THE ROUTLEDGE HANDBOOK OF INDIGENOUS ENVIRONMENTAL KNOWLEDGE

This volume provides an overview of key themes in Indigenous Environmental Knowledge (IEK) and anchors them with brief but well-grounded empirical case studies of relevance for each of these themes, drawn from bioculturally diverse areas around the world. It provides an incisive, cutting-edge overview of the conceptual and philosophical issues, while providing constructive examples of how IEK studies have been implemented to beneficial effect in ecological restoration, stewardship, and governance schemes.

Collectively, the chapters in the *Routledge Handbook of Indigenous Environmental Knowledge* cover Indigenous Knowledge not only in a wide range of cultures and livelihood contexts, but also in a wide range of environments, including drylands, savannah grassland, tropical forests, mountain landscapes, temperate and boreal forests, Pacific and Indian Ocean islands, and coastal environments. The chapters discuss the complexities and nuances of Indigenous cosmologies and ethno-metaphysics and the treatment and incorporation of IEK in local, national, and international environmental policies. Taken together, the chapters in this volume make a strong case for the potential of Indigenous Knowledge in addressing today's local and global environmental challenges, especially when approached from a perspective of appreciative inquiry, using cross-cultural methods and ethical, collaborative approaches which limit bias and inappropriate extraction of IEK.

The book is a guide for graduate and advanced undergraduate teaching, and a key reference for academics in development studies, environmental studies, geography, anthropology, and beyond, as well as anyone with an interest in Indigenous Environmental Knowledge.

Thomas F. Thornton is Dean of Arts and Sciences and Vice-Provost for Research and Sponsored Programs at University of Alaska Southeast, USA, and Associate Professor (part-time) at the Environmental Change Institute, School of Geography and the Environment, University of Oxford, UK.

Shonil A. Bhagwat is Professor of Environment and Development, and Head of the School of Social Sciences and Global Studies at the Open University, UK. His research focuses on the links between environment and development in the context of global challenges.

THE ROUTLEDGE HANDBOOK OF INDIGENOUS ENVIRONMENTAL KNOWLEDGE

Edited by Thomas F. Thornton and Shonil A. Bhagwat

LONDON AND NEW YORK

First published 2021
by Routledge
2 Park Square, Milton Park, Abingdon, Oxon OX14 4RN

and by Routledge
52 Vanderbilt Avenue, New York, NY 10017

Routledge is an imprint of the Taylor & Francis Group, an informa business

British Library Cataloguing-in-Publication Data
A catalogue record for this book is available from the British Library

Library of Congress Cataloging-in-Publication Data
Names: Thornton, Thomas F., editor. | Bhagwat, Shonil, editor.
Title: The Routledge handbook of indigenous environmental knowledge /
edited by Thomas F. Thornton and Shonil A. Bhagwat.
Other titles: Handbook of indigenous environmental knowledge
Description: Abingdon, Oxon ; New York, NY : Routledge, 2021. |
Includes bibliographical references and index.
Subjects: LCSH: Ethnoecology. | Traditional ecological knowledge.
Classification: LCC GF50 .R695 2021 (print) |
LCC GF50 (ebook) | DDC 304.2089–dc23
LC record available at https://lccn.loc.gov/2020021122
LC ebook record available at https://lccn.loc.gov/2020021123

ISBN: 978-1-138-28091-5 (hbk)
ISBN: 978-1-315-27084-5 (ebk)

DOI: 10.4324/9781315270845

Typeset in Bembo
by Newgen Publishing UK

The Open Access version of Chapters 10 and 23 was funded by Portland State University.

We dedicate this volume to all the Indigenous Peoples of the world and acknowledge their contribution to environmental knowledge, without which this volume would not have been possible.

CONTENTS

Contents

FIGURES

TABLES

CONTRIBUTORS

Ariell Ahearn is a departmental lecturer and the course director for the MSc/MPhil in Nature, Society and Environmental Governance at the School for Geography and the Environment at the University of Oxford, UK. She is a human geographer/social anthropologist focused on mobile pastoralism in Mongolia. Her current research aims to understand the relationship between social systems, resource distribution and governance frameworks particularly in areas affected by mining and land use change. ariell.ahearn@ouce.ox.ac.uk

Yildiz Aumeruddy-Thomas is Director of Research at the Centre for Functional and Evolutionary Ecology of the CNRS in Montpellier, France. She heads the Biocultural Interactions team and the CNRS national programme for Biodiversity in the Mediterranean region. As an ethnobiologist, she has worked on Indigenous and Local Knowledge in Indonesia, Nepal, Morocco and Sicily. She contributed on integrating ILK into the IPBES Global Evaluation as a lead author, and to the IPBES ILK Task Force. One of her favourite research themes relates to biocultural approaches to tree domestication and other trans-species relationships.

Shonil A. Bhagwat is Professor of Environment and Development, and Head of the School of Social Sciences and Global Studies, the Open University, UK. He is an environmental geographer with broad research interests at the cross-section between natural and social sciences. His research centres on the links between environment and development. In particular, his research engages critically with discussions on a variety of key environmental concerns: agriculture and food security, biodiversity conservation, climate change, and sustainability. Before joining the Open University as a Lecturer in Geography in February 2013, he directed an international and interdisciplinary master's programme in Biodiversity, Conservation and Management at the School of Geography and the Environment, University of Oxford, UK (2009–2013). He has also held post-doctoral research appointments at the University of Oxford (2006–2009) and at the Natural History Museum, London, UK (2003–2006).

Rochelle Bloom is a Research Associate with the Department of Anthropology at Portland State University, Oregon, USA. With an advanced degree in cultural heritage, she collaborates in ethnographic and ethnohistorical research focusing on U.S. national parks. In addition to working in support of several studies relating to Native American uses of national parks directed by Dr. Douglas Deur, Bloom has been stationed at Yosemite National Park, providing on-site research support to park staff and park-associated Native American tribes.

Eduardo Brondizio is Distinguished Professor of Anthropology in the Department of Anthropology, Indiana University, Bloomington, USA, where he is the Director of the Center for the Analysis of Social Ecological Landscapes (CASEL). Brondizio is the Co-Chair of the IPBES Global Assessment of Biodiversity and Ecosystem Services (2016–2019). He serves as co-Editor-in-Chief of *Current Opinions in Environmental Sustainability* (Elsevier).

Dawn Chatty is Emeritus Professor in Anthropology and Forced Migration and former Director of the Refugee Studies Centre, University of Oxford, and a Fellow of the British Academy. She is a social anthropologist with long experience in the Middle East as a university teacher, development practitioner, and advocate for Indigenous rights. Her research interests include nomadic pastoralism and conservation, gender and development, and coping strategies of refugee youth. Her most recent books include, *Syria: The Making and Unmaking of a Refuge State* (2018), *From Camel to Truck: The Bedouin in the Modern World* (2013), and *Displacement and Dispossession in the Modern Middle East* (2010).

Ugo D'Ambrosio is the Director of the Mediterranean Ethnobiology Programme of the Global Diversity Foundation and a research collaborator of the Etnbiofic research group (University of Barcelona and Botanical Institute of Barcelona). With a PhD. in Ethnobiology, Dr. D'Ambrosio's research interests revolve around rural and urban ethnobotany, especially food and medicinal plants, participatory methodology, Indigenous communities, human migration, as well as theory and practice in ethnobiology. The geographical areas where Ugo has specialized his fieldwork include lower Central America and the western Mediterranean Basin. His research includes the ethnobotany of Asian, African and Latin American origins sold in Barcelona's ethnic markets.

Douglas Deur is Professor in the Anthropology Department, Portland State University, and the University of Victoria School of Environmental Studies. He studies Indigenous North American ethnoecology; the meaning, contestation, and protection of Native sacred places; and communities' enduring attachments to landscapes within parks and other public lands. He has long served as the primary third-party researcher assisting Native American tribes and the US National Park Service in documenting enduring tribal connections to landscapes and resources in Western national parks – from Arctic Alaska to the desert Southwest. Adopted by Clan Chief Kwaxsistalla Adam Dick, he remains active in cultural and academic efforts in the Kwakwaka'wakw world. deur@pdx.edu

Sandra Díaz is a Professor of Community and Ecosystem Ecology at Córdoba National University, and a senior member of the National Research Council of Argentina. She is interested in plant functional trait diversity. She combines her plant ecology studies with interdisciplinary work on how different societies value and reconfigure biological communities and ecosystems. She has spearheaded the development of the concept of nature's contributions to people, and co-chaired the IPBES Global Assessment.

Chief Adam Dick, *Kwaxsistalla Wathl'thla* (Kawadillikalla Clan of Dzawatainuk Tribe of the Kwakwaka'wakw Nation in British Columbia) was a traditionally trained leader of his nation, and a celebrated teacher, interpreter, and preserver of his culture. Chosen by the chiefs of his youth at the age of 4, hidden and trained by these chiefs for years after, Adam became a keystone culture-bearer during a time when colonial authorities forced most other Indigenous children into residential schools. With specialized knowledge of all aspects of traditional Kwakwakwaka'wakw chiefly conduct, Adam became a leading authority to his people in the potlatch and in many other cultural and ceremonial domains throughout his adult life. His specialized knowledge on such topics as traditional resource harvests and management has been the focus of academic books, articles, dissertations and theses, and documentary films.

Chris S. Duvall is a Professor in the Department of Geography and Environmental Studies at the University of New Mexico, USA. His research examines people-plant interactions, principally in western Africa and in the African Atlantic Diaspora, from a range of academic perspectives, including historical geography, ecological biogeography, science studies, political ecology, cultural studies, and ethnobotany. He has published studies of Indigenous Knowledge in the journals *Geografiska Annaler*, *Journal of Biogeography*, and *Systematics and Geography of Plants*. His most recent book is *The African Roots of Marijuana* (2019). duvall@unm.edu

Roy Ellen is Emeritus Professor of Anthropology and Human Ecology at the University of Kent, Canterbury, UK. His main areas of current interest are ethnobiological knowledge systems, cultural cognition, social system resilience, and inter-island trade. He has conducted field research in archipelagic Southeast Asia over a period of 50 years, and his most recent book is *Kinship, Population and Social Reproduction in the 'New Indonesia'* (2018). He was elected to a fellowship of the British Academy in 2003, was President of the Royal Anthropological Institute between 2007 and 2011, and in 2017 received the Distinguished Economic Botanist award from the Society for Economic Botany. r.f.ellen@kent.ac.uk

Álvaro Fernández-Llamazares is a post-doctoral Research Fellow at the University of Helsinki, Finland. He is an ethnoecologist interested in biocultural approaches to conservation. Most of his research deals with the relations between biological and cultural diversity, and the role that Indigenous Peoples play in biodiversity conservation. He has more than 27 months of in-depth ethnographic field experience among Indigenous Peoples and other local communities across much of the Global South, working in countries such as Bolivia, Costa Rica, Kenya and Madagascar.

Joshua B. Fisher is Science Lead, ECOSTRESS Mission, at Jet Propulsion Laboratory, California Institute of Technology, USA, and Research Scientist in the Carbon Cycle & Ecosystems Group. He is also Associate Project Scientist at the Joint Institute for Regional Earth System Science and Engineering (JIFRESSE) University of California, Los Angeles (UCLA). Previously he was a post-doctoral researcher and departmental lecturer at the University of Oxford, Environmental Change Institute, School of Geography & Environment. He has a BSc and PhD from University of California, Berkeley, USA.

Jeff Gayman is Professor at the Hokkaido University Research Faculty of Media and Communication and Graduate School of Education, Japan. His main areas of research are Indigenous education, the human rights of Indigenous Peoples and intercultural relations between Indigenous Peoples and mainstream society. He has been working with the Ainu people for approximately 15 years, two of which were spent in the Ainu village of Nibutani. jeffry.gayman@imc.hokudai.ac.jp

Maximilien Guèze is a Science Officer in the Technical Support Unit for the IPBES Global Assessment and holds a PhD in Environmental Science from the Autonomous University of Barcelona, Spain. His research focuses on the distribution of tropical tree diversity and the relationship between plant diversity and Indigenous culture, including cultural change. He conducted over four years of fieldwork with Indigenous communities in the Bolivian Amazon and Indonesian Borneo.

Ariella Helfgott researches subjects spanning from conceptual and mathematical modeling of system resilience and adaptability, through to participatory approaches to building resilience and adaptive capacity on-the-ground. Within the later, she brings together stakeholders across and within multiple decision-making levels and engages them in programs that develop strategic capacity, for example exploratory scenarios, visioning and back-casting and resilience-based planning. She has extensive experience in the design, implementation and analysis of participatory planning, decision-aiding and capacity development programs in East and West Africa, South and South-East Asia, Latin America, Europe and Australia from individual, through community to national and international scales.

Eugene Hunn is Emeritus Professor at the University of Washington at Seattle, USA. He received his PhD in Anthropology from the University of California, Berkeley, in 1973 and has been on the UW faculty since. His scholarly interests are centered on ethnobiology, particularly the cognitive and ecological aspects of that field. He has conducted ethnobiological field research among Tzeltal Mayan speakers in Chiapas, Mexico (1971) (*Tzeltal Folk Zoology: The Classification of Discontinuities in Nature*, 1977). Since 1976, he has been studying the ethnobiology and cultural ecology of Sahaptin-speaking Indian people of the Columbia River basin in the Pacific Northwest of North America (*Nch'i-Wana, The Big River: Mid-Columbia Indians and Their Land*, 1990). He has also been involved in contract research for the National Park Service (U.S.) on subsistence issues in Alaska and has testified in court with regard to Pacific Northwest Native American resource and land rights. He is currently initiating a long-term ethnobiological/ethnoecological research project among Zapotec-speakers in southern Oaxaca, Mexico ("The utilitarian factor in folk biological classification", *American Anthropologist* 84: 830–847, 1982). enhunn323@comcast.net

Henry P. Huntington lives in Eagle River, Alaska. His research includes documenting Indigenous knowledge of marine mammals, examining Iñupiat and Inuit knowledge and use of sea ice, and assessing the impacts of climate change on Arctic communities. Huntington has been involved in several international research programs, was co-chair of the National Academy of Sciences Committee on emerging research questions in the Arctic and a member of the Council of Canadian Academies panel on the state of knowledge of food security in the North. Huntington has made long trips in the Arctic by dog team, small boat, and snowmobile. henryphuntington@gmail.com

Wendy Jackson is a part-time researcher and a full-time multilateral practitioner. She has 20 years of experience working for and with the United Nations system in various capacities. Her research interests include: the effectiveness of the multilateral system; Indigenous Knowledge and international decision-making; and the confluence of nature and culture, such as in urban sacred natural sites. She currently works for the New Zealand Ministry of Foreign Affairs and Trade as a relationship manager for several multilateral organizations. Previously she has worked for the International Institute for Sustainable Development, the United Nations Environment Programme, and the New Zealand Department of Conservation. She is a member of the Society for Conservation Biology and the IUCN's Commission on Environmental, Economic and Social Policy. wendy.jackson@gmail.com

Shiaki Kondo is Assistant Professor at the Center for Ainu and Indigenous Studies, Hokkaido University, Japan. He is a cultural anthropologist, who has conducted research in Alaska and Japan. His research interests include human-animal relations, Indigenous food sovereignty, and land-based pedagogy. He is a co-editor of *A Human History Through Canine Perspective* [In Japanese, 2019]. shiaki.kondo@gmail.com

Steve J. Langdon is Professor Emeritus of Anthropology at the University of Alaska Anchorage, USA, where he taught from 1976 until 2014. Over his 45-year career, Dr. Langdon has conducted research projects on many public policy issues impacting Alaska Natives. He has advocated for policies that enhance and promote rural Alaska Native communities and their cultures in such areas as fisheries, lands, tribal government, cultural heritage, customary trade and resource management. Dr. Langdon has specialized in research on the history and culture of the Tlingit and Haida peoples of southeast Alaska from precontact conditions through the historic period of nineteenth- and early twentieth-century US governance. He has conducted extensive research on archeological evidence of precontact Tlingit salmon fishing systems and on traditional ecological knowledge and uses of salmon by the Tlingit and Haida, demonstrating the complex and rich relations between the people and salmon that sustained their cultures for centuries. His book *The Native People of Alaska* is a widely used introduction to Alaska Native people. sjlangdon@alaska.edu

Phil Lyver has used ecological science and Indigenous and local knowledge to interpret changes in demographic trends and abundance of wildlife populations (e.g. terrestrial and marine birds). The key focus of his research has been on climatic and anthropogenic (e.g. harvest) drivers of population change, and the development of models to forecast population trends. He has expertise in interpreting the ecosystem structure and function within scientific and Indigenous world-views, in particular, the ways that different cultures "sense" the environment. Phil has experience interfacing both scientific and cultural frameworks to design and interpret biodiversity indicators for Māori resource users, scientists, policy-makers, and government and non-government officials. Strong cross-cultural partnerships with Indigenous people, especially Māori, are core to his research. Also of interest are the arrangements used by Māori and Indigenous People internationally to achieve governance and management over their lands and natural resources.

Nadezhda Mamontova received her PhD in Social Anthropology (*kandidat nauk* – domestic) from the Institute of Ethnology and Anthropology of the Russian Academy of Sciences in 2013. Currently she is doing her PhD studies at the School of Geography and the Environment and is a Research Fellow in the Department of Ethnology, Faculty of History, Lomonosov Moscow State University, Russia. She has been conducting research on Indigenous Russian minorities, mainly on the Ewenki living in Central Siberia and the Russian Far East, since 2007 and was involved in several projects aimed at documenting the Ewenki language, including the Ewenki digital text corpus. nadezhda.mamontova@ouce.ox.ac.uk

Gary Martin has a PhD in anthropology from the University of California, Berkeley, USA, and an undergraduate degree in botany. His applied research and teaching on conservation and ethnobotany has taken him to more than 60 countries over the last 30 years. He was a lecturer in the School of Anthropology and Conservation at the University of Kent from 1998 to 2011, and was a Fellow of the Rachel Carson Center for Environment and Society from 2010 to 2012. He created the Global Diversity Foundation in 2000, and launched the Global Environments Summer Academy and Global Environments Network in 2011.

Ashley Massey Marks is a science educator for girls in New York, USA. She has conducted field research in Borneo, Gambia, and South Africa, and assessed data from Ethiopia and Japan using GIS and remote sensing technologies. Ashley is Vice President of the Religion and Conservation Biology Working Group of the Society for Conservation Biology, and Co-Chair of the Theme on Culture, Spirituality and Conservation of the IUCN Commission on Environmental, Economic and Social Policy. She served as a Peace Corps volunteer in Guinea and Gambia. Ashley holds a DPhil in Geography and the Environment and an MSc in Biodiversity, Conservation and Management from Oxford, and an AB from Dartmouth College. ashley.massey@gmail.com

Pamela McElwee is Associate Professor of Human Ecology at Rutgers, the State University of New Jersey, USA. She has conducted research in Vietnam for the past 25 years on environment and development issues, particularly with Indigenous groups in the uplands including Ede, Mnong, Bru-Van Kieu, Katu, Pacoh, Hmong and Thai communities. She is the author of *Forests Are Gold: Trees, People and Environmental Rule in Vietnam* (2016). pamela.mcelwee@rutgers.edu

Armando Medinaceli is a Bolivian ethnobiologist with extensive research experience with Indigenous Peoples in Bolivian Amazonia, as well as southern Mexico and Guatemala. His research emphasizes collaborative approaches and topics in ethnobiology that range from traditional uses of plants and animals, community conservation, biocultural diversity, research ethics, natural resource management, collaborative video, and more. Armando currently teaches anthropology courses at Washington State University. He is a coordinator for the Ethics Committee for the Latin American Society of Ethnobiology and a member of the Ethics and Advocacy Committee for the Society of Ethnobiology. armando.medinaceli@wsu.edu

Zsolt Molnár is a botanist and ethnoecologist, and is the leader of the Traditional Ecological Knowledge Research Group at the MTA Centre for Ecological Research, Hungary. He has conducted research in Hungary, Romania, Serbia, and recently in Mongolia on traditional and local knowledge of herders and farmers on plants, vegetation, landscapes and their changes, particularly focusing on nature conservation issues and on knowledge co-production with locals in order to avoid or resolve conflicts with conservation and foster traditional land management. He is a coordinating lead author of the IPBES Global Assessment, and a member of the IPBES Indigenous and Local Knowledge Task Force.

Soufiane M'sou is the Field Scientist of Ethnobiology Programme of the Global Diversity Foundation, which currently focuses on the maintenance of High Atlas Cultural Landscapes as territories and areas conserved by Indigenous Peoples and local communities (ICCA), and research on policy and governance. He received his Bachelor's degree in Environmental Engineering & Management of Biodiversity and his Masters in Medicinal and Aromatic Plants: Protection, Conservation and Valorization and a PhD in Ecology and Phyto-chemistry from Cadi Ayyad University, where he focused on the study and monitoring of biodiversity, ethnobotany and cultural landscapes in the High Atlas, and phytochemistry.

Devi Mucina is an Associate Professor and the Director of the Indigenous Governance Program at the University of Victoria, Canada. His scholarship is focused on Indigenous Ubuntu knowledge systems and masculinities. At the center of this research are questions about how Indigenous communities are renewing and decolonizing Indigenous masculinities by grounding holistic relational engagements with intimate partners, families, communities, lands and bodies of waters. His academic interests and current research work focus on Indigenous men with reference to Indigenous men and fathering beyond imposed colonial masculinities; Indigenous men in elite sports; and contemplating how Indigenous spiritual mask dancing shapes Indigenous masculinities.

Hien T. Ngo is currently the Head of the Technical Support Unit for the Intergovernmental Science-Policy Platform on Biodiversity and Ecosystem Services' (IPBES) Global Assessment of Biodiversity and Ecosystem Services. Prior to this position she was also technical support person for the IPBES pollinator, pollination and food production assessment. She has experience working on the first Global Pollinator project of the UN-FAO and research involving bee pollination and coffee production in Costa Rica, Vietnam and the Philippines.

Alison A. Ormsby teaches Environmental Studies at the University of North Carolina Asheville, and is a graduate mentor in Environmental Studies at Prescott College, USA. She is a human ecologist with 25 years of experience working with people and protected areas, environmental education, and sacred natural sites. She is a member of the IUCN's Specialist Group for Cultural and Spiritual Values of Protected Areas, and has numerous publications, including the books *Asian Sacred Natural Sites: Philosophy and Practice in Protected Areas and Conservation* (2016), *Sacred Species and Sites: Advances in Biocultural Conservation* (2012), and *Sacred Natural Sites: Conserving Nature and Culture* (2010). ormsbyaa@gmail.com

Yoshitaka Ota is a social anthropologist specializing in global ocean governance. He is a Research Assistant Professor for the School of Marine and Environmental Affairs at the University of Washington, USA. With a BSc (1995), MSc (1998), and PhD (2006) in Anthropology from University College London, he has worked with coastal communities in various locations, such as Palau, Japan, the UK and Indonesia, where he investigated the social organization of human and marine environmental relationship.

Paul Porodong is a Senior Lecturer in Sociology and Social Anthropology in the School of Social Sciences, Universiti Malaysia Sabah, a post which he has held since 1998. He has more than 12 years' experience in collaborative work with government agencies, plantation industries, local and international NGOs, consultant firms in relation to rural communities in these three interrelated areas – socioeconomic development, community empowerment and environmental conservation. Additionally, he is involved in extensive community consultation involving over 30 villages to create an "eco link" between two protected areas in Sabah, namely, Kinabalu Park and Crocker Range Park. Dr. Porodong is also working with Sabah Parks, the Japanese International Cooperation Agency (JICA) and the Global Diversity Foundation (GDF), a UK-based NGO, to develop co-management protocols for community use zones (CUZ) inside the protected area in two different villages, namely, UluPapar, Penampang and UluSenegang, Keningau. He also contributes to consulting in Gaya Island, Taman Tunku Abdul Rahman Park with the aim of realizing the potential in getting the Island communities to participate in conservation.

Rajindra K. Puri is a Senior Lecturer in Environmental Anthropology and Director of the Centre for Biocultural Diversity, School of Anthropology and Conservation, University of Kent in Canterbury, UK. For over 30 years he has conducted interdisciplinary research on local knowledge systems, the dynamics of human-environment relations, and the application of anthropology to conservation social science in Southeast Asia. He is now working with forest ecologists on local adaptation to invasive Lantana camera in protected forests in southern India, using local knowledge to influence the contentious debates surrounding community-based forest management and climate change adaptation. R.K.Puri@kent.ac.uk

Kim Recalma-Clutesi, Ogwi'low'gwa (Kwakwaka'wakw/Pentlatch), is a cross-cultural interpreter, academic researcher, traditional artist, and award-winning videographer and producer. She is the daughter of Chief Ewanuxdzi (Buddy Recalma), sister and cultural advisor to Chief Klaqwagila (Mark Recalma), and the partner of the late Clan Chief Kwaxsistalla Wathl'thla Adam Dick. A noted expert on intellectual property rights and cultural repatriation, she has served as the elected Chief to the Qualicum Band of Indians, and has held positions in dozens of Aboriginal NGOs, as well as serving as a lead advisor to Canadian universities and government agencies on Indigenous cultural matters. Since 1989, she has administered the Ninogaad Knowledge Keepers Foundation, a nonprofit established by traditionally trained Kwakwaka'wakw cultural leaders to mentor younger leaders in training. kim_recalma-clutesi@shaw.ca

Victoria Reyes-García is ICREA Research Professor at the Institute of Environmental Science and Technology, Universitat Autònoma de Barcelona (ICTA-UAB), Spain. She has a PhD in Anthropology, 2001, from the University of Florida. Her research focuses on Indigenous and local knowledge, particularly regarding environmental issues. Reyes-García coordinates the Laboratory for the Analysis of Socio-Ecological Systems in a Global World (www.laseg.cat/en). Between 2010 and 2015, she directed an ERC Starting Grant to study the adaptive nature of culture using a cross-cultural approach. In 2017, she received an ERC Consolidator Grant to study the potential contribution of Indigenous and local knowledge to research on climate change impacts.

Samantha Russell was born and raised in Kenya, developing a love for wild places early on. Having completed schooling in Kenya, followed by a stint in the United Kingdom to study Zoology and Psychology at university, she returned to join the conservation world in Kenya at the age of 22. For the last (almost) two decades she has dedicated her life to working alongside the communities of Southern Kenya to better understand and protect their land, livelihoods and the wildlife that coexist with them. She lives in Shompole with her husband and two children.

Paul Sillitoe is Professor of Anthropology at Durham University, UK, and has a background in both cultural anthropology and agricultural science. His research interests focus on natural resources management. He has conducted extensive fieldwork in the Highlands of Papua New Guinea and has been involved in projects in South Asia and recently in the Arabian Gulf region. His recent books include *Sustainable Development: An Appraisal Focusing on the Gulf Region* (2014); *The Collaborative Moment: The Implications of Indigenous Studies for an Engaged Anthropology* (2015); and *Built in Niugini: Constructions in the Highlands of Papua New Guinea* (2017). paul.sillitoe@durham.ac.uk.

Tammy Steeves co-leads the Conservation, Systematics and Evolution Research Team (ConSERT) at Te Whare Wānanga o Waitaha/The University of Canterbury, New Zealand. In partnership with Māori tribes (*iwi* or *hapū*) and in collaboration with conservation practitioners and local communities, ConSERT combines genomic and non-genomic data to co-develop conservation genetic management strategies for some of Aotearoa/New Zealand's rarest *taonga* (treasured) species. Tammy was born and raised in Canada, and she has been living in Aotearoa/New Zealand since 2004.

Irene Teixidor-Toneu is a postdoctoral researcher at the University of Oslo, Norway, and guest researcher in Naturalis Biodiversity Center, the Netherlands, where she applies evolutionary thinking to ethnobiology. She holds a degree in biology from the University of Barcelona and a PhD in ethnobiology from the University of Reading, UK. Her PhD research focused on medicinal plant knowledge and its cultural transmission among Amazigh communities in the High Atlas region in Morocco. She has collaborated closely with the Global Diversity Foundation in studying biocultural diversity and its conservation.

Tarshish Thekaekara is a researcher-conservationist based in the Nilgiris. He holds a PhD in Geography from the Open University, and an MSc in Biodiversity, Conservation and Management from the University of Oxford, UK. He is a post-doctoral fellow at the National Centre for Biological Sciences, Bangalore, working on elephant genetics. He is also a trustee of The Shola Trust, a conservation organization he co-founded in 2008, that works on human-inclusive models of nature conservation. He is currently involved in two broad areas: better integrating human-elephant spaces, and community-based approaches to understanding and managing Lantana camara, an invasive plant. tarsh@thesholatrust.org

Jessica P. R. Thorn is a Namibian social-ecological systems scientist, with a background in ecology and human geography. She is a research associate in the Department of Environment and Geography at the University of York and Senior Researcher at the African Climate and Development Initiative at the University of Cape Town. Her research interests include climate adaptation, peri-urban resilience, China in Africa, development corridors, ecosystem services, agroecology, scenarios and urban green infrastructure. Jessica has been involved in various NSF, NERC, NRF, DFID, CGAIR, IDRC, ESRC, UNECA, WMO, ADB, and USAID-funded projects, conducting field research in 12 countries, and is a contributing author to IPCC and TEEB reports. Jessica received her DPhil from the University of Oxford. jessica.thorn@york.ac.uk

Thomas F. Thornton is Dean of Arts and Sciences and Vice-Provost for Research and Sponsored Programs at the University of Alaska Southeast, USA, and Associate Professor (part-time) at the Environmental Change Institute, School of Geography and the Environment, University of Oxford, UK. His research interests include Indigenous and local knowledge systems and human-environmental interactions, the political ecology of resource management in social-ecological systems, and human adaptation to environmental change in the North Pacific, especially Southeast Alaska. thomas.thornton@ouce.ox.ac.uk; tthornto@alaska.edu

Suzanne von der Porten is an independent consultant with expertise in Indigenous governance, water, marine policy, and environmental governance scholarship. Her postdoctoral research at Simon Fraser University, Canada, focused on the changing roles of Indigenous coastal nations, governments, and industry in relation to marine conservation. Suzanne has a PhD in Environment and Resource Studies from the University of Waterloo, and a BSc and an MBA from the University of Victoria. She is a Research Associate with the Nereus Program at the University of British Columbia and is Action Canada Fellowship Program Alumnus (2010/

2011). Both her consulting and research work toward improving the state of environmental decision-making and the inherent rights of Indigenous Peoples. suevonderporten@gmail.com

Krushil Watene is Associate Professor of Philosophy in the School of Humanities at Massey University, New Zealand. Krushil specializes in moral and political philosophy, with a particular commitment to Indigenous philosophies. She works closely with Māori communities to support the revitalization and sustaining of *mātauranga* Māori and writes about the contributions of Maori philosophies to contemporary social and global challenges. She is Ngāti Manu, Te Hikutu, Ngāti Whātua o Orākei, Tonga.

Priscilla Wehi is a government conservation biologist who works with Indigenous ecological knowledge at Manaaki Whenua Landcare Research. She is also Incoming Co-Director of Te Pūnaha Matatini, a Centre of Research Excellence in Complex Systems. A New Zealander of Scottish descent, she also has extended family responsibilities in Waikato-Tainui and Tūhoe. WehiP@landcareresearch.co.nz

Guy Western was raised and educated in Kenya and completed a BA in Environmental Studies from the University of California, Santa Barbara, USA, before going on to pursue an MSc in Biodiversity, Conservation, and Management at Oxford University, UK. He has since conducted his doctoral research at Oxford on understanding the ecological, socio-economic and cultural factors that promote coexistence between pastoralists and lions in human-dominated landscapes. Since 2010, he has worked in conjunction with the South Rift Association of Landowners to spearhead the establishment of Rebuilding the Pride, a community-based conservation programme aimed at enabling Maasai pastoral communities in Kenya's southern Rift to develop lion conservation initiatives rooted in traditional knowledge, sound science, and good governance. He continues to work with the Maasai communities in Kenya. guy.western.ke@gmail.com

Hēmi Whaanga is an Associate Professor in Te Pua Wānanga ki te Ao at the University of Waikato, Aotearoa. He has worked as a project leader and researcher on a range of projects centred on the revitalization and protection of Māori language and knowledge. He affiliates to Ngāti Kahungunu and Ngāi Tahu.

Kathy J. Willis is a biologist, who studies the relationship between long-term ecosystem dynamics and environmental change. She is Professor of Biodiversity in the Department of Zoology, University of Oxford, UK, and an adjunct Professor in Biology at the University of Bergen, Norway. In 2018, she was elected as Principal of St Edmund Hall. She held the Tasso Leventis Chair of Biodiversity at Oxford and was founding Director, now Associate Director, of the Biodiversity Institute Oxford, and was Director of Science at the Royal Botanic Gardens, Kew.

Victoria Wyllie de Echeverria is a doctoral candidate in the Environmental Change Institute, School for Geography and the Environment, at the University of Oxford, UK. Her research interests include ethnobotany, ethnoecology and folk taxonomy of the landscapes of the Pacific Coast of North America, people's perceptions and adaptations to a changing climate, and bridging knowledge between disciplines. victoria.wylliedeecheverria@ouce.ox.ac.uk

ACKNOWLEDGEMENTS

We dedicate this volume to the memory of those traditional knowledge-holders who contributed significantly to this volume but did not live to see the book in its final form. In particular, we wish to acknowledge the late Kwaxistalla Adam Dick, a Kwakwaka'wakw (Kwakiutl) clan chief who made vast contributions to the documentation and maintenance of Indigenous Environmental Knowledge throughout his life. He and his partner Kim Recalma-Clutesi (Oqwilowgwa) co-authored a chapter of this volume. In the cover photo we see them together, conducting a ceremonial blessing of a Bald eagle. Eagles hold tremendous cultural and spiritual significance in Kwakwaka'wakw tradition. To show due respect, clan chiefs look after eagles: Chief Kwaxistalla often brought food to wild eagles and honoured them in myriad ways, while Kim's father, the late Chief Ewanuxdzi, regularly donated a portion of his commercial salmon catch to support eagle protection and the rehabilitation of injured eagles. The eagle shown in this photo had been found injured, nursed back to health and given the noble name Ewanuxdzi, before being released back into the wild. The scene reminds us of the profoundly restorative aims and potentials of Indigenous Environmental Knowledge systems – carried by individuals such as Clan Chief Kwaxistalla Adam Dick, with fidelity, into modern times. Photo courtesy of Vim Roloff.

1
INTRODUCTION

Thomas F. Thornton and Shonil A. Bhagwat

Indigenous Knowledge and the organization of diversity
Knowledge and cultures

A simple definition of Indigenous Environmental Knowledge (IEK) is *knowledge generated by Indigenous Peoples about their surroundings, including relations with other beings, human and other-than-human, which is adapted and transmitted from generation to generation.* Yet the implications of this definition are not simple. While all human knowledge systems attempt to organize the profound diversity of life in the world to understand the boundaries and relations between its constituent beings, they do not do so in the same ways. Cultural, cognitive, and linguistic systems reflect both universal and disparate ways of classifying the world, including plants and animals (Berlin 1992; Ellen 2006), geographic places (Basso 1996; Feld and Basso 1996; Thornton 2008), landscapes (Hirsch and O'Hanlon 1995; Johnson and Hunn 2010), and other environmental phenomena. Form (in Linnaeus' taxonomic system) and function (in many Indigenous systems) are two major ways of classifying life in the world. Yet, the very nature of what is living (animate vs. inanimate), of what constitutes being-in-the world, or consciousness, and of what contributes to the biocultural diversity of life (Maffi 2005) is complex and relational. Indigenous Environmental Knowledge (IEK) necessarily touches on existential, ontological, epistemological, and teleological questions, which takes us into the realm of what anthropologist A.I. Hallowell termed "ethno-metaphysics."

In his famous study of "Ojibwa ontology, behavior and world view" (1960), Hallowell argued:

> Human beings in whatever culture are provided with cognitive orientation in a cosmos; there is 'order' and 'reason' rather than chaos. There are basic premises and principles implied, even if these do not happen to be consciously formulated and articulated by the people themselves. We are confronted with the philosophical implications of their thought, the nature of the world of being as they conceive it. If we pursue the problem deeply enough we soon come face to face with a relatively unexplored territory – ethno-metaphysics. Can we penetrate this realm in other cultures? What kind of evidence is at our disposal?

Hallowell suggested that language, myth, religion, and a variety of other cultural domains were the relevant "evidence" to understand Indigenous Knowledge systems, but most importantly,

DOI: 10.4324/9781315270845-1

one needed to see these knowledge systems deployed in the context of experience to understand how they really worked. Only through *culture in experience*—in the exigencies of living in a certain place—could one see how humans accessed, applied, refined, and transmitted knowledge of their environs, and how they reckoned various categories, boundaries, and relationships between beings, including human and other beings, and objects in the world.

Culture and experience in environmental knowledge-building

Understanding culture in experience typically requires fieldwork and participant observation, as well as interviews, linguistic and cultural domain and discourse analysis, and other techniques. One cannot simply gather environmental knowledge, as a number of authors in this volume point out, simply by presenting a Western (Linnean) taxonomy or plant or animal guide, and asking for "equivalent" names in the language of another cultural group. The categories may not be the same, the nature of the beings may differ in different contexts (some "rocks" speak, Hallowell famously found among the Ojibwa Anishinaabe), evolutionary and contemporary relations between species may differ, and so forth. In short, IEK is relational, contextual, contingent, and, ultimately, complex. In fact, so is Western scientific knowledge, which itself is descended from, reproduced, and represented in particular cultural systems. Thus, the genealogy of relations between the two knowledge systems is itself worthy of analysis (Zent 2009).

Despite this complexity and divergence, there are some ways that Indigenous and Western scientific environmental knowledge systems are similar, or at least can be analyzed comparatively. Consider that all knowledge systems are not only informed by their possessors' worldviews, but also have distinct processes of knowledge production, maintenance, and transmission.

As Table 1.1 shows, there are some broad contrasts that can be made between IEK and Western science in each of these areas. But it would be a mistake to over-emphasize these distinctions, for example, to argue, that IEK is only "particularistic" or concrete rather than general or abstract,

Table 1.1 How does IEK differ from Western science?

Aspects of knowledge	Western Scientific Knowledge (WSK)	Indigenous Environmental Knowledge (IEK)
World-view	Inanimate earth	Animate earth
	Nature/culture dichotomy	Nature/culture unity and reciprocity
	Material perspective	Material & spiritual perspective
	Mediated by technology	Mediated by technology & ceremony
Production/accumulation	Universalist, dispassionate/ amoral context	Cultured, passionate/moral context,
	Discrete "facts"/data	Integrated/ place-based "facts"
	Specialized approach, generalizing, objective centralized in institutions	Holistic approach Dispersed in individuals, esp. elders Reciprocal trust and sharing
Transmission	Written, quantitative analysis	Oral, mainly qualitative, *in situ*
	Transmission as duty (publish!)	Transmission as gift, or by protocol
Maintenance	Falsify hypotheses	Grasp wisdom of the past
	Controlled experiments	Refine IEK via continued lifeways
	Generate new knowledge	Conserve and adapt IEK to fit the
	Refine general theory, laws	exigencies of changing lifeworlds.

or that IEK does not address contradictions, or is more biased and less empirical than Western science (cf. Lévy-Bruhl [1926] 1985; Berkes 2012). Such assertions are simplistic and do not explain the unity and diversity of human thought systems. Indeed, there is evidence to suggest that all human knowledge systems, including science, are primarily defined by the relationships we maintain (Bateson 1972). Yet, this relational perspective is seen as a distinct advantage of IEK, which science has partly attenuated and devitalized by becoming too objectivist, mechanistic, and reductionist. Thomas Berry, echoing Hallowell, argues:

> The universe is a communion of subjects, not a collection of objects. The devastation of the planet can be seen as a direct consequence of the loss of this capacity for human presence to and reciprocity with the nonhuman world. This reached its most decisive moment in the seventeenth-century proposal of René Descartes that the universe is composed simply of "mind and mechanism." In this single stroke, he devitalized the planet and all its living creatures, with the exception of the human.

F.D. Peat, in his *Blackfoot Physics* (2002), makes a similar point, suggesting that Western science demands objectification of those conquered and of nature itself, in contrast to more integrative, holistic, and relational Indigenous world-views. Yet there are important exceptions to this tendency in Western science, such as Bateson's *Steps to an Ecology of Mind* (1972) and Lovelock's (2000) Gaia theory, which are richly relational and holistic in their approach.

Indeed, the development of science itself was never exclusively Western, but a multicultural sharing of ideas from the beginning (Weatherford 1988; Medin and Bang 2014). In fact, there is still a common foundation of characteristics that link science and IEK, as outlined in Box 1.1.

Box 1.1 How are IEK and Western science similar?

- Built on empirical, place-based inquiry
- Motivation to understand and successfully interact with environments
- Cumulative basis—knowledge accumulates over time and across generations
- Systemic character—knowledge systems are organized in particular ways (domains, etc.)
- Innovations and achievements enabled by knowledge breakthroughs are celebrated
- Protocols govern ownership, sharing, and use of knowledge
- Evolving—knowledge is continually refined, assessed and updated (not without loss)
- Relevance of knowledge is assessed in relation to exigencies of experience and needs

---

These common foundations lead us to potential answers to perhaps the most important question of this Handbook: How should IEK and Western science interact (Box 1.2)?

Box 1.2 How should IEK and Western science interact?

- Recognize and respect the differences, complementary nature, and validity of both systems of knowledge
- Accept that both systems of knowledge are informed by cultural values, beliefs, and ideological paradigms

---

- Define a protocol or agreement for relating (e.g., braiding or integrating) and maintaining the two systems of knowledge in research and policy, including: objectives and pathways; meaningful consultations at all phases of research (design, data collection, analysis, interpretation, dissemination); acquisition, handling, and transmission of IEK through culturally appropriate means; rights and ethical responsibilities; representing and crediting knowledge
- Re-evaluate protocol and research plans periodically

Recognizing the differences but also the complementary nature and validity of the two systems of knowledge is essential, as is the acceptance that both systems of knowledge are informed by cultural values, beliefs, and ideological paradigms.

Experience is key to all modes of knowledge building. Anthropologist Tim Ingold (2000: 146), an intellectual descendant of Hallowell, makes the point this way:

> The source of cultural knowledge lies not in the heads of predecessors but in the world that they point out to you—if, that is, one learns by discovery while following in the path of an ancestor—then words, too must gather their meanings from the contexts in which they are uttered.

From this standpoint, we must appreciate that all knowledge systems have a means of knowledge production that is tied not just to schooling or dissemination from elders but also to the contexts of experience. An example of this was presented to one of us (Thornton) recently by a Tlingit man, who wrote via email wondering about documentation of the "big storm" that comes after the herring spawn in Sitka Sound, Alaska, each spring. The big winds of the storm are said to help break up the herring spawn, which is laid on substrate (including Western hemlock boughs placed by Natives) in the intertidal zone, where the eggs eventually hatch out. The man remembered hearing about this wind from an elder but could not recall the Tlingit word or concept. Nevertheless, the *experience* of the wind after this (2019) herring season triggered the knowledge memory. According to documentation of Tlingit elder Al Martin (Thornton 2015: 219), this wind is called "'Wind of the washing of the spawn: L'uk' eeti.oosk,'" and it arises because "the herring had to be washed clean from the beach so the[y] ... can survive." This is a remarkable bit of IEK that many younger Tlingits are still aware of from experience, even if they are not familiar with the Tlingit term for this special wind.

Undoubtedly, the loss of Indigenous languages poses a serious threat to IEK, but an even greater threat may be the loss of lands and livelihoods that enable the generation, practice, and experience of such knowledge. If the Tlingit man in the above example did not have the opportunity—the right—to continue harvesting herring eggs each spring in his homeland, would he even remember the phenomenon of the "wind of the washing of the herring spawn?" Probably not, as without continuity of lifeways on traditional lands, IEK tends to atrophy. For this reason, it is important to link the maintenance of IEK to access to traditional lands and Indigenous rights to customary resources and livelihood practices (Thornton 2001; Nakashima and Roué 2002; UNESCO 2009; Maffi and Woodley 2010; McCarter et al. 2014).

Integrating knowledge systems

Since the millennium, the legitimation and "integration" of IEK into regional, national, and international scientific and governing bodies have become a priority. The United Nations, especially since the passage of the UN Declaration on the Rights of Indigenous People (2007), has worked to "integrate"Traditional Ecological Knowledge (TEK), including IEK, into its science and governance. Witness the 2018 establishment of the Local Communities and Indigenous Peoples Platform within the United Nations Framework Convention on Climate Change (UNFCCC). The creation of this platform represents an important step in valorizing IEK, as Indigenous Peoples, despite being a small minority of the world's population, dwell in some of the most biodiverse and climate vulnerable places in the world. Indigenous Peoples argue that they know their places best and thus their IEK should be considered as foundational in informing broader understandings of their lands as well as adaptive responses to the environmental changes affecting them. The Platform seeks to 'catalyze learning, engagement, and policy coordination that benefits local communities and Indigenous Peoples, as well as the international community', by embracing diverse knowledges and working toward their integration (see Shawoo and Thornton 2019).

Yet, Nadasdy (1999), Berkes (2012), and others (e.g., Hill et al. 2012; Ross et al. 2011) show that integration is not a straightforward process, but rather a political one, fraught with institutional conflicts, wherein the dominant partner, Western science and its institutions, tends to dictate the terms of IEK legitimation, integration, and application. This is particularly likely to occur when there are ontological or epistemological divides between scientific and Indigenous world-views and epistemologies. Such inequitable integration can be harmful, another manifestation of colonialism, and thus a range of scholars have called for decolonizing IEK "integration" in favor of more equitable and systematic collaboration in all phases of IEK documentation and use (e.g., Thornton and Scheer 2012; Spoon 2014; Velasquez Runk 2014).

Some scholars go further in arguing that constructive interactions between IEK and Western science cannot be achieved without first addressing the imbalance of power between the two systems (cf. Nadasdy 2005), or without deploying explicit decolonizing methodologies (Tuhiwai Smith 1999). These methodologies place Indigenous language, thought, and discourse at the center of the inquiry and analysis. From this perspective, extractive models of IEK gathering, seeking only to "mine" IEK systems for "data" that can be incorporated into non-Indigenous science models or databases, are viewed increasingly as a form of neocolonialism, which ultimately erodes the values of trust and reciprocity requisite for conducting collaborative scientific work on IEK systems. These issues power and hegemony are critically important to address in the context of globalization and international development (cf. Sillitoe 2007).

The organization of diverse knowledge systems

Given the reality of plural systems of environmental knowledge, the dominance of science over IEK, and the folly of extracting bits of IEK "data" for science without cognizance of the integrated whole of the Indigenous cultural system and experience from whence it originates, how can this diversity be organized? Can all knowledge systems be equally accommodated in research and policy? Can one kaleidoscopic perspective of diverse Indigenous Knowledge systems be successfully forged? These are challenging questions, and to date they have received little attention in comparison to the integrationist perspectives. Certainly, if places themselves are historical ecological products of human-environmental interactions, then, as Cuerrier et al. (2015) argue, IEK can provide long-term environmental perspectives that may be otherwise be inaccessible to science. Places as evolutionary human-environmental mosaics, or landscapes, also can provide a basis for understanding the structural and functional diversity of human environmental knowledge systems. In addition, the Indigenous biocultural diversity mosaic provides evidence of how Indigenous social-ecological systems may contribute lower carbon, environmentally-friendly, and resilient sustainable development solutions (Bohensky and Maru 2011; Mistry and Berardi 2016).

In reviewing the UNFCCC's Local Communities and Indigenous Peoples Platform, a framework for integrating IEK and science, Shawoo and Thornton (2019) offer three overarching recommendations to ensure it empowers Indigenous Knowledge systems: (1) reframe the platform toward rights-based recognition of the authority of IEK; (2) ensure final decision authority rests with Indigenous Peoples (IP) for actions affecting their lands and cultural lifeways; and (3) develop policies to ensure that the complex, integrated, and holistic context of IEK is not compromised, "braiding" IEK (Kimmerer 2013; Raygorodetsky 2017) or integrating IEK and Western science synergistically rather than simply incorporating relevant IEK bits into Western science frameworks.

One means of doing this is to conceptualize the embeddedness of IEK facts in knowledge-belief-practice complexes, nested within a set of concentric circles (representing levels of analysis), consisting of land and resource management systems, social institutions, and, at the widest level, world-view (Berkes 2012). Another is to consider the integrated processes of IEK production, accumulation, and transmission within their full cultural and situational context (Lauer and Aswani 2009), as Keith Basso (1972) does, for example, in explicating ice travel among the Fort Norman Slave (Dene, Athabaskan), in particular how hunters learn to make decisions about whether and how to cross a potentially dangerous expanse of ice. A third way is to start from the concept of Indigenous cultures existing in niches, unique roles within the environment that served to co-create biological and cultural diversity on earth. As Unangan (Aleut) leader Larry Merculieff (1990) suggests:

> We are disrupting them [ecosystems] because humans, unlike most life on earth, have no niche. Without a niche, we stumble into and disrupt natural systems all around us. The fact is cultural systems which have, as a paradigm, intimate interaction with the environment, are the closest humankind will come to having a niche. I suggest that the humane, morally proper and most efficient laws must complement, supplement and enhance rational local systems already in place seeking to do the same thing.

Nature, culture, and environmental knowledge are inextricably linked; they reflect and produce the rich diversity of life on earth. Monocultural approaches, be they of a universalizing science or a colonizing culture, can destroy this diversity and imperil the planet. As Lakota Chief Luther Standing Bear (1933: 255) said of Indigenous Knowledge ideals and practices, after having been educated in White boarding schools, more than a century ago:

> There were ideals and practices in the life of my ancestors that have not been improved upon by the present-day civilization; there were in our culture elements of benefit; and there were influences that would broaden any life. But that almost an entire public needs to be enlightened to this fact need not be discouraging. For many centuries the human mind labored under the delusion that the world was flat ...The human mind is not yet free from fallacious reasoning; it is not yet an open mind and its deepest recesses are not yet swept free of errors.

To open these recesses and reduce the errors that now threaten a healthy biocultural diversity on earth requires a more equitable, just, and comprehensive understanding and use of Indigenous Environmental Knowledges. The chapters in this book suggest how to achieve such an understanding of diverse human environmental knowledge systems for the sustainability of all humankind.

The structure of the volume

The chapters in this volume are divided into four Parts:

Part I: Concepts and context
Part II: Issues of perspectives, values, and engagement
Part III: Application of IEK for adaptation, conservation, and coexistence
Part IV: Governance and equity

However, it is important to note that the boundaries between these Parts are fluid and many chapters address themes that cross-cut different sections. As such, the division of chapters into a particular Part is only for the organizing of the volume and is not intended as a tool to pigeon-hole chapters in any way.

Part I Concepts and context

In Chapter 2, "Indigenous Ecological Knowledge: why bother?," Eugene Hunn unravels the complex terminology in describing Indigenous Knowledge. Reflecting on the discussions about the Indigenous Knowledge that began in 1983, he argues that the challenge has been to describe Indigenous Knowledge in terms that cannot be misconstrued by the Indigenous Peoples themselves or marginalized by science. Based on his own decades-long fieldwork in the Americas, Hunn argues against the notion that Indigenous Knowledge is magical, misguided, or misanthropic. Rather Indigenous communities are key allies in the efforts to protect the environment and their purpose and pride reflected in their knowledge should not be undermined by such bias and mischaracterizations.

Chris Duvall's Chapter 3, entitled "Context matters: The holism and subjectivity of environmental knowledge," is based on his fieldwork in West Africa and Central and North Americas. He argues that "environmental" knowledge constitutes more than the knowledge of a particular geographical location or biological feature. It encompasses a holistic understanding of the biophysical reality when Indigenous Peoples make decisions about resource use. He suggests that there is, however, no universal form of environmental knowledge and it is subjective and dependent on the local context. He goes on to argue that the holistic nature of environmental knowledge, together with its subjectivity, gives rise to new concepts that enter into environmental knowledge systems. Therefore, Indigenous Knowledge holds significant potential to transform "environmental science"—a project that will go a long way in decolonizing the institutions of education about and governance of the environment.

In Chapter 4, "Cultivar diversity and management as traditional environmental knowledge," Roy Ellen reviews the history of how Indigenous Peoples' management practices have generated enormous diversity in some of the world's staple crops today—including rice, potato, sweet potato, cassava, maize, and sago. He discusses the interconnection between ecological and cultural selection of cultivars, varieties, or landraces; knowledge of plant maturation and reproduction; and planting strategies of the landraces. He makes a case for social embeddedness of knowledge, highlights the need for dissemination and exchange of germplasm, and expresses concern about the consequences of diversity loss due to farming intensification. He concludes the chapter by arguing that preservation of cultural memory alongside conservation of biodiversity is necessary, such that the conservation of one can support the other.

In Chapter 5, "On serving salmon: an ethnography of hyperkeystone interactions in interior Alaska," Shiaki Kondo introduces IEK and conservation practices in the hunting, fishing, and gathering practices among Alaskan Athabascans. Based on his fieldwork in the community of Nikolai, Alaska, he discusses two examples of salmon-human interactions in the Upper Kuskokwim region. Engaging with the ecological and anthropological debates on "keystone species," Kondo proposes some solutions to bridge the gap between IEK and natural resource management regimes predominantly influenced by scientific management paradigms. He extends the keystone species concept further and considers the role of humans and some of the other non-human top predators in marine ecosystems as "hyperkeystone species"—those that affect multiple other keystone species across different habitats, and hence drive complex, potentially connected interaction chains. He argues that effective management of salmon in Alaska will need to consider Alaskan Native IEK and practices that are akin to the role of hyperkeystone species in ecosystems.

In Chapter 6, "Performance knowledge: uncovering the dynamics of biocultural diversity of Borneo's tropical forests through a Penan hunting technique," Rajindra Puri looks at the knowledge that emerges and evolves through subsistence activities such as collecting, hunting, fishing, herding, and various forms of agriculture. He calls this "performance knowledge" and distinguishes it from "declarative knowledge" (e.g., knowing the names and myths of animals) or "behavioral knowledge" (e.g., sharpening a knife or throwing a spear). He argues that performance knowledge is learned and transmitted largely through accumulated experiences as an observer or participant in those performances. He strongly advocates researchers embrace very acute observational techniques or apprenticeship with local expert practitioners in IEK communities in order to fully grasp the nuances of this kind of knowledge.

Paul Sillitoe's Chapter 7, "Soil ethnoecology," takes on the complexities in understanding Indigenous Knowledge about soils, or "ethnopedology." This is a type of IEK that has received relatively little attention, unlike ethnobotany or ethnozoology. His research focuses on the Wola speakers of the Southern Highlands Province of Papua New Guinea. The most widely cultivated soil types in Wolaland are Inceptisols and Andisols (derived from the sedimentary rock), dominated by volcanic ash, with some alluvial re-deposition. Elaborating on the Wola soil classification system, Sillitoe warns about the danger of uncritically imposing a scientific model of soil science on what we think we understand of others' knowledge, at the risk of distorting the IEK. He examines the contrasts between the "scientific" soil management and the management of soils according to the Wola farmers. He argues that the Wola soil management keeps cultivation in sustainable equilibrium with soil resources, so long as the appropriate fallow time is observed.

In Chapter 8, "Bridging paradigms: analyzing traditional Tsimane' hunting with a double lens," Armando Medinaceli proposes a new approach based on the combined use of traditional anthropological methodologies (e.g., participant observation, interviews) and Indigenous methodologies (e.g., *so'baqui*) searching for "true collaboration" between researcher and the local Indigenous communities. The chapter is set in the Bolivian Amazonia and focuses on comparing and contrasting the Tsimane' peoples' epistemologies with conventional ethnography. Medinaceli uses his own extensive experience of working with the Tsimane' people to bridge Indigenous epistemologies and conventional ethnography. He suggests that it is important to use locally known and understood customs (Indigenous methodologies) as part of the package of methods used for true collaboration through activities that the Indigenous Peoples are familiar with. In Medinaceli's case, this included a combination of *so'baqui* (visiting) with the use of participant observation, interviewing, and focus groups (based on local formats for discussion). This combination, in his experience, created a comfortable and familiar environment easy to understand, while maintaining the rigor and structure of academic research generating outcomes that are rewarding for both Indigenous Peoples and academic researchers.

Part II Issues of perspectives, values, and engagement

Ariell Ahearn and Dawn Chatty's Chapter 9, "Asian and Middle Eastern pastoralists," is set against a backdrop of attempts by governments worldwide to settle pastoralists, remove their access to open rangelands through land privatization or conservation schemes, forced settlement, and denial of citizenship. Yet, the Indigenous Environmental Knowledge of pastoralist groups in Asia and the Middle East has supported sustainable practices of resource use over many generations. Ahearn and Chatty argue that pastoralist knowledge that encompasses the relationships between social groups, between people and their livestock, and survival strategies in harsh dryland environments is relevant today in our plight to address global challenges.

In Chapter 10, "Balance on every ledger: Kwakwaka'wakw resource values and traditional ecological management," Douglas Deur, Kim Recalma-Clutesi, and Chief Adam Dick illustrate the core environmental values of the Kwakwaka'wakw (Kwakiutl) people on the Pacific coast of Canada, and explore how they manifest in the traditional management of coastal natural resources. Their survey of environmental values is based on the unique, authentic knowledge of Chief Adam Dick, a co-author of the chapter. They argue that extracting or speaking about Indigenous Knowledge without the broader context of environmental values can lead to serious scholarly misunderstandings, and these authors show how long-term collaborations between academic researchers and specialized knowledge holders from Indigenous communities are important to represent Indigenous Knowledge accurately.

Jeff Gayman's Chapter 11, "Challenges surrounding education and transmission of Ainu Indigenous ecological knowledge in Japan: Disparate valuations of a people and their IEK," examines the plight of the Ainu people, who are the Indigenous inhabitants of northern Honshu, the island of Hokkaido, southern Sakhalin, and the Kurile Islands in a geographical region that sits at the border between Japan and Russia. He argues that the environmental knowledge of the Ainu people can potentially provide valuable lessons for management of regional ecosystems and use of resources. However, this knowledge has been undervalued historically due to the cultural politics in and between Japan and Russia, leading to the development of the Ainu lands under legislation which neglected or denied traditional livelihoods. The chapter highlights some initiatives on Hokkaido in education, cultural transmission, and locally relevant research, and explores how the use of traditional mediums, such as Ainu oral tradition, might help with a greater recognition of Indigenous Knowledge in new Ainu "Cultural Promotion" legislation.

In Chapter 12, "Engaging with Indigenous Environmental Knowledge in the North American Arctic: moving from documentation to decisions in environmental governance," Henry P. Huntington argues that although Indigenous Environmental Knowledge has been documented in the North American Arctic, the engagement of that knowledge and that of knowledge-holders in making decisions about resource use is rare. He identifies three areas where change is needed in order to bring about more substantial engagement: First, an assessment by knowledge-holders and Indigenous leaders of the steps that need to be taken for such an engagement. Second, a thorough appraisal of resources and infrastructure required to achieve full engagement of IEK and its holders. Third, a careful consideration by IEK leaders and scholars of where IEK can and cannot contribute to the kinds of decisions needed for effective environmental governance. The chapter makes a strong case for greater leadership roles for IEK holders in shaping the nature of future engagement between IEK and resource management.

Nadezhda Mamontova's Chapter 13, "Taiga Forest reindeer herders and hunters, subsistence, stewardship," provides an ethnolinguistic account of terminologies used in hunting, reindeer herding, and fishing activities by the Ewenki people in Siberia. She observes that Ewenki peoples' IEK is closely related to and reflected in their endangered Indigenous terminologies used to encode local knowledge of land, vegetation, and animals, their traditional means of classification and land use, and their cultural values and beliefs toward their environment. The Ewenki native language is especially rich in terms that embed important IEK and there is no comparable Russian terminology to replace those terms. She argues that despite the prevalence of the Russian language and a decline in Ewenki native language, particularly among the younger members of the community, the Ewenki language is still important for its functional role in environmental cognition.

In Chapter 14, "Tlingit engagement with salmon: the philosophy and practice of relational sustainability," Steve Langdon shares his findings on 40 years of anthropological research on Tlingit knowledge of and relations with salmon in Southeast Alaska. Based on participant observation, in-depth interviews with knowledgeable elders and experts, and comprehensive examination of the ethnographic record and oral traditions, he argues that Tlingit engagement with salmon as non-human persons is one based on reciprocity and sustainment of positive relations. Further, Langdon shows that Tlingits' empathic identification with salmon is based on astute empirical observations of environmental conditions and how these conditions and specific human practices toward salmon affect salmon behavior, productivity, and returns.

In Chapter 15, "*Mātauranga* as knowledge, process and practice in Aotearoa New Zealand," Priscilla Wehi, Hēmi Whaanga, Krushil Watene, and Tammy Steeves explore the relationship between Māori ecological philosophies and mainstream ecological management. The chapter is set in Aotearoa New Zealand (ANZ), an archipelago of around 600 islands in the South Pacific. In Māori philosophies, the social, cultural, environmental, and ecological knowledge is collectively referred to as "*mātauranga*." A holistic approach adopted by *mātauranga* has helped the practitioners to develop cultural indicators that can be applied to monitor ecosystem health. Despite the merits of adopting a holistic approach to natural resource management, Wehi et al. argue that the relationship between Māori philosophies and Western science is uneasy, with many arguing that *mātauranga* and Western science are two different knowledge systems. This leads to the tension between romantic conceptualizations of Māori concepts and the reality of the current state of the environment in ANZ. Wehi et al. explore some of these tensions but also discuss the need for structures that create space for collaborations between *mātauranga* practitioners and ecosystem managers who value the potential of *mātauranga*.

Part III Applications of IEK for adaptation, conservation, and coexistence

In Chapter 16, "Integrating Amazigh cultural practices in Moroccan High Atlas biodiversity conservation," Irene Teixidor-Toneu, Gary Martin, Soufiane M'Sou, and Ugo D'Ambrosio analyze the cultural practices of conservation in the High Atlas Mountains of Morocco. They focus on two rural communities inhabiting the cultural landscapes of this mountainous region and find that most conservation practices are associated with agriculture, food, pastoralism, and soil and water management. These multifaceted practices help in developing the High Atlas cultural landscapes and support the biodiversity they harbor by: (1) creating mosaics of cultivated and grazing areas supported by communally-managed water sources; (2) managing plant resources in a way that enriches floristic diversity; and (3) regulating interactions between people and their local environment in ways that embody local values in access to lands and resources.

Alison Ormsby's Chapter 17, "Sacred groves of Sierra Leone: preserving Indigenous Environmental Knowledge," asks whether the sacred groves of Sierra Leone help preserve IEK. Sacred groves are community-managed forests that are protected because of cultural traditions and rituals these forests support. The research was conducted in the Tonkolili District in central

Sierra Leone and looked at residents' attitudes, use, and rules regarding local sacred groves, as well as the impact of war on the groves. This research suggests that societal traditions, knowledge, and rituals remain the most important aspects associated with the protection of sacred groves. Key informant interviews show that, even with modernization in Sierra Leone, the local communities are keen to protect the sacred groves provided that critical traditional beliefs, and social institutions are supported in their management.

In Chapter 18, "The role of biodiversity in the maintenance of ecosystem services in human-dominated landscapes: evidence from the Terai Plains of Nepal," Jessica Thorn, Thomas Thornton, Ariella Helfgott, and Kathy Willis present the results of ethnobotanical surveys conducted around rice farms in the Terai Plains of Nepal. They show that despite being situated in densely populated agricultural landscapes, these farms are reservoirs of biodiversity. They provide multiple provisioning, regulating, and supporting goods and services that sustain local Nepali livelihoods. They also observe that even today farmers maintain a rich ethnobotanical knowledge, and plants are being used for their wide-ranging benefits, including medicine, food, timber, fuel, fodder, soil enhancement, pesticides, as well as for spiritual or ritual purposes. They argue that low-cost, small-scale ethnobotanical practices enhance the overall resilience of these farming communities to the vicissitudes of social and environmental change in this Himalayan region.

Guy Western and Samantha Russell's Chapter 19, "Creating coexistence: traditional knowledge and institutions as a foundation for Maasai-wildlife coexistence in southern Kenya," demonstrates how Traditional Ecological Knowledge (TEK) of Maasai societies, combined with their social capital, facilitates coexistence between pastoralists and wildlife. This case study is situated in Kenya's southern Maasailand, on the group ranches of Olkiramatian and Shompole, an area inhabited by roughly 20,000 people in approximately 1,000 km^2 with the two ranches used as a single management area. The traditional seasonal livestock movements and herding practices in this area are planned and governed by local committees, which primarily dictate where settlement and grazing are allowed in a given season. Western and Russell report that in Shompole and Olkiramatian, a culture of coexistence with wildlife predators on livestock exists and is influenced by intimate knowledge and societal beliefs relating to predators such as lions. Their chapter provides novel insight into how Maasai pastoralists dwelling in the region are able to avoid wildlife conflicts and coexist with lions without unduly compromising their livestock.

Victoria Wyllie de Echeverria's Chapter 20, "Cultural keystone species as indicators of climatic changes," looks at the terminologies used to describe species that are considered important in ecosystems, often termed "keystone species." She looks at the advantages and disadvantages of focusing on single species in conservation management. The advantages include the preservation of other species alongside the one that is a focus of conservation. The disadvantages include contrasting the management needs of species, wherein the protection of one cannot always guarantee the protection of others. She argues that in order to acknowledge the anthropogenic nature of many of today's ecosystems, the concept of "cultural keystone indicator species" (CKIS) is more appropriate. Based on case studies from the temperate rainforest forest ecosystems of North-western North America, she finds that animal behavior or plant phenology of specific CKIS species and seasonal or annual changes in those regions can be important indicators of climate change and help us understand how climate change might impact on humans and other species important to their livelihoods and well-being.

Tarshish Thekaekara's Chapter 21, "Living with elephants: Indigenous world-views," focuses on the Kattunayakan and Bettakurumba peoples' relationships with, and strategies of living alongside, Asian elephants in Southern India. Kattunayakans are a forest-dependent community (Kattu: forest; Nayakans: rulers). Bettakurumbas are also forest people but are known also for their skillful handling of domesticated elephants. Thekaekara describes the complex, multiple conceptualizations of the elephant among these Indigenous communities. He argues that there are three underlying drivers of people's tolerance to elephants and the ability to share space with them: (1) elephant ontologies and the very conceptualization of what is an elephant; (2) a (non-competitive) mode of subsistence, including the kind of crops people choose to grow; and (3) a shared history of living together. Although these factors vary significantly between different Indigenous and local communities, Thekaekara suggests that they provide a heuristic tool to understand the ability and willingness of communities to share space with elephants and mitigate so-called "human-wildlife conflicts."

In Chapter 22, "Do dragons prevent deforestation? The Gambia's sacred forests," Ashley Massey Marks, Josh Fisher, and Shonil Bhagwat look at the cultural practice of forest protection in the Gambia in West Africa and ask whether this practice prevents the loss of forests to non-forest land uses. In West Africa, belief in *ninkananka*, translated by residents as "dragons" in English, spans ethnic groups from Senegal to Sierra Leone. It is said that if people see a *ninkananka*, they will die on the spot or very soon after. The research made use of 25 years of high-resolution Landsat satellite imagery in conjunction with *in situ* field campaigns to compare forest cover in dragon areas, formally protected areas (forest reserves and a national park), and unprotected surrounding areas. Massey, Fisher, and Bhagwat found that, like formally protected areas, dragon areas are significantly "greener" than surrounding areas, as indicated by the Mean Normalized Difference Vegetation Index (NDVI). These results suggest that in a state-owned common property regime, widespread belief in local cosmologies can conserve forest patches over time, even more consistently than formally protected areas.

Doug Deur and Rochelle Bloom's Chapter 23, "Fire, native ecological knowledge, and the enduring anthropogenic landscapes of Yosemite Valley," looks at the cultural practice of burning in the Yosemite Valley, USA. This practice goes hand-in-hand with other types of active management based on IEK, including selective harvesting, pruning, replanting, and many other techniques that produce signature anthropogenic plant communities. These anthropogenic plant communities have long sustained the food, material, medicinal, and spiritual needs of Yosemite Valley's Native residents. Deur and Bloom carried out extensive ethnographic fieldwork to understand the imprint of Indigenous Ecological Knowledge upon the Yosemite Valley landscape. This, combined with historical archival research, provides a resource of rare detail and comprehensiveness, containing the vast majority of the written corpus regarding Native American communities' relationships to the Yosemite landscape across time. Based on this knowledge, the chapter describes techniques that materially enhance the availability of culturally preferred plant communities in the Yosemite landscape. The chapter goes on to explore the tensions between these traditional management practices and those brought in by non-Native settlers and park managers. Deur and Bloom note that the park managers in Yosemite have begun incorporating the traditional practices, consulting with traditionally associated tribes, and welcoming their participation in unprecedented ways. Despite this, they argue, the quantity and quality of keystone plant habitats and species continue to decline and they make a strong case for multi-disciplinary research and cross-cultural dialogue to reverse this decline.

Part IV Governance and equity

Wendy Jackson and Phil Lyver in Chapter 24, "Who benefits? Indigenous Environmental Knowledge (IEK) in multilateral biodiversity agreements," focus on six international conventions on biodiversity and how IEK is represented in each. Together, these conventions comprise primary global agreements that constitute international law and policy on biodiversity. Jackson and Lyver conclude that there are very few examples where engagement between IEK and its holders is visible or credible in the international conventions on biodiversity. In the limited number of examples that exist, they suggest that the experience of IEK holders is not necessarily positive in relation to the conventions where they are represented. The chapter suggests how better-designed conservation policies and approaches may be able to engage Indigenous Peoples and local communities in problem identification and solutions, weave IEK into decision-making, and extend the repertoire of approaches required to address today's social-environmental challenges.

Pamela McElwee's Chapter 25, "The use and misuse of IEK in conservation in Vietnam," argues that IEK is not well understood as a complete cultural system or world-view, encompassing religion, ritual, and belief. In her case study from the forests of Vietnam, she observes that conservation projects have made little use of IEK and have seen Indigenous communities as obstacles to their work. The focus of studies on IEK is primarily on species' names and classifications with little regard for the knowledge systems underlying them. The Western-funded conservation projects in Vietnam are interested in knowledge that can be directly integrated into Western conservation paradigms with their primary focus on urging local people to "buy into" the idea that they can no longer use protected areas for agroforestry production or hunting. She argues that a fuller recognition of omens and taboos as local management practices can be beneficial for conservation and help improve the limited success of Western-funded conservation projects.

In Chapter 26, "Including indigenous and local knowledge in the work of the Intergovernmental Science-Policy Platform on Biodiversity and Ecosystem Services (IPBES): outcomes and lessons for the future," Pamela McElwee, Hien Ngo, Álvaro Fernández-Llamazares, Victoria Reyes-García, Zsolt Molnár, Maximilien Guèze, Yildiz Aumeruddy-Thomas, Sandra Díaz, and Eduardo Brondízio make a strong case for a greater inclusion of Indigenous and Local Knowledge (ILK) in global environmental policy fora and in science-policy interfaces. They specifically look at the IPBES Global Assessment, which has developed one of the first global-scale mechanisms for operationalizing ILK in sustainability decision-making. The types of knowledges that have been successfully integrated into this assessment include ways in which ILK can help (1) assess ecosystem change and health; (2) inform the achievement of global goals such as the Sustainable Development Goals and the Aichi Targets; and (3) inform policy-relevant options for decision-makers. They argue that other global initiatives seeking to engage ILK in their endeavors can learn from the ILK approach of the IPBES global assessment.

In Chapter 27, 'Indigenous Knowledge, knowledge-holders and marine environmental governance,' Suzanne von der Porten, Yoshi Ota, and Devi Mucina explore the perspectives on Indigenous Knowledge in the context of global marine governance. Indigenous Knowledge is important to environmental concerns, such as climate change, habitat destruction, pollution, and loss of biodiversity. These IEK systems align well to address today's environmental concerns because they are capable of interpreting ecological patterns in dynamic environments. Indigenous Knowledge in marine contexts is central to Indigenous Peoples' relationships to their coastal lands and oceans through activities such as fishing, navigating, subsistence, and trade. The chapter investigates Indigenous resurgence, or movements for reestablishing the autonomy of Indigenous nations, and discusses the political nature of Indigenous Knowledge in the context of contested coastal lands and resources. It argues that Indigenous Peoples' sovereignty and rights must be at the center of IEK policies applied to local and/or international oceans governance.

In Chapter 28, "Incorporating social-ecological systems into protected area networks: Territories and areas conserved by Indigenous Peoples and local communities (ICCAs) in Sabah, Malaysian Borneo," Ashley Massey Marks, Paul Porodong, and Shonil Bhagwat ask: "How can social-ecological processes beyond protected areas inform conservation policy and practice on a landscape scale?" The research was set in Sabah, Malaysian Borneo, and employed key informant interviews, oral and written questionnaire surveys, and participant observation on-site and at conservation planning and capacity-building workshops. Massey Marks, Porodong, and Bhagwat suggest that community-based forms of conservation are not static, but rather can be affected by events, including religious conversion, modernization, and the influence of outsiders on natural resource use. Successfully integrating ICCAs into the protected area network could provide landscape-scale rewards if the "proof of concept" engages neighboring communities. Just as protected area management has shifted toward adaptive management of ecological systems, ICCAs require adaptive governance of their social-ecological systems. Sacred forests and rivers can make valuable contributions in a new paradigm of community conservation—one that acknowledges conservation rooted in ecological understanding, supports conservation in the face of social and ecological changes, and recognizes the autonomy of local custodians in the process.

Collectively, the chapters in this volume cover Indigenous Knowledge not only in wide range of geographic cultures and livelihood contexts (hunting-gathering, pastoralist, horticultural, agricultural), but also in a wide range of environments, including drylands (Chapter 3); savannah grassland (Chapters 17, 18, 20); tropical forests (Chapters 6, 13, 16, 23, 26, 27, 28); montane landscapes (Chapters 12, 17, 24); temperate and boreal forests (Chapters 4, 11, 19, 22); Pacific and Indian Ocean islands (Chapters 7, 21, 25); and coastal environments (Chapters 8, 10) (Table 1.2). The chapters also discuss the complexities and nuances of Indigenous cosmologies and ethno-metaphysics (Chapters 1, 2, 5) and the treatment and incorporation of IEK not only

Table 1.2 Environmental settings of the book chapters

Chapter number and title	Drylands	Savanah grasslands	Tropical forests	Montane landscapes	Temperate and boreal forests	Pacific and Indian Ocean islands	Coastal	Multiple
Chapter 1: Introduction [Thornton and Bhagwat]								x
Chapter 2: Indigenous Ecological Knowledge: why bother? [Hunn]								x
Chapter 3: Context matters: The holism and subjectivity of environmental knowledge [Duvall]								x
Chapter 4: Cultivar diversity and management as traditional environmental knowledge [Ellen]			x					
Chapter 5: On serving salmon: ethnography of hyperkeystone interactions in Interior Alaska [Kondo]							x	

Table 1.2 Cont.

Chapter number and title	Environmental settings							
	Drylands	Savanah grasslands	Tropical forests	Montane landscapes	Temperate and boreal forests	Pacific and Indian Ocean islands	Coastal	Multiple
Chapter 6: Performance Knowledge: uncovering the dynamics of biocultural diversity of Borneo's tropical forests through a Penan hunting technique [Puri]			x					
Chapter 7: Soil ethnoecology [Sillitoe]				x				
Chapter 8: Bridging paradigms: analyzing traditional Tsimane' hunting with a double lens [Medinaceli]			x					
Chapter 9: Asian and Middle Eastern pastoralists [Ahearn-Chatty]	x							
Chapter 10: Balance on every ledger: Kwakwaka'wakw resource values and traditional ecological management [Deur et al.]					x		x	
Chapter 11: Challenges surrounding education and transmission of Ainu Indigenous Ecological Knowledge in Japan: disparate valuations of a people and their IEK [Gayman]						x		
Chapter 12: Engaging with Indigenous Environmental Knowledge in the North American Arctic: moving from documentation to decisions in environmental governance [Huntington]							x	
Chapter 13: Taiga Forest reindeer herders and hunters, subsistence, stewardship [Mamontova]					x			
Chapter 14: Tlingit engagement with salmon: the philosophy and practice of relational sustainability [Langdon]						x		

(continued)

Table 1.2 Cont.

Chapter number and title	Environmental settings							
	Drylands	*Savanah grasslands*	*Tropical forests*	*Montane landscapes*	*Temperate and boreal forests*	*Pacific and Indian Ocean islands*	*Coastal*	*Multiple*
Chapter 15: Mātauranga as knowledge, process and practice in Aotearoa New Zealand [Wehi et al.]						x		
Chapter 16: Integrating Amazigh cultural practices in Moroccan High Atlas biodiversity conservation [Martin et al.]				x				
Chapter 17: Sacred groves of Sierra Leone: preserving Indigenous Environmental Knowledge [Ormsby]		x						
Chapter 18: The role of biodiversity in the maintenance of ecosystem services in human–dominated landscapes: evidence from the Terai Plains of Nepal [Thorn et al.]		x		x				
Chapter 19: Creating coexistence: traditional knowledge and institutions as a foundation for Maasai–wildlife coexistence in southern Kenya [Western]		x						
Chapter 20: Cultural keystone species as indicators of climatic changes [Wyllie de Echeverria]					x			
Chapter 21: Living with elephants: Indigenous world-views [Thekaekara]			x					
Chapter 22: Do dragons prevent deforestation? The Gambia's sacred forests [Massey Marks et al.]			x					
Chapter 23: Fire, native ecological knowledge, and the enduring anthropogenic landscapes of Yosemite Valley [Deur and Bloom]				x				
Chapter 24: Who benefits? Indigenous Environmental Knowledge (IEK) in multilateral biodiversity agreements [Jackson and Lyver]								x

Table 1.2 Cont.

Chapter number and title	Environmental settings							
	Drylands	*Savanah grasslands*	*Tropical forests*	*Montane landscapes*	*Temperate and boreal forests*	*Pacific and Indian Ocean islands*	*Coastal*	*Multiple*
Chapter 25: The use and misuse of IEK in conservation in Vietnam [McElwee]			x					
Chapter 26: Including Indigenous and Local Knowledge in the work of the Intergovernmental Science-Policy Platform on Biodiversity and Ecosystem Services (IPBES): outcomes and lessons for the future [McElwee et al.]								x
Chapter 27: Indigenous Knowledge, knowledge-holders and marine environmental governance [von-der-Porten et al.]					x			
Chapter 28: Incorporating social-ecological systems into protected area networks: Indigenous and Community Conserved Areas (ICCAs) in Sabah, Malaysian Borneo [Massey Marks et al.]			x					

at local, national, and international levels of environmental policies (Chapters 9, 14, and others). Taken together, the chapters in this volume make a strong case for the potential of Indigenous Knowledge in addressing today's local and global environmental challenges, especially when approached from a perspective of appreciative inquiry, utilizing cross-cultural methods and ethical, collaborative approaches which limit bias and inappropriate extraction of IEK.

Acknowledgments

We are grateful to Dr. Alberta Jones, Associate Professor of Education at University of Alaska Southeast, for her early review and suggestions on this chapter.

References

Basso, K. H. 1972. Ice and travel among the Fort Norman Slave: Folk taxonomies and cultural rules. *Language in Society* 1(1): 31–49.

Basso, K. H. 1996. *Wisdom Sits in Places: Landscape and Language Among the Western Apache*. Albuquerque, NM: University of New Mexico Press.

Bateson, G. 1972. *Steps to an Ecology of Mind.* New York: Ballantine.

Berkes, F. 2012. *Sacred Ecology.* London: Routledge.

Berlin, B. 1992. *Ethnobiological Classification: Principles of Categorization of Plants and Animals in Traditional Societies.* Princeton, NJ: Princeton University Press.

Bohensky, E. and Maru, Y. 2011. Indigenous knowledge, science, and resilience: What have we learned from a decade of international literature on "integration"? *Ecology and Society* 16(4): 1–19.

Cuerrier, A., Brunet, N., Gérin-Lajoie, J., Downing, A., and Lévesque, E. 2015. The study of Inuit knowledge of climate change in Nunavik, Quebec: A mixed methods approach. *Human Ecology* 43(3): 379–394.

Ellen, R. 2006. *The Categorical Impulse: Essays in the Anthropology of Classifying Behaviour.* Oxford: Berghahn.

Feld, S. and Basso, K. (Eds.) 1996. *Senses of Place.* Santa Fe, NM: School of American Research Press.

Hallowell, A. I. 1960. Ojibwa Ontology, behavior, and world view. In S. Diamond, (Ed.), *Culture in History: Essays in Honor of Paul Radin.* London: Octagon Books.

Hill, R., Grant, C., George, M., Robinson, C. J., Jackson, S., and Abel, N. 2012. Typology of indigenous engagement in Australian environmental management: Implications for knowledge integration and social ecological system sustainability. *Ecology and Society* 17(1): 23.

Hirsch, E. and O'Hanlon, M. (Eds.) 1995. *The Anthropology of Landscape: Perspectives on Place and Space.* Oxford: Clarendon Press.

Hunn, E. 2007. Ethnobiology in four phases. *Journal of Ethnobiology* 27(1): 1–10. http://dx.doi.org/10.2993/0278-0771(2007)27[1:EIFP]2.0.CO;2

Ingold, T. 2000. *The Perception of the Environment: Essays on Livelihood, Dwelling and Skill.* London: Routledge.

Johnson, L. M. and Hunn, E. S. (Eds.) 2010. *Landscape Ethnoecology: Concepts of Biotic and Physical Space.* Oxford: Berghahn Books.

Kimmerer, R. W. 2013. *Braiding Sweetgrass: Indigenous Wisdom, Scientific Knowledge and the Teachings of Plants.* Minneapolis, MN: Milkweed Editions.

Laird, S. A. (Ed.) 2002. *Biodiversity and Traditional Knowledge: Equitable Partnerships in Practice.* London: Earthscan.

Lauer, M. and Aswani, S. 2009. Indigenous ecological knowledge as situated practices: Understanding fishers' knowledge in the western Solomon Islands. *American Anthropologist* 11(3): 317–329. http://dx.doi.org/10.1111/j.1548-1433.2009.01135.x

Lovelock, J. 2000. *Gaia: A New Look at Life on Earth.* Oxford: Oxford University Press.

Lévy-Brühl, L. [1926] 1985. *How Natives Think.* Princeton, NJ: Princeton University Press.

Maffi, L. 2005. Linguistic, cultural, and biological diversity. *Annual Review of Anthropology,* 34: 599–617.

Maffi, L. L. and Woodley, E. 2010. *Biocultural Diversity Conservation: A Global Sourcebook.* London: Earthscan.

McCarter, J., Gavin, M. C., Baereleo, S., and Love, M. 2014. The challenges of maintaining indigenous ecological knowledge. *Ecology and Society* 19(3): 39. http://dx.doi.org/10.5751/ES-06741-190339.

Medin, D. L. and Bang. M. 2014. *Who's Asking?: Native Science, Western Science, and Science Education.* Cambridge, MA: MIT Press.

Merculieff, L. 1990. Western society's linear systems and aboriginal cultures: The need for two-way exchanges for the sake of survival. Speech at Conference of Hunting and Gathering Societies. Anchorage, AK. May 30.

Mistry, J. and Berardi, A. 2016. Bridging indigenous and scientific knowledge. *Science* 352(6291): 1274–1275.

Nadasdy, P. 1999. The politics of TEK: Power and the" integration" of knowledge. *Arctic Anthropology* 36(1/2): 1–18.

Nadasdy, P. 2005. The anti-politics of TEK: The institutionalization of co-management discourse and practice. *Anthropologica* 47: 215–232.

Nakashima, D. and Roué, M. 2002. Indigenous knowledge, peoples and sustainable practice. In T. Munn et al. (Eds.), *Encyclopedia of Global Environmental Change,* vol. 5. Chichester: Wiley, pp. 314–324.

Peat, F. D. 2002. *Blac`kfoot Physics: A Journey into the Native American Worldview.* York Beach, ME: Red Wheel/Weiser.

Raygorodetsky, G. 2017. Braiding science together with Indigenous knowledge. Available at: https://blogs.scientificamerican.com/observations/braiding-science-together-with-indigenous-knowledge/.

Ross, A., Sherman, K. P., Snodgrass, J. G., Delcore, H. D., and Sherman, R. 2011. *Indigenous Peoples and the Collaborative Stewardship of Nature: Knowledge Binds and Institutional Conflicts.* London: Routledge.

Shawoo, Z. and Thornton, T. F. 2019. The UN local communities and Indigenous peoples' platform: A traditional ecological knowledge-based evaluation. *Wiley Interdisciplinary Reviews: Climate Change,* e00575.

Sillitoe, P. (Ed.). 2007. *Local Science vs Global Science: Approaches to Indigenous Knowledge in International Development.* New York and Oxford: Berghahn.

Smith, L. T. 1999. *Decolonizing Methodologies: Research and Indigenous Peoples.* London: Zed Books.

Spoon, J. 2014. Quantitative, qualitative, and collaborative methods: Approaching indigenous ecological knowledge heterogeneity. *Ecology and Society* 19(3): 33. http://dx.doi.org/10.5751/ES-06549-190333.

Standing Bear, L. 1933. *Land of the Spotted Eagle*. Lincoln, NE: University of Nebraska Press.

Thornton, T. F. 2001. Subsistence in northern communities: Lessons from Alaska. *The Northern Review* 23: 82–102.

Thornton, T. F. 2008. *Being and Place among the Tlingit*. Seattle, WA: University of Washington Press.

Thornton, T. F. 2015. The ideology and practice of Pacific herring cultivation among the Tlingit and Haida. *Human Ecology* 43(2): 213–223.

Thornton, T. and Scheer, A. 2012. Collaborative engagement of local and traditional knowledge and science in marine environments: A review. *Ecology and Society* 17(3): 8.

UNESCO 2009. *Learning and Knowing in Indigenous Societies Today*. Edited by P. Bates, M. Chiba, S. Kube, and D. Nakashima. Paris: UNESCO.

Velasquez Runk, J. 2014. Enriching indigenous knowledge scholarship via collaborative methodologies: Beyond the high tide's few hours. *Ecology and Society*, 19(4): 37. http://dx.doi.org/10.5751/ES-06773-190437.

Weatherford, J. 1988. *Indian Givers: How the Indians of the Americas Transformed the World*. New York: Ballantine.

Zent, S. 2009. A genealogy of scientific representations of indigenous knowledge. In S. Heckler (Ed.), *Landscape, Process, and Power: Re-evaluating Traditional Environmental Knowledge*. Oxford: Berghahn Books, pp. 19–67.

Zerner, C. (Ed.) 2000. *People, Plants and Justice: The Politics of Nature Conservation*. New York: Columbia University Press.

PART I

Concepts and context

2

INDIGENOUS ECOLOGICAL KNOWLEDGE

Why bother?

Eugene Hunn

Introduction

Why bother? This is for me a rhetorical question, as I've devoted the past 50 years to a series of ethnographic projects to document the sophistication of local understandings of natural history, from southern Mexico to Alaska. Today we must fight our way through a thicket of acronyms: IK/IEK/TK/TEK/LK/LEK/LTK/TKW/TRM in search of our goal: IK "Indigenous Knowledge," IEK "Indigenous Ecological/Environmental Knowledge," TK "Traditional Knowledge," TEK "Traditional Ecological/Environmental Knowledge," LK "Local Knowledge," LEK "Local Ecological/Environmental Knowledge," LTK "Local and Traditional Knowledge," TKW "Traditional Knowledge and Wisdom," and TRM "Traditional Resource Management." Much like the story of the six blind men describing an elephant, each scholar sees this subject matter from a personal perspective, thus perhaps we fail to appreciate the elephant in the room.

The impetus for these and allied acronyms dates to at least 1983 when a Working Group on Traditional Ecological Knowledge (TEK) was established by the Commission on Ecology of the International Union for Conservation of Nature and Natural Resources (IUCN), chaired by Gregory Baines (Williams and Baines 1993: 1). The goals of this working group emphasized the value of traditional knowledge of local ecosystems for biodiversity conservation and sustainable resource management. A parallel line of development (Brokenshaw, Warren, and Werner 1980) led to the founding in 1987 of CIKARD (Center for Indigenous Knowledge [IK] and Rural Development, at Iowa State University), focused on "preserving and using the local knowledge of farmers and other rural people around the globe" (www.ciesin.org/IC/cikard/CIKARD.html). However, our interest in Indigenous Ecological Knowledge has deeper roots, as we shall see.

Then, as now, isolated rural communities were widely judged in need of "development," which generally entailed converting their subsistence practices into commercially viable, scientifically managed, capitalist endeavors. If rural peoples should get in the way of such rural development plans, they must be encouraged to move to the nearest city. A hint of this attitude is current in pronouncements to the effect that global poverty has been drastically reduced in recent decades, coincident with rapid urbanization in the "developing world." This may well be the case. However, we must ask how is global poverty measured? Typically, analysts

DOI: 10.4324/9781315270845-3

average estimates of annual per capita cash income, now set at US$1.90 per day as the threshold for "extreme poverty" (*The Economist* 2015). By this measure, rural subsistence producers are relegated, by definition, to lives of "poverty and ignorance." This assessment seems to me to ignore the cultural and social capital of those rural communities that may still control the resources of their ancestral lands.

Those of us who have devoted our academic lives to documenting the complexities of Indigenous Knowledge – whether ecological, environmental, or otherwise – see a rather different picture. We have lived and worked in partnership with communities living "on the land," living as and where their distant ancestors lived, and in the process we have learned from the citizens of such communities to appreciate their intelligence and to *bear witness* to our common humanity. This testimony, I believe, is a key contribution of our studies of Indigenous Knowledge.

So what should we call this enterprise by which we affirm the value of Indigenous intellectual achievement? The challenge has been to find words that cannot be misconstrued by our audience (cf. Sillitoe 1998; Menzies and Butler 2006; Berkes 2008: 1–20; Heckler 2009). That may not be possible given the diverse audience we seek to address. "Traditional" has been my personal preference (Hunn 1993a, 1993b), but many insist that that will be (mis-)construed as implying a rigid attachment to an unchanging past. On the contrary, for me, traditions are dynamic collective products passed along from generation to generation within a community rooted in but not slave to a distant past. Other scholars prefer the term "Indigenous," which requires that such communities have deep roots – through centuries or millennia – in a particular home place. Indigeneity implies tradition, given the time depth of Indigenous territorial occupations. Furthermore, Indigenous peoples are by definition "local." However, there can be traditional communities that are not Indigenous to a particular place – such as gypsies – as traditions may travel. But such transient traditions may well lack the rich environmental knowledge that comes from roots in particular places, particular landscapes, particular soils (Johnson and Hunn 2010).

Likewise, there can be local communities, even local traditions, that are not Indigenous, for example, settler communities, which may involve several generations living in a place, but offspring of a mercantilist or capitalist colonial occupation. "Indigenous" demands a pre-colonial occupation. This creates a problem in ancient imperial societies, such as those of India and China, for which "indigeneity" loses much of its moral force (Agrawal 2002). The term "local" avoids the problems some associate with "traditional" or "Indigenous." Yet "local" is too bland for my taste, as it lacks the power of antiquity. In the final analysis, these semantic disputes solve nothing.

Ideally, the knowledge we honor is at once Indigenous, traditional, and local. We explore such knowledge systems to understand how that knowledge inspires Traditional Resource Management (TRM) practices that favor environmental sustainability, the goal that initially inspired this research enterprise. We also document the forces that disrupt such communities and do what we can to assure their survival for the future.

Epistemological issues

Ecological knowledge is *cultural* knowledge, that is, the collective mental product of the life of a *community* of individuals. Back in the 1950s and 1960s, cultural anthropologists debated the "true" nature of culture, judged to be the essential subject matter of anthropology. I found it then, as now, most congenial to consider culture to be "that which one must know or believe to act appropriately within one's society," to paraphrase Goodenough's succinct characterization

(1957). From this perspective, social relations, behavioral regularities, and artifacts are byproducts of cultural knowledge, are *generated* by that knowledge – which integrates an "image" of the world with "plans" to live in that world. This knowledge is by and large shared by the society's citizens as they go about the business of living in their local habitat. This knowledge need not be conscious nor entirely consistent, but is expressed in thought, memory, commentary, and conversation in the language of the community. Though our actions may speak louder than our words, we must have words to describe our actions. Thus, the *epistemological foundation* for the study of Indigenous Ecological Knowledge is cognitive, ideas and beliefs about the world, but consequent to an intimate, life-long immersion with a local habitat in pursuit of livelihood. Ingold appears to agree: we pursue "the study of how people perceive, act, think, know, learn and remember within the setting of their mutual, practical involvement in the lived-in world" (2000: 171).

Ethnoscience was an anthropological precursor to the study of Indigenous Knowledge. The term as used in the 1960s was ambiguous. On the one hand, it specified a scientific approach to ethnography whereby our investigations of cultural themes might be explicit and replicable. This proved an elusive goal, given the nature of the ethnographic encounter, at its best, intensely personal and not readily reduced to formulaic interactions. On the other hand, ethnoscience defined our subject matter as the science of the people: ethnobiology (or to avoid Latin, "folk biology"), ethnobotany, ethnozoology, ethnomycology, ethnomedicine, etc. Anthropology has argued that our Western academic topics – e.g., kinship, religion, art, law, medicine – are universal human capacities and concerns. We investigated kinship systems radically at odds with our contemporary notions, religions absent supreme beings; art in enacting daily life, justice without law books, judges, lawyers, or jails; medicine without doctors and pharmacists. So we were primed to acknowledge science without scientists, labs, and expensive instruments. Science in the most comprehensive sense is thus universal, the product of human curiosity, creative problem solving grounded in careful observation, reasoned analysis, while comparing notes with our neighbors.

Indigenous science is *natural history*, not far removed from Aristotelian science (Leroi 2014), yet a far cry from the juggernaut of modern, professional scientific endeavors, often presumed to be the "real science." It is "naked-eye" science, focused on describing the world around us first of all, with an eye to practical results, emphasizing the edible and medicinal plants, the insects that sting or bite, while duly appreciative of the aesthetic values in nature. If it is applied science rather than some imaginary "pure" theoretical science, it is not therefore less scientific, given that the professional science of today is likewise constrained by the requirements of funding agencies.

Ontological issues

The recent obsession with ontological relativism in anthropology, the so-called Ontological Turn (Daly et al. 2016), suggests that there is an unbridgeable gulf between the ontological grounding of "modern" societies and those of the "Others" that have to date remained uncontaminated by modernity.

Much has been made of our "Western" scientific objectification of life, attributed more-or-less accurately to Descartes. Are living organisms mere machines programmed by their DNA? By contrast, the perspective from Indigenous communities is at odds with this mechanistic ontology. Rather, "People, animals, plants, and other forces of nature – sun, earth, wind, and rock – are animated by spirit. As such they share with humankind intelligence and will, and thus have moral rights and obligations as PERSONS" (Hunn and Selam 1990: 230). Which side of this ontological fence are we? Anyone with a pet dog will attest to the intelligence and will of their beloved dog. Animism is alive and well. Meanwhile, ecologists are busy demonstrating how individual plants are able to communicate via chemical signaling systems. This suggests that even plants may "have agency." Yet, given the dominant evolutionary paradigm of modern biology, that agency is understood as the product of natural selection, not divine inspiration.

Is religion thus at odds with science by attributing an imaginary soul at the heart of the machinery of nature and ascribing a divine purpose to it all? I think not. There is awe and wonder in the varied pursuits of science, whether in the lab or the field, whether by professor or peasant. Furthermore, I believe there is an indispensable role for moral judgment beyond the purview of science, without which our engagement with the world would be quite flat.

Indigenous communities do not pit "science" against "religion" nor set them apart. There are likely no names for either in the Indigenous languages. Their dealings with their living environment are governed by what we might understand as a fusion of scientific, religious, and moral understandings. Koyukon hunters of bears (Nelson 1983) have an intimate knowledge of their prey and a profound respect bordering on fear of them, imagining that the bear can read their minds, detect any hint of arrogance, which to the bear is galling. Yet they must hunt to live, so they believe it is essential to show the bear, living and dead, the most meticulous courtesy and honor, by way of recompense for taking its life. Indigenous hunters often find the techniques of wildlife biologists – such as the darting and radio-tagging of animals to track their movements – offensive to the personhood of the animals. This is an ontological conflict, but not fatal, in my estimation, for respectful collaboration between university-trained wildlife biologists and Indigenous experts (contra Nadasdy 1999). Witness the productive collaboration between Inuit whaling-boat captains and government biologists in assessing the true status of bowhead whale populations in the Arctic Ocean near Barrow. The pessimistic estimates of the biologists proved unfounded after cooperative surveys employing aircraft, sophisticated sonar, and traditional expertise. A sustainable subsistence harvest was devised acceptable to both communities (Huntington 1992: 110–115).

Proponents of the "Ontological Turn" have argued that we cannot simply describe the beliefs of exotic cultures, such as their animistic assertion that natural elements are "persons," intelligent and moral beings. Rather, we must accept these assertions as literally true if we are to escape our own ontological prison (Kohn 2013). I would counter that one need not be a true believer to grasp the power of an alien belief to give life meaning. One need not believe that bears are mind-readers to appreciate the fact that presuming that they are can motivate wise action.

I have argued that the epistemological and ontological impediments to the recognition of Indigenous Ecological Knowledge as a valuable scientific contribution have been exaggerated. However, there remain political barriers to this "meeting of minds." The careful ethnographic documentation of IEK requires a considerable dedication of time, effort, and funding. If such research is considered trivial or controversial by academic research institutions and funding agencies, the traditional knowledge may be lost, as the knowledgeable elders pass on and younger generations are moved to abandon traditional subsistence harvests or are forced off the land. IEK is undeniably fragile. Much of the detail of IEK is acquired in the doing – whether by explicit instruction or more simply by participant observation. "Use it or lose it." Already much has been lost which might have been documented by previous generations of ethnographers.

Political barriers are encountered on both sides. Indigenous activists are intensely aware of past exploitation by outsiders and demand that those who would study their communities – no matter how "pure" the motives from the researchers' perspective – justify their project as beneficial not only to the researcher and "the public" but also to the Indigenous community and engage with the local community on a truly collaborative basis (Smith 1999; Posey and Plenderleith 2004). Researchers must provide a detailed account of how they intend to do their work and obtain "prior informed consent," from individual consultants and responsible community authorities (as well as state agencies where the research is proposed, University Human Subjects Review Committees, and publishers). It is no longer acceptable to "parachute in," then cut and run with the data. Often Indigenous communities become second homes to the researcher, their understanding of the local language and perspectives on the world maturing year by year. I suspect that few students of IEK pursue this line of work casually, given the effort required, but rather will be inspired by a profound respect for the local people and committed to faithfully communicating that appreciation in their subsequent writings and teaching.

An Indigenous Ecological Knowledge Cook's Tour

I would like to share some of my favorite accounts of Indigenous Ecological Knowledge. Several predate the invention of the term. Conklin's dissertation on "The relation of Hanunóo culture to the plant world" (1954) established the ethnobiological enterprise as a worthy endeavor. Conklin's attention to detail, his "fine description" (Kuipers and McDermott 2007), set the methodological standard for all subsequent efforts to document Indigenous botanical knowledge. Berlin, inspired by Conklin's example, collaborated with two distinguished botanists in a comprehensive study of Tzeltal Mayan botanical classification and nomenclature (Berlin, Breedlove, and Raven 1973). He built on this foundation an elaborate perceptual/taxonomic framework for comparative analysis (Berlin 1992). Though initially skeptical of the concordance of Linnaean and Tzeltal recognition of genera and species (Berlin, Breedlove, and Raven 1966), he subsequently revised that view, in part with the benefit of more appropriate measures of taxonomic correspondence (Hunn 1975). Berlin's search for "universal principles" has fallen on hard times, as our emphasis has shifted to appreciating the complex conceptual ecology of particular communities (Hunn 2007).

Ralph Bulmer partnered with a young Kalam hunter, Ian Saem Majnep, to describe in meticulous detail the bird life of Saem Majnep's New Guinea highland community (Saem Majnep and Bulmer 1977). The text alternates between Saem Majnep's Kalam descriptions and Bulmer's academic reflections on the Kalam accounts, which provides a unique binary perspective. An interesting aside: Bulmer reported that after a number of years of collaborative study with the Kalam, a geologist colleague dropped in for a visit at Bulmer's field site. This geologist friend soon was absorbed in a lively conversation with a local. When Bulmer asked what was so interesting, the geologist exclaimed at the fascinating geological details they had discussed. Bulmer was slightly annoyed, asking why, after all these years, this information had never been shared with him. The reply, "You never asked!" In short, you learn best when you share a love of the subject. Many ethnographers, perhaps most, lack the background knowledge necessary to engage in such conversations, and thus fail to appreciate the sophistication of Indigenous Knowledge.

Consider Thomas Gladwin's study of Micronesian navigation and canoe design (Gladwin 1970). Lacking magnetic compasses and our contemporary GPS satellite technology, Micronesian master navigators guided their sophisticated outriggers across hundreds of miles of trackless ocean to make landfall on tiny atolls, to visit relatives or hunt delicacies. These navigators relied on star paths, patterns of interacting oceanic swells, the flight of seabirds, and a somewhat mysterious mnemonic process to tack into prevailing winds. (They also put great store by what appear to be magical encounters *en route*.) Recent revivals of this endangered Indigenous Knowledge by native Hawaiians have retraced the 2,000-plus mile journey that first brought humans to Hawaii (Finney 1994). This traditional technology is not likely to replace our reliance on GPS. However, it is an inspiring example of human perspicacity and rationality.

The !Kung San people of the Kalahari in what is now Botswana and Namibia are descended from ancestral populations who have lived by hunting and gathering in this harsh savannah country for at least 50,000 years. In 1970, a leading expert on the ethology of large African mammals, Nicholas Blurton Jones, visited anthropologist Melvin Konner at his field site. They organized "seminars" with local San hunters at three local San villages, sitting together around evening campfires. The "Western" scientist posed questions from his studies of animal behavior, which Konner translated into the local language for the half-dozen hunters gathered at each camp (Blurton Jones and Konner 1976). The hunters were eager participants, never tiring of discussing among themselves and with Konner their often quite personal conclusions to the riddles posed by Blurton Jones. By and large, their observations squared with those documented in the scientific literature. On occasion, they argued points that Blurton Jones found hard to believe, such as their obstinate assertion that lions were fastidious eaters, carefully removing and burying the intestines of prey, even rejecting meat that had been soiled by feces when innards were spilled during a kill. However, subsequent investigations proved the San hunters to be correct in nearly all these respects (ibid.: 331–332). In other instances, the hunters vigorously disputed with one another the reliability of some inferred conclusion, such as that elephants "bury their babies up to their necks in sand," as one young hunter had deduced but which his fellows considered laughable (ibid.: 331). The authors conclude: "That !Kung have an advanced ability to observe and assemble facts about behavior and to discriminate facts from hearsay and interpretation. In this ability they surpass lay observers and many professionals in western society." Furthermore, "Their motivation for acquiring knowledge about what animals do goes far beyond the immediate, momentary needs of hunting" (ibid.: 344). However, !Kung hunters seemed rather disinterested in explaining their observations, at least in terms comparable to the theoretical emphases of professional biologists. Lions do what lions do, might sum it up. This "seminar" approach, akin to contemporary "focus groups," seems a most productive strategy for IEK "rapid assessment," but clearly requires that the interlocutors share mutual trust, a deep appreciation for the common subject matter, and fluency in a common natural language.

!Kung San hunters are hardly unique in their meticulous observation and sophisticated analysis of the ecological and behavioral constancies of their prey. Richard Nelson's collaboration research with Inuit and Koyukon Indigenous hunters in Arctic Alaska is most informative.

Nelson first set out to study the survival strategies at Wainwright, an Inuit village on Alaska's North Slope, funded by the United States Airforce and the Office of Naval Research (1969: xvii). His funding was based on the assumption that Arctic hunters would have developed over centuries of Arctic living superior strategies for living on the Arctic sea ice. However, as Nelson's academic advisor, William S. Laughlin, notes in his Foreword, "Hunting behavior cannot be studied in the warmth of a kitchen; it must be studied where and when it takes place." Furthermore,

> The observer must be trained by the Eskimos [now better known as Inuit in this part of the Arctic], he must travel with Eskimo hunters and be able to help rather than hinder the efforts of his more knowledgeable Eskimo colleagues. This in turn requires a broad background amenable to Eskimo tuition, in addition to physical agility, endurance, and a substantial sense of humor adequate to withstand and enjoy the joking and ridicule which the Eskimos may use as part of their instructional system.
>
> *(ibid.: xv)*

These requirements for Nelson's research are no doubt extreme, but suggestive of what is necessarily involved in a convincing study of IEK. Nelson describes a strategy for hunting polar bears, an always dangerous adversary on the sea ice. The hunter cannot afford to miss, or the wounded bear may well kill the hunter first. To be sure of a fatal shot, the hunter first shoots at the bear's haunch knowing that the bear will react by biting the wound, which thus exposes the bear's neck where a second bullet will stop it in its tracks (ibid.: 201–202). Meanwhile, in the Bering Sea, Savoongan walrus hunters name and carefully distinguish 99 varieties of sea ice, detailing in their native Yupik language when, where, and how to recognize and utilize each (Oozeva et al. 2004).

Robert Johannes – a student of tropical fish biology – engaged Palauan fishermen in an extended conversation about the lives of reef fishes (Johannes 1981). He notes that this Indigenous Knowledge significantly advanced our scientific understanding of tropical fisheries, as local observers could rely on consecutive lifetimes of close daily monitoring of local reefs. Their lives depended on it. He details also traditional conservation practices disrupted by colonial occupation. Some of these traditional conservation strategies are being successfully revived (Johannes 2002).

Let me cite four of my own ethnographic studies. Mayan peasants living on their own resources on traditional lands in the highlands of Chiapas, Mexico, distinguish nearly 50 species of social hymenoptera, ants, bees, and wasps, by Tzeltal name. They have learned to differentiate many of these species by their nest architecture, anticipating a key strategy of professional entomologists (Wilson 1971; Hunn 1977: 267). Elders of the Indigenous bands and tribes of the Columbia River basin of the Pacific Northwest, native speakers of the Sahaptin language, name over a dozen species of "desert parsleys" of the genus *Lomatium* of the carrot family. Many are key starchy staples in their traditional diet, others are recognized for their medicinal value or used to poison fish (Hunn and French 1981). Meanwhile, professional botanists struggle to make sense of this diverse genus.

My Zapotec natural history project in San Juan Mixtepec, Oaxaca, Mexico, revealed the astounding fact that children as young as 6 learned to recognize several hundred local species of plants and to name them in their native Zapotec language (Hunn 2008). They did not gain this botanical expertise in school, as their teachers, speaking Spanish and at home in the city, knew nothing of the local flora. Rather, they soaked it up eagerly from their families working and playing in the gardens, fields, and forests of their town. Their parents value "their forests" and resist efforts to open them to commercial harvest and in the process conserve habitat critical to the lives of nearly 200 bird species (Alcántara-Salinas et al. 2015). I recorded some 270 species of local plants used to treat some 100 illnesses, again named in Zapotec. This expert knowledge was not esoteric shamanistic lore, as popular accounts might suggest (cf. Plotkin 1994), but widely shared within the community.

I participated in a study commissioned by Glacier Bay National Park at the behest of the local Huna Tlingit community to evaluate the impact of their traditional harvest of seagull eggs from a colony now within the park boundaries (Hunn et al. 2003). We learned that the Huna valued this spring egg harvest for its culinary and social contribution. We also learned that the Huna managed these harvests by carefully selecting only eggs from nests with just one or two eggs, leaving nests with three to hatch. This proved an effective conservation strategy, given the gull's reproductive strategy and modal clutch size of three. This traditional resource management strategy now contributes to a cultural and linguistic renaissance in these Indigenous communities. In the Huna case, documenting their strategic harvest led directly to a modicum of legislative relief for the Huna, now authorized to resume, with strict limits, their traditional egg harvests within the National Park.

Micronesians are fisher folk home to islands scattered across the western equatorial Pacific. The Hanunóo, the Kalam, and the Mayans are subsistence farmers who have carved out fields from the tropical and montane forests of Mindanao, New Guinea, and Central America. Columbia Plateau and Huna Tlingit communities of the Pacific Northwest harvest wild salmon and suckers, roots and berries, deer and elk, traveling by season to key familial harvest sites. Let us expand our coverage.

Traditional pastoralists manage their herds by moving to fresh pasture as the seasons dictate. James Mandaville's *Bedouin Ethnobotany* (2011) demonstrates the depth of the pastoralists' botanical expertise. His historical investigations show furthermore the impressive persistence of this "traditional ecological knowledge." An Arab scholar, Abū Ḥanīfah ad-Dīnawarī, authored a treatise dated to c. AD 895 on desert plants, relying heavily on Bedouin sources. He notes a key ecological distinction between two types of plants in terms of the nutritional requirements for camels, precisely as contemporary Bedouin herders distinguish plants that are *ḥamḍ* "like meat" versus those that are *khillah* "like bread," the first salty, the second lacking salt (ibid.: 339). They search for the appropriate plants to properly provision their camels. Krohmer (2010) shows how finely detailed an Indigenous pastoralist image of the local landscape can be in the African Sahel.

Zsolt Molnár (2012) likewise captures this complex environmental knowledge for the cattle herders of the Hortobágy steppe of the Great Hungarian Plain. What might appear a seemingly endless and monotonous expanse of short grass prairie is shown to be a complex mosaic of discriminable microhabitats harboring nearly 200 species of plants, each affording local herders critical resources through the seasons. The author offers his ethnographic contribution – the product of years of collaborative research – "to help herdsmen to be proud of their knowledge and be willing to pass it on to the next generation, and the young generation willing to accept this knowledge without hesitation" (ibid.: 139).

The question of Indigenous conservation

A key question is whether Indigenous Peoples infer from their detailed knowledge of their local environment the necessity to conserve those resources for the future. If not, our efforts to document such knowledge would seem misguided, as that was our starting point. Skeptics have argued to the contrary, that despite their environmental expertise they are as likely as the rest of us to squander their resources in the pursuit of short-term payoffs. For example, Diamond (1986), Krech (1999), and Smith and Wishnie (2000) are skeptical of Indigenous conservation in practice. This issue is critically important in that our judgment of the evidence may well determine whether those in power will decide if Indigenous communities are allies or antagonists in pursuit of biodiversity conservation. If judged a threat to conservation goals, they could be removed from their traditional homes in an attempt to set aside ecologically rich areas in the interest of conserving threatened biodiversity or, more likely, to facilitate resource extraction or some other development project. If judged to be effective stewards of their local territories, they may be granted legal protection for the occupation of their homelands, without which it is unlikely their communities and the traditions of ecological knowledge that have sustained them for centuries will survive. (As students of Indigenous Knowledge, we must do all we can to support these communities as they struggle to survive.)

The issue should not be reduced to a choice between opposites, nor should we judge the question from ideological first principles. Rather, we should let the facts speak. Richard Nelson confronted this controversy by comparing his experience with Inuit, on the one hand, and Athabaskan Indians, on the other, both Arctic hunting peoples with whom he had long-term personal engagement (Nelson 1982). Inuit, it seemed, were less explicitly concerned to husband their local resources than the Koyukon Athabaskans. Were the Inuit on the tundra therefore profligate despoilers while the Indians of the forests to the south dedicated conservationists? Nelson argues rather that the Koyukon are in a better position to assess the impact of their hunting pressure on their local prey populations and thus to reason that one should not "take too much out of it right now or you'll get nothing later on," as one elder remarked. His wife added that, "People never kill animals for no reason, because there's times when they'll really need to kill anything they can find" (Nelson 1983: 200). Nelson details the complex rituals that restrain Koyukon hunting, their fear of offending the spirits of their prey, and the retribution they expect for such disrespect. Thus, they conserve their prey for both practical and, if you will, spiritual reasons. Nelson suggests that this restraint might be attributable to the fact that key Koyukon resources are relatively sedentary, such as moose and beaver, and that hunters may readily judge the impact of their hunting on their prey populations. By contrast, Inuit hunters pursue highly mobile prey, such as caribou, whales, and seals. The impact of their harvest of such prey is thus far more difficult to judge, and therefore the necessity to restrain their harvests far less obvious.

It is tempting to assume that conservation is a moral imperative everywhere and at all times a good thing. However, it should be obvious that one does not conserve without good reason, given that there will be a cost involved. It may be easier or more profitable to take more than one needs, as we see in the notorious cases of the slaughter of the American bison for their hides, leaving the carcasses to rot, or the slaughter of African elephants for their ivory tusks – in high demand on the global black market. Is it simply "human nature" to maximize profit, to minimize care if there is a cost incurred? This is the logic behind the "Pleistocene overkill" hypothesis promoted by the late paleontologist, Paul S. Martin (1973). However, the evidence to support his analysis is extraordinarily weak (Grayson and Meltzer 2003). There is neither time nor space here to properly refute this popular account, but the assumptions on which it is based are without merit. One key issue is the demand factor. If a resource is harvested to feed global markets, demand is for all practical purposes infinite. If the resource is harvested for subsistence, demand is strictly finite. A key factor that will constrain subsistence harvests by Indigenous communities in most times and places is the concern to preserve a way of life for the future, the *longue durée* of indigeneity. Given that, while strategic choices may well obey an economic imperative such as "least effort for maximum payoff," scarce resources will likely be subject to a countervailing conservative principle. This is not designed to conserve biodiversity for its own sake, but rather to conserve the life of one's family and community, "to the seventh generation," according to the Great Law of the Iroquois, with the incidental outcome of conserving local biodiversity (cf. Hunn et al. 2003).

I would like to include one final example of IEK, this regarding TRM, that is, Traditional Resource Management, the systematic use of fire by Indigenous Peoples from the Australian outback and by North American Indians before Smokey the Bear embarked upon his fire-fighting crusade. Krech dismisses the evidence for the sophisticated application of Indigenous fire technology for habitat modification as mere pyromania (Krech 1999). The evidence that fire was a powerful tool applied with care for a range of practical purposes – e.g., driving game, initiating ecological succession for useful plants, clearing campsites, harvest areas, and routes of travel – and that the repeated burning at short time intervals enhanced the health of forests and grasslands, is a lesson modern-day fire ecologists are finally coming to appreciate, after a century or more of mismanagement motivated by the narrow interests of timber harvesters and range management for cattle. However, Indigenous fire management was (since it is rarely feasible in the present context) quite complex, often ritualized, and highly localized. The Australian aboriginal case is illustrative of the difficulty of reviving pre-settlement burning practices today. In Australia's Northern Territory, Kakadu National Park is now jointly managed by professional resource managers and representatives of the original Aboriginal Owners (Lewis 1989). In their tradition, specific clans were spiritually obligated to keep their particular clan lands and sacred sites "clean" by means of periodic burning. However, the times and places of burning depend on both ritual considerations and a detailed understanding of how fire behaves within each micro-habitat. Thus, one cannot simply transfer the local knowledge of one's home territory to the terrain of a neighbor's clan. Given the disruption of traditional territorial occupation, it is now rarely possible to match knowledgeable elders of one clan or another with the lands for which they have a spiritual obligation and with that the detailed knowledge required to burn appropriately. In short, we may now be able to appreciate the sophistication of Aboriginal fire technology only in the abstract, with the fine detail of its practical application no longer accessible.

Conclusion

In conclusion, we must reject the argument that Indigenous Knowledge is necessarily magical, misguided, or misanthropic. I doubt that IEK will provide us with some magic bullet to cure all ills or to avoid global ecological catastrophe. IEK is tied to particular places and addresses small-scale challenges. Yet, Indigenous communities today occupy, and in many cases have effective control of, habitats that provide shelter for much critical biodiversity (Maffi 2001; Alcántara-Salinas et al. 2015). Such communities are key allies in our efforts to protect the planet for the future. IEK is also instructive with regard to the value of a "natural" diet, the effectiveness of certain herbal cures for common ailments, the use of targeted burning to managed fuel loads in our national forests, or the most efficient and sustainable use of our agricultural lands. Most important, I believe, is that Indigenous Knowledge sustains a sense of purpose and pride for Indigenous Peoples. But only if we grant that we share with our Indigenous colleagues a common human intelligence in the service of a meaningful life.

References

Agrawal, A. 2002. "Indigenous knowledge and the politics of classification." *International Social Science Journal* 54(3): 287–297.

Alcántara-Salinas, Graciela., Eugene S. Hunn and Jaime E. Rivera-Hernández. 2015. "Avian biodiversity in Two Zapotec communities in Oaxaca: The role of community-based conservation in San Miguel Tiltepec, and San Juan Mixtepec." *Human Ecology* 43(5): 735–748.

Berkes, Fikret. 2008. *Sacred Ecology: Traditional Ecological Knowledge and Resource Management*. 2nd edn. New York: Routledge.

Berlin, Brent. 1992. *Ethnobiological Classification: Principles of Categorization of Plants and Animals in Traditional Societies*. Princeton, NJ: Princeton University Press.

Berlin, Brent, Dennis E. Breedlove, and Peter H. Raven. 1966. "Folk taxonomies and biological classification." *Science* 154: 273–275.

Berlin, Brent, Dennis E. Breedlove, and Peter H. Raven. 1973. *Principles of Tzeltal Plant Classification: An Introduction to the Botanical Ethnography of a Mayan-Speaking Community of Highland Chiapas*. New York: Academic Press,

Blurton Jones, Nicholas and Melvin J. Konner. 1976. "!Kung knowledge of animal behavior (or: The proper study of mankind is animals)." In Richard B. Lee and Irven DeVore (Eds.), *Kalahari Hunter-Gatherers: Studies of the!Kung San and Their Neighbors*. Cambridge, MA: Harvard University Press. pp. 325–348.

Brokenshaw, David, David M. Warren, and Oswald Werner. 1980. *Indigenous Knowledge Systems and Development*. Lanham, MD: University Press of America.

Conklin, Harold C. 1954. "The relation of Hanunóo culture to the plant world." Unpublished PhD dissertation, Yale University, New Haven, CT.

Daly, Lewis, Katherine French, Theresa L. Miller, and Liúsiach Nic Eoin. 2016. "Integrating ontology into ethnobotanical research." *Journal of Ethnobiology*, 36(1): 1–9.

Diamond, Jared. 1986. "The environmentalist myth." *Nature* 324: 19–20.

Finney, Ben. 1994. *Voyage of Rediscovery: A Cultural Odyssey through Polynesia*. Berkeley, CA: University of California Press.

Gladwin, Thomas. 1970. *East Is a Big Bird: Navigation and Logic on Puluwat Atoll*. Cambridge, MA: Harvard University Press.

Goodenough, Ward H. 1957. "Cultural anthropology and linguistics." In *Report of the Seventh Annual Roundtable Meeting on Linguistics and Language Study*. Georgetown University Monography Series on Language and Linguistics No. 9.

Grayson, Donald K., and David J. Meltzer. 2003. "A requiem for North American overkill." *Journal of Archaeological Science* 30: 585–593.

Heckler, Serena. 2009. "Introduction." In Serena Heckler (Ed.), *Landscape, Process, and Power: Re-evaluating Traditional Environmental Knowledge*. New York: Berghahn Books, pp. 1–18.

Hunn, Eugene S. 1975. "A measure of the degree of correspondence of folk to scientific biological classification." *American Ethnologist* 2: 309–327.

Hunn, Eugene S. 1977. *Tzeltal Folk Zoology: The Classification of Discontinuities in Nature*. New York: Academic Press.

Hunn, Eugene S. 1993a. "What is TEK?" In Nancy M. Williams and Graham Baines (Eds.), *Ecologies for the 21st Century: Traditional Ecological Knowledge, Wisdom for Sustainable Development*, Report of the Traditional Ecological Knowledge Workshop, Centre for Resources & Environmental Studies, Australian National University, Canberra, Australia, pp. 11–17.

Hunn, Eugene S. 1993b. "The Ethnobiological Foundations for TEK." In Nancy M. Williams and Graham Baines (Eds.), *Ecologies for the 21st Century: Traditional Ecological Knowledge, Wisdom for Sustainable Development*, Report of the Traditional Ecological Knowledge Workshop, Centre for Resources & Environmental Studies, Australian National University, Canberra, Australia, pp. 18–29.

Hunn, Eugene S. 2007. "Ethnobiology in four phases" *Journal of Ethnobiology* 27: 1–10.

Hunn, Eugene S. 2008. *A Zapotec Natural History: Trees, Herbs, and Flowers, Birds, Beasts, and Bugs in the Life of San Juan Gbëë*. Tucson, AZ: University of Arizona Press.

Hunn, Eugene S., and David H. French. 1981. "*Lomatium*: A key resource for Columbia Plateau native subsistence." *Northwest Science* 55: 87–94.

Hunn, Eugene S., Darryll Johnson, Priscilla Russell, and Thomas F. Thornton. 2003. "Huna Tlingit traditional environmental knowledge and the management of a 'wilderness' park." *Current Anthropology* 44(S5): 79–104.

Hunn, Eugene S., with James Selam and Family. 1990. *Nch'i-Wána 'The Big River': Mid-Columbia Indians and Their Hunn Land*. Seattle, WA: University of Washington Press.

Huntington, Henry P. 1992. *Wildlife Management and Subsistence Hunting in Alaska*. Seattle, WA: University of Washington Press.

Ingold, Tim. 2000. *The Perception of the Environment: Essays in Livelihood, Dwelling and Skill*. London: Routledge.

Johannes, Robert E. 1981. *Words of the Lagoon: Fishing and Marine Lore in the Palau District of Micronesia*. Berkeley, CA: University of California Press.

Johannes, Robert E. 2002. "The renaissance of community-based marine resource management in Oceania." *Annual Review of Ecology and Systematics* 33: 317–340.

Johnson, Leslie Main, and Eugene S. Hunn (Eds.) 2010. *Landscape Ethnoecology: Concepts of Biotic and Physical Space*. New York: Berghahn Books.

Kohn, E. 2013. *How Forests Think: Towards an Anthropology Beyond the Human*. Berkeley, CA: University of California Press.

Krech, Shepard, III. 1999. *The Ecological Indian: Myth and History*. New York: Norton.

Krohmer, Julia. 2010. "Landscape perception, classification, and use among Sahelian Fulani in Burkina Faso." In Leslie Main Johnson and Eugene S. Hunn (Eds.), *Landscape Ethnoecology: Concepts of Biotic and Physical Space*. New York: Berghahn Books, pp. 49–82.

Kuipers, Joel, and Ray McDermott (Eds.) 2007. *Fine Description: Ethnographic and Linguistic Essays by Harold C. Conklin*. New Haven, CT: Yale University Press.

Leroi, Armand Marie. 2014. *The Lagoon: How Aristotle Invented Science*. New York: Viking.

Lewis, Henry T. 1989. "Ecological and technological knowledge of fire: Aborigines versus park managers in northern Australia." *American Anthropologist* 91: 940–961.

Maffi, Luisa (Ed.) 2001. *On Biocultural Diversity: Linking Language, Knowledge, and the Environment.* Washington, DC: Smithsonian Institution Press.

Mandaville, James P. 2011. *Bedouin Ethnobotany: Plant Concepts and Uses in a Desert Pastoral World.* Tucson, AZ: University of Arizona Press.

Martin, Paul S. 1973. "The discovery of America." *Science* 179: 969–974.

Menzies, Charles R., and Caroline Butler. 2006. "Understanding ecological knowledge ." In Charles R. Menzies (Ed.), *Traditional Ecological Knowledge and Natural Resource Management.* Lincoln, NE: University of Nebraska Press, pp. 1–17.

Molnár, Zsolt. 2012. *A Hortobágy Pásztorszemmel: A Puszta Növényvilága* [Traditional Ecological Knowledge of Herders on the Flora and Vegetation of the Hortobágy]. Debrecen, Hungary: Hortobágy Természetvédelmi Közalapítvány.

Nadasdy, Paul. 1999. "The politics of TEK: Power and the 'integration' of knowledge." *Arctic Anthropology,* 36(1–2): 1–18.

Nelson, Richard K. 1969. *Hunters of the Northern Ice.* Chicago: University of Chicago Press.

Nelson, Richard K. 1982. "A conservation ethic and environment: The Koyukon of Alaska." In Nancy M. Williams and Eugene S. Hunn (Eds.), *Resource Managers: North American and Australian Hunter-Gatherers.* Washington, DC: American Association for the Advancement of Science, pp. 211–228. Reprinted by the Australian Institute of Aboriginal Studies, Canberra, Australia.

Nelson, Richard K. 1983. *Make Prayers to the Raven: A Koyukon View of the Northern Forest.* Chicago: University of Chicago Press.

Oozeva, Conrad, Chester Noongwook, George Noongwook, Christina Alouwa, and Igor Krupnik. 2004. *Watching Ice and Weather Our Way.* Washington, DC: Arctic Studies Center, National Museum of Natural History, Smithsonian Institution.

Plotkin, Mark J. 1994. *Tales of a Shaman's Apprentice: An Ethnobotanist Searches for New Medicines.* New York: Penguin Books.

Posey, Darrell A. with Kristina Plenderleith (Eds.) 2004. *Indigenous Knowledge and Ethics: A Darrell Posey Reader.* London: Routledge Harwood Anthropology.

Saem Majnep, Ian and Ralph Bulmer. 1977. *Mnmon Yad Kalam Yakt: Birds of My Kalam Country.* Auckland, NZ: Auckland University Press.

Sillitoe, Paul. 1998. "Defining indigenous knowledge: The knowledges continuum." *Indigenous Knowledge and Development Monitor* 6(3): 14–15.

Smith, Eric A. and Mark Wishnie. 2000. "Conservation and subsistence in small-scale societies." *Annual Review of Anthropology* 29: 493–524.

Smith, L. Tuhiwai. 1999. *Decolonizing Methodologies: Research and Indigenous Peoples.* London: Zed Books.

The Economist, 2015 "The tricky work of measuring falling global poverty." October 12.

Williams, Nancy M. and Graham Baines (Eds.) 1993. "Ecologies for the 21st century: traditional ecological knowledge, wisdom for sustainable development." Canberra, Australia: Report of the Traditional Ecological Knowledge Workshop, Centre for Resources & Environmental Studies, Australian National University.

Williams, Nancy M. and Eugene S. Hunn (Eds.) 1982. *Resource Managers: North American and Australian Hunter-Gatherers.* Washington, DC: American Association for the Advancement of Science. Reprinted by the Australian Institute of Aboriginal Studies, Canberra, Australia.

Wilson, E. O. 1971. *The Insect Societies.* Cambridge, MA: Belknap Press.

3

CONTEXT MATTERS

The holism and subjectivity of environmental knowledge

Chris S. Duvall

Introduction

In this chapter, I want to make three points. First, I want to underscore the distinction between what I call "environmental" knowledge from knowledge of particular geographic or biological features, such as plants, animals, or soils. This distinction is crucial. People, regardless of language or culture, tend to rely upon holistic—that is, environmental—understanding of biophysical reality when making decisions about resource use, even though academic researchers have tended to emphasize particularistic aspects of Indigenous Knowledge systems. Second, there is no objective or universal form of environmental knowledge, ultimately because most geographic features exhibit continuous variation, and do not form discrete units that might be distinguished cross-culturally. Since there is no singular way to divide continua into discrete units, environmental knowledge is inevitably subjective. Finally, the importance and subjectivity of environmental knowledge together create remarkable possibilities for new concepts to enter into environmental knowledge systems. Among these possibilities, Indigenous Knowledge has the potential to transform environmental science, within an overarching project to decolonize institutions of education and governance.

The theoretical bases of my argument are twofold. I am convinced that similar processes of knowledge production and application exist across human cultures, including ethnolinguistic and scientific cultures (Agrawal 1995). Different knowledge cultures include divergent worldviews, but the processes through which new ideas are formed and enacted are not strongly divergent, despite strong cosmological differences between peoples. Of course, my understanding is constrained by my personal experience as a non-Indigenous Westerner and trained academic (Radcliffe 2017; Whitson 2017). I approach my topic based upon scholarly literature, particularly the fields I'll label ethnoscience (that is, analyses of Indigenous Knowledge systems) and science studies (analyses of the philosophy and practice of formal, academic research). I will discuss Indigenous Knowledge I have encountered in Mali, as well as scientific knowledge I have encountered in my professional life.

DOI: 10.4324/9781315270845-4

The *whole* context matters

I began to think about Indigenous Environmental Knowledge as a doctoral student, when I found that my own environmental knowledge did not work in the West African landscape where I was doing research. As a PhD student, I was not completely new to the middle Bafing River valley in western Mali (Figure 3.1). I had done my Master's thesis research there, after having lived in the valley for two years as a U.S. Peace Corps Volunteer. It would have been presumptuous to think that these experiences made me an expert on the knowledge that Maninka farmers have about the environments where they live, but I was oblivious rather than presumptuous.

I discovered my problem when I began trying to understand why my hosts decided to farm in some locations, but not others. I figured that this knowledge could be accessed by translating words between my language, English, and theirs, Maninkakan. Yet bilingual dictionaries had no translations for "slope," "valley," "forest," or so many other words I thought must certainly be relevant in agricultural decision-making. Further, when I asked someone what their word was for that thing over there—as I pointed toward what looked to me like a hill, or some other particular feature—I never seemed to get the same answer. In the end, I learned that the translation of words does not yield knowledge, at least not directly nor necessarily. The seemingly inconsistent answers I got were perfectly consistent with a system that I did not know, or, more precisely, that I did not know existed. Even more precisely, I had not imagined that there were different systems of environmental knowledge, only different sets of words for plants, animals, soils, and so on.

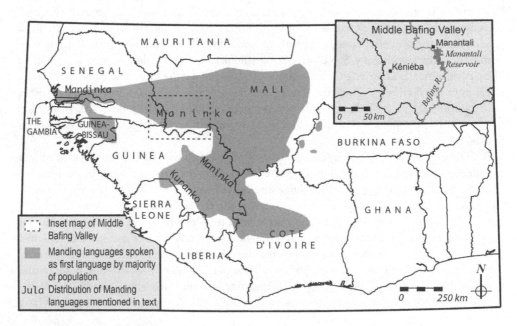

Figure 3.1 West Africa, showing the Middle Bafing River Valley, and the distributions of linguistic groups mentioned in the text

Note: Maninka, Mandinka, and Kuranko are examples of the widespread Manding language family.

Source: Adapted from Figure 1 in Duvall (2008).

Colonial attitudes can be deeply ingrained in Western minds, and neo-colonial institutions of education and governance perpetuate these attitudes. My personal environmental knowledge originated in rural Wyoming (a state in the High Plains region of the U.S.), where my recent ancestors have been mostly ranchers and wage laborers, often with incomes supplemented by gardening, hunting, fishing, and berry-picking. The public education system I entered established a presumption that my geographically specific knowledge was globally applicable. I learned nothing about how Indigenous Peoples understood the environments in which I lived. The U.S., after all, had taken those lands from the Oglala Lakota in the triumphal expansion of settler colonialism. In settler schools like mine, if Indigenous Knowledge existed, it existed only in history or in the distant Indian Reservations spattered across the national map. From kindergarten to college, I learned instead of the universality of my budding scientific knowledge, of the power of scientists to know objectively the patterns and processes of biophysical reality.

After college, in the Peace Corps—a roundly neo-colonial institution, a branch of the U.S. government that sends American volunteers to advance "development" in less-developed countries (Moseley and Laris 2008)—I learned Maninkakan, but not that there existed Maninka environmental knowledge. Other volunteers were doubtlessly more insightful, but I simply translated my message to try and get it across. *My* message; following institutional guidance from the Peace Corps, I determined the message and the words to express it. Since I was then concerned just with plants and animals—*stop cutting down trees! stop hunting!*—dictionaries were helpful. I was dealing with biological nomenclature. Even across language divides, people often similarly perceive different groups of organisms (Berlin 1992; Ellen 1993), so that dictionary-writers have discovered strong, cross-cultural correspondences among plant and animal names. Maninkakan *lè*, French *phacochère*, English *warthog*, and the scientific binomial *Phacochoerus africanus* effectively refer to the same entity. Such pat translations, however, can sustain—and may arise from—the presumption that scientific knowledge is universally valid: if I point at a photo of *Phacochoerus africanus* and you say *lè*, you must have an untrained inkling of my (universal) zoological taxonomy. Of course, it's possible to presume universality for systems of Indigenous Knowledge, but such presumption is characteristic of Western science, which dominates in political-economic and educational institutions worldwide.

In shifting my concern from Malian plants and animals to the decisions of farmers, I moved from thinking about biological to environmental knowledge. Understanding a different culture's biological knowledge is not easy, but understanding environmental knowledge poses a different set of challenges.

To begin, it is necessary to establish what "environment" might be, and why the concept is useful. In academic geography, "environment" has been formally defined as the biophysical conditions, aesthetic values, and human activities that exist in a given location (Mayhew 2009: 159). Environment is a holistic concept, useful for understanding the integration of biophysical and sociocultural, objective and subjective, material and non-material processes in a location (Radcliffe et al. 2009; Gregory 2017). By describing and categorizing environmental conditions—using labels like forest, periglacial, urban, sedimentary, and such—people can characterize and compare locations. Environmental knowledge is emergent knowledge, arising from detailed understanding of particular features yet greater than the sum of its parts.

Concepts analogous to "environment" exist in many non-English-speaking cultures (Burenhult 2008; Johnson and Hunn 2010; Mark et al. 2011). Indeed, awareness of what surrounds us—the ground, the sky, more-than-human social-ecological relations, and so on—is unavoidable and inevitable (Ingold 2000). This awareness is expressed partly via knowledge of particular features, including biological entities like trees, birds, and reptiles, as well as non-biological things like soils, waters, and clouds. Many academic researchers have studied particular aspects of Indigenous Knowledge systems, under umbrella terms like "ethnobotany" (relating to plants), "ethnozoology" (animals), and "ethnopedology" (soils). However, environmental knowledge is relational and embodied, not mechanistically descriptive or coldly intellectual. By focusing on particularistic knowledge, academic researchers may overlook holistic awareness that is central to decisions about resource use (Duvall 2008). People depend upon holistic awareness when deciding where and how to use particular resources, whether for farming, fishing, or some other activity. Hunters do not simply go out and kill warthogs, for instance, but they think about where, why, and when they might find animals they want, and whether they can allowably (according to human and non-human beings) take the animals in any particular location. Further, hunters sense corporeally and without thinking the conditions encompassing their actions. Humans occupy "taskscapes" (arrays of related activities) as well as landscapes (arrays of related features) that are perpetually in flux, because social-ecological relations are constantly changing (Ingold 1993). Humans require holistic awareness to exist in these scapes; more accurately, human existence within taskscapes and landscapes inevitably produces holistic knowledge. "Ethnogeography" and "ethnoecology" have been proposed as the study of such knowledge (Blaut 1979; Toledo 1992), but relatively few researchers have studied in detail how people perceive "environment" within broader world-views.

In Mali, as I began to think less naïvely about the decisions of Maninka farmers, I learned that my hosts paid less attention to soil, vegetation, or other particular features than I thought they would. Certainly, these farmers examined particular features to understand the arability of a site, but they relied much more upon a synthetic view of all characteristics. They favored certain soil types for agriculture, yet found many sites with favorable soil to be non-arable because of topography, vegetation, non-soil characteristics of the ground, human history, or non-human (animal and spirit) activities. Similarly, some sites with less favored soils were arable because the total biophysical conditions were suitable for certain crops. A challenge I had in recognizing the importance of holistic knowledge was that I found no equivalent word to "environment," even though I encountered at least 34 terms that each represent different configurations of biophysical conditions (Duvall 2008: 340). The closest I came to a general term for 'environment' was *dan*, a concept that encapsulates a great range of conditions (Figure 3.2). However, *dan* excludes any sets of conditions that are considered anthropogenic, not because these conditions are necessarily distinct in biophysical terms but because their human past suggests social meanings that limit individual use rights. Sites characterized as *manyang* ("fallow field") and *lemukan* ("arable woodland with loamy soil") often look identical, but only *lemukan* is a type of *dan*; a person's knowledge of local social history was often more important than knowledge of plants or soil in distinguishing *manyang* from *lemukan*. Soil and vegetation characteristics varied within most of the *dan* categories I identified, but not enough for informants to recognize soil- or plant-based subcategories. The only categories for which a single substrate was diagnostic were non-arable, but not all non-arable categories were associated with one substrate.

My hosts were not unusual. Farmers elsewhere in West Africa make decisions based on a synthetic view of soil, vegetation, slope, hydrology, microclimate, human history, and other social conditions, and the activities of animals and spirits (Osunade 1988; Croll and Parkin 1992; Fairhead and Leach 1996; Laris 2002; Osbahr and Allan 2003; Fraser, Leach, and Fairhead 2014). The importance of "environment" in classifying site arability is a widespread but underemphasized aspect of Indigenous Knowledge systems. Published information on Indigenous Environmental Knowledge is limited, because researchers have often defined their study focus narrowly, and seemingly independently of Indigenous thought. With regard to farmers in particular, researchers have focused on soil and characterized what appears to be holistic knowledge as pertaining only to the substrate of a site (Duvall 2008). I'll offer

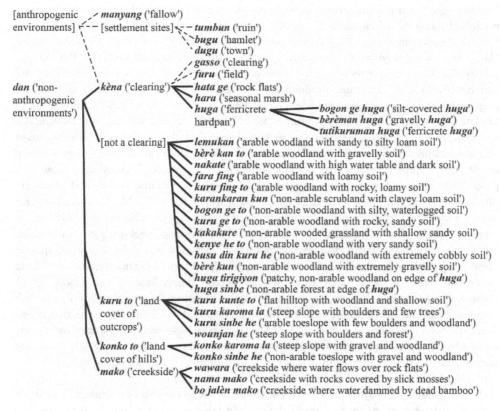

Figure 3.2 Taxonomy of "environment" categories for Maninka speakers in the Middle Bafing Valley, Mali

Note: Brackets indicate covert categories, which are not named but serve to organize thought. Dashed lines connect categories that include human-made features; solid lines connect categories that do not include human-made features. See Duvall (2008), which should be consulted for more information about taxonomic and diagrammatic conventions.

Source: Adapted from Figure 10 in Duvall (2008).

two examples from groups whose languages are related to Maninkakan. First, Carney (1991) describes how Mandinka farmers in The Gambia recognize "micro-environments" based on hydrology, topography, vegetation, and soil. Soil plays a minor role in distinguishing these "micro-environments," yet other researchers have categorized Carney's paper as describing soil knowledge (WinklerPrins 1999; Barrera-Bassols and Zinck 2000). Second, Tabor (1993: 47) translates the Maninka term *huga* as a specific soil type (ferricrete), although this is a type of *dan* (Duvall 2008). Western scientists have noticed ferricrete—hardened, iron-rich soil—for over a century in West Africa, fixating upon it as an indicator of environmental change (Duvall 2011). My Maninka hosts noticed this unusual substrate, which they called *tutikuru*. However, substrate does not define the *huga* concept; *huga* are locations that are not seasonally flooded, and where grasses grow but trees do not, so that sweeping views are possible. I encountered three types of *huga*, in which the substrate is either poorly drained silt, densely packed gravel, or ferricrete; further, two other types of *dan* were associated with ferricrete, both being types of location where trees grow in the hardened substrate.

Although I'm using the term "environment" here, the concept labeled "land" in several published accounts of Indigenous Knowledge seems to be similar (Ingold 2000). For instance, Barrera-Bassols and Zinck (2003) show how Purhépecha farmers in Central Mexico conceive "land" as an integrated whole composed of water, climate, relief, and soils. The Purhépecha classify "land" according to how these four components interact at a given site; they are variables that jointly determine productive potential (ibid.). However, "land" and "soil" are frequently confounded, particularly when researchers claim examples of holistic knowledge to validate their interests in particular features. Thus, Carney's "micro-environments" became soil knowledge to researchers interested in soil knowledge. Soil characteristics are an important and salient aspect of site arability, yet farmers use knowledge about more than soil in selecting arable sites. Similarly, studies that describe Indigenous "vegetation" knowledge show that people rely upon holistic assessments when deciding about resource use, even if plants are often most salient (Fleck and Harder 2000; Verlinden and Dayot 2005). My Maninka hosts, for instance, identified trees and grasses as indicators of several types of *dan*, but the only "environment" with a plant in its name—*bò jalèn mako* ('dried bamboo creekside')—was identified by its topographic location, its soil texture, and by the difficulty of walking easily and quietly through a jumble of bamboo stalks. I don't know whether I should say my hosts were looking at "environment" or "land," but I do know they weren't simply looking at bamboo or soil. Plants are salient in *bò jalèn mako*, because it's difficult to hunt while stumbling noisily through bamboo stalks, and creeksides are valued as hunting grounds. The salience of any particular feature depends upon human activities (Ingold 1993). *Bò jalèn mako* was unimportant and poorly known by the people with whom I spoke, who were not hunters. The people who were not farmers, but teachers or merchants or laborers, didn't know as much about *dan* categories that centered on site arability.

For my Maninka hosts, the biophysical environment always expresses something about social and spiritual relations, partly because in their world-view—at least in what I understand of it—human and non-human beings continuously seek to capture and control valuable resources. Environmental knowledge offers one route to identify the activities of other beings, to avoid social-spiritual dangers, and to gain individual power. For instance, the tree species that I identified in scientific terms as *Dombeya senegalensis* indicated unhealthy conditions produced by spirit activity under the ground surface. Such unhealthiness can exist within *dan* categories generally preferred for farming, showing again that the whole context matters. Farmers who know the significance of this tree can better protect themselves, their families, and their crops from spiritual malevolence. Holistic knowledge is the fundamental basis for decisions about resource use, not particularistic knowledge plants, soil, or other features.

The subjectivity of environmental knowledge

I found that hunters knew more about *bò jalèn mako* than non-hunters, because this type of environment impacted their activities. Women who rarely traveled beyond their village and their fields knew little about *dan*, because they lived almost entirely within anthropogenic environments. Subjective experience shapes individuals' understandings of the world. However, subjectivity is ultimately inescapable in environmental knowledge not because of differences between people, but because of the characteristics of the particular geographic features that are encompassed in holistic views of reality.

To begin, reconsider biological knowledge. When I began working in Mali, I was interested in plants and animals, and found a good correspondence between English, scientific, and Maninka names for organisms. If one language recognized an organism with a distinct word, the others usually did too. More broadly, ethnobiological research has shown that humans classify plants and animals in highly predictable ways across cultures, when the purpose of classification is to identify generally different sets of organisms (Brown 1984). A pan-environmental taxonomy may exist because the plants and animals in any given landscape do not represent continua of variation, but comprise discrete groups of morphologically distinctive organisms separated by objective discontinuities in the observed range of variation (Hunn 1976; Malt 1995).

In contrast, physical features such as landforms, soils, and vegetation are less discrete (WinklerPrins 1999; Ingold 2000). There are objective discontinuities in the biophysical environment—sky/earth, elevation/depression, verdant/senescent—but within the framework suggested by these discontinuities there are few obvious breaks in the range of variation exhibited by particular feature classes. Few features are discrete objects, even if they are conceptualized as such; few have natural boundaries that correspond to objective discontinuities on the Earth's surface, such as rivers, oceans, or cliffs. Nearly all features are some part of a field of continuous variation, such as slope, soil texture, or plant density. The subjectivity of environmental knowledge arises from this fundamental property of geographic features (B. Smith 2001; B. Smith and Mark 2003). Any boundaries defined for geographic features within continuous fields of variation are inherently arbitrary (Bennett 2001; Johnson and Hunn 2010).

The boundaries between individual geographic features are socially constructed, based upon whatever criteria is most important to the people whose thoughts include a particular feature (B. Smith 2001; Smith and Varzi 2000). The constructedness of physical geographic knowledge is evident in comparing how different groups of people understand the same set of features. Williams and Ortiz-Solorio (1981) show that a Nahuatl soil taxonomy from Central Mexico differs from that of technical soil science because the Indigenous taxonomy classifies surface soils (important to Nahuatl farmers) rather than soil profiles (important to soil scientists). Similarly, Zimmerer (1996, 2001) describes how Quechua-speaking Andean farmers in Peru and Bolivia classify "landscape units," some of which correspond to widely used, general geographic terms—like "valley" or "hill"—while others derive from use-values in potato farming—like locations used for early planting or others farmed with ox-plows. Zimmerer further shows that "landscape" knowledge structures potato nomenclature, taxonomy, and horticulture for his hosts. Their particularistic knowledge of the crop arises from their holistic knowledge of the various arrays of biophysical features in which they exist (Zimmerer 1998).

Defined geographic features are merely one among many possibilities for differentiating portions of continuous fields of variation. Concepts like savannah, *manyang*, moraine, or *lemukan* are historically contingent: they emerged at particular moments and have changed within broader sociocultural change (Hacking 2002). Very few physical geographic features are natural kinds, or entities recognized pan-culturally (B. Smith and Mark 2001, 2003). Instead, the knowledge of different sociocultural groups includes unique sets of geographic features that seem objectively self-evident to group members, but are in fact specific to each group. Geographic features become real through social consensus; in turn, environmental categories are as socially constructed as the features that compose them. Again, though, some features are recognized cross-culturally so that strict constructivism is not appropriate for ethnoscientific analyses. In terms of physical geography, broad features like the ground surface, topographic elevations, topographic depressions, and water bodies are widely recognized; social constructions serve to differentiate narrower, meaningful features within these broad and broadly shared categories. Resource-use decisions are based principally upon knowledge of narrower, constructed features.

Changing environmental knowledge

Environmental knowledge is necessary in interacting with biophysical reality, and inescapably social. Consequently, environmental knowledge bears historically contingent political-economic meanings. Indigenous geographic and environmental concepts have histories, but these are, in general, poorly known. Among my Maninka hosts, the idea of *koti* ("slope") traces to the experiences of men who worked on state-sponsored construction projects in which the French word *côte* ("slope") was useful. Their word for quartz is *jaman*, which comes from the French *diamond*, which I believe many Maninka speakers first encountered via colonial prospectors in the early 1900s. I'm sure there was an earlier word for quartz, but I didn't encounter it.

Environmental knowledge is powerful because it relates to ideas shared widely across society; this broad basis is the essence of holistic thought. The concept of "tropical rainforest," for instance, arose within nineteenth-century European botany but reflected contemporaneous belief in Social Darwinism, imperatives for colonial expansion, and imagined geographies of "the tropics" (Stott 1999). The concept helped enable colonial powers to gain and exert political-economic authority, because "tropical rainforest" was constructed as nature's untouched and thus unclaimed bounty. Indigenous Peoples worldwide did not share this concept, even in narrow situations where people recognized something physically analogous to "tropical rain-forest." An example comes from Southern Guinea, among people whose Kuranko language is closely related to Maninkakan. The Kuranko vegetation term *tu* might be translated as "rain-forest," but in most instances this vegetation occurs in an "environment" called *tòmbòndu* ("the land of ruined [i.e. abandoned] villages") (Fairhead and Leach 1996: 129). In its social-cultural context, *tu* vegetation is evidence of past human efforts to improve the agricultural fertility of a site, through the manipulation of plants, soil, animals, fire, spirits, socially determined use-rights, and other factors. The biophysical distinctiveness of dense forest with tall trees might be recognized across language divides, but the whole meaning of such plant assemblages differs between human groups.

The inevitable subjectivity of environmental knowledge makes it a slippery subject of thought. There is no inherently correct or incorrect way to describe or categorize environments. The usefulness of an environmental concept depends upon social context (Robbins 2001a, 2001b; Simon 2010, 2016). Critical thinking is possible within environmental knowledge systems, if it is possible to question why a particular environment might or might not be considered to exist in a location. My Maninka hosts thought critically about environmental description, such as I encountered once when two different men separately gave me different "environment" terms for a single location. The first said *lemukan* (a category of *dan*, meaning it lacks a human past); the second said *manyang* (fallow land). When I asked the second man about this discrepancy, he guessed that the first man wanted to farm the site, and was trying to obscure its long-past history of cultivation in order to claim it as land that might be freely claimed. The essence of holistic knowledge is that it entails many bits of particularistic knowledge. Environmental concepts are useful because they describe locations as hosting a defined array of particular features. However, since the qualities of all particular features vary continuously over space and time, the existence of any environment can be questioned simply by selecting a different array of features as more salient or relevant in a given context.

The contingent and potentially strategic character of environmental knowledge creates remarkable possibilities for new concepts to gain power. The history of science offers many examples of new concepts becoming fundamentally important in understandings of biophysical reality. For instance, "savanna" is one of the principal environmental categories in academic thought about Africa nowadays, but this idea has a readily traceable history in Western thought

(Duvall, Butt, and Neely 2018). A traveler first saw "savanna" in Africa—in Senegambia—in 1738 (Moore 1738), but scientists began seeing it only gradually over the following centuries. By 1880, savanna covered 21.3 percent of Africa, 37 percent by 1923, and 65 percent by 1995 (White 1892: 53; Shantz and Marbut 1923: 57; Archibold 1995: 61). Biophysical conditions on the continent did not change as drastically as these numbers suggest. Instead, observers increasingly found the array of features they called "savanna" to be more salient and relevant than the different arrays that they had previously noticed. Examples like "savanna" and "tropical rainforest" do not show that environmental scientific knowledge is especially arbitrary, but only that environmental knowledge in general is subjective and contingent upon historical context. Science is merely better documented than Indigenous Knowledge, so that illustrative cases are easy to find.

One possibility that could be pursued by exploiting the openness of environmental knowledge systems is to work toward the decolonization of educational institutions. Scholars and activists have made efforts to "indigenize the academy" by developing culturally responsive pedagogies and curricula (L.T. Smith 1999; Stauss 2002; Mihesuah and Wilson 2004). This broad project does not entail the rejection of science, but emphasizes cultural sovereignty in educational theory and practice. A potential desire that may exist is to bring Indigenous perspectives into the theory and practice of environmental science. From my perspective, settler-society schools in the U.S. would benefit students by accepting cultural plurality in environmental science curricula. The scientific method, for instance, is an epistemology and does not depend upon a single set of environmental concepts. If environment concepts are taught as arbitrarily defined arrays of particular features, an education system can help students understand how and why people might think differently about biophysical reality. Teaching people that environmental concepts are inescapably subjective can strengthen scientific knowledge by weakening facile characterizations, that "tropical rainforest" is virgin nature, that "savanna" covers 65 percent of Africa, and so on. Facile environmental characterizations have widely enabled colonial and neo-colonial governance (Duvall, Butt, and Neely 2018). Challenging the objectivity of environmental knowledge opens space for Indigenous ideas within dominant education systems.

References

Agrawal, Arun. 1995. "Dismantling the divide between indigenous and western knowledge." *Development and Change* 26: 413–439.

Archibold, O.W. 1995. *Ecology of World Vegetation*. London: Chapman & Hall.

Barrera-Bassols, Narciso, and Alfred Zinck. 2000. *Ethnopedology in a Worldwide Perspective: An Annotated Bibliography*. Enschede, The Netherlands: International Institute for Aerospace Survey and Earth Sciences.

Barrera-Bassols, Narciso, and Alfred Zinck. 2003. "Ethnopedology: A worldwide view on the soil knowledge of local people." *Geoderma* 111: 171–195.

Bennett, Brandon. 2001. "What is a forest? On the vagueness of certain geographic concepts." *Topoi* 20: 189–201.

Berlin, Brent. 1992. *Ethnobiological Classification: Principles of Categorization of Plants and Animals in Traditional Societies*. Princeton, NJ: Princeton University.

Blaut, James M. 1979. "Some principles of ethnogeography ." In Stephen Gale and Gunnar Olsson (Eds.), *Philosophy in Geography*. Dordrecht, The Netherlands: D. Reidel, pp. 1–7.

Brown, Cecil H. 1984. *Language and Living Things: Uniformities in Folk Classification and Naming*. New Brunswick, NJ: Rutgers University Press.

Burenhult, Niclas. 2008. "Language and Landscape: Geographical Ontology in Cross-Linguistic Perspective." Special issue of *Language Sciences* 30(2–3): 135–382.

Carney, Judith. 1991. "Indigenous soil and water management in Senegambian rice farming systems." *Agriculture and Human Values* 8: 37–48.

Croll, Elisabeth and David Parkin. 1992. *Bush Base: Forest Farm: Culture, Environment and Development.* London: Routledge.

Duvall, Chris S. 2008. "Classifying physical geographic features: The case of Maninka farmers in Southwestern Mali." *Geografiska Annaler: Series B, Human Geography* 90: 327–348.

Duvall, Chris S. 2011. "Ferricrete, forests, and temporal scale in the production of colonial science in Africa." In Mara J. Goldman, Paul Nadasdy, and Matthew D. Turner (Eds.), *Knowing Nature: Conversations between Political Ecology and Science Studies*. Chicago: University of Chicago Press, pp. 113–127.

Duvall, Chris S., Bilal Butt, and Abigail H. Neely. 2018. "The trouble with savanna and other environmental categories, especially in Africa." In Rebecca Lave (Ed.), *Critical Physical Geography*. Basingstoke: Palgrave-Macmillan.

Ellen, Roy. 1993. *The Cultural Relations of Classification*. Cambridge: Cambridge University Press.

Fairhead, James and Melissa Leach. 1996. *Misreading the African Landscape: Society and Ecology in a Forest-Savanna Mosaic*. Cambridge: Cambridge University Press.

Fleck, David W. and John D. Harder. 2000. "Matses Indian rainforest habitat classification and mammalian diversity in Amazonian Peru." *Journal of Ethnobiology* 20: 1–36.

Fraser, James Angus, Melissa Leach, and James Fairhead. 2014. "Anthropogenic dark earths in the landscapes of Upper Guinea, West Africa: Intentional or inevitable?" *Annals of the Association of American Geographers* 104(November): 1222–1238. https://doi.org/10.1080/00045608.2014.941735.

Gregory, Kenneth J. 2017. "Putting physical environments in their place: The next chapter?" *The Canadian Geographer/Le Géographe Canadien* 61: 11–18. https://doi.org/10.1111/cag.12333.

Hacking, Ian. 2002. *Historical Ontology*. Cambridge, MA: Harvard University Press.

Hunn, Eugene. 1976. "Toward a perceptual model of folk biological classification." *American Ethnologist* 3: 508–524.

Ingold, Tim. 1993. "The temporality of the landscape." *World Archaeology* 25(2): 152–174.

Ingold, Tim. 2000. *The Perception of the Environment: Essays on Livelihood, Dwelling and Skill*. Hove: Psychology Press.

Johnson, Leslie M. and Eugene S. Hunn. 2010. *Landscape Ethnoecology: Concepts of Biotic and Physical Space*. Oxford: Berghahn Books.

Laris, Paul. 2002. "Burning the seasonal mosaic: Preventative burning strategies in the wooded savanna of Southern Mali." *Human Ecology* 30: 155–186.

Malt, B.C. 1995. "Category coherence in crosscultural perspective." *Cognitive Psychology* 29: 85–148.

Mark, David M., Andrew G. Turk, Niclas Burenhult, and David Stea. 2011. *Landscape in Language: Transdisciplinary Perspectives*. Amsterdam: John Benjamins Publishing Company.

Mayhew, Susan (Ed.) 2009. *Oxford Dictionary of Geography*, 4th ed. Oxford: Oxford University Press.

Mihesuah, Devon Abbott, and Angela Cavender Wilson. 2004. *Indigenizing the Academy: Transforming Scholarship and Empowering Communities*. Lincoln, NE: University of Nebraska Press.

Moore, Francis. 1738. *Travels into the Inland Parts of Africa*. London: E. Cave.

Moseley, William G. and Paul Laris. 2008. "West African environmental narratives and development–volunteer praxis." *Geographical Review* 98: 59–81. https://doi.org/10.1111/j.1931-0846.2008.tb00288.x.

Osbahr, Henny and Christie Allan. 2003. "Indigenous knowledge of soil fertility management in Southwest Niger." *Geoderma* 111: 457–479.

Osunade, M.A. Adewole. 1988. "Nomenclature and classification of traditional land use types in South-Western Nigeria." *Savanna* 9: 50–63.

Radcliffe, Sarah A. 2017. "Geography and indigeneity i: Indigeneity, coloniality and knowledge." *Progress in Human Geography* 41: 220–229. https://doi.org/10.1177/0309132515612952.

Radcliffe, Sarah A., Elizabeth E. Watson, Ian Simmons, Felipe Fernández-Armesto, and Andrew Sluyter. 2009. "Environmentalist thinking and/in geography." *Progress in Human Geography* 33: 1–19.

Robbins, Paul. 2001a. "Fixed categories in a portable landscape: The causes and consequences of land-cover categorization." *Environment and Planning A* 33: 161–179.

Robbins, Paul. 2001b. "Tracking invasive land covers in India, or why our landscapes have never been modern." *Annals of the Association of American Geographers* 91: 637–659.

Shantz, Homer LeRoy and Curtis Fletcher Marbut. 1923. *The Vegetation and Soils of Africa*. American Geographical Society Research Series, No. 13. New York: National Research Council and American Geographical Society.

Simon, Gregory L. 2010. "The 100th meridian, ecological boundaries, and the problem of reification." *Society & Natural Resources* 24(November): 95–101. https://doi.org/10.1080/08941920903284374.

Simon, Gregory L. 2016. "How regions do work, and the work we do: A constructive critique of regions in political ecology." *Journal of Political Ecology* 23: 197–203.

Smith, Barry. 2001. "Fiat objects." *Topoi* 20: 131–148.

Smith, Barry. and David M. Mark. 2001. "Geographic categories: An ontological investigation." *International Journal of Geographic Information Science* 15: 591–612.

Smith, Barry., and David M. Mark. 2003. "Do mountains exist? Towards an ontology of landforms ." *Environment and Planning B* 30: 411–427.

Smith, Barry, and Achille Varzi. 2000. "Fiat and bona fide boundaries." *Philosophy and Phenomenological Research* 60(2): 401–420.

Smith, Linda Tuhiwai. 1999. *Decolonizing Methodologies: Research and Indigenous Peoples.* London: Zed Books.

Stauss, Joseph H. 2002. *Native American Studies in Higher Education: Models for Collaboration between Universities and Indigenous Nations.* Walnut Creek, CA: Altamira Press.

Stott, Philip. 1999. Tropical Rainforest: A Political Ecology of Hegemonic Myth-Making. London: Institute of Economic Affairs.

Tabor, Joe. 1993. *Risk and Tenure in Arid Lands: The Political Ecology of Development in the Senegal River Basin.* Tucson, AZ: University of Arizona Press.

Toledo, Victor M. 1992. "What is ethnoecology? Origins, scope and implications of a rising discipline." *Etnoecologica* 1: 5–21.

Verlinden, A. and B. Dayot. 2005. "A comparison between indigenous environmental knowledge and a conventional vegetation analysis in North Central Namibia." *Journal of Arid Environments* 62: 143–175.

White, Arthur Silva. 1892. *The Development of Africa: A Study in Applied Geography.* 2nd edn. London: George Philip & Son.

Whitson, Risa. 2017. "Painting pictures of ourselves: Researcher subjectivity in the practice of feminist reflexivity." *The Professional Geographer* 69(April): 299–306. https://doi.org/10.1080/00330124.2016.1208510.

Williams, Barbara. J. and Ortiz, C. A. 1981. Middle American folk soil taxonomy. *Annals of the Association of American Geographers* 71(3): 335–358.

WinklerPrins, Antoinette M.G.A. 1999. "Local soil knowledge: A tool for sustainable land management." *Society & Natural Resources* 12: 151–161.

Zimmerer, Karl S. 1996. *Changing Fortunes: Biodiversity and Peasant Livelihoods in the Peruvian Andes.* Berkeley, CA: University of California Press.

Zimmerer, Karl S. 1998. "The ecogeography of Andean potatoes" *BioScience* 48: 445–454.

Zimmerer, Karl S. 2001. "Report on geography and the new ethnobiology" *The Geographical Review* 91: 725–734.

4

CULTIVAR DIVERSITY AND MANAGEMENT AS TRADITIONAL ENVIRONMENTAL KNOWLEDGE

Roy Ellen

Introduction

Techniques for recognizing, maintaining and developing cultivar diversity have existed since the beginnings of human plant domestication and cultivation. For example evidence for early cereals in the Near East around 8000 BCE indicates the hybridization of wild grasses, such as *Triticum tauschii* with cultivated *Triticum dicoccum* (emmer) to give *Triticum aestivum* subsp. *spelta* (spelt wheat), and thus suggesting that the management of wild and cultivated varieties has long overlapped (Harlan 1995: 101). The earliest documentation of local variation within named cultigens is from 1000–600 BCE in a Chinese text (Ho 1969). Although pre-modern European sources are less helpful on the matter of cultivar diversity, archaeological samples from English medieval roof thatch indicate various landrace mixtures of *T. aestivum* (Letts 2001), and more recent written sources for the 1830s list some 150 named wheat varieties in circulation (Ambrose and Letch 2009). However, it is to the work of early anthropological pioneers such as Conklin (1954a) that we owe our first appreciation of the extensiveness of such local knowledge among contemporary traditional farmers, and especially how this is encoded in language (see Box 4.1). Ethnographic work on cultivar management using specialized investigatory methods, together with the application of such knowledge, begins effectively with the research of Boster (1984a, 1984b), which has inspired many subsequent studies of major crop species, by ethnobotanists in particular. The rise of the Green movement in the West and the knowledge sovereignty movement among traditional farmers have led to increased attention, in particular in the context of erosion of genetic diversity and knowledge under modern farming conditions.

DOI: 10.4324/9781315270845-5

Box 4.1 Harold Conklin: pioneer of linguistic ethnobotany

Harold C. Conklin (1926–2016) was an anthropologist and ethnobotanist whose work marked a fundamental shift in methodology and theory, away from conventional economic botany towards a serious engagement with people's detailed knowledge of biota as linguistically expressed and culturally embedded. Rather than focusing simply on lists of local plant uses, Conklin sought to understand how people conceptualized their plant worlds, while vernacular plant terminologies and categories were not seen simply as clues to the identification of scientific taxa, but of interest in themselves (perhaps especially) where they diverged from scientific taxa. Remarkably, these insights were achieved through a PhD thesis on Hanunóo ethnobotany (1954a) that was never published, but widely praised and emulated.

In particular, Conklin did the following:

- Provided a set of analytic terms and concepts to describe and analyse folk nomenclature of plants: how to make sense of the meaning of plant terms and how to distinguish genuine terms from other words for plants that local people often provide.
- Produced the first comprehensive and analytic description of the plant world of a group of small-scale farmers: the Hanunóo of Mindoro in the Central Philippines. A later study was to move the focus to "secondary" landscape categories, but adopting a broadly similar approach, this time through a long-term study of the Ifugao, a people of the Luzon cordillera (e.g. Conklin 1980).
- Demonstrated the extent of local knowledge, its rationale and why it should be taken seriously, for example in his work on swidden cultivation (e.g. Conklin 1954b).
- Introduced a model (e.g. Conklin 1962) to describe the folk classifications of local populations, specifying types and levels, and distinguishing 'basic' and 'terminal' categories, which was foundational for the later work of Berlin (e.g. Berlin, Breedlove and Raven 1973) and others.
- Insisted that plant knowledge and expertise should be understood in its rich cultural context, through intensive and linguistically informed fieldwork.
- Outside ethnobiology, Conklin is perhaps more widely associated with the 'Nida-Conklin' hypothesis, which predicts a positive correlation between lexical elaboration of a semantic domain and its cultural importance. It is a prediction not always realized, but has the virtue of testability.

For more on Conklin's ethnobotany, see Hunn 2007.

This chapter reviews some indicative current work on our understanding of diversity management among local and traditional peoples, in relation in particular to rice (*Oryza sativa* L.), potato (*Solanum tuberosum* L., and related species), sweet potato (*Ipomoea batatas* (L.) Lam.), cassava/manioc (*Manihot esculenta* Crantz), maize (*Zea mays* L.) and sago (*Metroxylon sagu* Roetboll). I explore the conceptual issues arising from the use of terms such as 'cultivar', landrace, varietal and folk-variety, and how these are treated in ethnobiological classification. I discuss the interconnection between ecological and cultural selection, knowledge of plant maturation and reproduction (both seed and vegetative propagation), and planting strategies. I look at how farmers use different cultivar qualities, and thereby influence their responses to innovation and change. I emphasize the social embeddedness of knowledge, dissemination and exchange of germplasm, and the consequences of diversity loss, due to farming intensification, environmental challenges, and, finally, I address the introduction of hybrid strategies combining modern crop breeding and local knowledge.

Terminological and conceptual issues in relation to folk classification

It is common enough to find farmer recognition of biological diversity within crops without the variants being systematically named, but as soon as such variants are transmitted between farmers, they generally acquire names. A common proxy for the measurement of diversity within a crop, therefore, becomes the number of distinct names attached to variants below the species level, for example the 89 names for kinds of rice reported for the Baduy of West Java (Iskandar and Ellen 1999), or the 43 names reported for Debut cassava in the Kei islands (Ellen and Soselisa 2012: 24). However, there is debate in the literature as to how these should be conceptualized and described. Taxonomists use the term 'variety' to describe a taxon below the species or sub-species level that can be distinguished using general morphological indicators and (increasingly) DNA bar-codes. However, agricultural and horticultural scientists have long used the term 'cultivar' to refer to variants of planted crops, usually emphasizing those properties of practical farming significance (e.g. early maturing, late maturing) or factors important in consumption (e.g. glutinous, non-glutinous). This means that the concept of cultivar does not map comfortably onto the taxonomic notion of 'variety'. Agriculturalists looking at the traditional practices of farmers also speak of 'landraces', which are simply the types of a crop identified by farmers themselves without necessarily having been subjected to the scrutiny of agricultural scientists or professionals. The development of studies of ethnobiological folk classification led to the introduction of the term 'varietal' (Berlin, Breedlove and Raven 1973) in the expectation of achieving greater conceptual rigour ('varietals' being contrasting linguistic segregates that divide a conceptual 'specific', and the lowest level of folk classification). However, most cultivars are conceptual folk-specifics in Berlin's terms, leading to an element of confusion. Some writers prefer to use the term 'folk variety' which does not assume the technical characteristics attributed to 'varietal' or 'variety', and also avoids the slippage between 'cultivar' and 'landrace', stressing that these are based on criteria used by a local population rather than having some objective international scientific status.

To avoid the constant interchange of terms, I will here use 'cultivar' throughout, except where it is necessary to distinguish the technical sense of this term from other terms. Cultivar (Bailey 1923; Gibson 2009) is usually defined as a sub-specific variant of a cultigen (itself generally defined as a cultivated species). However, cultigens are typified by a high degree of hybridization, which makes their representation in taxonomies awkward. Moreover, some folk-varieties may cut across several conventional scientific species. Thus, potato in the Andes comprises a complex of eight related species, the four most widely distributed taxa of which are *Solanum stenotomum* Juz. & Bukasov, *S. goniocalyx* Juz. & Bukasov, *S. x chauca* Juz. & Bukasov, *S. tuberosum* subsp. *tuberosum antigenum* Juz. & Bukasov Hawkes, which alone contain thousands of native cultivars (Zimmerer 1991: 29; Brush 1992).

The largest numbers of locally recognized cultivars have been reported for rice (e.g. Conklin 1954a [Hanunóo: 90], 1988 [Ifugao: 85–7]; Iskandar and Ellen 1999 [Baduy: 89]), though there are also large numbers reported for cassava (Heckler and Zent 2008). In calculating figures for comparative purposes, it is necessary to avoid confusion with synonyms, dialectal variants for the same type of germplasm, and variants at a regional rather than local level, or for an entire ethno-linguistic group rather than for a local socially interacting population exchanging plant material. Where there are large numbers of cultivars distinguished, the question arises as to internal classificatory organization. The Berlin folk classificatory scheme, which delineates levels of life form, intermediate, folk generic and folk specific in terms of increasing specificity, would suggest a downward extension of the two-dimensional taxonomic logic of contrast and division over time (to what Berlin calls folk varietal), and this may work in some cases.

Conklin (1988), for example has suggested that the Ifugao divide rice on the basis of several further sub-specific levels of contrasting categories. However, the continuous and overlapping distribution of qualities often makes this difficult. In sexually reproducing grains such as rice, it is often more accurate to represent the groupings of cultivars that farmers distinguish as non-taxonomic cross-cutting categories, based on multiple criteria, with no fixed two-dimensional hierarchy (Iskandar and Ellen 1999). Clonally propagated crops, such as cassava and sago (e.g. Ellen 2006), pose special problems for folk classification. Local populations often separate cassava cultivars into 'bitter' and 'non-bitter', as in the Kei island distinction between *enbal* and *kasbi*, though even in such cases the fine gradation in hydrogen cyanide toxicity across different named cultivars, and variation in toxicity according to micro-ecology, make even such a distinction problematic (Ellen, Soselisa and Wulandari 2012). It is also important to differentiate between all locally known named cultivars and those actually grown in particular locations at any one time. Thus, Nazarea (1998) in her work on Bukidnon sweet potato cultivation in Mindanao in the Philippines reports two sites, one at which 29 cultivars were planted and 26 others known about but not planted, and another site where 23 cultivars were planted and only five known about but not planted.

The names applied locally to cultivars generally take a binomial form (e.g. Nuaulu *kasipii putie*, 'white cassava'), where the adjectival qualifier may describe a morphological feature (colour, shape of fruit or seed, height), a quality of processing (glutinous, bitter), a growing characteristic (rate or time of maturation, resistance to pests or growing condition), place of origin, or some other word that identifies place of origin (e.g. Ellen and Soselisa 2012). Zimmerer (1991: 32, 39), however, notes that in Andean potato cultivation not only are cultivars uninomials rather than binomials, but the Berlin model is not appropriate for a folk classification which, below the cultigen level, tends to involve a use category-cultivar-subcultivar hierarchy. In some cases the name is not a binomial and highly idiosyncratic, for example 'Kentish fire', for a cultivar introduced from commercial chili seed (*Capsicum anuum* L.) purchased by Kentish recreational gardeners as food and subsequently successfully bred in allotments (Ellen and Platten 2011: 570). But while names given to cultivars can be misleading, and over-reliance on them as guides to genetic difference, and over-interpretation of folk etymologies, unwise, they can provide important clues about significant properties, or how cultivars originate and are distributed. Moreover, for farmers themselves, names (that is lexical knowledge) often serve as proxies or trigger-words that point to other (often un-lexicalized) qualities or repositories of complex 'substantive' knowledge related to the cultivar concerned. Thus, for Kei Islanders to call a cassava plant *enbal* rather than *kasipii*, is to point not only to the toxicity of the root, but to other requirements related to its ecology, planting, management and processing.

The interconnection between ecological and cultural selection

Looked at biologically, local cultivars are a consequence of adaptations to local conditions that improve their fitness. That fitness may be in relation to local ecologies (soil type, pests, climatic and weather conditions) or in terms of qualities which human cultivators find significant (productivity, taste, morphological characters, size). Where this happens in cultivated contexts without human intervention, or where human intervention is inadvertent, we can speak of ecological selection. However, where local cultivators identify variation and deliberately manage it, we may talk of cultural selection. Both ecological and cultural selection interact, and the occurrence of diversity is a fluid and evolving process (Elias, Rival and McKey 2000). Processes of selection apply equally to plants reproducing sexually or vegetatively, but with different evolutionary consequences. It is important to remember that small-scale, traditionally based farmers select germplasm as part of an integrated system of production and consumption that distorts classic Darwinian selection pressures (Cleveland and Soleri 2007). Where farmers deliberately choose particular strains that arise through sexual reproduction, the process usually involves selection of seed and its preferential subsequent planting. Where selection is through vegetative reproduction, it is, for example roots, suckers or stems that are removed and replanted. Effective management implies skills in recognizing useful variation, separating out new germplasm and subsequent planting and monitoring strategies for maturation and reproduction.

Brush and others (1981) have tended to favour a model of selection in which farmers seek to match cultivar qualities to micro-environmental conditions. However, Boster (1985; also Gibson 2009) introduced a model of selection based on perceptual difference, by which visible features are seen to serve as proxies for other characteristics valued by farmers because of growth or consumption factors. In his assessment of the two models among Andean farmers, Zimmerer (1991) notes that potato selection tends to be for perceptual difference, while maize selection is for direct consumption and production traits. Zimmerer prefers a 'perception-selection-maintenance-dispersal model', and concludes (following Clawson 1985) that colour and other non-adaptive characters serve as morphological markers for other things, and that the perceptual distinctiveness model does not account for why agriculturalists undertake such practices in the first place.

The degree of variation between cultigens in terms of whether cultivars are selected on the basis of sexual recombination or vegetative propagation to a large extent depends on the evolved biology of the crop, the ecological conditions and farmers' preferences. Thus, in parts of Indonesia, cassava does not flower predictably, and elsewhere although farmers recognize volunteer seedlings resulting from sexual recombination, they are not motivated to breed new cultivars on this basis (Ellen and Soselisa 2012: 28–32). Similarly, sago palms will often be harvested before they fruit, but may be propagated by the replanting of suckers. Potato – like cassava – is mainly propagated through vegetative recombination. In the Andes, altitude is a strong selective factor differentiating the composition of potato fields, associated with ecological habitat and productive zone use-categories, that is whether or not the tubers are developed for freeze-drying, boiling, commerce, or soup-making (Brush 1992). By contrast, grains such as rice and maize are selected mainly on the basis of sexual recombination. In the Andean study by Zimmerer (1991), potatoes appeared to be selected for diversity, while maize was selected for specific characters.

In some cases, genetic diversity within a crop may be enhanced by natural crosses with wild-growing populations of the same species. Often farmers are aware of this but attach little importance to it, though in other cases it may be deliberately encouraged by maintaining wild groves in close proximity, sometimes ritually supported, as in the cultivation of Abyssinian banana (*Ensete ventricosum* (Weiw.) Cheeseman) by Ari farmers in Ethiopia (Shigeta 1996).

How farmers use cultivar diversity in response to changing conditions

There are many good examples of how local management of cultivars reflects successful adaptation to varying environmental conditions. In fact, the entire history of domestication is the history of traditional management methods to extend the usefulness of particular crops under varying conditions, of which the process whereby various potato cultivars adapted to different altitudinal zones and uses in the Andes is among the most dramatic. New cultivars may emerge or be deliberately selected for many reasons. Among the Baduy of upland West Java (Iskandar and Ellen 1999), different rice cultivars are maintained and encouraged not only because of their productivity, but as likely because of resistance to disease, rate of maturation, tolerance to water shortage, ease of harvesting, and due to several consumption factors, such as glutinousness and colour. Given the large number of characteristics that farmers seek to keep in play at any one time, it is not surprising that hyper-diversity of planting itself is a feature of traditional farming systems.

Over and above the qualities of particular cultivars, the maintenance of high levels of diversity is characteristic of many traditional farming systems. The extent of diversity in any one plot depends on cultigen as well as ecological and cultural factors. Thus, Zimmerer (1991) found in the Andean fields that he studied 21 cultivars per field for potato, but only 2.9 cultivars per field for maize. High levels of diversity have the effect of buffering adverse short-term ecological conditions. As conditions change, so farmers vary the proportion of different cultivars in their fields, and how groups of cultivars are arranged within a field. Maintenance of high levels of diversity is achieved through planting a wide range of cultivars in a given year, in the same field or over a number of fields, but also in the case of seed crops by long-term storage of germplasm. Thus, Baduy rice barns (Iskandar and Ellen 1999: 121) contain many more types of rice than are necessarily used in any one year, some bunches of which have been stored for up to 90 years and yet still maintain their viability.

Where germplasm cannot easily be stored, as in the case of many clonally reproducing crops, diversity is maintained through live-storage in fields, by periodically supplementing planted cultivars from wild stock, or by social storage: either relying on others to plant cultivars or by keeping germplasm in constant circulation. While much selection and the incorporation of individual cultivars in a local inventory are calculated and deliberate, it has been widely reported that many cultural populations appear to have a shared aesthetic of diversity, encouraging variation for its own sake, with longer-term inadvertent beneficial qualities. Thus, the maintenance of diversity itself can be a key factor in long-term adaptation, reflected and supported through distinctive moral regimes (Ellen 2017).

Varying the number, combination and proportions of different cultivars in any one plot, or over a number of plots, is a risk-averse strategy for buffering against uncertainty. For example during El Niño years, and in particular in 1998, the year of ecological, economic and political crisis in Indonesia, and a year which saw major farmer suffering in lowland Java due to the rigid adoption of a simplified Green Revolution high-input model favoured by state agricultural extension services (Ellen 2007), Baduy upland farmers coped well using polydiverse, rain-fed swiddens and traditional cultivars. However, risk aversion does not necessarily motivate high cultivar diversity in itself, or specialization to climatic conditions or soil conditions (Zimmerer 1991: 28).

A major environmental hazard influencing diversification of crop cultivars has been aridification. One of the reasons why cassava spread so widely out of its area of endemism in north-west Amazonia, was its flexibility as a crop, and particularly its tolerance of dry conditions. A key feature that makes this possible is high levels of HCN (hydrogen cyanide). This serves to combat competing pathogens and thus confers an advantage in dry zones. The range of HCN toxicity is wide in cassava, but, in dry areas, toxicity tends to be higher and the ecology selects for cultivars with high HCN concentration. Farmers in Africa, Indonesia and elsewhere have learned to take advantage of this and deliberately favour high toxicity cultivars. Thus, in the Kei islands there are as many high toxicity cultivars (*enbal*) as low toxicity cultivars (*kasbi*) (Ellen and Soselisa 2012). However, high toxicity is also a danger to human health, and where diets are poor can be life-threatening. As a result, farmers have developed methods for detoxification, not only for cassava but for other roots and tubers, such as yams (Johns 1990). In a comparative study of Kei and Nuaulu cassava diversity, Ellen, Soselisa and Wulandari (2012) used DNA evidence to show that the close genetic relatedness between most of the larger number of Kei cultivars and a distant genetic relatedness between all of the smaller number of Nuaulu cultivars, strongly indicated that Kei farmers were much more active in selecting propagative material than Nuaulu, who – living in a less arid area – were far less dependent on the crop.

Dissemination and exchange of germplasm

The redistribution of domesticate germplasm, and hence cultivar variations, is inevitably related to human movement, whether inadvertent or deliberate, but it also depends on evolved forms of plant reproduction and their different properties. Seed is the most resilient form of germplasm, and different forms of vegetative propagule vary greatly in their ability to move effectively through human systems (Ellen and Platten 2011). For example Zimmerer (1991: 39) found that, in his Andean study, maize seed was distributed more frequently, easily and widely than potato tubers.

In the case of deliberate distribution, it is not the germplasm alone that moves but often the knowledge associated with it, while both knowledge of cultivars and actual germplasm pass between cultivators in ways consistent with wider social norms and practices. This import-antly includes division of labour by gender, depending on whether it is males or females who have the predominant role in farming. Thus, Boster (1986) has described how cassava cultivar stem cuttings and the knowledge associated with them move between female cultivators along kinship lines. Among allotment-keepers in East Kent, bean cultivars move through friendship networks and those renting contiguous plots (Platten 2013). Such forms of dissemination pro-vide a robust means of social storage, redistribute both germplasm and knowledge diversity, and provide a useful reservoir as conditions change. In traditional societies, most management knowledge rests in individuals, who transmit this through distributed kinship links. However, knowledge is not always distributed equally, and in addition to a strong gender bias in favour of women in developing world agricultural economies (Howard 2003), there is some additional specialist division of labour, with examples of cultivar management operating at a village or societal level. For example Kasepuhan peoples of upland West Java have a specialist role for those who vet newly identified rice types, and who accept or decline them by keeping a register of named cultivars in circulation (Soemarwoto 2007).

As well as local circulation, cultivars have travelled over wide geographic areas under trad-itional cultivation regimes, as evidenced not only by physical location, but in cultivar names. However, the age of European expansion initiated by the 'Great Columbian Exchange' redistributed germplasm (particularly that of native American domesticates) on a previously unprecedented global scale, providing new opportunities for adaptation by local peoples, but – as with invasive species generally – often threatening previously important local domesticates. However, the physical transfer was often accompanied by knowledge loss, followed by rapid rediscovery and diversification of cultivars to suit new environments, as was the case with cas-sava in eastern Indonesia (Ellen, Soselisa and Wulandari 2012). Deliberate retention of one kind of local knowledge (associated with the germplasm itself), on the assumption that it could be unproblematically transferred to a new location, has sometimes proved misplaced, and what has proved more important has been local people's ecological knowledge of the recipient environ-ment. The story of the introduction of *Hevea* rubber among Malay planters and smallholders is a good case in point here (Dove 2000).

The evidence for germplasm exchange and dissemination demonstrates vividly the import-ance of acknowledging that cultivar management skills are not free-floating, but are everywhere socially embedded and rooted in the practices of individual persons. Not only are these skills organized through local divisions of labour, transmitted generationally and spatially through culturally-specific arrangements, but are valorized and even sanctified, as when glutinous rice and millets hold a special place in ritual. The diversity of cultivars is often closely associated with the identity of particular ethnic groups, for example among Quechua with potato diversity (Zimmerer 1991), and, as we have seen, an 'aesthetic of diversity' itself may be an important factor underpinning risk-averse management strategies.

Diversity loss and its consequences

As farming intensifies, due to government policy initiatives, market penetration, commodification and other local economic factors, so traditional farmers are under pressure to relinquish traditional cultivars and reduce their diversity, both in the subsistence and commercial sectors, making them more vulnerable. Traditional management techniques and levels of diversity have been widely seen as incompatible with models of farming modernization and Green Revolution (GR) strategies, which have strongly prioritized a small number of commercially bred high-yielding varieties (HYVs) and more recently genetic modification (GM). The consequences of this have not simply been a reduction in diversity (leading to genetic erosion), and therefore ability to buffer against changing conditions, but have included loss of valuable cultivars due to their incompatibility with introduced new technologies, with implications for the restructuring of power relations. The GR impacted on gender relations where innovations were channeled through males as household heads rather than females as the principal farmers, and on land distribution (with richer farmers generally being better placed to take advantage of new technologies, and to buy out smaller farmers who then either become agricultural labourers or migrate to towns). These are all processes that are well recorded for rural areas of Java (Ellen 2007). More specifically, a major problem encountered with the introduction of HYV rice in Java was that their shorter stems and panicles were difficult to harvest using customary methods, and in particular the hand-held finger knife (Stoler 1977). There were also cultural consequences, as when glutinous cultivars, important in ritual and feasting, were lost. In Bali (Lansing 1991), state attempts to 'modernize' irrigation systems through centralization and simplification interfered with traditional forms of local management that had effectively optimized water delivery and restricted pest outbreaks by limiting contiguous areas at the same stage in the agricultural cycle.

In some cases these changes have been actively resisted, as among Baduy, where it has sometimes been simply put down to 'cultural conservatism' (Iskandar and Ellen 1999). Elsewhere there have been compromises. In Brunei, Dusun swidden farmers rely on imported rice for most food, but retain some swiddens to ensure availability of cultivars required to support traditional religious and festive practices (Ellen 2012: 27), Kasepuhan in upland west Java maintain a hybrid system in which HYVs have been incorporated and traditional landraces kept (Soemarwoto 2007), while Brush (1992) has shown that it is simplistic to assume that modern varieties of potato simply replace traditional cultivars in the Andes.

Paradoxically, despite the erosion of cultivar diversity under the pressures of farming modernization, there has been scientific and non-governmental organization (NGO) recognition for some decades of the importance of conserving traditional germplasm, and globally there are research centres specializing in the *ex situ* conservation of germplasm ('gene banks') for particular cultigens (Harlan 1995: 245–248), such as at the International Rice Research Institute at Los Baños in the Philippines and CIP, the International Potato Center in Lima, Peru, both run under the auspices of the Consultative Group for International Agricultural Research (CGIAR). But while such seed banks are good at preserving physical germplasm, they have tended to ignore the knowledge of the people who created the diversity and the practices that assisted this, and which we now understand to be integral in preserving effective biocultural systems.

Moreover, there have been ethical risks and political controversy associated with 'biopiracy', and violation of the intellectual property rights (IPR) of those who developed them (Baumann, Bell, Koechlin and Pimbert 1996). By contrast, *in situ* conservation (not only, but especially for vegetatively reproduced cultivars), has the advantage of preserving cultural memory alongside biodiversity, where the conservation of one can support the other. Such community-based memory banking protocols have been developed, for example for sweet potato, in a way that is sympathetic to social norms and supports the interests and values of local populations themselves (Nazarea 1998). It would be a mistake, though, to assume that *in situ* conservation simply freezes the biological composition of existing stock, alongside farmer knowledge. Indeed, it is rather an open and dynamic process in which change can continue to occur (Perales, Brush and Qualset 2003).

Hybrid strategies

A false dichotomy and a fallacy underlie some conceptions of traditional knowledge, which opposes it with respect to modern or scientific knowledge. We have already seen that the dynamic character of traditional cultivar diversity management depends on its openness to new variations and geographic transfer. Certainly, there are examples – such as the Baduy (Iskandar and Ellen 1999) – where there is cultural resistance to the introduction of new crops, but this has to be understood in the context of general cultural resistance to the outside world. Many traditional communities welcome scientifically developed inputs to their body of existing knowledge, and have done so for over 100 years (e.g. Perales, Brush and Qualset 2003). In many cases, germplasm and knowledge from outside sources have been progressively introduced over a long period of time, such that local farmers' knowledge can already be said to be 'hybrid'. Nevertheless, germplasm and knowledge introduced from outside continue to be tested against the local situation and ecology, and therefore are dynamic and rightfully considered local, contributing to adaptation and crop development in important ways. Kasepuhan, for example combine traditional cultivars with high yield varieties (Soemarwoto 2007).

Where there is resistance, it is generally the other way round, as when commercial firms attempt to control the terms by which local farmers use patented seed, and where government and NGO and science-driven efforts to increase productivity insist on replacing existing cultivar ranges with single scientific varieties, either because of blanket assumptions about the unsuitability of existing cultivars, very tight monocultural models of the Green Revolution type, or biosecurity fears, including genetic contamination. Usually, however, failure to consider existing cultivars is simply based on ignorance of what is already there, and what might be adapted. A more strident version of this argument – how globalized corporate agriculture undermines the lives, knowledge and environments of small farmers – is found in the work of Vandana Shiva (2000).

The development of *ex situ* collections of local cultivars has to some extent been linked to scientific plant breeding programmes, though generally with less attention being paid to local farmer knowledge. Modern crop-breeding based on genetics and evolutionary biology, like pharmaceutical research based on synthetic drugs, has over a 100-year period sought to separate itself from traditional knowledge systems, whether through ignorance, a sense of superiority, or because of policy, commercial and legal barriers that have entrenched the division, most saliently reflected in the confrontational neo-liberal politics of the Green Revolution and of the genetically-modified (GM) crop industry. There is now increasing evidence of a 'third way', in the form of a body of research on the active engagement between farmers and crop-breeders linked to attempts to promote it (Cleveland and Soleri 2002). This acknowledges the dangers of separating professional plant breeding and seed supply systems from more traditional methods and farmer knowledge, and is encouraging 'participatory plant breeding'. Moreover, it acknowledges that plant breeding systems consist not only of genetic material and growing environments, but also of the social context in which the breeding is undertaken and the local knowledge of farmers. It is not always the conservation of genetic resources that is important, but preservation of knowledge among farmers that permits continuing flexible adaptation through selecting diverse cultivars (Soleri and Smith 1999).

The synergy between local knowledge and scientifically driven agriculture is fundamentally no different in the agricultural and horticultural sectors of advanced industrial societies than among small and subsistence farmers in less-developed countries. Farmers continue to use local knowledge to manage cultivar diversity, while recreational gardeners and small producers actively contribute to maintaining diversity and developing new strains (Ellen and Platten 2011; Platten 2013). The constraint here, however, is less that scientific agriculture has undermined a role for local knowledge, than that the bureaucratic weight of standardization legislations, de-skilling of labour, biosecurity measures and commercial protection prevent widespread dissemination through sale of heritage varieties. In response, private clubs have been formed to maintain and circulate germplasm or it is freely given away, or redistributed through informal exchange (as in allotment associations).

On East Kent allotments (Ellen and Platten 2011: 570), several gardeners make a habit of growing 'Leo's peas'. This is a brand name for dried peas intended for direct consumption rather than planting, purchased from a local supermarket. In this example, a commercial variety, grown by a farmer for market sale and consumption, has been re-branded by a food packager and sold to a supermarket retailer as a separate comestible. The allotment plotholder has then purchased the peas and put them to use outside the context envisaged by either farmer, packager or retailer. It is such ingenious reworking through different and informal social channels that suggests that the forces of local generation of cultivar diversity hold some prospect of resisting the relentless pressure for monoculture and ecological simplification.

References

Ambrose, M. and S. Letch. 2009. Thatching with longstraw wheat in relation to onfarm conservation in England. In M. Veteläinen, V. Negri and N. Maxted (eds), *European Landraces: On-Farm Conservation, Management, and Use*. Biodiversity International. Available at: www.bioversityinternational.org/uploads/tx_news/European_landraces__onfarm_conservation__management_and_use_1347.pdf

Bailey, L. H. 1923. Various cultigens, and transfers in nomenclature. *Gentes Herbarum* 1(3): 113–136.

Baumann, M., J. Bell, F. Koechlin and M. Pimbert (eds) 1996. *The Life Industry: Biodiversity, People and Profits*. London: Intermediate Technology Publications.

Berlin, B., D. Breedlove and P. Raven. 1973. General principles of classification and nomenclature in folk biology. *American Anthropologist* 75(1): 214–242.

Boster, J. 1984a. Classification, cultivation, and selection of Aguaruna cultivars of *Manihot esculenta* (Euphorbiaceae). *Advances in Economic Botany* 1: 34–47.

Boster, J. 1984b. Inferring decision making from behaviour: An analysis of Aguaruna Jivaro manioc selection. *Human Ecology* 12(4): 347–358.

Boster, J. 1985. Selection for perceptual distinctiveness: Evidence from Aguaruna cultivars. *Economic Botany* 39: 310–325.

Boster, J. 1986. Exchange of varieties and information between Aguaruna manioc cultivators. *American Anthropologist* 88(2): 428–436.

Brush, S. B. 1992 Ethnoecology, biodiversity and modernization in Andean potato agriculture. *Journal of Ethnobiology* 12(2): 161–185.

Brush, S. B., J. H. J. Carney and Z. Huaman 1981. Dynamics of Andean potato agriculture. *Economic Botany* 35(1): 70–88.

Clawson, D. L. 1985. Harvest security and intraspecific diversity in traditional tropical agriculture. *Economic Botany* 39: 56–67.

Cleveland, D. A. and D. Soleri (eds) 2002. *Farmers, Scientists and Plant Breeding: Integrating Knowledge and Practice.* Wallingford: CABI Publishing.

Cleveland, D. A. and D. Soleri 2007. Extending Darwin's analogy: Bridging differences in concepts of selection, between farmers, biologists, and plant-breeders. *Economic Botany* 61: 121–136.

Conklin, H. C. 1954a. *The relation of Hanunóo culture to the plant world.* PhD thesis, Yale University.

Conklin, H. C. 1954b. An ethnoecological approach to shifting agriculture. *Transactions of the New York Academy of Sciences* 17: 133–142.

Conklin, H. C. 1962. Lexicographical treatment of folk taxonomies. *International Journal of American Linguistics* 28: 119–141.

Conklin, H. C. 1980. *Ethnographic Atlas of Ifugao: A Study of Environment, Culture and Society in Northern Luzon.* New Haven, CT: Yale University Press.

Conklin, H. C. 1988. Des orientements, des vents, des riz...: pour une étude lexicologique de savoirs traditionnels. *Journal d'Agriculture Traditionelle et de Botanique Appliquée* 33: 3–9.

Dove, M. 2000. The life-cycle of indigenous knowledge, and the case of natural rubber production. In R. Ellen, P. Parkes and A. Bicker (eds), *Indigenous Environmental Knowledge and Its Transformations: Critical Anthropological Perspectives.* Amsterdam: Harwood, pp. 213–251.

Elias, M., L. Rival and D. McKey. 2000. Perception and management of cassava (*Manihot esculenta* Crantz) diversity among Makushi Amerindians of Guyana (South America). *Journal of Ethnobiology* 20: 239–265.

Ellen, R. 2006. Local knowledge and management of sago palm (*Metroxylon sagu* Rottboell) diversity in south central Seram, Maluku, eastern Indonesia. *Journal of Ethnobiology* 26(2): 83–123.

Ellen, R. 2007. Introduction. In R. Ellen (ed.), *Modern Crises and Traditional Strategies: Local Ecological Knowledge in Island Southeast Asia.* Oxford: Berghahn, pp. 1–45.

Ellen, R. 2012. Studies of swidden agriculture in Southeast Asia since 1960: An overview and commentary on recent research and syntheses. *Asia Pacific World* 3(1): 18–38.

Ellen, R. 2017. Is there a connection between object diversity and aesthetic sensibility?: A comparison between biological domesticates and material culture. *Ethnos* 82(2): 308–330.

Ellen, R. and S. J. Platten 2011. The social life of seeds: The role of networks of relationships in the dispersal and cultural selection of plant germplasm. *Journal of the Royal Anthropological Institute* 17: 563–584.

Ellen, R. and H. L. Soselisa 2012. A comparative study of the socio-ecological concomitants of cassava (*Manihot esculenta*) diversity: Local knowledge and management in eastern Indonesia. *Ethnobotany Research and Applications* 10: 15–35.

Ellen, R., H. L. Soselisa and A. P. Wulandari 2012. The biocultural history of *Manihot esculenta* in the Moluccan islands of eastern Indonesia: Assessing the evidence for the movement and selection of cassava germplasm. *Journal of Ethnobiology* 32(2): 157–184.

Gibson, R. W. 2009. A review of perceptual distinctiveness in landraces including an analysis of how its roles have been overlooked in plant breeding for low-input farming systems. *Economic Botany* 63: 242–255.

Harlan, J. R. 1995. *The Living Fields: Our Agricultural Heritage.* Cambridge: Cambridge University Press.

Heckler, S. and S. Zent 2008. Piaroa manioc varietals: Hyperdiversity or social currency. *Human Ecology* 36: 679–697.

Ho, P.-T. 1969. The loess and the origin of Chinese agriculture. *American Historical Review* 75: 1–36.

Howard, P. (ed.) 2003. *Women and Plants: Gender Relations In Biodiversity And Management.* London: Zed Books.

Hunn, E. 2007. Commentary: Conklin's ethnobiological contribution. In J. Kuipers and R. McDermott (eds), *Fine Description: Ethnographic and Linguistic Essays by Harold C. Conklin.* New Haven, CT.: Yale University Press, pp. 191–195.

Iskandar, J. and R. Ellen 1999. In situ conservation of rice landraces among the Baduy of west Java. *Journal of Ethnobiology* 19(1): 97–125.

Johns, T. 1990. *The Origins of Human Diet and Medicine.* Tucson, AZ: University of Arizona Press.

Lansing, J. S. 1991. *Priests and Programmers: Technologies of Power in the Engineered Landscape of Bali.* Princeton, NJ: Princeton University Press.

Letts, J. 2001. Living under a medieval field. *British Archaeology* 58 (April). Available at: www.archaeologyuk.org/ba/ba58/feat1.shtml

Nazarea, V. D. 1998. *Cultural Memory and Biodiversity.* Tucson, AZ: University of Arizona Press.

Perales, H. R., S. B. Brush and C. O. Qualset 2003. Dynamic management of maize landraces in central Mexico. *Economic Botany* 57: 21–34.

Platten, S. 2013. Plant exchange and social performance: Implications for knowledge transfer in British allotments. In R. Ellen, S. J. Lycett and S. E. Johns (eds), *Understanding Cultural Transmission in Anthropology: A Critical Synthesis*. Oxford: Berghahn, pp. 300–319.

Shigeta, M. 1996. Creating landrace diversity: The case of the Ari people and ensete *(Ensete ventricosum)* in Ethiopia. In R. F. Ellen and K. Fukui (eds), *Redefining Nature: Ecology, Culture and Domestication*. Oxford: Berg, pp. 233–268.

Shiva, V. 2000. *Stolen Harvest: The Hijacking of the Global Food Supply*. London: Zed Books.

Soemarwoto, R. 2007. Kasepuhan rice landrace diversity, risk management and agricultural modernization. In R. Ellen (ed.), *Modern Crises and Traditional Strategies: Local Ecological Knowledge in Island Southeast Asia*. Oxford: Berghahn, pp. 84–111.

Soleri, D. A. and S. E. Smith 1999. Conserving folk crop varieties: Different agricultures, different goals. In V. D. Nazarea (ed.), *Ethnoecology: Situated Knowledge/Located Lives*. Tucson, AZ: University of Arizona Press, pp. 133–154.

Stoler, A. L. 1977. Rice harvesting in Kali Loro: A study of class and labor relations in rural Java. *American Ethnologist* 4(4): 678–698.

Zimmerer, K. S. 1991. Managing diversity in potato and maize fields of the Peruvian Andes. *Journal of Ethnobiology* 11(1): 23–49.

5

ON SERVING SALMON

An ethnography of hyperkeystone interactions in Interior Alaska

Shiaki Kondo

Politics of fish weirs and king salmon in Salmon River

Salmon River, or *Hotoleno'* in the local Athabascan language, is known as one of the good fishing places for king salmon (*Oncorhynchus tshawytscha*) in the Upper Kuskokwim region. Even though the king salmon population was in a period of sharp decline in the 2010s, one still sees vast numbers of salmon going upstream for spawning during June and July. Dichinanek' Hwt'ana (also known as Upper Kuskokwim) Athabascans have caught the migrating salmon for food for hundreds and possibly thousands of years. King salmon is a prime food source for humans, while chum salmon (*Oncorhynchus keta*) is locally called "dog salmon" and considered to be a good food for sled dogs because of its abundance. Coho salmon (*Oncorhynchus kisutch*) is in-between in terms of status.

While there is still an ongoing debate over when Alaskan Athabascan started to "depend" on salmon as a resource (Osgood 1936; Hosley 1977; cf. Halffman et al. 2015), various methods of catching salmon signify the continued utilization and cultural importance of salmonid species. Salmon fishing techniques can be roughly divided into two, based on the physical characteristics of a river. In muddy rivers, people can use fishing nets (made from plants in pre-contact times and from synthetic fibers nowadays) and the rather old-fashioned fishwheels, which were said to be introduced by non-native miners during the Gold Rush period (Collins 2004: 28). In clear-water rivers, fishing nets and fishwheels would be ineffective because fish can see the devices. Instead, people built fish weirs to intercept and catch the migrating fish. Other Indigenous technologies include fish spears. Ignatti Petruska, an Athabascan elder, originally from the Upper Kuskokwim region, told me that people used to climb out on the trees that lean almost horizontally over the river and speared the fish from these perches.

King salmon spawning in the Upper Kuskokwim region have been conserved by local Indigenous fishers. As in many Alaska Native cultures, people stop fishing when they think they have enough for their family. Respectful behavior toward king salmon has also been taught to the younger generation. For example, Miska Deaphon, a traditional chief of Nikolai, narrated the story called "Tr'aha Goya Łuk'a Ił Ghedo' Nin" (The Girl who Stayed with the Fish), where a girl turned into a king salmon and learned about the importance of the proper behavior toward the salmon such as cleaning the poles for fish camp before the fishing season (Deaphon 1980: 2–8). It is important to note that traditional stories are not "just stories"

DOI: 10.4324/9781315270845-6

for entertainment: they teach young people important lessons in life (Cruikshank 1990). In Nikolai, it was even said that children who fall asleep during the storytelling would not live a long life.

In 1889, the U.S. Congress passed a law that forbade "the erection of dams, barricades, or other obstructions in any of the rivers of Alaska" for the purpose of protecting the migrating salmon populations. At that time, law enforcement aimed at large fishing companies was not so effective (Arnold 2008: 82). However, in 1959, the State of Alaska made it illegal to use fish traps for commercial purposes. During the 1950s, there was an intense conflict between small-scale fishers (gillnetters) and large-scale (non-Native) fish trap operators. Against this background, the ban on fish trap was considered by the State government as a part of process to take over the control of natural resources from outside large businesses and the federal government (Arnold 2008: 158).

However, from the Upper Kuskokwim people's perspective, the ban on fish traps actually meant the loss of "local" control. State natural resource managers only told Upper Kuskokwim people about the new regulations to ban fish traps (i.e., fish weirs) sometime during the 1960s. Due to the increased fishing pressure, use of traditional fish weirs, including the one operated in Salmon River, was discontinued after the late 1960s. The Upper Kuskokwim people think that the ban on fish weirs is one of the many attempts by the non-Natives to attack the subsistence ways of life in Native Alaskan societies. While the State government's reason for such regulation is that the fish trap is too effective and likely to overfish, one local resident pointed out the fact that the traditional fish weirs cannot be used continuously during the run because people needed to repair the wooden weirs frequently (Holen, Simeone, and Williams 2006: 95). Traditional fish weirs seem to be an effective device to catch enough salmon for subsistence, while fish escapement is ensured by the characteristics of the materials used.

Around ten years after the ban on fish weirs, the Upper Kuskokwim people took up rod and reel fishing gear and have continued to fish for food at Salmon River since the late 1970s. They consider rod and reel fishing at Salmon River to be subsistence fishing (Stokes 1985: 224–225). Nowadays, Salmon River is a venue for a culture camp where local youth and elders gather to learn ways of living in the bush (Kondo 2016a). Unfortunately, fish weirs are still forbidden as fishing gear, but it is crucial for the next generation to maintain connections with the traditional land and water. Some of Upper Kuskokwim people also have a plan of re-establishing a summer trail from Nikolai to Salmon River (*Hotoleno' hwdazgwn' shanh tin*), which was used until the late 1960s.

In the midst of king salmon decline in the 2010s, Salmon River has hosted a scientific investigation. The Alaska Department of Fish and Game and MTNT Ltd. (an assembly of four Native Village Corporations) decided to conduct research on the king salmon population in the Pitka fork of Salmon River. In 2015, they built a weir for scientific purposes and an accommodation unit adjacent to it. They hired two people from Nikolai as research technicians to count the fish that went through the weir over the summer. The numbers of returning fish and other related data are sent to the scientists every day during the run through a wi-fi system established at the accommodation unit.

The site of the research weir is a three-minute walk away from one of the local family's fish camps, where the culture camp was held. Locally-hired research technicians took some of their relatives to their accommodation unit and showed them the operation of the weir when they were visiting Salmon River for the culture camp. One of the older Nikolai residents mentioned that her older brother would be happy to see the research weir because it would remind him of the old days when they used to operate a fish weir for salmon, which required continuous maintenance. She is proud of her son and grand-niece who are using a fish weir in their traditional fishing place, even though it is strictly for a scientific purpose and does not involve catching fish for food.

In the case of Salmon River, we see signs of collaboration between Indigenous fishermen and scientists for the shared goal of protecting the king salmon population in the region. Hopefully, this will lead to more advanced biocultural restoration where the Upper Kuskokwim people can use fish weirs again to catch abundant king salmon for their own use, not just to count them for scientific research. It is interesting to note that the abundance of king salmon in Salmon River makes it an ideal and crucial place for the production of scientific and Indigenous Environmental Knowledges. Fish weirs are still one of the most sophisticated gears for both scientific and Indigenous practices.

Interspecies interactions in coho salmon spawning streams

Salmon spawning streams are a transient home for many people and wildlife. There is a coho salmon spawning stream which is located upstream from Nikolai. Hunters in Nikolai go there to hunt brown bears that feed on the fish. Getting to the place takes a few hours by motorboat and involves navigating a shallow stream where partially submerged logs may imperil watercraft. However, once you get close to the destination, you will be unable to miss the feast that is going on there. Golden eagles (*Aquila chrysaetos*) and ravens (*Corvus corax*) hover in the sky, while brown bears left partially devoured salmon carcasses below. People set up a camp at the edge of the forest and visit the nearby stream to hunt bears.

On one of these hunting trips, three Nikolai hunters and I went to take a walk along the stream after an unsuccessful bear hunt. One of the hunters spotted a beaver (*Castor canadensis*) dam built across the stream, and we decided to make an opening because the hunters thought that it was blocking the pathway of migrating salmon. It took us around an hour to make an opening in the dam by using a long stick to remove the entangled branches. Soon, water started to flow from the opening, and the water level gradually changed. After a while, we started to see red fish swimming upstream: it was coho salmon resuming their migration. Earlier, I had observed more than ten salmon in the stream. They became active again, swam over to the opening and proceeded upstream. Seeing off some salmon that went upstream, one of the hunters said that salmon would come back next year because of our efforts. He then added that the Alaska Department of Fish and Game should hire the villagers to do this kind of river maintenance activities.

From what I observed in that particular salmon spawning stream, it may be possible to argue for the ecological benefits of the human interventions described above. Salmon migration attracts a plethora of wildlife, including bears that feed on the fish. Athabascan hunters put modest pressures on the bear population by hunting at the stream, which then reduces the number of salmon preyed upon by bears.[1] Further, when the beaver dams are observed to be blocking the stream, partial destruction of the dams ensures the safe travel of the migrating salmon upstream. In this sense, activities at the stream can be described as Indigenous "service to ecosystems" (Comberti et al. 2015), which is a complementary concept to "ecological services" (i.e. the benefits humans receive from the ecosystem).

According to the Upper Kuskokwim people, beaver dams also adversely affect whitefish populations in the region. Along with low water levels, the migration of whitefish may be impeded by beaver dams. One respondent of a study conducted by the Alaska Department of Fish and Game said:

> Whitefish, we don't have any more up here. Beaver mess it up. Like, here, where it's spawning area, it's messed up with beaver dam, [whitefish will] never get out or

something, or never come in or whatever, you know ... But nobody take care of it. They have to break the dam out and get it out ... Some don't know what to do with that, cause they don't hunt beavers. Yeah, long time ago, there was hunters, you know? I guess but they just don't want to go anymore for skin. Been true that nobody use beaver like they use to ... All dying out, and all the fish dying out ...

(Holen, Simeone, and Williams 2006: 92)

However, marine biologists tend to downplay the possibility of beaver dams negatively affecting the passage of migrating fish. One time, I mentioned my observation in a discussion with an Alaskan marine biologist, but his answer is that beaver dams should not be destroyed because they provide benefits to the fish. Actually, a recent study demonstrated that many species, including coho salmon, benefit by beaver activities. Beavers build dams to flood the area, creating ponds to protect their nests. These beaver-engineered ponds provide excellent habitat for juvenile salmon, thus helping their survival (Naiman 1988; Pollock et al. 2004).

These studies have been put forth as a direct counter-argument against early resource management policies. In the first half of the twentieth century, early Alaskan managers conducted what they called a "stream improvement" project, completely removing log jams, beaver dams, and other obstructions in some of the Alaskan rivers and streams. Following the same logic used when banning the fish weir, any impediment in the passage of the fish was pinpointed as a reason for salmon decline while apparent overfishing by industry was left untouched. This "friendly" gesture might have helped to partially ease the problem of the fish passage, but it nevertheless destroyed the fish habitat and is thought to have caused more damage than benefits (Arnold 2008: 89). Overhunting of beavers in the fur trade period is a likely factor contributing to the slow recovery of the fish. It is no wonder that my marine biologist interlocutor is reluctant to revisit the issue of beaver dams blocking the salmon migration.

My opinion is that the above-mentioned Indigenous practices and the scientific research on the benefit of beaver dams do not necessarily contradict each other. Beaver dams can provide long-term ecological services, especially to the juvenile fish, as many studies show, but it could be the case that they potentially have a negative effect during the upstream migration for spawning (cf. Gard 1961; Schlosser 1995). Then, the minor destruction of the dams in a crucial time of the salmon migration eases the negative effects. As I observed when following a hunting trip to the same stream a week later, beaver dams can be repaired fairly quickly as long as a healthy beaver population is maintained. Thus, by the time the juvenile salmon need their suitable habitats, the beaver dams and ponds are there for them to use. It is important to note that what the bear hunters did was a minor destruction, instead of complete removal, of the beaver dam. Indigenous "service to the ecosystem" as described in this section maximizes the "ecosystem service" by balancing the interspecies interactions to make it possible for the multispecies community to sustain themselves. Furthermore, these trade-offs are recognized implicitly in the knowledge-practice-belief complexes (Berkes 2012) that characterize Indigenous Environmental Knowledge (IEK).

From the beaver's point of view

In the previous two sections, I have described two contrasting examples of salmon–human entanglements in the Upper Kuskokwim region. In Salmon River, despite the colonial legacy of the continued ban on subsistence fish weirs, scientists and Indigenous fisher-hunters began working together to better manage the declining king salmon population. However, both parties do not seem to agree on how beavers influence the survival of salmon populations.

At first sight, this reminds us of the classic studies of the conflict between Indigenous and scientific knowledge practices (Nadasdy 2003; Blaser 2009). I have discussed the salmon management issue in Interior Alaska from the perspective of Indigenous/scientific knowledge(s) elsewhere (Kondo 2016b). However, here I want to complicate the picture by highlighting non-human agency. Beavers are known to possess highly sophisticated engineering capabilities that enable them to build dams. Lewis H. Morgan (1986 [1868]: 91) writes of beaver dams,

> [i]t is generally asserted that the introduction of a curve, with its convexity up stream, was the result of intelligence and design on the part of the architects; and that its use at the precise point where the pressure of the water is the greatest, affords conclusive evidence that the beavers understood its mechanical advantages.

Both beavers and humans are ecosystem engineers.

By stressing non-human animals' capacities as engineers, we are now faced with not just the two types of human knowledge practices (e.g. those of Athabascan and wildlife biology) but at least three world-making practices, namely, those of Athabascan hunter-fishers, wildlife biologists, and beavers. All three actors are "ecosystem engineers" in one respect or the other. The Athabascan hunter-fishers used to make fish weirs for fishing, while wildlife biologists install weirs for research purposes, especially to count fish. Beavers make dams to protect themselves from predators, such as wolverines and bears.

Beavers' capacities as builders won them the status of "keystone species" (Paine 1969). A keystone species can be defined as "one whose effect is large, and disproportionately large relative to its abundance" (Power et al. 1996: 609). Based on this definition, Mary E. Power and colleagues (1996: 611) point out that top-level predators, key food resources, ecosystem engineers, and seed dispersers are likely to be candidates for keystone species in their particular ecosystems.

Thomas Thornton (2015: 221) mentions that humans can be a keystone species. Describing Tlingit and Haida practices of herring "cultivation" (e.g. transplantation of eggs and enhancement of spawning grounds), Thornton argues that their "service" is integral to the maintenance of the Southeast Alaska coastal environment. In this sense, the Tlingits and Haidas occupy keystone status in the particular Pacific Northwest Coast biocultural contexts.

My analysis of Athabascan relations with boreal forest environment points to a similar conclusion: Athabascan hunter-fishers can be said to occupy the keystone status in the Interior Alaskan boreal forest environment. Athabascans not only have practiced sustainable salmon fishing through their traditional technology, but also have engaged in "service" to salmon in the form of bear hunting and making a passage route for trapped fish. They eat salmon as part of their traditional food and also labor for the benefit of them, thus the double-meaning of my chapter title "On Serving Salmon."

Ironically, while beaver activities have been increasingly appreciated and put into practice in ecosystem restoration projects (Woelfle-Erskine and Cole 2015), those of a human keystone player, Athabascan hunter-fishers, remain neglected, silenced, and criminalized. Use of fish weirs for fishing is still illegal, and damaging beaver dams is technically a violation of fish and wildlife regulations. All-too-human politics of salmon management revolves around who should build obstacles (e.g. dams and weirs) that may intercept the migrating salmon and for what purposes.

Hyperkeystone interactions

It is interesting to note that ecologists have discussed many of the species so far mentioned in this chapter as examples of keystone species or keystone interactions: beavers (Power et al.

1996), salmon, and brown bears (Helfield and Naiman 2006). Salmon in Southwestern Alaska transport oceanic nutrients to spawning grounds, while bears take the nutrients further inland in the form of scat and carcasses, which then will be decomposed by insects and other micro-fauna (Helfield and Naiman 2006). This chapter has been engaged in the ethnography of multiple keystone species and the interactions among them.

Recently, Boris Worm and Robert Paine (2016: 601) coined the term "hyperkeystone species" in order to denote "a species that affects multiple other keystone species across different habitats, and hence drives complex, potentially connected interaction chains." They define humans as a hyperkeystone species, since their activities have significantly influenced orcas (*Orcinus orca*), sea otters (*Enhydra lutris*), star fish (*Asteroidea spp.*), salmon, and brown bears, all of which are keystone species in their respective ecosystems in the Pacific Northwest according to Worm and Paine's criteria (see Figure 5.1).

Anthropologists would be tempted to criticize the concept of "hyperkeystone species" for its apparent anthropocentrism. Because Worm and Paine (2016) steer clear of the fundamental contribution of non-human keystone species in the maintenance and well-being of diverse human societies. In Figure 5.1, the human (H) as one homogeneous group stands independently from other species and influences them in a unidirectional manner. Figure 5.1 looks like a contemporary version of the Great Chain of Beings, where humans stand at the top of the ladder, seconded by large predatory mammals (Lovejoy 1936).

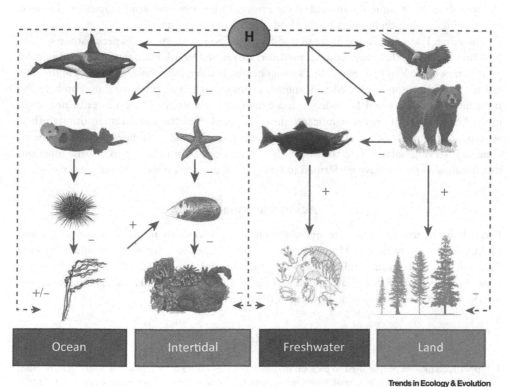

Trends in Ecology & Evolution

Figure 5.1 Humans as hyperkeystone species

Source: Worm and Paine (2016).

Despite its shortcomings, Worm and Paine's idea of looking at the interactions among key-stone species can be developed to describe how Indigenous practices serve the environment. I was surprised when I first looked at Figure 5.1, as it is supposed to be describing the position of humans in the Anthropocene[2] (likely as a destroyer). However, the right-hand side (i.e. Freshwater and Land) happens to depict the very interspecies interactions I observed in a salmon spawning stream in the Upper Kuskokwim region (though Figure 5.1 omits the beavers!). Athabascan hunter-fishers' activities there are carefully crafted to intervene in interactions among several keystone species in order to optimize salmon's survival in the lifecycle. In this sense, Athabascan hunter-fishers not only take up the role of a *keystone* player but also possibly perform a *hyperkeystone* function.

Another interesting point Worm and Paine (2016) raise is the possibility of non-human hyperkeystone species. Worm and Paine say, "Far-ranging generalist predators, such as orcas for example, might upon closer study turn out to qualify as hyperkeystone species" (ibid.: 603). We may need to consider the possibility of treating brown bears as hyperkeystone species in the Interior Alaskan boreal forest.

Interestingly, Athabascan hunter-fishers have shown deep reverence toward the top-level predators as well. I will summarize what Ignatti Petruska learned from Miska Deaphon, a traditional chief of Nikolai who spent a lot of time fishing in Salmon River: People used to pay a special respect to wolves (*Canis lupus*), wolverines (*Gulo gulo*), black bears (*Ursus americanus*), and brown bears. All these four animals have a type of spirit called *tsanza'*, which is different from *yeja'*, a type of spirit shared by all living beings. When butchering these animals with *tsanza'*, children should not come downwind of the carcass. Otherwise, the wind may carry the spirit of the animal, which then enters his/her body and causes sickness.

Of course, I do not mean to argue that all four top-level predators are hyperkeystone species. Nor am I suggesting that only top-level predators are revered by Athabascan people, who actually posit that one needs to pay respect to all living beings. Rather, the Athabascan belief in the special spirits of these top-level predatory animals seems to suggest an interesting overlap between Indigenous Environmental Knowledge and conservation ecology: These powerful neighbors in the boreal forest can exert significant effects on people and the environment through their activities (either as keystone or hyperkeystone species). If Alaskan wildlife managers hope to continue "serving" salmon into the future, they should learn from the hyperkeystone functions that Athabascan people have performed in service to ecosystems since time immemorial.

Acknowledgments

Early drafts of this chapter were orally presented at workshops in Rikkyo University and Hokkaido University, both in December 2017. I thank David Koester, Keiichi Omura, Heather Swanson, Hiroki Takakura, and the editors for their helpful comments and criticism. My gratitude also goes to my Athabascan friends and mentors in Nikolai, Alaska, who patiently taught me the importance of salmon and fishing in their culture.

Notes

1 Upper Kuskokwim people used to pick up spawned-out salmon for food while it was still fresh (Collins 2004: 23). Hunting bears at salmon spawning streams might have meant eliminating competitors for the same salmon resources.
2 Worm and Paine's argument unwittingly reproduces and reinforces anthropocentrism in their efforts to identify the position of the human in Anthropocene. I do not have enough space here to develop a

more sophisticated criticism and theoretical reflections on Worm and Pain's idea, but I plan to address it in the near future.

References

Arnold, David F. 2008. *The Fishermen's Frontier: People and Salmon in Southeast Alaska*. Seattle, WA: University of Washington Press.

Berkes, Fikret 2012. *Sacred Ecology*. London: Routledge.

Blaser, Mario 2009. The threat of the Yrmo: The political ontology of a sustainable hunting program. *American Anthropologist* 111(1): 10–20.

Collins, Raymond L. 2004. *Dichinanek' Hwt'ana: A History of People of the Upper Kuskokwim Who Live in Nikolai and Telida*. Edited by Sally Jo Collins. McGrath, AK: National Park Service.

Comberti, C., Thornton, Thomas F. Wyllie de Echeverria, Victoria, and Patterson, T. 2015. Ecosystem services or services to ecosystems? Valuing cultivation and reciprocal relationship between humans and ecosystems. *Global Environmental Change* 34: 247–262.

Cruikshank, Julie 1990. *Life Lived Like a Story: Life Stories of Three Yukon Native Elders*. Lincoln NE: University of Nebraska Press.

Deaphon, Miska 1980. *Nikolai Hwch'ihwzoya': Stories from Nikolai*. Anchorage, AK: National Bilingual Materials Development Center.

Gard, Richard 1961. Effects of beaver on trout in Sagehen Creek, California. *Journal of Wildlife Management* 25(3): 221–242.

Halffman, Carrin M., Potter, Ben A., McKinney, Holly J., Finney, Bruce P., Rodrigues, Antonia T., et al. 2015. Early human use of anadromous salmon in North America at 11,500 y ago. *Proceedings of the National Academy of Sciences of the U.S.A.* 112(40): 12344–12348.

Helfield, James M. and Naiman, Robert J. 2006. Keystone interactions: Salmon and bear in riparian forests of Alaska. *Ecosystems* 9: 167–180.

Holen, Davin L., Simeone, William E., and Williams, Liz 2006. Wild resource harvests and uses by residents of Lake Minchumina and Nikolai Alaska, 2001–2002. ADFG Technical Paper 296. Juneau, AK: Alaska Department of Fish and Game.

Hosley, Edward 1977. A reexamination of the salmon dependence of the Pacific drainage culture Athapaskans. In J. Helmer, S. Van Dykes, and F. J. Kense (Eds.), *Problems in the Prehistory of the North American Subarctic: The Athapaskan Question*. Calgary, AB: Department of Archaeology, University of Alberta, pp. 124–129.

Kondo, Shiaki 2016a. Hunting and fishing education and recurrence of the past: revitalization of subsistence activities in Interior Alaska. In Shinjilt Okuno and Katsumi Okuno (Eds.), *Ethnography of Killing Animals*. Showado, pp. 293–326. [In Japanese].

Kondo, Shiaki 2016b. An ethnography of knowledge production in the issue of declining salmon populations in Alaska: What should researchers do? *Annual Papers of the Anthropological Institute, Nanzan University*, 6: 78–103. [In Japanese].

Lovejoy, Arthur O. 1936. *The Great Chain of Being: A Study of the History of an Idea*. Cambridge, MA: Harvard University Press.

Morgan, Lewis H. [1868] 1986. *The American Beaver: A Classic of Natural History and Ecology*. New York: Dover Publications.

Nadasdy, Paul 2003. *Hunters and Bureaucrats: Power, Knowledge, and Aboriginal-State Relations in the Southwest Yukon*. Vancouver, BC: UBC Press.

Naiman, Robert J. 1988. Alteration of North American streams by beaver. *BioScience* 38(11): 753–762.

Osgood, Cornelius 1936. *The Distribution of the Northern Athapaskan Indians*. New Haven, CT: Yale University Press.

Paine, Robert T. 1969. A note on trophic complexity and community stability. *American Naturalist* 103(929): 91–93.

Pollock, Michael M., Pess, George R., and Beechie, Timothy J. 2004. The importance of beaver ponds to coho salmon production in the Stillaguamish River basin, Washington, USA. *North American Journal of Fisheries Management* 24: 749–760.

Power, Mary E., Tilman, David, Estes, James A., Menge, Bruce A., Bond, William J. et al. 1996, Challenges in the quest for keystones. *BioScience* 46(8): 609–620.

Schlosser, Isaac J. 1995. Dispersal, boundary processes, and trophic-level interactions in streams adjacent to beaver ponds. *Ecology* 76(3): 908–925.

Stokes, Jeff 1985. Natural resource utilization of four Upper Kuskokwim communities. Technical Paper #86. Juneau, AK: Alaska Department of Fish and Game.

Thornton, Thomas 2015. The ideology and practice of Pacific herring cultivation among the Tlingit and Haida. *Human Ecology* 43: 213–223.

Woelfle-Erskine, Cloe, and July Cole 2015. Transfiguring the Anthropocene: Stochastic reimaginings of human-beaver worlds. *Transgender Studies Quarterly* 2(2): 297–316.

Worm, Boris. and Robert T. Paine 2016. Humans as a hyperkeystone species. *Trends in Ecology & Evolution* 31(8): 600–607.

6

PERFORMANCE KNOWLEDGE

Uncovering the dynamics of biocultural diversity of Borneo's tropical forests through a Penan hunting technique

Rajindra K. Puri

Introduction

This chapter illustrates a kind of local ecological knowledge that develops and is applied in the course of resource use activities, in part, to deal with contingency and the dynamics of biocultural diversity (Bridgewater and Rotherham 2019). Specifically, what I call *performance knowledge* (Puri 1997a, 2005) emerges and evolves through the practice of such subsistence activities as collecting, hunting, fishing, herding, and various forms of agriculture, and is procedural in form and strategic in function, allowing a practitioner to "manage" people, wildlife, tools, materials, and various contexts of a particular activity and to respond to change in these components and contexts in the course of that activity. It contrasts in form with both *declarative knowledge* (verbalizable, lexical knowledge, e.g., knowing the names and folklore/myths of prey animals) and another type of procedural knowledge, which I call *behavioral knowledge* (tacit, practical skills, e.g., sharpening a knife, making a blowpipe, throwing a spear), thus nuancing the commonly referred to dichotomy of "knowledge and skills" or "knowing and doing" (Ellen and Harris 2000; Anderson 2011). As tacit procedural knowledge, performance knowledge is learned and transmitted differently, largely through accumulated experiences (as an observer or a participant), including the recounting of such experiences by others. As researchers, we are most likely to come upon such knowledge through very acute observational techniques such as *active participant-observation* (*sensu* Nelson 1969) or apprenticeship (Puri 2015).

A close equivalent is the classical Greek concept of "metis" (combining wisdom and cunning), made popular by James Scott (1998), which he describes as a kind of local "know how" that is gained from long practical experience (in traditional farming, for instance), and is contrasted with epistemic knowledge, the more abstract form that is lexical. In my conception, performance knowledge includes, even requires, "metis" in the sense of the wisdom of how something works, such as "rules of thumb," gained from experience, that is then used strategically to manage activities and events to produce consistent success despite contingencies that arise naturally. "Metis" may be a cultural institution, in the sense of transmitted socially, but performance knowledge is often discovered, and re-discovered, spontaneously by

DOI: 10.4324/9781315270845-7

individuals in the course of their practice as they have to solve problems, some of which are known and some of which are surprises. To the extent these solutions are consciously knowable, successful, and in line with a local world-view and value system, they will be transmitted and become cultural. But whether or not such knowledge will be used will depend on the particular circumstances that individuals and groups find themselves in; hence, performance knowledge is a type of *meta-knowledge*, in practice, it both reveals and instantiates biocultural diversity.

The Penan hunters of Borneo

To illustrate performance knowledge, I will describe a hunting technique known as *nedok*, performed today only by a very few expert hunters of the Penan Benalui people of Indonesian Borneo (Puri 2005). Two generations ago, they were semi-nomadic hunters and gatherers in North Kalimantan's rainforest which covered the mountainous interior. Despite long periods in the forests—hunting wild pigs, processing sago palms for their starchy pith, and collecting wild plants and fruit—they were seldom very far from rice swidden farming communities with which they interacted socially and economically—trading a variety of forest products, such as meat, skins and other animal parts, rattan, damar, aloeswood, and medicinal plants, for rice, steel, cloth, and manufactured trade store goods. I lived and conducted research with the Penan and their Kenyah Badeng neighbors in the Pujungan district of North Kalimantan for many years between 1990 and 2000. Between 1991 and 1993, I conducted 21 months research on hunting, using *active participant observation* to be able to observe closely how hunting works, how children learn to hunt, and to learn to hunt myself Penan-style (Puri 1997a).

At that time, the Penan and their Kenyah neighbors mostly hunted with dogs and spears (*ngasu*), primarily for the wild bearded pig (*Sus barbatus*). Occasionally, they caught animals with homemade shotguns (*nyalapang*) or by ambush with blowpipes (*ngeleput*). Usually once a year (starting in May), there was a mass migration of wild pigs into the area to fatten up on oil-rich fruits and acorns of Dipterocarp and Oak species respectively, and then another migration out of the area (in November) once the fruit season was finished (Caldecott 1990). Residents, men and women of all ages, from nearby villages would line the banks of the larger rivers and wait in ambush for the pigs to swim across (*mabang satong*), and then kill them, in the river itself from boats, or when they emerged onto the riverbank. This was a season of great abundance, of meat and fat, and fruit too. When the fruit season ended and the pigs departed the area, rice swiddens were still being weeded with several months to go before harvest in February and March. This was the start of a leaner part of the year: a cooler rainy season where wild meat gave way to fish, traps provided a little meat from gardens and fields and most people relied on stored rice, root vegetable crops (manioc, yams, taro), village fruit, such as bananas, and forest and garden greens.

It was during this time of the year that I first observed the use of *nedok*, used to track down what few pigs remained in the surrounding forests. It also proved to be useful when dogs became scarce, as they often did following epidemics, and when Penan and Kenyah unexpectedly encountered pigs, or evidence of their presence, while traveling through the forest for other reasons. But it is very difficult to pull off such opportunistic hunts successfully; the conditions have to be manageable, not necessarily perfect, and the hunter has to have considerable skills that can be flexibly applied, given the particular circumstances. In short, this is a technique for experts, and it ranks number one for difficulty and, thus, for prestige gained.

Nedok

Nedok in Penan means "to dok," *modok* or *dok* being the vernacular for the pig-tailed macaque (*Macaca nemestrina*). Thus, the hunter or hunters must become a pig-tailed macaque, by mimicking the sounds and movements of an individual or a group of macaques traveling through the forest. The bearded pig, hearing these sounds, moves toward the "monkeys," thinking they will lead him to food, specifically fruit that has been dropped or shaken free from the forest canopy as the macaques feed or travel from tree to tree. The hunters, hearing a pig approaching, or coming across one near their path, then lure the animal closer with a set of different calls. When they get within range, which can be as close as a few meters, the lead hunter will launch his spear or fire his gun. They prefer the former, which is quieter, so if they miss they can possibly get another try. Macaques travel in quite noisy groups, with young ones often throwing dead branches around, presumably for fun. When hunters travel as macaques, they too will throw branches around and make a bit of a racket. Thus, a spear whizzing by an animal can be somewhat disguised as branch throwing, giving the hunter another attempt. The whole ruse is dependent on the pigs not detecting the humans by their smell, the whites of their eyes, or their normal voices and behaviors. If the wind shifts in their favor, the pigs will bolt away in a flash; if a hunter does not cover his eyes with a leafy branch, the pigs may see the whites, know it can only be a human, and flee; if young hunters drop out of character and start chatting or stop making macaque calls, the pigs will register the change and take flight.

There are many more variables to consider: the wind direction and the way it changes during the day; whether it is overcast or sunny or about to rain; the path taken through the forest, and whether to travel on ridge tops, or follow a route past certain fruiting trees, such as the hill sago palm (*Eugeissona utilis*); the number and types of companions along, their skills and abilities; past sightings or catches; the tools available for the day, etc. From the moment the hunters enter the forest, they need to be in character, and so the performance begins and continues on for hours and hours if need be, traveling fast and far, until a kill or two or even three is made. During the day, circumstances may change, it may rain, or get very windy, prey animals may be wounded and need to be tracked, equipment may malfunction or break and need to be replaced or repaired, demands at home may drive a hunter to pursue prey well into the evening and perhaps even force him to stay overnight in the forest, someone may get injured and need to be cared for. Hunters have to respond to such exigencies and changing conditions continuously in real time in order to be successful, day in and day out, and it is in these responses where their performance knowledge, their ability to manage, to make decisions, and apply various kinds of knowledge, comes to the fore.

Those few hunters that use *nedok* regularly have very high success rates; between 80 and 100 percent of day trips result in a catch. There is less success when trying to use the technique spontaneously in a chance encounter, but still it was surprising how often really good hunters could get into character, get close enough and use whatever was at hand, a machete or even rocks, to make a kill (Puri 2005).

Palms, pigs, and primates

The Penan and their neighboring farmers can also be said to *manage* the forest to promote the success of *nedok*, revealing a deep understanding of the ecology of Borneo's tropical forests (Colfer et al. 2005). During the off-season, some of the most reliable sources of fruit, for animals and humans, are forest palms of all sorts, from climbing rattans to tree palms, which fruit asynchronously (Dransfield et al. 2008; Whitmore 1985). The hill sago, *nanga* or *uvut* (*Eugeissona utilis*), formerly the Penan's staple source of starch, is now a famine food that grows in very dense monospecific stands (known as *birai*) on dry ridge tops, often very close to, and overlooking, villages (Puri 1997b). These stands are encouraged to expand, vegetatively, by cutting new shoots for palm cabbage, and sexually, by allowing several of the larger stems to go to seed, and not harvesting them for their sago flour. There is, therefore, a tradeoff between pith production and seed production, managed by Penan and farmers to insure that there are stems for pith if a rice crop should fail and food is needed, and also fruits for animals that can be hunted, such as the pig-tailed macaque and the bearded pig. The former eats the outer green fruit, or ectocarp, the latter consumes the remaining coconut-like oily seed. So, it's not surprising that hunters *nedok* to and from these hill sago stands when fruit and pigs elsewhere are in short supply.

Conclusion

As a type of meta-knowledge, the Penan's performance knowledge both integrates and thus reveals an ecological and dynamic complex of animal and plant species, habitats, seasonal changes, local environmental knowledge, and practices. In this sense, performance knowledge is *biocultural diversity in practice*. While it does hold some of the accumulated wisdom of "metis," it is always dynamic, always responding to whatever the context is that hunters or fishers or farmers find themselves in; incorporating and generating new ideas, understandings, innovations in order to solve the daily problems of survival in an ever-changing environment. As global environmental change comes to the rainforests of Borneo, it is performance knowledge that will register these changes and capture human responses, indicating both successful and not so successful adaptations to change (Puri 2015). Paying attention to the way that biodiversity-dependent people, like the Penan, perform those everyday tasks that make up their livelihoods, and in doing so manage biocultural diversity, promises a better understanding of local change, local responses, and ways to better integrate such understandings across larger scales of response and management. For this to happen, the Penan and others must have the space and freedom to evolve their performance knowledge, their biocultural diversity, to fit the times in a manner that suits them.

References

Anderson, E. N. 2011. Ethnobiology: overview of a growing field. In E. N. Anderson, D. M. Pearsall, E. S. Hunn, and N. J. Turner (Eds.), *Ethnobiology*. Hoboken, NJ: Wiley-Blackwell, pp. 1–14.

Bridgewater, P. and Rotherham, I. D. (2019). A critical perspective on the concept of biocultural diversity and its emerging role in nature and heritage conservation. *People and Nature*, 1, 291–304. https://doi.org/10.1002/pan3.10040.

Caldecott, J. O. 1990. Eruptions and migrations of bearded pig populations. *Bongo* 18: 2–12.

Colfer, C. J. P., Colchester, M., Joshi, L., Puri, R. K., Nygren, A., and Lopez, C. 2005. Traditional knowledge and human well-being in the 21st century. In G. Mery, R. Alfaro, M. Kanninen, and M. Lobovikov (Eds.), *Forests in the Global Balance: Changing Paradigm*, vol. 17. Helsinki: IUFRO, pp. 173–182.

Dransfield, J., Uhl, N. W., Asmussen, C. B., Baker, W. J., Harley, M. M. and Lewis, C. E. 2008. *Genera Palmarum*. Chicago: University of Chicago Press, pp. 410–442.

Ellen, R. and Harris, H., 2000. Introduction. In R. Ellen, P. Parkes, and A. Bicker (Eds.), *Indigenous Environmental Knowledge and Its Transformations: Critical Anthropological Perspectives*. Amsterdam: Harwood Academic Publishers, pp. 1–34.

Nelson, R. K. 1969. *Hunters of the Northern Ice*. Chicago: University of Chicago Press.

Puri, R. K. 1997a. *Hunting knowledge of the Penan Benalui of East Kalimantan, Indonesia*. PhD thesis, Department of Anthropology, University of Hawai'i, Honolulu, HI.

Puri, R. K. 1997b. Penan Benalui knowledge and use of treepalms. In K. W. Sorensen and B. Morris (Eds.), *People and Plants of Kayan Mentarang*. London: WWF-IP/UNESCO, pp. 194–226.

Puri, R. K. 2005. *Deadly Dances in the Bornean Rainforest: Hunting Knowledge of the Penan Benalui.* Leiden: KITLV Press.

Puri, R. K. 2015. The uniqueness of the everyday: Herders and invasive species in India. In J. Barnes and M. Dove (Eds.), *Climate Cultures: Anthropological Perspectives on Climate Change*. New Haven, CT: Yale University Press, pp. 249–272.

Scott, J. C. 1998. *Seeing like a State*. New Haven, CT: Yale University Press.

Whitmore, T. C. 1985. *Palms of Malaya*. Oxford: Oxford University Press.

7

SOIL ETHNOECOLOGY

Paul Sillitoe

Introduction

The majority of us, urban dwellers distant from land and farming, take the soil for granted and if it is brought to our attention, it stirs little interest, equated in our minds with clods of mud and mouldy rotting vegetation, but see Box 7.1. The study of others' soil knowledge or 'ethnopedology' (Tabor 1992; Sillitoe 1996), the subject of this chapter, reflects this popular attitude, having received relatively little attention compared to other ethnoscientific fields, notably ethnobotany and ethnozoology (see Winklerprins 1999 and Barrera-Bassols and Zinck 2003 for literature reviews).[1] This is strange, for even some early ethnographers noted the soil's importance, such as George Forster who, together with his father, acted as scientists on Cook's second voyage, and who, in an essay on the breadfruit tree, observed that the 'produce of the soil is deeply entwined in the fate of humankind' (Forster 1843: 347); and his acolyte von Humboldt (1822: 525), the renowned eighteenth-century polymath, who referred to the 'influence exercised by the composition of the soil on agriculture, trade and the more or less slow progress of society'.

Box 7.1 Soil

Soil is a remarkable substance upon which all life depends. It is the medium in which plants root themselves and it supplies the inorganic nutrients they use, along with water, sunlight and atmospheric gases, to synthesise the organic materials upon which the trophic pyramid of animal, including human, life stands. As Russell comments in the Preface to his classic soil science textbook: 'At first sight the subject appears very simple; in reality it is highly complex' (1912: vii). It is challenging because of the complexity of soil systems, their components interrelated in complicated feedback relationships, and the soil constantly changing with chemical and physical processes going on continuously.

Human beings contribute to natural changes when they disturb the soil in cultivation, sometimes skilfully managing it to maintain fertility, other times degrading it. It adds a further

DOI: 10.4324/9781315270845-8

dimension of complexity to consider the ethnopedological knowledge and management of soil by populations elsewhere, such as communities in the mountains of Papua New Guinea who are the ethnographic focus of this chapter. The relevant issues include how people think of, classify and appraise, their soils, how they manage them under cultivation, and the impact of their activities on soil conditions, notably what happens to soil processes, particularly productivity, under various cultivation regimes.

The Wola and their soils

The Wola speakers of the Southern Highlands Province in Papua New Guinea live in small houses scattered along the sides of their valleys (Figure 7.1), in areas of extensive cane grassland, the watersheds between which are heavily forested (Sillitoe and Sillitoe 2009). Dotted across the landscape are their neat gardens. They depend almost exclusively on horticulture to meet their subsistence needs, living on a predominantly vegetable diet in which sweet potato (*Ipomoea batatas*) is the staple. They keep pig herds of considerable size. They hand these creatures, together with other valuable things, around to one another in unending series of ceremonial exchanges, which mark all important social events. These transactions are a notable force for order in their fiercely egalitarian acephalous society.

The most widely cultivated soil types in Wolaland are Inceptisols and Andisols, which are derived from the sedimentary parent materials, variably affected by volcanic ash (dominated by it, to no evident effects), with some alluvial re-deposition. Some soils are subject to wet conditions and are gleyed, others are peaty. They feature dark topsoils, high in organic matter, with a porous crumb structure and indeterminate volcanic ash contents, overlying a firm, moist, aerobic, bright-brown clay subsoil, sometimes stony and faintly mottled (Bleeker 1983: 67–70).

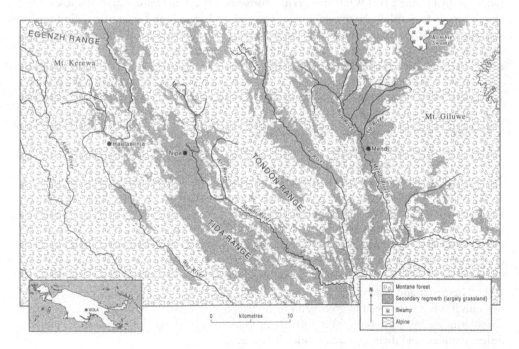

Figure 7.1 Map of Wola region

These young soils, experiencing several rejuvenating episodes of volcanic ash fall, are fairly productive with appropriate management.

Knowledge and cultural context

A challenge that faces the ethnosciences is how to interface with their eponymous natural science counterparts, which tend to dominate. There is a danger of uncritically imposing a scientific model on what we think we understand of others' knowledge and distorting it. This was evident from the start in the ethnosciences with the categorisation of subject matter according to supposedly synonymous scientific subjects, such as ethnobotany, ethnozoology and ethnopedology. It is an error to treat local understandings of the natural world as culturally decontextualised knowledge amenable to investigation according to the conceptual framework of the matching science. The assumption that there are delimited bodies of knowledge independent of socio-cultural context is dubious, leading to ambiguous science-like representation that focuses on aspects thought to mirror science, whereas their status may be different from a local viewpoint.

It is necessary to acknowledge the 'cultural construction of the environment' (Simmons 1993; Atran and Medin 2008), that understanding of the natural world is culturally mediated. People may codify the intelligence amassed over many generations of trial-and-error in ways quite different to science. Furthermore, the imposition of scientific categories not only threatens to misrepresent knowledge, but also to limit analysis by prompting us to ignore issues that may be important to local understanding and experience, notably the non-empirical such as supernatural, symbolic, socio-political and kin considerations, albeit expressed in an alien idiom, perhaps involving spirits, mythical events, and so on.

We see the consequences in scientific soil and land resource surveys that have limited relevance locally (Young 2017). While farmers may use some of the same information as soil scientists to assess soils (e.g. colour, permeability and texture), Indigenous and scientific definitions of soils and land types, and associated soil classifications, often compare poorly (Sikana 1993). Soil scientists have long used local soil names. The pioneering Russian soil scientist Dokuchaev is an early example, using such vernacular terms as *chernozem*, *solonetz* and *gley* in his nineteenth-century soil classification, some of which continue in scientific use to this day (Krasilnikov and Tabor 2003: 201). A subsequent twentieth-century example concerns the incorporation of local soil names in a scientific soil survey in Sukumuland, Tanzania (Milne 1947), which have replaced those of scientific soil classification in the region (Acres 1984).[2] But it amounts here to little more than equating local names with soil science catenae, although it may promote interaction between scientific and Indigenous Knowledge, highlighting their potentially dynamic relationship. It is necessary to exercise care in scientifically recodifying and interpreting local understandings. It may isolate for analysis certain resource use practices from the broader circumstances that critically inform them.

Scientists identify classes by a range of technically assessed properties, whereas farmers frequently look for a dominant property which may have further associations, and their approach is frequently more holistic, incorporating exotic social and cultural aspects in distinguishing land and soils according to various uses and associations (Niemeijer 1995; Talawar and Rhoades 1998; Niemeijer and Mazzucato 2003). The manner in which New Guinea Highlanders classify soils illustrates the need to go beyond the conceptual framework of Western science when considering pedological issues, for it is necessary to know something about their tribal political order to understand their approach to soil classification.

Issues of soil classification and representation

In the Highlands of Papua New Guinea, people have names for different types of soils (Box 7.2), and we can match these terms against those of soil science in a Dokuchaev-cum-Milne-like manner. Table 7.1 presents such a correlation for the Wola speakers. We might use their terms to name the soils mapped in a land resources survey. But this would be a distortion of their ideas, using their terms in an alien context because these people traditionally have no notion of maps. (This criticism applies equally to participatory soil mapping exercises that seek to draw on local farmers' knowledge while imposing foreign cartographic representations; Barrera-Bassols, Zinck and Van Ranst 2009.) It amounts to plucking their words out of cultural context and using them to gloss our classes; so instead of talking about Psamments, for example we substitute *iyb muw*. But does this matter because we presumably see the same soil 'out there', to which we just give different names? The rub is that while we may see the same soil, we may think about it in quite different ways.

Box 7.2 Wola soil classification

Regarding naming of soils, Wola people have no concept equivalent to the soil profile. They do not classify soils as entire profile sequences. They conceive of soils as a series of named horizons. The Chimbu people likewise name and classify soils by horizons (Brookfield and Brown 1963: 35–36). Thus, the classes in Table 7.1 refer to the predominant subsoil at any location. When they talk about soils, they name each horizon, adding qualifying terms as necessary to specify particular characteristics. So they might refer to a Humitropept, which is a common soil type overlying

Table 7.1 Soil names of the Wola correlated with those of USDA and the FAO

Wola		USDA		FAO
Major group	Minor groups	Sub-order	Great group	
haenbora		Orthent	Troporthent	Lithosols
haen hok		Rendoll		Rendzinas
hundbiy	payhonez	Tropept	Humitropept	Cambisols
	kas	Humult	Tropohumult	Acrisols
	tongom			
tiyptiyp	kolbatindiy	Andepts	Hydrandept	Andosols
			Dystrandept	
iyb dor tilai		Fluvent	Tropofluvent	Fluvisol
iyb muw		Psamment	Troposamment	Arenosol
pa tongom		Aquepts	Andaquept	Gleysols
			Tropaquept	
		Aquent	Fluvaquent	
iyb uw damiy		Saprist	Troposaprist	Histosols
		Hemist	Tropohemist	
			Cryohemist	
waip		Folist	Tropofolist	
		Fibrist	Tropofibrist	

Figure 7.2   Soil profile pit: Hydrandept; *tiyptiyp*

limestone in their region, as *suw pombray diy hundbiy* 'black soil and bright-orange-clay' (Figure 7.2), and they may qualify this further by specifying notable features of each horizon *suw pombray hemem iyba wiy hundbiy araytol omb hul haeruw* 'black soil fertile with grease and stony bright-orange-clay with ferralitic nodules' (Figures 7.3 and 7.4).

   If we think of spatial relationships between soils, as in land resource surveys, and dig pits at intervals across an area, and ask Wola people to comment on the soil exposures, they will name and may describe each horizon separately. Their manner of naming soils prompts them to treat each exposure independently, constructing an appropriate label using various terms to qualify what they see. There is no compulsion to organise classes so that they relate one to another, to form catenae or landscape-related sequences. While they are fully aware that topography can influence soil type, that wet gleyed soils, for example will more likely occur in low-lying areas and depressions, they do not relate their soil classification systematically to landscape. In part, this reflects their disinterest in soil genesis, which informs and orders the classifications of soil science. They name what they see with no such theory to structure their classes, which are consequently diverse and variable.

Figure 7.3   Soil profile pit: stony Humitropept; *hundbiy araytol*

Figure 7.4   Soil profile pit: Humitropept; three horizons of *pombray, hundbiy sha* and *hundbiy*

If we focus on the boundaries between soil classes, as opposed to the centrally defined modal categories, and ask what is going on, we see the Wola approach in an interesting light. Soils in the field naturally comprise a continuum usually without sharp discontinuous breaks between one class and another. They intergrade gradually one into another, whereas, in soil mapping, a firm line distinguishes sharply between them. The catena conveys the continuity, conceived as a series of defined soil types that change in some predictable manner across the landscape, as demarcated on a map. But the associated categorisation and spatial representation face particular challenges not overly to distort nature's arrangements. The Highland Papua New Guinea system of soil naming handles the boundary conundrum better by doing away with it. Lacking any idea of representing soils spatially, the Wola have no need to draw boundaries between one soil class and another. They can accommodate to any observable changes in soil type they think significant as they move across the landscape by modifying their descriptor terms, so the soil here may be 'some of this and a little of that' whereas that one over there has 'less of this and more of that' and so on, with no thought to specify a boundary between them (Figure 7.5).

Figure 7.5   Soil profile section, showing blocky structure

An intriguing consequence of the soil classification is that it is more straightforward than those of soil science and more faithfully reflects observed reality. These Highlanders have no rigidly defined classes into which they have to sort soils, only names subject to whatever qualifying terms are necessary to further specify the soil, horizon by horizon. They avoid the difficulties soil scientists often face in deciding exactly how to classify soils according to precisely defined classes and modal types; the monumental USDA scheme exemplifies the complexity (Soil Survey Staff 2014). And scientists go further in soil surveys, drawing boundaries between different soil classes, which require them to decide that somewhere on the ground one soil type ends and another begins, represented with a line on maps, with different keyed colour codes either side for the two soil types distinguished.

A further consequence of doing away with the need to draw boundaries is that the classification scheme is inherently flexible and better able to handle the wide range of soils 'out there'. The Wola have no need of a hierarchical classification scheme of the sort needed to accommodate evolutionary hypotheses, lacking, as noted, any concern for soil genesis. The problems that arise from scientific schemes structured in this way include sometimes putting agronomically similar soils into different, sometimes widely separated, soil classes according to their genetic criteria. The absence of a hierarchical scheme dispenses with the need rigidly to define classes and divide soils. If they wish, people can associate soils one with another, in any way that seems appropriate to them at the time.

The flat and fluid classification scheme reflects the culture in which it occurs. Briefly, Wola society is, as mentioned, aggressively egalitarian, albeit with a marked gender differentiation in social life. It lacks, as an acephalous or stateless order, any political offices that vest authority in certain persons over others. The ethos prompts the subversion of hierarchies that place some above others, and part of this compulsion is conceptually to obfuscate the existence of boundaries because their existence implies an embryonic hierarchy, the centre more important than the periphery (Sillitoe 1999). The stateless cultural context has further ramifications for any comparison of local and scientific perceptions of soils. There is no recognition of any expert authority to validate the correct identification and naming of soils. The upshot is that unexpected levels of disagreement characterise the system. Different persons may name the same soil somewhat differently, particularly at the level of descriptor terms, which suggests that they 'see' different features as characteristic of the soil. And no one can adjudicate over differences between them about the naming of soils. They are all 'correct'. This relates to Indigenous Knowledge being fragmentary and dynamic, not centralised and static. The fluidity of the local versus the rigidity of the scientific soil classifications starkly reveals the problems of comparing and correlating them, and questions the validity of such comparisons, when they depend on quite different criteria, one concerning everyday practical management and the other landscape evolutionary processes. Local Indigenous Knowledge is by definition parochial and culturally relative, its local embedded character is intrinsic to its success, whereas science strives for a generic and global perspective.

Issues of soil cultivation and recultivation

The classificatory schemes of people only reflect part of their soil knowledge, which also has a practical dimension, notably in respect of management under cultivation (Figure 7.6). The farming regime found in the rugged Wola highland region defies characterisation according to any classification (Sillitoe 1983, 2010). The idea of stationary cultivation under a shifting regime appears a contradiction in terms, but this antilogy characterises the farming system. On some plots, people practise classic shifting cultivation, clearing for one, or possibly two cropping cycles, and then abandoning to natural regrowth for many years. Other plots they keep under almost permanent cultivation for decades, with occasional, brief periods of grassy fallow, which the Wola call *em hul* or 'bone gardens', after their bone-like durability.[3] They farm all plots, whatever their productive life, using the same methods and technology. The result is a continuous spectrum of agricultural land use, comprising a single farming system (Figure 7.7).

The maintenance of some gardens under crops long term, within the broad context of a shifting cultivation strategy, contradicts widely held suppositions about subsistence farming in

Figure 7.6 A view across a valley side showing the mosaic of gardens and forest

Figure 7.7 View of a slope showing gardens in varying stages of cultivation

the tropics (Ruthenberg 1976; Norman 1979), which assume brief periods under cultivation followed by abandonment due to declining productivity (Nye and Greenland 1960: 73–126; Sanchez 1976: 374–380) with long intervals under natural fallow to allow soil recovery from nutrient losses, weed proliferation, disease build-up, erosion damage, and so on.[4] What is it about the soils and crops of this region that allows such a cultivation regime to exist, featuring minimal or no fallow breaks and no outside amendments (such as manure), without catastrophic declines in productivity, resulting in a degraded landscape like other tropical regions, such as Amazonia, following extended forest clearance? Indeed, how come the reverse sometimes occurs and crop yields improve?

If you put such questions to the Wola, they are likely to respond '*em hul wiya ngora*' (bone gardens just exist), which is sufficient to their minds and for their purposes. Encouraged to elaborate further, they may comment '*nao shumbaen em nguwbiy bismiyuwp*' (our ancestors cultivated such gardens), implying that time and tradition have proved the practices effective. Some individuals might mention observations of some soil conditions under cultivation, such as '*em ngo suw iyba bawiy*' (soil 'grease' remains in such gardens). In short, the Wola follow a cultivation system evolved over many generations[5] of experimentation without apparent need to explain the soil processes that underpin it. Indeed, they maintain that when they cultivate a new garden, they often do not know for how long it will remain under cultivation,[6] soils changing and sometimes improving with use. They plant the plot and see how their crops fare (Sillitoe 1996: 310–311). They may continue to cultivate it indefinitely if yields are respectable and the garden's location convenient.

The validity of people's assertions about their soil knowledge is difficult to assess. It is arguable that soil conditions enter indirectly into their calculations, because some of the landscape factors that inform their choice of site – like slope, terrain, and so on – also influence soils. Furthermore, while they say that they do not inspect the soil before cultivating, they may do so unwittingly. They know their local regions intimately and may have no need to look closely at the soil before deciding to cultivate. They already know its status by virtue of constantly walking across the land during their everyday lives, and doing so barefoot, they are well aware of soil texture and structure – they 'know' the soil through their feet. And the vegetation growing on a site may indicate soil fertility, by its luxuriance, even the presence of certain species above others, although farmers largely deny that this is so.[7] Whatever, the epistemological challenges that attend the documentation of such local environmental understandings are large.

If you are a Wola, you just know, you are not used to being asked how you know. The awareness that you have of the soil, for instance, is an accumulation of experiences, built up over the years, cultivating the soil and hearing many comments from others on it. These people live rather than reflect on their environmental knowledge. It has a marked practical aspect to it. When asked, for instance, to assess the soil at a particular site, individuals may inspect it, and even handle it, before passing judgement. If you then ask them to justify their assessment, they will look somewhat bewildered. They will probably tell you to look at the soil for yourself, maybe even pass you a handful to feel, the implication being that surely you can judge for yourself.

The knowledge is passed on by informed experience and practical demonstration, more likely shown than spoken, being as much skill as concept, conveyed as required in everyday life. It is difficult for an outsider convincingly to gain some understanding of such knowledge. If actors know by doing, can we learn to know too, which presumes going barefoot, for instance, to make the tactile connection with the earth necessary to knowing? And if we can gain such tacit understanding, can we express it in words (Sillitoe 2017: 300–303)?

Local knowledge and natural science

There is another avenue, which reverses the foregoing critique and endorses soil science. While I can give an account of Wola horticultural practices (Sillitoe 2010: 253–329) and report their comments on soils and their behaviour under cultivation – about 'grease' levels, and so on, so far as I apprehend them – the question remains: what is it about the soils and crops of this region that allows the anomalous cultivation regime to exist? No systematic answer is forthcoming locally. People follow practices that impact on the environment, without articulating a theory of why, which is understandable given the pragmatic character of their soil knowledge, it not being codified but diffuse and communicated piecemeal, even disjointedly from our perspective.

The theory and concepts of natural science allow us to address the question. The implication is *not* that we can translate Wola conceptions about the environment into scientific discourse. It is entirely foreign to them, with no ideas equivalent to nutrient ions, gas exchange, physical functions, and so on. Nor is the aim to assess the veracity of local ideas against scientific ones, both are relative. It is to further our understanding of environmental interactions within the cultural context. Although taken up with different issues expressed in quite different idioms, both concern the *same* natural environment 'out there' and together can further our understanding of it, people's place within it and the impact of their activities on it. (For similar attempts, which consider both Indigenous and scientific knowledge of soils in African and MesoAmerican cultural contexts, see Krogh and Paarup-Laursen 1997; Ericksen and Ardón 2003; and Gray and Morant 2003.) Furthermore, combining natural science with ethnographic enquiry responds to post-modern criticism, distancing such work from any pretense about achieving an understanding of a foreign population as it understands itself.

It is necessary to know something about the changes that occur in the nutrient status of soils under cultivation for varying periods of time to understand how the Wola can cultivate sites semi-permanently with no soil amendments. Briefly, the properties of the region's soils (see Table 7.2)[8] that limit crop nutrition and production are: (1) low levels of phosphorus availability and high rates of phosphate fixation (reflecting the strong immobilising capacity of volcanic ash-derived minerals; Parfit and Mavo 1975);[9] (2); sub-optimal pH, these acid conditions interfering with the supply of some nutrients, notably reducing total base saturation; (3); depressed cation exchange capacities and lowered availability of exchangeable cations, which is particularly problematic with potassium; and (4); low levels of available nitrogen, which is probable with the high organic matter contents (as high C:N ratios indicate). The physical properties of the soils are, by contrast, generally favourable to crop production, with their high organic matter contents, low bulk densities, and good topsoil aeration and drainage.

The following soil-related processes occur with the establishment of a garden. The burning of cleared vegetation returns nutrient elements to the soil (except for that fraction lost as gas to the atmosphere) via ash that is rapidly broken down further for plant uptake, as documented for swidden regimes elsewhere. It increases pH and gives a critical, though short-lived boost to the availability of several elements. The boost is particularly significant for phosphorus, and also potassium and nitrogen, three major plant nutrients. The increased availability of these limiting nutrients is sufficient to allow the cultivation of a wide variety of crops (Figure 7.8), several of them annuals, including a range of beans (e.g. *Lablab niger, Phaseolus vulgaris*), green leafy vegetables (e.g. *Rorippa* sp., *Dicliptera papuana*), aroids (e.g. *Colocasia esculenta, Xanthosoma sagittifolium*) and cucurbits (*Lagenaria siceraria, Cucumis sativus*), plus some longer-term crops such as bananas (*Musa* sp.) and sugar cane (*Saccharum officinarum*) (Sillitoe 1983: 29–136). But the increase is short-lived (Figure 7.9). There is a decrease in the variety of crops cultivated after two or three plantings, reflecting a change in soil nutrient status, with nutrient availabilities

Table 7.2 Measures of topsoil chemical fertility compared with site land use status

	Site status: virgin through cropped to fallow vegetation												Statistical significance		
	Virgin Sites	Cropped sites: times cultivated						Fallow sites: years under fallow						
		1	2	3	4	>5	>10	<1	1–5	5–10	>10	F	P	
pH	5.02 (0.66)	5.39 (0.92)	5.08 (0.47)	4.94 (0.40)	5.01 (0.67)	5.32 (0.53)	5.25 (0.28)	5.28 (0.61)	5.22 (0.41)	5.1 (0.31)	4.86 (0.27)	1.14	0.34	N.S.
Phosphorus (ppm)	17.1 (12.1)	17.3 (16.2)	12.2 (13.4)	7.1 (4.5)	7.8 (6.3)	7.2 (5.5)	5.7 (5.2)	13.9 (17.4)	7.2 (7.9)	6.6 (6.0)	7.2 (6.8)	1.71	0.09	N.S.
Potassium (me/100g)	1.04 (0.57)	0.56 (0.20)	0.68 (0.30)	0.63 (0.36)	0.44 (0.49)	0.27 (0.13)	0.28 (0.17)	0.87 (0.96)	0.58 (0.28)	0.79 (0.97)	0.46 (0.25)	2.09	0.03	S.
Calcium (me/100g)	13.9 (11.2)	12.2 (6.4)	14.1 (9.4)	8.7 (7.4)	15.3 (11.7)	16.7 (15.8)	8.1 (3.7)	19.7 (23.8)	12.4 (9.4)	8.0 (5.3)	7.3 (7.7)	1.23	0.29	N.S.
Magnesium (me/100g)	3.04 (2.53)	3.36 (2.25)	3.25 (2.23)	1.90 (1.31)	2.11 (1.54)	2.14 (0.70)	2.35 (0.98)	2.65 (2.48)	2.04 (0.92)	1.66 (0.96)	2.45 (2.39)	0.95	0.49	N.S.
CEC (me/100g)	31.3 (5.2)	27.4 (9.4)	31.0 (8.3)	27.1 (5.6)	27.4 (12.1)	29.0 (4.9)	22.6 (7.4)	28.0 (7.4)	31.6 (6.4)	23.3 (7.1)	26.9 (6.9)	1.58	0.13	N.S.
Carbon (%)	25.9 (6.6)	17.6 (8.6)	17.3 (6.4)	14.9 (5.3)	16.8 (10.9)	12.8 (5.9)	11.6 (6.4)	14.5 (6.8)	16.7 (5.2)	11.9 (4.1)	17.5 (6.0)	3.55	0.001	S.
Nitrogen (%)	1.41 (0.62)	0.96 (0.29)	1.06 (0.41)	0.86 (0.39)	0.99 (0.32)	0.93 (0.28)	0.74 (0.26)	0.93 (0.42)	1.12 (0.43)	0.79 (0.29)	1.08 (0.46)	2.23	0.02	S.

Notes: values in brackets = standard deviations; n = 110 sites; degrees of freedom = 10, 99.

Figure 7.8   New garden with taro, greens, sugar cane and beans (trained up poles)

Figure 7.9   Mature garden of sweet potato with bananas and some sugar cane

falling to pre-burn levels due to soil processes largely, with some removed in harvested crops. Organic matter and nitrogen decline significantly with time under cultivation together with potassium, but it is the latter, together with phosphate (the availability of which is relatively low throughout) that fall to levels below those required by many crops (other nutrients show no significant variation).[10] They decline to new equilibrium points that remain relatively constant, even after years under cultivation.

After the first couple of cropping cycles, a markedly narrower range of crops occurs, with many gardens passing under a virtual monocrop of sweet potato, perhaps with a few longer-term crops, and the occasional patch of pumpkin (*Cucurbita maxima*), edible pitpit (*Setaria palmifolia*) and acanth greens (*Rungia klossii*). The sweet potato (Bourke 1985) occupies a central place in this farming system. It is the staple crop, comprising about 75 per cent of all food consumed by weight (Sillitoe 1983: 239), and makes up by far the largest area under crops. It has the capacity to continue producing tolerable yields under these nutritionally constrained conditions with its relatively low phosphorus requirements and preference for fairly high potassium to nitrogen ratio.[11] Contrary to expectations, these changes in soil fertility do not necessarily lead to reduced staple sweet potato yields. Farmers maintain that the soil on some sites improves with use, becoming better for tuber production with time. Measurement of crop yields confirms their assertions, the harvest of tubers from newly cleared garden areas being some 40 per cent below that from established ones. Far from experiencing a decline in yields, as the accepted model of low-input subsistence agriculture predicts, the reverse occurs on some sites and they increase under cultivation.

Managing soil resources

While sweet potato may yield adequately on soils relatively low in extractable phosphorus, even potassium, so long as its ratio relationship with nitrogen remains favourable, farming arrangements must nonetheless maintain minimal levels for tolerable tuber production to continue. How do Wola farmers sustain them and keep their 'bone gardens' under near continuous cultivation without adding external amendments to the soil? The answer is by cultivating sweet potato in earth mounds, composted with the weeds and grasses that colonise sites during brief fallows (Figure 7.10). The plano-convex mounds of soil, called *mond* locally, vary in size, between 2–3 metres or so in diameter (Figure 7.11). This soil management technique is characteristic of subsistence agriculture across the Central Highlands of Papua New Guinea.[12]

A key feature of mounding is the incorporation of all weedy regrowth and any remaining crop residues into them as compost, or coarse grasses[13] and herbaceous vegetation, either green or burnt, if a site is left fallow for any time (Box 7.3). The vegetation soon rots down into a soft compost, the general Wola term for which is *paenpaen* (e.g. *iysh shor paenpaen*, literally 'tree leaf bedding-accumulated', for tree leaf compost, *gaimb paenpaen*, literally 'cane-grass bedding-accumulated', for cane grass compost, and so on). It allows farmers to manage soil fertility constraints by incorporating nutrients stored in plant residues into the soil,[14] particularly with organic matter playing a notable role in maintaining the fertility of soils containing volcanic ash.[15] The supply of those nutrients identified as the probable major constraints on yield – namely, potassium, phosphate and possibly nitrogen – is more or less sufficient from grassy regrowth to meet the demands of sweet potato. Although phosphorus availability may remain below that necessary for optimal yields (Parfitt and Mavo 1975), the effect may be relatively insignificant, given the crop's ability to manage on low phosphate supply (Nicholaides et al. 1985). The boost in available potassium is significant to the success of composted mounds in

Figure 7.10　Mound recently planted with sweet potato cuttings

Figure 7.11　View across garden area previously under coarse grass (in the foreground) showing mounds in various stages of cultivation

cultivating established gardens where levels are low, grasses being well known for their high potassium contents.

Box 7.3 The efficacy of mounds

The benefits of mounds are several (Waddell 1972: 150–166) (Figures 7.12–7.14). The composted mounds are particularly suitable for the management of soils containing volcanic ash, the mechanism of nutrient uptake which they afford being especially effective at overcoming phosphate fixation and poor base saturation, uptake occurring directly from the decomposing vegetation concentrated at the centre of the mound as roots grow through it (Floyd et al. 1987).[16] The delay in the release of nutrients from the compost doubtless permits more opportunity for uptake, allowing time for initial root growth, and more intimate contact between roots and nutrients as they enter solution. Further benefits of compost include improved water-holding capacity. The microbial decomposition of the compost also increases temperatures, further encouraging vigorous root growth. Composting mounds also helps control weeds, by burying them so that sweet potato has a head start in the competition for light. Another benefit mentioned by local people is that it reduces the incidence of disease and rotten tubers,[17] although care is needed, especially using crop residues, not to spread disease.

The farmers also mention the physical results, pointing out that mounding ensures that the soil is friable, the compost giving mounds a soft centre favourable to root penetration and enlargement into long and straight, regular-shaped tubers. The breaking up of the soil with mound tillage certainly avoids compaction and ensures favourable bulk densities regardless of the time soils are kept under cultivation. It also promotes soil aeration and drainage by encouraging a loose friable structure, which is significant in a climate where high rainfall is usual (Clarke and Street 1967). The building of mounds also increases effective topsoil depth, which is noteworthy in a region where they commonly cultivate steep slopes having shallow topsoils. Local farmers observe that tuberisation is poor with plants that have to root into the denser, clayey subsoil.

Figure 7.12 Starting mound: heaping crumbled soil over uprooted grass (newly planted mounds to left)

Figure 7.13 Finishing mound: scooping fine tilth over it (mature garden to rear)

Figure 7.14 Firming sweet potato cuttings into soil mound

The incorporation of compost in mounds has a significant effect on sweet potato yields. A series of trials in the Southern Highlands Province, on soils similar to those discussed here, showed a linear relationship between rate of compost application and increases in mean tuber yield (Floyd et al. 1988).[18] The extent to which compost augments soil fertility varies with the vegetation mix; although the Wola do not manage fallow sites to promote the growth of certain species above others. They use whatever vegetation comes naturally available. They manage regrowth by the time they leave gardens fallow. The immediate cultivation of sites repeatedly (incorporating crop residues and pioneer weeds as compost, or after short-term fallow intervals with more advanced herbaceous colonisation) will result in declining sweet potato yields. The yield of fresh vegetation in such gardens ranged from 28–35 t/ha. But it only requires a longer fallow interval dominated by coarse grasses between some cultivations to sustain available nutrient supplies indefinitely at levels sufficient to assure adequate tuber yields.[19] When left fallow under grassy regrowth for several months, compost yields ranged from 32 t/ha after nine months, to 85 t/ha after two years plus, with a mean rate of 62 t/ha.[20]

Another source of fertiliser for small cultivations near to homesteads is *dowhuwniy* (sweepings of everyday refuse from inside houses) and *showmay iy hiym* (pig manure), which people regularly toss onto such gardens (Sillitoe 2010: 16–17). They acknowledge that such waste promotes fertility, the small enriched areas supporting a variety of high-yielding crops. They exploit these nutrient sources when they abandon houses too, cultivating small mixed vegetable gardens after the building materials have rotted or are burnt. These sites can initially prove particularly fertile, acidity approaching neutral, which favours phosphate availability, and they have substantial exchangeable cation availabilities and nitrogen levels too, although the topsoils are usually thin and cannot sustain these nutrient levels for long.

Conservation versus degradation

Shifting cultivation has a maligned reputation, regardless of attempts to rehabilitate it (see, e.g. Dove 1983; Aweto 2013). According to one commentator, 'shifting cultivation activities destroy 50,000 km2 ... of tropical rainforest a year', part of the problem being that 'underlying soils are inherently infertile' (Park 1992: 46–47). Such conceptions derive in large part from Latin America and Equatorial Africa where people cultivate old soils, such as Oxisols and Ultisols that occur on ancient land surfaces (Nye and Greenland 1960; Spencer 1966; Watters 1971).[21] Some commentators argue that the horticultural activities of New Guinea Highlanders, including the Wola, result in similar land degradation, portending an ecological crisis (Brookfield and Brown 1963: 117–124; Wood 1979, 1982; Allen 1984; Allen and Crittenden 1987). We need to be wary of allowing our current concerns about the destruction of natural resources – deforestation, biodiversity loss, industrial pollution, global warming, etc. – to distort our view of the activities of others. While in some tropical regions, such as Amazonia and the Congo, a degraded environment results from extended cultivation, it is an error to generalise from them for all the tropics. In view of the wide variation in land resources across tropical latitudes (Sanchez 1976: 52–87; Juo and Franzluebbers 2003: 131–237), we should anticipate different environmental responses to cultivation. After all, it is common knowledge that volcanic fallout may rejuvenate soils, keeping them youthful, although it leads to certain nutrient availability problems.

The claims of Wola farmers that some soils progressively improve and staple crop yields increase the longer they are under cultivation, and that the only way to ascertain this is to garden them, are the exact reverse of the widespread image of soils quickly exhausted and forcing a change of site. The evidence suggests that the Wola have a remarkably conservational horticulture regime within the constraints of their technology. The flexibility afforded by fallow options keeps cultivation in sustainable equilibrium with soil resources, so long as the appropriate fallow time elapses. When sites show signs of reduced yields, farmers leave them longer under grass to recover their fertility status, whereas when they crop well, they may cultivate them again immediately. It is a stable management system. The character of the region's soils, their response to clearance and cropping, together with local people's understandings and their management of them under cultivation, are central to understanding the processes that allow the continuance of this agricultural regime, which allow farmers to maintain semi-permanent plots of non-perennial crops by overcoming the soil constraints reported elsewhere in the tropics that oblige subsistence farmers to shift their cultivations frequently.

Notes

1 Publications that contribute to ethnopedological knowledge in various regions of the world include: Africa and Asia (Fairhead et al. 2017), Niger (Lamers and Feil 1995; Osbahr and Allan 2003), Tanzania (Ostberg 1995), Nigeria (Kundiri et al. 1997; Adewole Osunade 1995), Ivory Coast (Birmingham 2003), Burkina Faso (Krogh and Paarup-Laursen 1997; Gray and Morant 2003), Rwanda (Habarurema and Steiner 1997), Nepal (Scott and Walter 1993), Latin America (WinklerPrins and Barrera-Bassols 2004), Brazil (Hecht 1989; De Queiroz and Norton 1992), Mexico (Barrera-Bassols, Zinck and Van Ranst 2009), Peru (Sandor and Furbee 1996), Honduras (Ericksen and Ardón 2003) and Papua New Guinea (Sillitoe 1996, 1998).

2 In Sukumuland, for example the Ukiriguru Agricultural Research Institute continues to use Milne's (1947) names that feature in Lake Zone farming systems research.

3 These are not annual replantings but extend over one to two years usually, depending on the plants intercropped.

4 Some of the data and arguments presented in this chapter have appeared in P. Sillitoe, *A Place Against Time*, published by Harwood Academic in 1996, and I acknowledge permission to reproduce them.

5 Note that archaeological finds indicate that farming started in the New Guinea Mountains some 9,000 years ago (Denham 2017; Golson et al. 2017).

6 Unless they intend to establish a stand of taro, a crop which they say rapidly reduces soil productivity; such a garden, usually on a wet site, will be a one-off cultivation.

7 The Wola deny that they look systematically for the presence or absence of certain associations to assess soil fertility. If the vegetation, whatever the succession, obviously looks spindly and poor, they may take this as evidence of poor fertility.

8 For details of analytical methods and discussion (including statistical analysis), see Sillitoe (1996: 339–364; 1998).

9 Phosphorus sorption isotherms, which give accurate assessments of phosphorus fixation, constructed for similar soils elsewhere in the Southern Highlands region confirm these observations of very high phosphate fixation capacities (Radcliffe 1986: 114–118; Floyd et al. 1988).

10 These parallel findings elsewhere (Wood 1979; Goodbody and Humphries 1986; Floyd et al. 1988).

11 Regarding sweet potato, a wide range of factors may limit yields, among them deficiencies in soil fertility, virus diseases, root-knot nematode infestation, leaf scab, insect damage, and the genetic potential of cultivars. The balance of environmental evidence, general observations of many gardens over several years, and the comments of local people, suggest that soil fertility is the major constraint on sweet potato production in the Wola region, the incidence of disease and pest infestation rarely rising to levels sufficient seriously to limit yields.

12 On the role of mounding elsewhere in subsistence farming, see Dubois (1957), Stromgaard (1990), Siame (2006), McKey, Renard and Comptour (2017).

13 Notably *Ischaemum polystachum*.

14 The incorporation of herbaceous and grassy regrowth as compost shows the role weeds may play in the management of soil fertility in shifting cultivation contexts (Swamy and Ramakrishnan 1988; Lambert and Arnason 1989).

15 Adding crop residues or straw is successful on a range of soils under shifting cultivation, giving yield responses as good, or higher than fertiliser or manure applications (Lal 1975; Sanchez 1976). In 12 out of 13 trials, e.g. conducted throughout Papua New Guinea, organic soil amendments increased sweet

potato yields, these consistently positive yield responses suggesting that organic materials may offer more scope in this country for increasing food productivity than inorganic fertilisers (D'Souza and Bourke 1982;Velayutham et al. 1982).

16 Floyd et al. (1987) also take the absence of any residual composting effect on the next crop as further evidence that it is direct nutrient uptake from decomposing compost that makes this manuring method particularly effective. Under traditional cultivation, however, women add new compost to mounds every time they cultivate a garden.Although there may be no evident residual effects on yield, the addition of organic matter in composting makes a long-term contribution to the maintenance of soil fertility – carbon content decline being one of the few significant effects evident under ongoing cultivation.

17 In Enga Province in the Central Highlands, where people say compost helps prevent tubers from rotting, black rot (caused by *Ceratocystis fibriata*) was observed four to five times (2.5–13.6 per cent) less often on tubers from composted mounds (Preston 1990).

18 The observed response to the improvements in soil fertility was an increase in tuber initiation over tuber bulking, with more tubers per plant, with considerably more tubers reaching marketable size (>100g).

19 A long-term soil exhaustion experiment at Aiyura Highlands Agricultural Experiment Station in the Eastern Highlands Province, which was first planted in November 1955, demonstrates the effectiveness of grass fallows between crops in maintaining sweet potato yields (Kimber 1974).

20 There was a marked increase in the amount of decomposing grass litter, the longer sites were under fallow, adding considerably to their compost yields. The dry weight of the vegetation was about one-third of its green weight.

21 Recent research on these old soils has revealed that people can manage soil fertility in previously unforeseen ways, notably through traditional cultivation practices involving biochar, such as the anthropogenic dark earths of Amazonia (*terra preta*), recently reported in Africa too (Glaser et al. 2001; Glaser and Birk 2012; Frausin et al. 2014; Fairhead et al. 2017).

References

Acres, B. D. 1984. Local farmers' experience of soils combined with reconnaissance soil survey for land use planning: An example from Tanzania. *Soil Survey and Land Use* 4: 77–86.

Adewole Osunade, M. A. 1995. Identification of crop soils by small farmers of South-western Nigeria. *Journal of Environmental Management* 35: 193–203.

Allen, B. J. 1984. Land use, population and nutrition. In B. J. Allen (ed.), *Agricultural and Nutritional Studies on the Nembi Plateau, Southern Highlands*. Port Moresby: UPNG Geography Department. Occasional Papers (New Series) No. 4 (pp. 89–95).

Allen, B. J. and Crittenden, R. 1987. Degradation and a pre-capitalist political economy: The case of the New Guinea Highlands. In P. Blaikie and H. Brookfield (eds), *Land Degradation and Society*. London: Methuen, pp. 143–156.

Atran, S. and Medin, D. 2008. *The Native Mind and the Cultural Construction of Nature*. Cambridge, MA: MIT Press.

Aweto, A. O. 2013. *Shifting Cultivation and Secondary Succession in the Tropics*. Wallingford: CAB International

Barrera-Bassols, N. and Zinck, J. A. 2003. Ethnopedology: A worldwide view on the soil knowledge of local people. *Geoderma* 111(3–4): 171–195.

Barrera-Bassols, N., Zinck J. A. and Van Ranst, E. 2009. Participatory soil survey: experience in working with a Mesoamerican indigenous community. *Soil Use and Management* 25(1): 43–56.

Birmingham, D. M. 2003. Local knowledge of soils: The case of contrast in Côte d'Ivoire. *Geoderma* 111(3–4): 481–502.

Bleeker, P. 1983. *Soils of Papua New Guinea*. Canberra: CSIRO and Australian National University Press.

Bourke, R. M. 1985. Sweet potato (*Ipomoea batatas*) production and research in Papua New Guinea. *Papua New Guinea Journal of Agriculture, Forestry and Fisheries* 33(3, 4): 89–108.

Brookfield, H. C. and Brown, P. 1963. *Struggle for Land: Agriculture and Group Territories among the Chimbu of the New Guinea Highlands*. Melbourne: ANU and Oxford University Press.

Clarke, W. C. and Street, J. M. 1967. Soil fertility and cultivation practices in New Guinea. *Journal of Tropical Geography* 24: 7–11.

Denham, Tim 2017. The antiquity of agriculture. In J. Friede, T. Hays and C. Hellmich (eds), *New Guinea Highlands: Art from the Jolika Collection*. San Francisco and Munich: De Young Fine Arts Museum and Prestel Publishing DelMonico Books, pp. 57–61.

De Queiroz, J. S. and Norton, B. E. 1992. An assessment of an indigenous soil classification used in the Caatinga Region, of Ceara State NE Brazil. *Agricultural Systems* 39: 289–305.

Dove, M. 1983. Theories of *swidden* agriculture and the political economy of ignorance. *Agroforestry Systems* 1: 85–99.

D'Souza, E. J. and Bourke, R. M. 1982. Compost increases sweet potato yields in the highlands. *Harvest* 8(4): 171–175.

Dubois, J. 1957. Semis forestiers sur buttes incinères. Leur importance dans les travaux de déforestation des savanes du Bas-Congo. *Bulletin d'Information de 'l'Institut national et Agronomique du Congo Belge (INEAC)* 6(1): 21–30.

Ericksen, P. J. and Ardón, M. 2003. Similarities and differences between farmer and scientist views on soil quality issues in central Honduras. *Geoderma* 111(3–4): 233–248.

Fairhead, J., Fraser, J. A., Amanor, K., Solomon, D., Lehmann, J. and Leach, M. 2017. Indigenous soil enrichment for food security and climate change in Africa and Asia: A review. In P. Sillitoe (ed.), *Indigenous Knowledge: Enhancing Its Contribution to Natural Resources Management*. Wallingford: CAB International, pp. 99–115.

Floyd, C. N., D'Souza, E. J. and LeFroy, R. D. B. 1987. Composting and crop production on volcanic ash soils in the Southern Highlands of Papua New Guinea. Technical Report 87/6. Port Moresby: Department of Primary Industry.

Floyd, C. N., Lefroy, R. D. B. and D'Souza, E. J. 1988. Soil fertility and sweet potato production on volcanic ash soils in the highlands of Papua New Guinea. *Field Crops Research* 19: 1–25.

Forster, G. 1843. *Saemtliche Schriften*. Vol. 4. Leipzig: F.A. Brockhaus.

Frausin, V., Fraser, J. A. Narmah, W., Lahai, M. K., Winnebah, T., Fairhead, J. and Leach, M. 2014. 'God made the soil, but we made it fertile': Gender, knowledge and practice in the formation and use of African dark earths in Liberia and Sierra Leone. *Human Ecology* 42(5): 695–710.

Glaser, B. and Birk, J. 2012. State of the scientific knowledge on properties and genesis of Anthropogenic Dark Earths in Central Amazonia (*terra preta de Índio*). *Geochimic et Cosmochimica Acta* 82: 39–51.

Glaser, B., Haumaier, L., Guggenberger, G., and Zech, W. 2001. The *terra preta* phenomenon: A model for sustainable agriculture in the humid tropics. *Naturwissenschaften* 88(1): 37–41.

Golson, J., Denham, T., Hughes, P., Swadling, P. and Muke, J. (eds) 2017. *Ten Thousand Years of Cultivation at Kuk Swamp in the Highlands of Papua New Guinea*. Canberra: ANU Press (Terra Australis 46).

Goodbody, S. and Humphries, G. S. 1986. Soil chemical status and the prediction of sweet potato yields. *Tropical Agriculture* 63(2): 209–211.

Gray, L. C. and Morant, P. 2003. Reconciling indigenous knowledge with scientific assessment of soil fertility changes in southwestern Burkina Faso. *Geoderma* 111(3–4): 425–437.

Habarurema, E. and Steiner, K. G. 1997. Soil suitability classification by farmers in southern Rwanda. *Geoderma* 75: 75–87.

Hecht, S. B. 1989. Indigenous soil management in the Amazon Basin. Some implications for development. In J. Browder (ed.), *Fragile Lands of Latin America*. Boulder, CO: Westview Press, pp. 166–181.

Humboldt, Alexandre de 1822. *Voyage aux Régions Equinoxiales du Nouveau Continent, fait en 1799, 1800, 1801, 1802, 1803 et 1804 par A. de Humboldt et A. Bonpland*. Vol. VIII. Paris: chez N. Maze, p. 525.

Juo, A. S. R. and Franzluebbers, K. 2003. *Tropical Soils: Properties and Management for Sustainable Agriculture*. New York: Oxford University Press.

Kimber, A. J. 1974. Crop rotations, legumes and more productive arable farming in the highlands of Papua New Guinea. *Science in New Guinea* 2(1): 70–79.

Krasilnikov, P. V. and Tabor, J. A. 2003. Perspectives on utilitarian ethnopedology. *Geoderma* 111(3–4): 197–215.

Krogh, L. and Paarup-Laursen, B. 1997. Indigenous soil knowledge among the Fulani of northern Burkina Faso: Linking soil science and anthropology in analysis of natural resource management. *Geojournal* 43: 189–197.

Kundiri A. M., Jarvis, M. G. and Bullock, P. 1997. Traditional soil and land appraisal on Fadama lands in northeast Nigeria. *Soil Use and Management* 13: 205–208.

Lal, R., 1975. Role of mulching techniques in tropical soil and water management. *International Institute of Tropical Agriculture Bulletin* 1.

Lambert, J. D. H. and Arnason, J. T. 1989. Role of weeds in nutrient cycling in the cropping phase of milpa agriculture in Belize, Central America. In J. Proctor (ed.), *Mineral Nutrients in Tropical Forest and Savanna Ecosystems*. Oxford: Blackwell, pp. 301–313.

Lamers, J. P. A. and Feil, P. R. 1995. Farmers' knowledge and management of spatial soil and crop growth variability in Niger, West Africa. *Netherlands Journal of Agricultural Science* 43: 375–389.

McKey, R. and Comptour, M. 2017. Will the real raised-field agriculture please rise? Indigenous knowledge and the resolution of competing visions of one way to farm wetlands. In P. Sillitoe (ed.), *Indigenous Knowledge: Enhancing its Contribution to Natural Resources Management*. Wallingford: CAB International, pp. 116–129.

Milne, J. A. 1947. A soil reconnaissance journey through parts of Tanganyika Territory, December 1935 to February 1936. *Journal of Ecology* 35: 192–265.

Ngailo, J. A. and Nortcliff, S 2007. Learning from Wasukuma ethnopedology: An indigenous well established system for transfer of agro-technology in Tanzania. *Indilinga: African Journal of Indigenous Knowledge Systems* 6(1): 64–75.

Nicholaides, J. J., Chancy, H. F., Mascagni, H. J., Wilson, L. G. and Eaddy, D. 1985. Sweet potato response to K and P fertilization. *Agronomy Journal* 77: 466–470.

Niemeijer, D. 1995. Indigenous soil classifications: Complications and considerations. *Indigenous Knowledge and Development Monitor* 3: 20–21.

Niemeijer, D., and Mazzucato, V. 2003. Moving beyond indigenous soil taxonomies: Local theories of soils for sustainable development. *Geoderma* 111(3–4): 403–424.

Norman, M. J. T. 1979. *Annual Cropping Systems in the Tropics*. Gainsville, FL: University of Florida.

Nye, P. H. and Greenland, D. J. 1960. The soil under shifting cultivation. Technical Bulletin No. 51. Harpenden: Commonwealth Bureau of Soils.

Osbahr, H. and Allan, C. 2003. Indigenous knowledge of soil fertility management in southwest Niger. *Geoderma* 111(3–4): 457–479.

Ostberg, W. 1995. *Land Is Coming Up: The Burunge of Central Tanzania and Their Environments*. Stockholm: Almqvist and Wiksell International.

Parfitt, R. L. and Mavo, B. 1975. Phosphate fixation in some Papua New Guinea soils. *Science in New Guinea* 3: 179–90.

Park, C. C. 1992. *Tropical Rainforests*. London: Routledge.

Preston, S. R. 1990. Investigation of compost-fertilizer interactions in sweet potato grown on volcanic ash soils in the highlands of Papua New Guinea. *Tropical Agriculture* 67(3): 239–242.

Radcliffe, D. J. 1986. The land resources of Upper Mendi, 2 vols. Research Bulletin No. 37, AFTSEMU Tech. Report No. 8. Konedobu: Department of Primary Industry).

Russell, E. J. 1912. *Soil Conditions and Plant Growth*. London: Longmans, Green and Co. (Monographs on Biochemistry).

Ruthenberg, H. 1976. *Farming Systems in the Tropics*. Oxford: Clarendon Press.

Sanchez, P. A. 1976. *Properties and Management of Soils in the Tropics*. New York: Wiley.

Sandor, J. A. and Furbee, L. 1996. Indigenous knowledge and classification of soils in the Andes of Southern Peru. *Soil Science Society of America Journal* 60, 1502–1512.

Scott, C. A. and Walter, M. F. 1993. Local knowledge and conventional soil science approaches to erosional processes in the Shivalik Himalaya. *Mountain Research and Development* 13: 161–172.

Siame, J. A. 2006. The Mambwe mound cultivation system. *LEISA Magazine* 22(4): 14–15.

Sikana, P. 1993. Mismatched models: How farmers and scientists see soils. *ILEA Newsletter* 9(1): 15–16.

Sillitoe, P. 1983. *Roots of the Earth: The Cultivation and Classification of Crops in the Papua New Guinea Highlands*. Manchester: Manchester University Press.

Sillitoe, P. 1996. *A Place Against Time: Land and Environment in the Papua New Guinea Highlands*. Amsterdam: Harwood Academic.

Sillitoe, P. 1998. It's all in the mound: Fertility management under stationary shifting cultivation in the Papua New Guinea highlands. *Mountain Research and Development* 18(2): 123–134.

Sillitoe, P. 1999. Beating the boundaries: Land tenure and identity in the Papua New Guinea Highlands. *Journal of Anthropological Research* 55(3): 331–360.

Sillitoe, P. 2010. *From Land to Mouth: The Agricultural 'Economy' of the Wola of the New Guinea Highlands*. New Haven, CT: Yale University Press.

Sillitoe, P. 2017. *Built in Niugini: Constructions in the Highlands of Papua New Guinea*. London: Royal Anthropological Institute. Monograph Series, vol. 1.

Sillitoe, P. and Sillitoe, J. A. 2009. *'Grass-Clearing Man': A Factional Ethnography of Life in the New Guinea Highlands*. Long Grove, IL: Waveland Press.

Simmons, I. G. 1993. *Interpreting Nature: Cultural Constructions of the Environment*. London: Routledge.

Soil Survey Staff. 2014. *Keys to Soil Taxonomy* (12th edn). Washington, DC: USDA-Natural Resources Conservation Service.

Spencer, J. E. 1966. *Shifting Cultivation in Southeastern Asia*. Berkeley, CA: University of California Press.

Stromgaard, P. 1990. Effects of mound-cultivation on concentration of nutrients in a Zambian *miombo* woodland soil. *Agriculture, Ecosystems and Environment* 32: 295–313.

Swamy, P. S. and Ramakrishnan, P. S. 1988. Ecological implications of traditional weeding and other imposed weeding regimes under slash and burn agriculture (*jhum*) in north-eastern India. *Weed Research* 28: 127–136.

Tabor, J. A. 1992. Ethnopedological surveys: Soil survey that incorporate local systems of land classification. Soil Survey Horizons (Spring).

Talawar, S. and Rhoades, R. E. 1998. Scientific and local classification and management of soils. *Agriculture and Human Values* 15(1): 3–14.

Velayutham, K. S., Pondrilei, K. S. and Natera, E. A. 1982. Effects of soil native nutrients and applied nutrients on sweet potato. In R. M. Bourke and V. Kesavan (eds), *Proceedings of the Second Papua New Guinea Food Crops Conference*. Konedobu: Department of Primary Industry, pp. 382–390.

Waddell, E. 1972. *The Mound Builders: Agricultural Practices, Environment, and Society in the Central Highlands of New Guinea*. Seattle, WA: University of Washington Press.

Watters R. F. 1971. *Shifting Cultivation in Latin America*. Rome: Food and Agriculture Organization of the United Nations.

Winklerprins, A. M. G. A. 1999. Local soil knowledge: A tool for sustainable land management. *Society and Natural Resources* 12: 151–161.

WinklerPrins, A. and Barrera-Bassols, N. 2004. Latin American ethnopedology: A vision of its past, present, and future. *Agriculture and Human Values* 21(2): 139–156.

Wood, A. W. 1979. The effects of shifting cultivation on soil properties: An example from the Kirimui and Borrai Plateaux, Simbu Province, Papua New Guinea. *Papua New Guinea Agricultural Journal* 30: 1–9.

Wood, A. W. 1982. Food cropping systems in the Tari Basin. In R. M. Bourke and V. Kesavan (eds), *Proceedings of the Second Papua New Guinea Food Crops Conference*. Konedobu: Department of Primary Industry, pp. 256–267.

Young, A. 2017. *Thin on the Ground: Soil Science in the Tropics*. 2nd edn. Norwich: Land Resources Books.

8

BRIDGING PARADIGMS

Analyzing traditional Tsimane' hunting with a double lens

Armando Medinaceli

Introduction

At the beginning of the twenty-first century the need for Indigenous voices in ethnography and ethnobiology is greater than ever, due increasing ethical concerns over balance of power in research, and for the improved reliability of ethnographic data. Two related paradigms have emerged to help address these needs: Collaborative ethnography and Indigenous methodologies. Despite common philosophical underpinnings, the two approaches are largely treated in different bodies of literature. The aim here is to examine some points of contact between these two approaches for studying traditional hunting practices in Bolivian Amazonia. The challenge is to create a space for Indigenous research and researchers within academia, while also working on meaningful, useful and ethical research with the Indigenous Peoples with whom my research takes place.

In my case, after working in several ethnobiological projects with Indigenous Peoples of Latin America, gaining a better understanding of their realities, curiosities and necessities, I often confronted the challenge of having to frame my work within a certain academic paradigm for research. I have decided to do that in order to maintain the legitimacy and relevance of my work for an academic audience—for my academic tribe, as it were—while at the same time portraying an approach to research that reflects local realities and that is well accepted and understood locally. More importantly, I want my research to respond to the local regulations, traditions and beliefs of Indigenous cultures.

Traditional anthropology, historically the academic discipline most concerned with Indigenous Environmental Knowledge, has focused on observing "the other" (Lévi-Strauss, 1966), from an *etic* perspective. Etic (from the linguistic term phon*etic*) understood as a viewpoint that studies cultural behavior from the outside, thus offering a representation created by the researcher (like the unified way that linguists represent the sounds of all the worlds languages for researcher comprehension and comparison; cf. Pike, 1967). But anthropology has been evolving through time, creating new approaches and paradigms for anthropological research. The notion of collaboration within anthropology finds its roots at an early stage. Malinowski stated that anthropologists should move "off the veranda" in order to engage with the everyday life of natives, through the use of participant observation, as a tool to systematically document natives' lifestyles while participating in them (Lassiter, 2005a, 2005b). This also leads to changing strategies for fieldwork and those subjects of study became friends, informants, and teachers.

DOI: 10.4324/9781315270845-9

Franz Boas's approach to collaboration was noticeable while doing fieldwork and sharing not only the fieldwork experience but also his writing of results with his native interlocutors, making sure to represent the nature of his informants' participation (Lassiter, 2005b). At the beginning of the 1900s, Leenhardt, following Boas, taught his informants to transform their traditions into text, but going even further, he often negotiated the final interpretation of the results with his collaborators.

Following this historical tendency toward dialogue and involvement with "informants" (now often called collaborators, consultants, or interlocutors in recognition of this co-production of knowledge), some academics have contributed to raising the voices of the natives and acting on behalf of their subjects in political activism. More recently, Lassiter, based on his own experiences, proposes collaborative ethnography, explaining that it is the natural next step in the collaboration between researchers and subjects, and a continuation of the works of classical anthropologists such as Malinowski, Boas, and Mead (Lassiter, 2005b, 2008).

Collaborative ethnography

Collaborative ethnography is a paradigm of research described by Lassiter (2005a, 2005b). It focuses on the importance of collaboration between the researcher and the "subjects" in an attempt to disrupt the traditional view of studying "the other" (Lassiter, 2005b). Collaborative ethnography emphasizes the importance of complete collaboration between the researcher and the "subjects," where collaboration should start from the conception of the research idea, and accompany the research until the interpretation and dissemination of results, usually in written ethnography (Lassiter, 2005b, 2008; Sillitoe, 2006; Rappaport, 2008). Lassiter and others argue about the importance of responding to the local demands, interests or necessities, in order to be true collaboration (Sillitoe, 2006; Rappaport, 2008; Schensul et al., 2014).

The encouragement for researchers to collaborate and interact on equal ground with "subjects," proposed by the proponents of collaborative ethnography, is based on the premise of being honest and vulnerable to improve the accuracy of research. Therefore, this means engaging with other cultures focusing on dialogue instead of approaching the "subjects" based on a text metaphor, as proposed by Geertz (1973). The use of dialogue provided for some anthropologists a new platform to represent the joint production of knowledge between themselves and their informants/collaborators in their ethnographic production (Lassiter, 2005; see also Marcus and Fisher, 1986).

As Rappaport argues, there are situations where collaborative research is equated with citizenship. She explains that, in Colombia, anthropologists consider themselves as "citizen researchers" under the premise that "the exercise of the profession is simultaneously the exercise of citizenship" (Rappaport, 2008). This view of research gives a collaborative approach that is embedded in what it really means to be an anthropologist, particularly in the case of Colombia and possibly Latin America in general. Furthermore, when analyzing and distinguishing differences between "metropolitan" anthropology and anthropologies from the periphery (Cardoso, 1995), such as Latin American anthropology, Cardoso argues that metropolitan (or traditional) anthropology, carried out by the pioneers in the field such as Boas and Malinowski among others, had as an objective the learning and describing of traditional cultures. But in the anthropology of the periphery (i.e. Latin America), the focus for research was transformed when local anthropologists change the focus to research entering sociopolitical grounds, topics relevant to the national society, thus transforming the position of the researcher from a foreigner to the communities to a member of the same nation that houses both Indigenous communities and the researchers themselves (Cardoso, 1995, 1998; Jimeno, 2000).

Indigenous methodologies and epistemology

On a different spectrum of anthropology, Indigenous scholars developed a distinct approach to anthropological research in response to the historical mistreatment of native peoples. They argue that Western imperialism, colonization, and capitalism have had a deleterious effect on Indigenous customs, traditions, and beliefs (Smith, 1999). Smith discusses how history in the Western eye is described as a glorious set of actions and conquests that lead to a better and more globalized world. But for many Indigenous and native peoples, history involves devastating destruction. The Western conquerer-heroes are seen as the carriers of such burdens as capitalism, predatory individualism, and disease, which devastated entire populations of local peoples of the Americas, and elsewhere (Smith 1999).

With the advances in academia and the global aim for equality for all people in the world, reflected in the Universal Declaration of Human Rights of 1948, some Indigenous Peoples gained access to certain resources of large-scale, developed societies, such as education and health care. Global discussions of local Indigenous Peoples' self-determination offered a space for Indigenous scholars to raise their voices and contest some of the theories and traditions of the academic world. Thus, through Indigenous scholars, such as Smith (1999), Wilson (2008, and Kovach (2009), a new paradigm for research involving Indigenous Peoples emerged under the banner of Indigenous Epistemologies. In this paradigm, Indigenous scholars aim to assert responsibility for telling their own stories and documenting their own knowledge and practices, following their own customary regulations and formats (Smith, 1999; Wilson, 2008; Kovach, 2009). The proponents of this paradigm are Indigenous scholars with vast experience working and collaborating within their own cultures. This paradigm emerges from reflecting on the complicated relationship between Indigenous Peoples and Western researchers, especially anthropologists who, in the past, were mostly focused on describing and deciphering Indigenous ways of life (Smith,1999; Wilson, 2008). This historical relationship has led to some Indigenous Peoples to think of anthropologists as "all that is bad with academics" (Smith, 1999).

The Indigenous Epistemologies movement suggests that research should emerge from Indigenous People, and should be carried out, as much as possible, by the Indigenous People themselves. This paradigm also encourages research in which ethical considerations are more central to the project, one that responds to local problems, demands, or necessities, and that should be implemented following the Indigenous traditions and customs. Research should use methodologies that are not strictly "academic," but locally accepted and adequate for gathering knowledge. For instance, these methods could be in the form of rituals, talking circles, or any other methods that are locally practiced and culturally respected and understood (Wilson, 2008). An important point is that approaches to mutual understanding should involve local tradition and custom to ensure that Indigenous voices are made clear. This approach requires close attention to local relevance of social processes for shared understanding. With this motion, research maintains academic rigor and also takes into account and respects traditions, customs, and beliefs from the local people. Therefore, approaching research this way, there is no gap between researcher and subject, as it is the "subject" who is in charge of the research.

A bridge between paradigms?

Here I argue that both approaches have many similarities that might be framed in a single approach, though each brings its own theory and concepts for respectful, appropriate, and ethical research. Is it possible to build a bridge between Indigenous methodologies and collaborative ethnography? To answer the question, I use my research focusing on the traditional practices

Table 8.1 Comparison of paradigms

Indigenous methodologies	Collaborative ethnography
Supports an *emic* or local perspective	Supports an *etic-emic* perspective
All research must respond to local interests, necessities★	Research should respond to local interests, necessities★
Research should be done by locals (collaboration with committed outsiders)★	Research done in constant collaboration between researcher and local participants★
Emphasize the importance of local customs and traditions★	Emphasize the importance of local customs and traditions★
Use of local methods (formal and non-formal)	Use a combination of academic methodologies (mixed methods)
Results from research should be beneficial for the local population★	Results from research should be beneficial for the local populations and for the researcher★
Maintain the insider-outsider frame of mind★	Ethically present the results back to locals and elsewhere★

Note: ★ represent similarities.

of hunting and fishing of the Tsimane' in Bolivian Amazonia, which I carried out over several field trips to local Tsimane' communities between 2013 to 2015 for a total of 10 months of fieldwork.

Reflecting on my work with the Tsimane' in Bolivia, I came to realize that, in general terms, it seems possible and desirable to develop research consistent with both paradigms. Thus, in attempt to develop a new approach to research bridging-related paradigms, I considered each paradigm individually to assess similarities between them. Table 8.1 shows, in detail, a description of the characteristics and similarities proposed for each paradigm.

In general terms, both paradigms attempt to demonstrate the importance of the integration of local people, and put emphasis on local needs and desires at every step of research. Both approaches address similar situations but from distinct angles, possibly due to the status of the proponents of each paradigm. On one side, Western academics propose collaboration, and, on the other hand, Indigenous scholars propose Indigenous methodologies. These are two fronts of academic researchers that, for no apparent reason except their own ethnic background, are separated, even though they both belong to an academic realm and also participate with Indigenous, native, and/or local populations. For both approaches, a central theme is to redress imbalances in power between researchers and their "subjects."

The main challenges that I perceive for this separation are from academia itself. Traditionally, this means maintaining an authoritative, and at times dominant, stance in the conduct of research. For the implementation of research that is collaborative and locally engaged, it is important to cultivate personal humility, aiming for the creation of a horizontal and balanced relationship between researcher and collaborators. This avoids hierarchical statuses created by the current social, political, and academic structures. We engage as individuals with the same capacities yet distinct skills that, when put together, can be beneficial for everyone involved.

When planning to implement cultural ecological research utilizing *etic* concepts and categories, I consider that those concepts should be framed in terms of the local communities where research will take place. Engaging in discussion and reflection between Western concepts and local epistemologies as a first step for research could be immensely beneficial for both Western and local interests. This first step could also be used to apply collaborative principles for research questions driven largely by outside interests in medicine, economics, among other.

For instance, when engaging with the Maya Q'eqchi' from Guatemala, a new research region where I started a collaboration in 2016, I initiated the process of obtaining the local consent for research by asking a Guatemalan colleague, with experience working with that Maya Q'eqchi' village, to present a letter I wrote to the village explaining my intentions for research collaboration with the locals. After a positive response to my letter, I planned my first visit where I engaged in meetings with local authorities and the community's general assembly. After dialoguing about my intentions and interests and making sure that my research was aligned with the local interests and curiosities, I was asked to create a 12-point written agreement to be evaluated and approved, based on their local customs. Once my research was approved locally and I was granted the local consent, I initiated my collaborative research. (For details see Medinaceli, 2018[BIB-014].)

Even though this might seem very similar to what should be standard practice for any anthropological and ethnobiological research, this approach (taking largely into consideration the local interests and curiosities) was a surprise for the villagers and local authorities who said they were not used to researchers engaging in dialogue and collaboration at such an early stage of the research. Much of their experiences were negotiating agreements for already established research proposals that were not open to allow much change in their implementation, focus, etc.

Large organizations, as well as small and local ones, are slowly creating a new trend in research on the conservation of the biocultural diversity in the world; a good example of such organizations is the Global Program of the Christensen Fund which aims to fund research on issues regarding Indigenous Peoples' biocultural rights, resilience and agrobiodiversity, and food sovereignty (see www.christensenfund.org). Processes of empowerment and self-determination of Indigenous and local communities are becoming more common and local peoples are demanding reciprocal and fair treatment regarding access to and benefit from resources that are the focus or results of research. Indigenous Peoples and local communities are more involved in global issues (i.e., climate change, exploitation of fossil resources), and they are raising their voices in order to safeguard their own traditions and practices. One of many examples is the recent COMBIOSERVE program, based on a co-enquiry approach to community conservation in Latin America (www.combioserve.animalared.org).

An important step that can contribute to facilitating the incorporation of collaborative and Indigenous approaches to large-scale research is to respond to the international and national regulations that are mandating fairer treatment of peoples involved in and impacted by research (i.e., CBD, 1992; Universal Declaration of Human Rights, 1948; United Nations Declaration for the Rights of Indigenous Peoples, 2007). Additionally, research should make use of codes of ethics from academic and Indigenous organizations that offer guidelines for better implementation of research in communities (i.e., Code of Ethics of the International Society of Ethnobiology 2006; Latin American Society of Ethnobiology Code of Ethics, 2016).

From my own point of view, these approaches to research are of great importance for current ethnobiological and anthropological research, including studies enquiring about Indigenous Environmental Knowledge. This is because too much ethnographic research is still strictly academic, creating a strong divide between researcher and subjects. In fact, research results are still often not made available to local communities. In Lassiter's words: "ethnographers are still writing only for their academic peers, and not for their collaborators" (2005b). This also relates to non-academic and professional (i.e. non-profit organizations, development agencies) research.

In order to realize research that is both locally and academically relevant, I engage with communities that I collaborate with based on an open dialogue and reflection process with community representatives (selected by the local peoples following their own traditional systems), in order to identify and pinpoint the research topic. Once identified, I begin with the process of obtaining the permission and consent of local peoples following the guidelines I propose in a different publication (Medinaceli, 2018). After consent, we discuss the terms of our collaboration and participation in the process, the aim is to carry out the research based on a co-enquiry approach, understood as a research process where I, as an external researcher, become a facilitator in community-led research that is responding to local necessities and curiosities (Caruso et al., 2015).

Identifying research interests of local people, obtaining permission in locally acceptable ways, returning results to locals in appropriate formats, etc. will help operationalize the bridge of paradigms, merging Indigenous methodologies with true collaboration. The intention of this approach is to support research that is beneficial for academic purposes while also benefiting local communities and to support, as Smith (1999) proposes, the decolonization of the research practice while supporting the self-determination of indigenous and local communities.

Research collaboration with the Tsimane': whose research?

The Tsimane' are an Indigenous group from Bolivian Amazonia with a population of approximately 6,000 people, based on the 2012 national census. Traditionally the Tsimane' were semi-nomadic, but with the adoption of agriculture and the influence of western urbanization, political pressures, and missionary activities, they have settled in small villages mostly along rivers or, more recently, some villages settled near roads (Reyes-Garcia et al., 2014). They are heavily dependent on the forest and surrounding biodiversity; their main activities are hunting, fishing, gathering and, depending on the location of the villages, and to a lesser extent, agriculture.

Traditionally Tsimane' people used a variety of hunting weapons, from bows and arrows to traps, machetes, and their bare hands. While they still make use of most of those (except traps), bows and arrows were by far the preferred and most used hunting weapon. The introduction of firearms into the Tsimane' hunting activities has been slowly but gradually impacting the traditional hunting. While there are not many firearms available for use in the villages studied here, they are slowly becoming the weapon of choice.

I have been collaborating with the Tsimane' since 2003. This allowed me, among other things, to build friendly relationships, and I became committed to collaboration with them in a manner that put ethical standards and the balance of power at the center of the research methodology. The personal relationship I built with the local population is what helped me first, to understand that this research is not "my research" anymore, but a collaborative effort between the Tsimane' and me; a research that aims to respond local demands while maintaining sufficient academic form, while maintaining and respecting the local traditions, beliefs, and practices.

The collaboration discussed here began with informal discussions with representatives (local authorities and non-authorities) from two Tsimane' villages where together we identified and agreed on the research topic: To document the knowledge related to the practice of fabricating bows and arrows. This topic came as a result of an attempt to understand and explore the possibility of controlling the western-imposed trading/selling activities that take place in their villages, in which the Tsimane' values and traditions are usually overstepped (for an example of this, see Medinaceli and Quinlan 2018).

Once we identified the research topic, I started a process of obtaining the authorization for my research following ethical and customary guidelines from codes of ethics relevant to ethnobiological and anthropological research (ISE's code of ethics (2006); SOLAE's (2016) code of ethics, United Nations Declaration on the Rights of Indigenous Peoples, the Bolivian national Constitution) and also following the nine-step guidelines that include a discussion on the potential benefits and risks from the study, the role of researchers and all actors involved, the dissemination and property rights of the results, among others that I propose in a separate publication (Medinaceli, 2018). After consent was obtained, we agreed on the schedule for work and elaborated the research protocol, always giving most importance to the local traditions and also respecting the time allocation for everyday activities of each participant. That is, I did not wish to waste their time, knowing that research involves what ecologists call "opportunity costs" (Hames and Vickers 1983). We decided that work for the project should be carried out only when participants had time for it, or when time for participation was agreed in advance. This process—based on discussion sessions, informal conversations, collaboratively identifying the research topic and creating a research schedule—comes under the premises of both paradigms analyzed here (Indigenous methodologies and collaborative ethnography). This process also includes negotiating time to discuss the research, because this is not something an outside researcher can simply expect. Assuming collaborators have time for us and our projects is to assume an imbalance of power.

Methodology

While I was visiting Tsimane' villages at the beginning of 2012, I engaged in discussions with the local Tsimane' about life in the villages and the, mainly, economic conflicts with the western world that most villagers were dealing with. During these discussions, several topics came to light that were of local importance and great concern for many of the inhabitants of the two villages I visited. As a result of these informal discussion sessions, we decided to embark on a research project with a holistic approach, that would be developed as a collaboration between Tsimane' and me. As part of the collaborative approach in one of the Tsimane' villages, three men were chosen in the village's general assembly to participate as active collaborators accompanying the research process in its entirety. Even though this was an excellent initiative from the Tsimane' village in order to represent their commitment to the collaborative approach, it also represented a new challenge for the research process, because research then had to be organized based on times where the three community researchers were able to participate. Then it was decided that when timeframes were not compatible as a team, we would hold a series of short meetings in which I would report all activities that I had carried out.

Before starting with gathering information, after the research topic had been identified ("traditional ways for hunting and fishing among the Tsimane'"), I held meetings with the local population of both villages as well as with the two regional Tsimane' institutions that act as the Tsimane' legal authorities. After the first meeting with the regional authorities, a first official visit to the villages took place to present once more the proposal for research, the implications of participating in it, and the expected results and other issues. Even more importantly, during these visits I obtained the free prior and informed consent of the Tsimane' for the implementation of this study following the codes of ethics from both the Latin American Society of Ethnobiology and the International Society of Ethnobiology (Código de Etica-SOLAE, 2016; ISE-CoE 2006) while also creating a new set of guidelines for engaging Indigenous communities (Medinaceli, 2018).

We discussed and negotiated the scope of the project, my interests and theirs, and agreed on some clear terms for conduct of the research, giving most importance to the local traditions and also respecting participants' time for everyday activities. The protocol also highlights that

Table 8.2 Collaboration agreement with the Tsimane'*

Some points of the collaboration agreement	
Summary	Brief description of the research process
Property rights	Agreement of formats for results, issues about authorship and co-authorship (i.e., intellectual property)
Community benefits	How results would benefit local Tsimane' villages
Role of researcher and participants	Detail of what is expected from researcher and participants from the local villages
Payments	Detail of the payments agreed for different items (i.e., translators, meals, labor, etc.)

Note: *For details on this and other agreements, see Medinaceli (2018).

work for the project should be carried out when participants offer the time for it, or when time for participation was agreed in advance. These conversations for creating the research protocol fall into the framework of both paradigms, Indigenous methodologies and collaborative ethnography. Table 8.2 shows the most relevant points identified and agreed as part of the consent agreement.

Methods

We decided on a qualitative approach for our research. Based on the premise that research would be focused on the knowledge of the participants, therefore all analysis of results would be better represented qualitatively.[1] Qualitative information was collected from all participants who voluntarily agreed to collaborate as part of the research. The range of ages of the participants was between 12 to 70 years old, both women and men. When children (below 16 years old according to local customs) were participating, I obtained the parent's consent prior to their participation.

Anthropological methods

I proposed the use of several classical anthropological methods during research, the selection of these methods was part of my contribution as an external researcher with knowledge about organizing, structuring, implementing the methods, and analyzing the results. Once accepted, I held sessions (usually as part of community meetings) where I explained, to the villagers, the intention and applicability of the methods and that, as a researcher, I would be implementing them:

Participant observation

As proposed by Bernard (2006) and Newing (2011), participant observation was the method that accompanied the whole process of research. It was used to first gain the trust (build rapport) of the local participants, also to create a better understanding of the local context and customs, to identify the patterns of daily activities and behavior of the Tsimane' and to frame all the other methods to be utilized.

Informal interviews

Based on descriptions from Bernard (2006), and Newing (2011), we used informal interviews during the entire time of fieldwork. These interviews were mostly carried out during regular daily activities for the Tsimane' (such as working in their crop fields, gathering, hunting, fishing, visiting, cooking, among other tasks). All informal interviews were carried out with people who had decided in advance to participate in the research, and including both men and women.

Semi-structured interviews

The Tsimane' and I engaged in semi-structured interviews in situations that were arranged in advance or when it was possible for the participant to take some time for the interview and also spontaneously when it caused no disruption to any other activity.

Questions and prompts for the interview guide were discussed and agreed in advance in discussions among the research team.

Focus groups

Focus groups were arranged and implemented several times during fieldwork and were used to discuss several topics. For instance, one focus group served to elaborate a seasonal calendar showing activities related to traditional hunting and fishing within the Tsimane' traditional seasons.

Some focus groups took place in two or more sessions. For instance, we discussed a qualitative cost-benefit analysis regarding the implementation of firearms into the Tsimane' traditional hunting techniques in several sessions because that was a mutually interesting topic.

Participatory video

As part of the agreement for the implementation of this research, we decided on the elaboration of a video documentary that would portray the fabrication and use of traditional bows and arrows used for hunting and fishing. We also used video to document some of the meetings, and sessions for discussion and reflection during the research.

Indigenous methodologies

Conversations/dialogue (informal and formal)

For the Tsimane', the word "*so'baqui*" (or visiting), is of great importance. It is an important aspect of their culture. The Tsimane' are traditionally semi-nomadic hunter-gatherers who used *so'baqui* as a way of interacting among different families, and to meet your potential "*fom*" (a person who is marriageable in Tsimane' kinship). Also *so'baqui* was used as a way to get important information about the environment, other Tsimane' villages and the world in general, and to make friends and to maintain contact with your extended family.

Nowadays, the Tsimane' are settled in small villages, but the importance of *so'baqui* persists. *So'baqui* is still an important characteristic of their own identity, and it helps them maintain communication with neighboring and non-neighboring villages, other Indigenous groups, and their extended families.

During these "visits," the Tsimane' share experiences, engage in conversations to discuss news and information relevant to their community, and life in general. Understanding and recognizing the importance of this conversation/communication, I decided to adopt it as a method to collect information needed for our research.

Discussion and reflection

At times, the Tsimane' engage in discussion (based on *so'baqui* or not), in which more than two people gather to discuss particular or general topics of interest, usually in response to a particular issue that is relevant for the participants. We incorporated group discussions as another method of gathering information that Tsimane' traditionally used and with which they feel comfortable. These are similar to the academic focus groups, but are more flexible. Also, they are usually spontaneous and discussion is open-ended.

Results and discussion

The collaborators also discussed results or "products" of this research and how they would be prepared for sharing outside the community. For focus groups and informal discussions we decided to collaborate on: (1) the development of a video documentary showing the fabrication and use of traditional bows and arrows; (2) a booklet to be used at the local and regional schools; (3) the creation of a mobile exhibit (i.e. for museums); (4) creation of posters and brochures; and (5) the development of research publications (academic papers) and presentations at national and international conferences to share our results and experiences, thus sharing part of the Tsimane' lifestyle with a wider audience.

We decided that given our different cultural positions, I would be responsible for the initial development of materials for dissemination with the proviso that the Tsimane' authorities and villagers would review, comment, and ask for editions or changes if needed, before submission for publication.

A new paradigm

Our collaboration is based on trust and friendship that has a foundation in several years working together and getting to know each other. This process has enabled us to coordinate and carry out research that is meaningful and beneficial for both the Tsimane' and me as an academic researcher.

Indigenous methodologies aim to decolonize research in a way that is acceptable by academics while at the same time is accepted in the local communities. The idea of being "insider-outsider" proposed by Brayboy and Deyhle (2000), demonstrates that Indigenous scholars have an advantage when it comes to discussing their research. They have a foot in the academic world, therefore are able to understand and use academic epistemologies, while at the same time have another foot within the Indigenous world, therefore, they are carriers of a cultural understanding and meaning of the Indigenous epistemologies.

On the other hand, some research is done with Indigenous communities where, for political or cultural reasons, the level of education is minimal (middle school level at most for the Tsimane'). Therefore, they do not have the knowledge to understand western, academic epistemologies, and have to rely on committed "outsiders" who are willing to collaborate at a deeper level. Outsiders should implement western research ways while learning and gaining an acceptable understanding of the traditional Indigenous epistemologies.

As I presented in the example of collaborating with the Tsimane', most of the work has been done based on strong commitment and ties of friendship and trust built over several years of collaborating and getting to know one another. This has enabled us to coordinate and carry out research that is meaningful and beneficial to both the local Tsimane' and me.

In general terms, the implementation of "formal" academic methodologies along with Indigenous ones was well received by the local population, in part, because I believe that by combining formal (anthropological) and familiar (Indigenous) methodologies, searching to answer a locally relevant topic, the Tsimane' people became more relaxed, thus more eager to participate and contribute in the research process. I argue then, that when using a less intimidating approach, such as collaborative ethnography, participants become more interested and invested in the research process and they seemed to be entertained by engaging in conversations or meetings (interviews, focus groups, etc.) to discuss issues that are relevant and of interest to many people in the villages. At times I also was interviewed about my own views of certain topics relevant to the local context. This demonstrates to me that engaging in formal and informal conversations (interviews or not), were inherently beneficial for both the local Tsimane' and me as a researcher.

Conclusion

Using the double lens suggested here, I am not able to frame my own research in terms of either paradigm described (Indigenous methodologies or collaborative ethnography), but rather through a hybrid approach. Some components of my research fall within the framework of one paradigm, while other components fall within the other. For instance, the process of agreeing on the topic for research, scheduling, and consent, came out of discussion and conversations that happened locally between the Tsimane' and me. This scheme is framed on the Indigenous methodologies paradigm.

On the other hand, the use of an approach, combining academic methodologies while also identifying and using locally familiar methods such as sessions following the *so'baqui* tradition, and encouraging the continuous local participation, relies on a framework incorporating both collaborative ethnography and Indigenous epistemologies.

The double lens (both paradigms) suggests the possibility of the inverse of Brayboy and Deyhle's (2000) "insider-outsider" concept. I propose the "outsider-insider", that is: an outsider (academic) researcher who combines the two paradigms. This maintains the focus on the local communities and their benefit from research while also aiming for academic validity. In this approach, the researcher (outsider) should have strong ties with the local communities (trust/rapport) and be committed to the research in order to well represent the insider's perspective, thus sharing a voice with the Indigenous collaborators. Based on this example, I believe that this approach is an interesting and necessary innovation, especially in situations such as the remote Tsimane' villages where literacy and access to education are not common.

The hybrid, "dual-lens" approach represents a more complete collaboration with Indigenous Peoples, incorporating their own methods and traditions as part of the methodologies implemented for research. Total collaboration means that researchers no longer work independently, but in research teams formed between academics and their collaborators (local communities). The outsider, who belongs to the academic, non-profit organization, or development agency spheres, will find ways to offer academic or professional rigor to the findings of the study, while also prioritizing the return of the information resulting from the research to the local communities in formats that are accepted and understandable for the locals.

I believe that it is time for academic research to become even more flexible and researchers should be discussing the new formats of representation in ways that acknowledge and respect local traditions, support the self-determination of the communities, and at the same time are academically recognized. Thus, supporting the necessary correction of the historical imbalance of power when engaging in academic research with Indigenous Peoples and local communities.

This rebalance is key for a fully acceptable, beneficial, and exciting program of IEK research, that offers benefits for both researchers and Indigenous Peoples.

Note

1 I believe it possible to also represent the data quantitatively, but it would be important to analyze and possibly incorporate new steps into the process, since quantitative analysis might be a completely foreign representation of information for communities with limited or no schooling.

References

Bernard, H. R. 2006. *Research Methods in Anthropology.* 4th edn. Oxford: Altamira Press.

Brayboy, B. and Deyhle, D. 2000. Insider-outsider: Researchers in American Indian communities. *Theory into Practice* 39(3), 163–169.

Cardoso de Oliveira, R. 1995. Notas sobre uma estilística da antropologia. In R. Cardoso de Oliveira and G. Raul Ruben (Eds.), *Estilos de antropologia.* Campinas: Editora da UNICAMP, pp. 177–196.

Cardoso de Oliveira, R. 1998. *O trabalho do antropólogo.* Brasilia: Paralelo 15 Editora da UNESP.

Caruso, E., Camacho, C., del Campo, C., Roma, R., and Medinaceli. A. 2015. Co-enquiry and participatory research for community conservation: A methods manual. Combioserve. Available at: www.global-diversity.org/co-enquiry-and-participatory-research-for-community-conservation-a-methods-manual

Convention on Biological Diversity (CBD). 1992. United Nations. Available at: www.cbd.int/doc/legal/cbd-en.pdf

Geertz, C. 1973. *The Interpretation of Cultures.* New York: Basic Books.

Hames, R. and Vickers. W. 1983. Introduction. In R. Hames and W. Vickers (Eds.), *Adaptive Responses of Native Amazonians.* New York: Academic Press, pp. 1–24.

International Society of Ethnobiology 2006. ISE Code of Ethics (with 2008 additions). Available at: http://ethnobiology.net/code-of-ethics/

Jimeno, M. 2000. La emergencia del investigador ciudadano: estilos de antropología y crisis de modelos en la antropología colombiana. In J. Tocacinpá (Ed.), *La formación del Estado Nación y las disciplinas sociales en Colombia.* Popayán: Taller Editorial, Universidad del Cauca, pp. 157–190.

Kovach, M. 2009. *Indigenous Methodologies: Characteristics, Conversations, and Contexts.* Toronto: University of Toronto Press.

Lassiter, E. 2005a. *The Chicago Guide to Collaborative Ethnography.* Chicago: University of Chicago Press.

Lassiter, E. 2005b. Collaborative ethnography and public anthropology. *Current Anthropology* 46(1), 83–106.

Lassiter, E. 2008. Moving past public anthropology and doing collaborative research. *NAPA Bulletin* 29, 70–86.

Lévi-Strauss, G. 1966. *The Savage Mind.* Chicago: University of Chicago Press.

Marcus, G. and Fisher, M. 1986. *Anthropology as Cultural Critique: An Experimental Moment in the Human Sciences.* Chicago: University of Chicago Press.

Medinaceli, A. 2018. Taking an early step to ethnobiological research: A proposal for obtaining prior and informed consent from indigenous peoples. *Ethnobiology Letters* 9(1), 76–85. DOI:10.14237/ebl.9.1.2018.1054.

Medinaceli, A. and Quinlan., R. 2018. Firearms effects on Tsimane' hunting and traditional knowledge in Bolivian Amazonia. *Ethnobiology Letters* 9(2), 230–242.

Newing, H. 2011. *Conducting Research in Conservation: A Social Science Perspective.* London: Routledge.

Pike, K. L. 1967. Etic and emic standpoints for the description of behavior. In K. L. Pike, *Language in Relation to a Unified Theory of the Structure of Human Behavior,* 2nd edn. The Hague: Mouton & Co., pp. 37–72.

Rappaport, J. 2008. Beyond participant observation: Collaborative ethnography as theoretical innovation. *Collaborative Anthropologies* 1, 1–31.

Reyes-Garcia, J., Paneque-Galvez, P., Bottazzi, A. C., Luz, M., Gueze, M.J. et al. 2014. Indigenous land reconfiguration and fragmented institutions: A historical political ecology of Tsimane' lands (Bolivian Amazon). *Journal of Rural Studies* 34, 282–291.

Schensul, S., Schensul, J., Singer, M., Weeks, M. and Brault, M. 2014. Participatory methods and community-based collaborations. In R. Bernard and C. Gravlee (Eds.), *Handbook of Methods in Cultural Anthropology*. Lanham, MD: Rowman & Littlefield, pp. 185–212.

Sillitoe, P. 2006. Ethnobiology and applied anthropology: Rapprochement of the academic with the practical. *The Journal of the Royal Anthropological Institute* 12, 119–142.

Smith, L. 1999. *Decolonizing Methodologies: Research and Indigenous Peoples,* 2nd edn. London: Zed Books.

Latin American Society of Ethnobioloby-SOLAE (Sociedad Latinoamericana de Etnobiología) 2016. Código de Ética. *Revista Etnobiología* 14(1). Distrito Federal, Mexico: SOLAE.

United Nations Declaration for the Rights of Indigenous Peoples. 2007. United Nations. Available at: www.un.org/development/desa/indigenouspeoples/wp-content/uploads/sites/19/2018/11/UNDRIP_E_web.pdf

United Nations Universal Declaration of Human Rights. 1948. Available at: www.un.org/en/universal-declaration-human-rights/

Wilson, S. 2008. *Research Is Ceremony: Indigenous Research Methods*. Black Point, Nova Scotia: Fernwood Publishing.

PART II

Issues of perspective, values, and engagement

9

ASIAN AND MIDDLE EASTERN PASTORALISTS

Ariell Ahearn and Dawn Chatty

Introduction

Mobile pastoralists inhabiting the drylands of Asia and the Middle East provide a lens through which to examine the increasing importance of Indigenous Environmental Knowledge in protecting landscapes which have become sites of large-scale nature reserves, extractive industry activity (mining and petroleum extraction), land degradation and the effects of climate change (Sternberg and Chatty 2016). Expertise in mobility, animal husbandry and pasture conditions (dryland environments) are the foundations of pastoralist Indigenous Environmental Knowledge. Asian drylands span a vast territory from China to the Arabian Peninsula, and both the lands and people of these regions have changed considerably over the course of the last century. The construction of new infrastructure and adoption of modern technology have opened up new forms of mobility and communication. The Soviet and Chinese communist legacies of state planning have reshaped institutions and contributed to desertification in the Inner Asia region (Sneath and Humphrey 1999). The Middle East has seen continued urbanization and in-migration from the wider region. Thus, when we read UNESCO's definition of Indigenous Knowledge, "the understandings, skills and philosophies developed by societies with long histories of interaction with their natural surroundings," it must be read with these ongoing changes and histories in mind. Indigenous Knowledge, as can be seen through long-term ethnographic research, is not static or unchanging. It includes new innovations and ways of adapting traditional knowledge to new circumstances. It also involves new ways of teaching and learning traditional knowledge. This is especially true for mobile pastoralist groups who have always maintained relations with settled communities and adapted their lifeways to new economic, social and political circumstances.

In particular, over the course of history, pastoralist knowledge and skills have been subject to discrimination and discourses of deficit (Krätli and Schareika 2010). These discourses have discounted Indigenous Knowledge, particularly those associated with mobile lifeways, as incompatible with modernity. Common myths are the idea that pastoralists are "backward" or a "throw-back to a past era" (Chatty 2006, p. 1; Humphrey and Sneath 1999) and engage in irrational economic practices. Other myths (Dyer 2014, p. 23) include the idea that "pastoralists need to settle to benefit from services," and "most rangelands are degraded as a result of pastoral over-grazing." Colonial and neo-liberal governments have sought to settle pastoralists or

DOI: 10.4324/9781315270845-11

remove their access to open rangelands through land privatization or conservation schemes, forced settlement and denial of citizenship. Emily Yeh, in her work on Tibetan pastoralists in China (2003, 2013) writes of Chinese television programmes which declare the "need to 'smash the traditional pastoral ideas' and 'liberate the herders' thoughts' in order to 'force herders to learn to turn their assets into market goods'" (2003, p. 499). Despite the ongoing discrimination of pastoralists against their Indigenous Knowledge and skills, there has been a growth in institutions which support Indigenous Knowledge and seek to apply this knowledge to address complex environmental and social challenges. The Dana Declaration on Mobile Peoples and Conservation, which was formalized in 2002 at the Wadi Dana Nature Reserve in Jordan and reaffirmed in 2012 at the Dana+10 Congress is one example of this effort. An excerpt from the declaration (2002) states:

> [T]hrough their traditional resource use practices and culture-based respect for nature, many mobile peoples are still making a significant contribution to the maintenance of the earth's ecosystems, species and genetic diversity – even though this often goes unrecognised. Thus the interests of mobile peoples and conservation converge, especially as they face a number of common challenges. There is therefore an urgent need to create a mutually reinforcing partnership between mobile peoples and those involved with conservation.

The Dana Declaration highlights the specialist knowledge and skills of mobile pastoralists and their role as stewards of dryland ecosystems, and the need for governments and international agencies to value these livelihoods. It also reveals the politicized narratives and conflicts of interest around Indigenous Knowledge and land use practices in developing countries where multiple interests compete for resource access and control. The following sections of this chapter will discuss pastoralism in dryland environments and highlight pastoralist attitudes and practices around land tenure and livestock husbandry. It will provide two cases to showcase how Indigenous Environmental Knowledge can be employed to advance conservation efforts in drylands and maintain fragile ecosystems. Finally, this chapter will suggest possible ways to mainstream pastoralist Indigenous Environmental Knowledge in conservation and environmental monitoring programmes.

Pastoralism as an adaptive system in dryland environments

Mobile pastoralism is practiced in dryland environments around the world. Drylands make up approximately 41 percent of the world's land and are home to over two billion people, the majority of whom live in developing countries (Pravalie 2016, p. 260). These regions are defined by very low levels of precipitation with high temporal and seasonal variability (Goudie 2012, 2013). Vegetation cover and quality also vary with the amount and seasonality of rainfall. In some regions, fog provides the moisture necessary to sustain plant life (Goudie 2013). These environments include diverse ecosystems; some drylands "rival tropical rainforests for their species richness" (Middleton 2016).

The dryland region of East Asia includes the contemporary states of Mongolia and the northwest region of China. This area is approximately 4.81 million square kilometers in size (Chen et al. 2014, p. 3) characterized by mountain grasslands, shrub lands, steppe, and temperate forests as well as gravel and dune desert lands. The East Asia dryland region is a high altitude zone, with elevations up to 7,929 meters above sea level (ibid., p. 5). In Asia, China has the largest dryland territory, while Saudi Arabia is unique for its hyper-aridity. Beyond East Asia, the Asian drylands

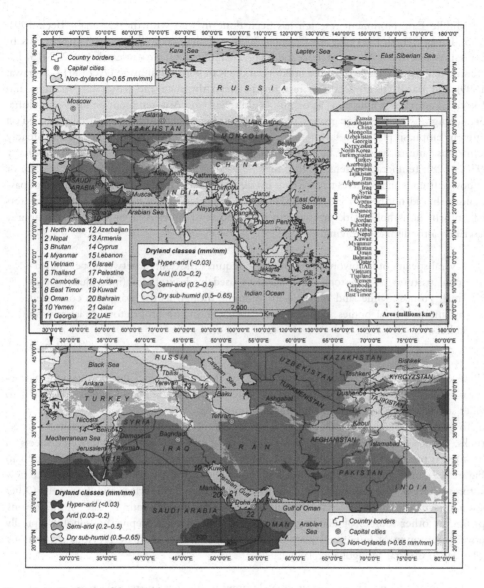

Figure 9.1   Drylands of the world

Source: Used with permission from author.

include the Arabian Desert, the Syrian Desert, Karakum, Kyzylkum, and the Thar located in the countries of Iraq, Iran, Jordan, Syria, Turkmenistan, Kazakhstan, Uzbekistan, India and Pakistan (Pravalie 2016) (Figure 9.1). These regions may also experience wide fluctuations in daily and seasonal temperatures. The Central Asian steppe has temperatures more than 40 degrees Celsius in the summer months and below 40 in the winter months. Drylands are also the sites of particular climatic and environmental hazards, including "drought, heat waves, extreme rainfall events, dust storms, wildfire and *dzud* (extreme snow and cold)" (Pravalie 2016, p. 262).

In these environments, mobile pastoralism is a central livelihood practice. Rough figures indicate that there are 100–200 million pastoralists in the world today (Dyer 2014, p. 22). In the Middle East, prominent research has been done in Iran with the Qashqa'i (Beck 1991, 2015), Basseri (Barth 1961), Baluch (Salzman 2000) and Shahseran (Tapper 1979) pastoralists. In Central Asia between Western China and Kyrgyzstan, Kreutzmann's research (2012, 2015) has documented social change since the building of the Karakorum Highway. Dyer (2014), Agrawal (1999) and Barth (1959) focus on relations between the state and pastoralists in India and Pakistan. Pastoralist populations across the Middle East and Asia represent an array of cultural forms, religions and linguistic groups. There are a number of definitions of pastoralism, based on different emphasis placed on factors such as mobility, social, and economic forms. It is broadly defined as "animal husbandry by natural graze with some access to crop cultivation" (Chatty 1972, p. 28). Temporary sedentarization may be used as a strategy to re-invest in livestock or empower friends and relatives to take over livestock holdings. In this way, the definition of pastoralism should not be so rigid as to exclude the important relationships with settled centers that inform household work practices. In the context of Inner Asia, Sneath and Humphrey describe pastoralism as involving

> several species of herbivorous livestock and requires movement between specific seasonal pastures … Livestock was owned or used by households (individually or in small groups), while land was held in common, though its use was often controlled by 'higher' social bodies such as lords, patrons, monastic proprietors or latterly by collectives
>
> *(1999, p. 2)*

These definitions reveal three main categories of skills and knowledge involved in pastoralism: mobility, livestock husbandry, and the environment. This knowledge is connected to wider social systems and decision-making institutions.

To enable mobile livelihoods, technology such as mobile housing (yurts, caravans, or tents), trucks, water bowsers, motorcycles, carts and draft animals are used. In Mongolia, sheets of thick felt is produced from livestock wool and provides insulation for yurts (*ger* in Mongolian) (Figure 9.2). In more arid regions, some herder households move once every two weeks to new pastures where dew or rain is harvested, or water is obtained from surface and deep wells. Other pastoralists in mountainous areas typically move seasonally and source water from streams, springs, or other sources of standing water. On the Mongolian steppe, households typically move a minimum of four times per year, returning to camps used by previous generations. A herder in Central Mongolia explained:

> I go following my father's places. In early times when there were collectives, he used to move these places … So these *nutag* (homelands), they [his parents] inherited these winter and spring camps from their ancestors and then I received and then I will pass to my children. This way we will pass and inherit.

In the Gobi Desert, where water is scarce, herders can move more than 20 times per year in order to provide adequate pasture and water for livestock. Additionally, each type of livestock requires different herding strategies and pasturage conditions. Grasses which are palatable for camels are unsuitable for goats and sheep. In Mongolia, livestock are grouped into "large" breeds and "small breeds." Goats and sheep are actively tended as a mixed group, returning to the camp every evening. Horses, cattle/yaks, and camels have a longer range and are left to graze independently while milking females and newborns are kept close to the home during certain

Figure 9.2 Traditional Mongolian *ger* with solar panel and satellite TV

Source: Photo by Ariell Ahearn.

times of the year. One example of traditional milking practices in rural Mongolia is the establishment of a *sahaalt ail* arrangement. In this arrangement, the milking goats and sheep of one household are switched with a nearby household (known as the *sahaalt*). Switching these groups allows households to control when the goat kids and sheep ewes are allowed to nurse. This both maximizes milk yields and its weaning technique.

Pastoralist knowledge encompasses the relationships between social groups, livestock and particular environments. Knowledge of animal husbandry includes how to graze different species of animals, livestock health, the palatability of plant species, skills around milking, wool and leather production, training draft animals and producing a variety of dairy products. Specialized skills include survival in sustained extreme hot and cold temperatures which requires conservation of key resources such as water. Gendered division of labor characterizes household production, usually with women responsible for milking small livestock, such as goats and sheep. In particular, the development of specialized livestock breeds and training methods to aid survival, milk and meat production, and the reproduction of livestock in challenging environmental conditions involve close socialization between humans and animals. Krätli and Schareika have argued that successful pastoralists practice an "intelligent harvesting of unstable concentrations of nutrients on the range" (2010, p. 606). They emphasize the situated knowledge gained from working closely with animals in uncertain environmental systems and the training involved in teaching animals how to graze on certain types of grasses in order to maximize nutritional gains (Figure 9.3). In this sense, pastoralism represents specialized knowledge of animal physiology, nutrition, behavior and training techniques. In Mongolia, herders use specialized techniques to communicate with livestock when they are grazing and to encourage livestock to feed their

Figure 9.3 Training horses in rural Mongolia using a traditional willow lasso pole, summer 2014

Source: Photo by Ariell Ahearn.

young. Certain vocalizations and song are used when training horses and directing them to move in certain directions. Fijn writes:

> The prominent use of song in Mongolian herders' relationships with their animals exemplified the use of vocal signals to elicit not only a desired response but to elicit an emotive response from the animal. Mongolians relate how singing is 'a sound that touches the heart of the animal'.
>
> *(2011, p. 111)*

Pastoralist livestock are specially adapted to the different extremes of dryland regions and typically include breeds of yak and cows, goats, sheep, camels and horses. Mongolian livestock have features such as thick skin, undercoats and layers of fat to survive the extreme winters and intense sunlight (Fijn 2011). According to Fijn: "Mongolian herding families do not make written records of the pedigrees of the herd animals, but they are able to mentally recall the genealogy of their horses and cattle for a number of generations" (2011, p. 83). In Mongolia, livestock are identified by their physical features and behavior as well as color. Horse, cattle, sheep and goats are identified additionally by their age, with different terms utilized for each of the first five years of each breed. Pastoralist livelihoods rely on the production of meat, dairy products, skins, and fiber (wool and cashmere). The household production of fermented mare's milk throughout Central and Inner Asia has been identified both as a source of valuable nutrition and a cultural tradition (Bat-Oyun et al. 2015). Nutrition from camel and horse milk often goes unexplored in development literature on health and household economies (Figure 9.4). Livelihood strategies also rely on relations with sedentary communities, markets and technology. As Barth (1961) has discussed in his work with the Basseri in Southern Iran, pastoralist

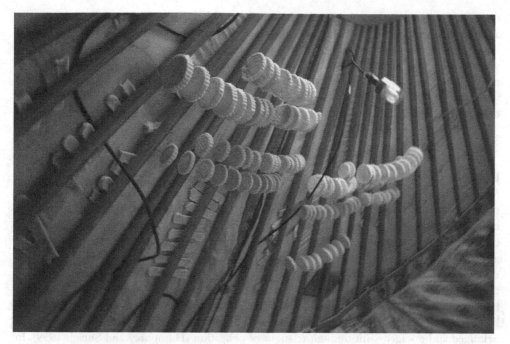

Figure 9.4   Dairy products drying on the roof poles of a Mongolian *ger*, summer 2014

Source: Photo by Ariell Ahearn.

households not only maintain active relationships with people in settled areas, but invest in land and do settle for temporary or long-term periods of time (p. 106). Pastoralist use of cellular and satellite phones, satellite television, trucks, wells, electric generators, solar power and other tools illustrates the innovative use of modern technology to enable continued mobility as well as communication over vast distances (Chatty 1986). Thus, while pastoralism involves a set of practices and skills around livestock husbandry, it is also embodied in distinct cultural forms which vary widely over the Middle East and Asia.

Pastoralist practices and knowledge have been at the center of debates around land degradation and desertification, which remains an important concern in dryland regions. Desertification refers to the loss of vegetation in dryland regions which may lead to the loss of top soils and increased erosion. While it has been acknowledged that both human and climatic factors contribute to land degradation, the effects of livestock grazing on rangelands has received disproportionate attention and blame. This view sees grazing as a primary cause of degradation and draws upon the more static equilibrium model's 'tragedy of the commons' and carrying capacity concepts (Middleton 2016). Studies of degradation and pastoralism have revealed a far more complex and dynamic picture regarding dryland ecosystems. A study of the legacy effects of livestock grazing on grasslands in Inner Mongolia (Han et al. 2014), for example, has shown that grazing increased the resilience of certain plant species and promoted carbon sequestration. The legacy effects of crop cultivation or fire (frequently used in Soviet-era collective farming) on contemporary grazing areas, in combination with climatic conditions, have also been identified as a critical component in degradation (Middleton 2016). Although climate hazards such as drought and forms of extreme weather such as dust and cold have always been a feature of drylands, trends in increasing temperatures and the volatility of weather events pose new challenges for pastoralists (Sternberg and Chatty 2016). As Fernandez-Gimenez has discussed, herders in Mongolia have developed specialized grazing strategies and classify pasture according to season, "nutritional quality and suitability for different types of livestock, topography and elevation, aspect, ecological zone and plant community, color, soil characteristics, water quality and quantity, distance from camp and degree of utilization by livestock" (2000, p. 1320). Additionally, close observation of rodent behavior, such as marmots, provide another indication of pasture quality and changes in weather, which herders can use to make pasturing choices and contribute to grassland resilience.

Mobility and pastoralist land tenure systems

Pastoralists across Asia and the Middle East have adapted their lifestyles to enable a variety of forms of both daily and seasonal mobility. Mobility, however, is enabled by institutions and open access lands. David Sneath (2003) for example, identifies the rural economy of Mongolian households as being historically characterized by "yield-focused or specialist" and "subsistence or domestic" forms of production (Sneath 2003, p. 446, 2014), which reflect the role of wider governance institutions in providing risk management services to herders. Yield-focused production coordinated by wealthy institutions or governments managed risk by overseeing long migrations, providing markets for agricultural goods, and emergency supplies or transport for households in cases of natural disaster or other threats. Environmental risks were often absorbed by large institutions (such as Buddhist monasteries or socialist collectives) to which households belonged and engaged with through relations of obligation (Humphrey and Sneath 1999). In this context, pastoralism is as much a cultural institution as an economic livelihood. Although exclusive or private ownership of land is not a traditional practice among pastoralists, social hierarchies based on concepts of kinship and age-based hierarchies constitute informal rules by which households establish seasonal range territories. Among Mongolian pastoralists, where pastureland is protected as open access in the national constitution, the occupancy or frequent use of seasonal camp sites structures land use rights (Ahearn 2016; Sneath 2003). The non-exclusive use of land has been described as "custodial" by scholars of Mongolia and emphasizes humans as occupants or inhabitants of a shared natural landscape rather than owners or exclusive resource users (Humphrey, Mongush and Telengid 1993). These institutional frameworks are supported by Indigenous Environmental Knowledge which considers nature to be shared with spiritual entities (Figure 9.5). For example, each year the Mongolian President with members of Parliament conduct sacred mountain worship rituals which acknowledge and respect the presence and power of these forces. As Sneath (2014, p. 461) explains, "The ceremonies reflected a cosmology in which humans do not hold land as they do other mundane possessions, but must enter into relations with the spiritual powers of the locality to ensure favorable conditions." These rituals occur annually on a smaller scale as part of the summer Naadam festival. This festival also features traditional games which showcase Indigenous Knowledge, such as horse racing, wrestling, ankle bone games and archery. Dairy products such as *airag*, fermented horse milk, is shared widely as part of these festivities across rural Mongolia.

In other regions of Asia and the Middle East, pastoralist tenure has become insecure and mobility curtailed. In Tibet and Lebanon, for example, many pastoralist households have to pay rental fees to utilize formerly communal or open access pasture (see Gongbuzeren, 2016; Obeid 2006). Chatty also highlights the complex relationship between state administrations and the Bedouin community in Lebanon and Syria. Nearly four decades after her initial contact, the Bedouin community in Lebanon was not able to secure citizenship from the Lebanese government due to their decision not to participate in a French colonial census to avoid enforced settlement (2013, p. 154). This decision left them undocumented, without citizenship, and with

Figure 9.5   Sacred site in South Gobi, Mongolia, summer 2016

Source: Photo by Ariell Ahearn.

no power to prevent or modify the state's efforts to transform the semi-arid common grazing land into private agricultural holdings. Across the border in Syria on the other hand, Chatty writes that the Bedouin community was much more aware of the dangers which their common use rights and intimate knowledge of their landscapes faced. Beginning at the end of the nineteenth century and intensified in the twentieth century, these Bedouin worked at integrating their leadership into the official state political system and thus maintaining a voice in how their traditional grazing lands were used or transformed (ibid.). These examples point to the different placement of people who are members of pastoralist groups within formal practices of government, sometimes marginalized and sometimes central to them, but in all cases firmly placed in the landscapes of power and the politics of territoriality where Indigenous Knowledge and ideas of homeland and place are rooted through social forms historically connected to pastoralist livelihoods. Chatty's body of work and focus on the resilience of pastoralists in the Middle East and their adaptation and change strategies in contemporary conditions continue to bear fruit. Her argument that pastoralists are "constantly changing their way of life to best suit their needs" (Chatty 1996) illustrates the ways in which traditional pastoralist knowledge is combined with new types of housing, income diversification and mobility. Mobile pastoralists in Lebanon, for example, adapt their livelihood strategies to changing economic, political and environmental conditions while still maintaining practices, such as hospitality and adherence to a certain kinship structure, which makes their social forms distinctly Bedouin, though they may not be living in their traditional homelands of the Badia (Chatty 2013, p. 4). Pastoralist Indigenous Knowledge constitutes socio-technical systems which are highly suited to extreme environments and constantly changing conditions. This flexibility has allowed pastoralists to weather periods of economic and political uncertainty, demonstrating forms of resilience which continue to allow pastoralist to adapt to changing conditions.

Conservation schemes in Oman

Conservationists – both national and international – have regarded the central desert of Oman as their own backyard, ignoring the presence and environmental knowledge of its local human inhabitants. International experts held significant 'scientific' knowledge of the deserts of Oman shaped by plants and animals, not people. Early in the 1970s, at about the time that the Arabian Oryx was declared extinct in Oman, they became concerned to restore a balance to this land-scape. The international flagship conservation effort promoted by the IUCN (International Union for the Conservation of Nature) and the WWF (World Wildlife Fund) set up the Arabian Oryx Re-introduction Project in the Jiddat il-Harasiis – the traditional homeland of the Harasiis camel and goat pastoral tribe. This process was envisaged from abroad and created in the offices of His Majesty, the Sultan's Advisor for the Environment without any consultation with the local tribesmen in the desert. Between 1980 and 1996, 450 Arabian oryx were either returned to the wild or were born in the Jiddat il-Harasiis with Harasiis tribesmen hired to track these animals as only they had the expert environmental knowledge of the Jiddat and the skills to follow these animals and to observe in minute detail these animals' social habits – as they were accustomed to doing prior to their near extinction from over-hunting by urban Emirati and Saudi elites. Throughout this period, the Harasiis' broad knowledge of the oryx was hardly recognized other than what the conservation scientists demanded be recorded on their movement charts. Furthermore, the relationship of this tribe with surrounding tribes was also undermined as jealousies emerged between rival tribes and the conservation project. In 1994, Oman succeeded in getting this conservation project recognized formally as the UNESCO World Heritage Arabian Oryx Sanctuary. But ongoing and constant friction between the Western managers of the conservation project and the local Harasiis tribesmen regarding their "rights" to graze their domestic herds in large parts of their territory – then officially a UNESCO nature reserve – eventually resulted in a distancing from the project by the Harasiis and a lack or diminution of any sense of 'ownership' or commitment to the conservation project. Harasiis herders had insisted that there was no competition for grazing between oryx and their own herds of camels. In fact, they insisted that camels kept the thorny scrubs under control and hence complemented oryx grazing habits. Although there has been no systematic study of graze competition between camels and oryx, the experience of the Shaumari Oryx Reserve in Jordan is telling. When the reserve was created in 1975, the Bedouin were forced to remove their camels from a special small enclosure earmarked for the reintroduction of 12 oryx. Bedouin tribesmen warned the Jordanian conservationists that this removal would result in the enclosure eventually becoming over-run with unpalatable woody shrub. Several years later, that is exactly what happened and the entire enclosure had to be bulldozed. By 1996, poaching and live capture from competing and rival tribes resulted in nearly two-thirds of this wild herd being captured or killed. In order to save the remaining 100 or so female oryx, the Oman army was called in to herd these animals into a small fenced enclosure. Soon thereafter the UNESCO World Heritage recognition was withdrawn (Chatty 2002, pp. 228–234). Unwillingness to recognize the broad environmental knowledge of the local tribal community and the delicate power balance between tribes and their environment hastened the decline and eventual failure of the international conservation effort to restore the Arabian oryx to the Jiddat il- Harasiis.

Mining in Mongolia

As mentioned earlier, the dryland regions where pastoralists live have become sites for a variety of mineral, oil and gas extraction activities. These industries have become a primary source of revenue for national governments across Asia and the Middle East and are framed as essential to national development. Indeed, for many of these countries, the majority of exports consists of minerals such as copper, coal, gold, rare earth, oil or natural gas. This is the case for one of the largest gold and copper mines in the world, which is located in the South Gobi Desert in Mongolia. The mine license area occupies approximately 86.2 km² in Omnogovi Province's Khan Bogd county (*soum*). The total area of the county is approximately 15,200 km² and also features major infrastructure related to the mining activities of other companies in the region as well as supporting industry (MDT/IEP Final Report 2017). The growth of roads, railroads, water pipelines, and immigration to the area has transformed this region from one largely occupied by pastoralists to a global mining production center. At the same time, local herders continue to graze camels, goats, sheep, and horses throughout the county and utilize water points such as wells or surface water (Jackson 2015). Although the scale of Oyu Tolgoi is massive, many of the governance issues at the interface of mining and pastoralism across Mongolia involve common issues around land tenure and use rights. Rights to practice traditional livelihoods feature strongly in pastoralist campaigns to protect their homelands from mining licenses. These issues are also commonly faced by reindeer herders in Siberia who face oil and gas infrastructure development (Forbes 2013).

After a resettlement program in 2004 and a compensation program in 2010, in 2012, the herders of two districts in Khan Bogd submitted a complaint to the Compliance Advisor Ombudsman of the International Finance Corporation regarding the lack of attention to traditional livelihoods given in the initial compensation packages for some herders with winter pastures within the mine license area. One section of the complaint reads:

> We consider ourselves as indigenous to the area, as well as carriers of the ancient tradition of nomadic herding. We are mobile pastoralists dependent on pasture for our livelihoods. These pastures are ours as recognized under the customary law. The Company, however, does not recognize our rights, justifying their decision only by the fact that we are not an ethnic minority. The compensation does not include mitigation or remedy for the loss of opportunity to carry on our traditional nomadic herding lifestyle and the related loss of property and cultural heritage to be passed on to our descendants.
>
> *(CAO 2012)*

This complaint illustrates the place of custodial land rights within the framework of traditional knowledge – it is where practices of mobility, social customs of respect, attitudes toward nature and moralities come together into an accepted code of practice. As discussed previously, Mongolian pastoralists consider nature to be shared by spiritual entities and thus not open to exclusive land ownership and exploitation. These ways of knowing and using the environment are jeopardized through government actions which allow entities like mining companies to have exclusive land rights without consultation with local communities or a form of legal process. This kind of tenure insecurity combined with climatic hazards risk along with wider social discourses around modernity, development and formal education has led many herders in Mongolia to invest in non-pastoralist futures for their children.

Despite these wider trends, fieldwork conducted by one of the authors over the course of 2016 on the Oyu Tolgoi CAO Complaint revealed a host of problems related to environmental monitoring and the capacity of the local government and national government to regulate water and land use in Khan Bogd. Herding households, through their close interaction with livestock, pasture and water points discovered a series of leaking bore holes which

Figure 9.6   Camels congregating around water source, South Gobi, Mongolia, 2016

were constructed during the initial research of water availability in the region. The scarcity of water in this region means that sounds of flowing or dripping water attracts camels. In this particular case, a herder discovered water cascading from unsealed bore holes when a herd of camels were found congregating around the capped pipes (Figure 9.6). Observing livestock behavior, health, milk yields and quality of wool and meat are tools herders use to assess long- and short-term changes in the environment. In Khan Bogd, herders also observe changes in the mucus of sheep and goat lungs which indicate consumption of dust and other airborne particles. These observations have led to a call for the Oyu Tolgoi mine to do more rigorous monitoring of their production of dust, water use and potential seepage of an open air tailings storage facility.

Currently, the Mongolian government lacks the capacity to conduct environmental monitoring of the rural areas with operating mine sites. Despite the presence of small-scale teams connected to the River Basic Authority under the Ministry of Environment and Green Development, the Mongolian government lacks the capacity to monitor mining sites across the country due to the huge scale of the country. In Khan Bogd, for example, the local government is given the responsibility for environmental monitoring across the whole *soum* territory. This is an overwhelming task for an understaffed and under-resourced team. In this absence, there is a strong potential for herders, who live in proximity to mining sites, to monitor environmental conditions and collect data on water levels, dust, pasture disturbances and other changes. The Mongolian government is highly dependent on revenue from mineral exports, which indicates that this industry will continue to operate in rural pastures for the foreseeable future. During the life course of the mines, herders can play an important role as environmental stewards, and are in a good position to rehabilitate pastureland after a mine has been closed. Mining companies, environmental NGOs and the federal government should engage with herders and value their environmental knowledge and close interaction with the land. Rather than exclude herders from pastures, these entities need to create policies which protect herder rights to mobility and access to environmental resources in order to promote both rangeland and human health for future populations.

Conclusion

Given the widespread changes across political, economic and environmental systems in Asia and the Middle East, there is an opportunity for pastoralists to play a more central role in managing their own environments and participating in decision-making forums. Mobile pastoralism and the unique Indigenous Environmental Knowledge that supports these systems, has been and continues to be discriminated against and devalued by government policies, despite the continued resilience of pastoralists for centuries preceding these interventions. Ethnographic research on pastoralist livelihoods has revealed the innovations involved in pastoralism and the potential for pastoralists to contribute to environmental monitoring and large-scale conservation projects based on their practices of mobility and close observation of pasture conditions and livestock behavior. What is required is the political will from governance and development institutions to value pastoralist skills and knowledge and to partner with them as equals.

Additionally, while pastoralist knowledge includes a range of specific skills, these are enabled by institutions which grant pastoralists rights to practice mobility on open access range. Mobility is a central practice which allows pastoralists to take advantage of variable rainfall and vegetation across large ranges. The effects of curtailing mobility can be seen in satellite images of Inner Asia, where the border between Russia and Mongolia can clearly be seen. Institutions that promoted large-scale mobility in Mongolia promoted grassland re-growth, while the enclosure systems of the former Soviet Union heavily degraded the region (Sneath 1998). Mainstreaming pastoralist environmental knowledge would involve investment in rural infrastructures and forms of tenure which promote mobile livelihoods. Innovative schemes like seasonal mobile schools for children would allow households to practice pastoralism without being marginalized from public services.

The dryland landscapes where pastoralists live are largely located in developing countries, many of which are largely dependent on revenue from the extractive industries and lack government capacity to monitor expansive territories. Pastoralists are in a position to fill this gap and can be equipped with data collection tools to record and measure water levels, dust, vegetation as well as numbers and types of wildlife. As pastoralists are quick to adopt new technology to facilitate mobile lifeways, their use of satellite technology and mobile phones is highly feasible. This citizen-science format of environmental monitoring allows for the collection of longitudinal data sets. In addition to serving as environmental monitors, pastoralist knowledge can be mainstreamed in projects which seek to protect ecosystems and promote social adaptability. The Dana Declaration provides a platform for pastoralists, environmental organizations and conservation agencies to develop a shared agenda around maintaining fragile dryland environments. The failure of conservation agencies to forge a shared agenda with the Harasiis around the UNESCO World Heritage Sanctuary illustrates the importance of valuing Indigenous Knowledge and engaging with local populations to design projects. Such engagement will not only increase the chances of project success and sustainability, but has the potential to expand current ways of thinking about the environment and conservation. The Indigenous Environmental Knowledge of pastoralist groups in Asia and the Middle East has supported sustainable practices over the course of many generations. The close relationships between humans and livestock in these regions involves forms of communication, practices and ways of relating to the landscape which have adapted to changing economic, political and environmental conditions. The adaptation and resilience strategies used by pastoralists across the region should be considered a valuable resource to address today's global environmental challenges.

References

Agrawal, A. 1999. *Greener Pastures: Politics, Markets, And Community Among A Migrant Pastoral People*. Durham, NC: Duke University Press.

Ahearn, A. 2016. The role of kinship in negotiating territorial rights. *Inner Asia* 18: 245–264.

Barth, F. 1961. *The Nomads of South Persia: The Basseri Tribe of the Kamseh Confederacy*. Boston: Little Brown.

Barth, F. 1959. *Political Leadership Among Swat Pathans*. London: Athlone Press.

Bat-Oyun, T., Erdenetsetseg, B., Shinoda, M., Ozaki, T., and Morinaga, Y. 2015. Who is making *airag* (fermented mare's milk)? A nationwide survey of traditional food in Mongolia. *Nomadic Peoples* 19: 7–29.

Beck, Lois. *Nomad: A Year in the Life of a Qashqa'i Tribesman in Iran*. Univ of California Press, 1991.

Beck, Lois. 2015. *Nomads in Postrevolutionary Iran: The Qashqa'i in an era of change*. London: Routledge.

CAO 2012. Mongolia/Oyu-Tolgoi-01/Southern Gobi, Letter of Complaint. Available at: www.cao-ombudsman.org/cases/document-links/documents/OyuTolgoiCAOComplaint_Oct122012_Redacted.pdf. (accessed 15 April 2017).

Chatty, Dawn. *From Camel to Truck : The Bedouin in the Modern World*. New York: Vantage, 1986.

Chatty, D. (1972) Pastoralism: Adaptation and optimization. *Folk* 14–15, 27–38.

Chatty, D. (1996) *Mobile Pastoralists: Development Planning and Social Change in Oman*. New York: Columbia University Press.

Chatty, D. (2006) Boarding schools for mobile people: The Harasiis in the Sultanate of Oman. In C. Dyer (Ed.), *The Education of Nomadic Peoples: Current Issues, Future Prospects*. New York: Berghahn Books, pp. 212–230.

Chatty, D. (2013) *From Camel to Truck: The Bedouin in the Modern World*. 2nd ed., Cambridge: White Horse Press.

Chatty, D. and Colchester, M. (Eds.) (2002). *Conservation and Mobile Indigenous Peoples: Displacement, Forced Settlement, and Sustainable Development*. New York: Berghahn Books.

Dana Declaration. 2002. Dana Declaration on Mobile Peoples and Conservation. Available at: http://danadeclaration.org/main_declarationenglish.shtml. (accessed 13 April 2017).

Dyer, C. (2014). *Livelihoods and Learning: Education for All and the Marginalization of Mobile Pastoralists*. New York: Routledge.

Fernandez-Gimenez, M. 2000. The role of Mongolian nomadic pastoralists' ecological knowledge in rangeland management. *Ecological Applications* 10 (5): 1318–1326.

Fijn, N. 2011. *Living with Herds: Human-Animal Coexistence in Mongolia*. New York: Cambridge University Press.

Forbes, B. C. 2013. Cultural resilience of social-ecological systems in the Nenets and Yamal-Nenets Autonomous Okrugs, Russia. A focus on reindeer nomads of the tundra. *Ecology and Society* 18(4): 36.

Galvin, K. A., Reid, R. S., Behnke, R. H., and Hobbs, N. T. 2008. *Fragmentation in Semi-Arid and Arid Landscapes*. Dordrecht: Springer.

Gongbuzeren. 2016. Herder participation in modern markets: The issues of the credit loan trap. In A. Ahearn, T. Sternberg and A. Hah (Eds.), *Pastoralist Livelihoods in Asian Drylands*. Cambridge: White Horse Press, pp. 91–109.

Goudie, A. 2013. *Arid and Semi-Arid Geomorphology*. Cambridge: Cambridge University Press.

Han, J., Chen, J., Han, G., Shao, C., Sun, H., & Li, L. (2014). Legacy effects from historical grazing enhanced carbon sequestration in a desert steppe. *Journal of Arid Environments, 107*, 1–9.

Humphrey, C., Mongush, M. and Telengid, B. 1993. Attitudes to nature in Mongolia and Tuva: A preliminary report. *Nomadic Peoples* 33: 51–61.

Humphrey, C. and Sneath, D. 1999. *The End of Nomadism? Society, State and the Environment in Inner Asia*. Durham, NC: Duke University Press.

Ingold, T. (2000). *The Perception of the Environment: Essays on Livelihoods, Dwelling and Skill*. London: Routledge.

Jackson, S. 2015. Dusty roads and disconnections: Perceptions of dust from unpaved mining roads in Mongolia's South Gobi Province. *Geoforum* 66: 94–105.

Krätli, S. and Schareika, N. 2010. Living off uncertainty: The intelligent animal production of dryland pastoralists. *European Journal of Development Research* 22: 605–622.

Kreutzmann, H.(Ed.) 2012. *Pastoral Practices in High Asia.* Berlin: Springer Science & Business Media.

Kreutzmann, H. 2015. *Pamirian Crossroads: Kirghiz and Wakhi of High Asia.* Wiesbaden: Harrassowitz.

MDT/IEP Final Report 2017. Available at: www.cao-ombudsman.org/cases/document-links/links-196.aspx (accessed 22 April 2017).

Menzies, C. R. 2006. *Traditional Ecological Knowledge and Natural Resource Management.* Lincoln, NE: University of Nebraska Press.

Middleton, N. 2016. Rangeland management and climatic hazards in drylands: Dust storms, desertification and the overgrazing debate. *Natural Hazards* DOI:10.1007/s11069-016-2592-6.

Middleton, N., Rueff, H., Sternberg, T., Batbuyan, B., and Thomas, D. 2015. Explaining spatial variations in climate hazard impacts in Western Mongolia. *Landscape Ecology* 30: 91–107.

Obeid, M. 2006. Uncertain livelihoods: Challenges facing herding in a Lebanese village. In D. Chatty (Ed.), *Nomadic Societies in the Middle East and North Africa.* Leiden: Brill, pp. 463–495.

Pravalie, R. 2016. Drylands extent and environmental issues. A global approach. *Earth-Science Reviews* 161: 259–278.

Salzman, P. C. 2000. *Black Tents of Baluchistan.* Washington, DC: Smithsonian Institute Press.

Sneath, D. 1998. State policy and pasture degradation in Inner Asia. *Science, New Series* 281(5380): 1147–1148.

Sneath, D. 2003. Land use, the environment and development in post-socialist Mongolia. *Oxford Development Studies* 31(4): 441–459.

Sneath, D. 2014. Nationalising civilisational resources: Sacred mountains and cosmopolitical ritual in Mongolia. *Asian Ethnicity* 15(4): 458–472.

Sternberg, T. and Chatty, D. 2016. Marginality, climate and resources in pastoral rangelands: Oman and Mongolia. *Rangelands* 38(3): 145–151.

Tapper, R. 1979. *Pasture and Politics: Economics, Conflict, and Ritual Among Shahsevan Nomads of Northwestern Iran.* London: Academic Press.

UNESCO n.d. What is local and indigenous knowledge? Available at: ww.unesco.org/new/en/natural-sciences/priority-areas/links/related-information/what-is-local-and-indigenous-knowledge/

Yeh, E. T. 2003. Tibetan range wars: Spatial politics and authority on the grasslands of Amdo. *Development and Change* 34(3): 499–532.

Yeh, E. T. 2013. *Taming Tibet: Landscape Transformation and the Gift of Chinese Development.* Ithaca, NY: Cornell University Press.

10

BALANCE ON EVERY LEDGER

Kwakwaka'wakw resource values and traditional ecological management

Douglas Deur, Kim Recalma-Clutesi, and Chief Adam Dick

On the Pacific coast of Canada, where the temperate rainforests meet the island-studded sea, the Kwakwaka'wakw have lived since before remembered time. There, the Kwakwaka'wakw (often, if inaccurately, called the "Kwakiutl") have dwelt in coastal villages, each community organized into clans overseen by specially trained chiefs who administer both worldly and spiritual life. These communities are world-renowned for their wealth: their abundance of natural resources and resource harvesting skill; their rich artistic and oral traditions; their ceremonials tradition that engages domains both tangible and intangible, seeking balance in the human and natural worlds. Clan chiefs (*Oqwa'mey* or *Tla'qwa'mey* in Kwak'wala) who oversee this wealth are traditionally aided by an entourage of key assistants: talented public speakers, shamans, woodcarvers, and many others, all working with chiefs to advance community interests, drawing on specialized training received from elder specialists in those fields. Traditionally, the clan chief and his assistants continuously invoked, shared, and applied core values in relationships to the lands and resources of the Kwakwaka'wakw world. In turn, these values have made a discernible imprint upon their home landscapes, fostering resilient human–environment relationships that have long sustained both human and biotic communities.

Here, we seek to summarize some of these core environmental values of the Kwakwaka'wakw (Kwakiutl) people, and to explore how they manifest in the traditional management of coastal natural resources. In doing so, we draw especially from the teachings conveyed by Chief Adam Dick, Kwaxsistalla, chief of the *Qawadiliqalla* [wolf] clan of the *Dzawada'enuxw* [Tsawataineuk] Kwakwaka'wakw—from Kingcome Village, on the mainland coast of British Columbia. One of the last chiefs to have been fully trained in the traditional way, Kwaxsistalla was isolated from the non-Native world for many years in his youth, being systematically educated in all aspects of chiefly knowledge—from the most profound to the mundane. His teachers consisted of a core group of clan chiefs born in the nineteenth century, who resisted cultural erosion by entrusting their core teachings to Kwaxsistalla and a small number of other children, urgently hoping they might carry forward these teachings into the coming century and beyond. His training was directed by three main clan chiefs: *Kodi* (Dick Webber), *Giyu'sti'stalathl* (Herbert Johnson), and Adam's grandfather, who held the title of Kwaxsistalla at that time; these three men, in turn, arranged to have other clan chiefs provide lessons over the course of several years, providing young Adam with specialized training in their particular areas of expertise. The traditional ecological knowledge conveyed to Kwaxsistalla through this intensive traditional

DOI: 10.4324/9781315270845-12

education, and passed on by him in turn, has provided a wealth of detail omitted from past writings on Northwest Coast cultures. In recent decades, this methodically transmitted knowledge has launched a revolutionary reinterpretation of cultural ecologies along the entire coast (Deur and Turner 2005; Mathewes and Turner 2017). It is from that authority, rooted in the teachings of the deep past, that we offer comment and clarification on many points of enduring concern.

Outside of the Kwakwaka'wakw world, in the academic domain, much has been written regarding the connection between traditional environmental values and the resource management practices of Native American communities—a topic of perennial interest to both academic and popular audiences. While these accounts generally accept that "resource conservation" has long been integral to Native American cultural practices in various ways, written treatments tend to be speculative, imprecise, and romanticized—at once mischaracterizing entire cultures and misleading those who might wish to learn from their examples. This is of much interest to the Kwakwaka'wakw, who have occupied a unique position in the literature and logic of academic anthropology. Being the focus of nearly five decades of research by Franz Boas, starting in the late nineteenth century, as well as his students and later generations of researchers, "Kwakiutl" cultural practices have been a source of inspiration, and a subject of considerable speculation on human-environment themes, since the beginnings of American anthropology (e.g., Boas 1921, 1966; Curtis 1915).

Of particular interest to anthropologists has been the institution of the "potlatch" (*pa'sa* in Kwak'wala). The term is so woven into academic discourse, and so contested, that we might start with a conventional definition taken from Merriam-Webster dictionary. There we find "potlatch" defined as "a ceremonial feast of the American Indians of the Northwest Coast marked by the host's lavish distribution of gifts or sometimes destruction of property to demonstrate wealth and generosity with the expectation of eventual reciprocation." Much of what was originally written of the potlatch in academic contexts resulted from Franz Boas' collaboration with George Hunt. A man of mixed Tlingit and English ancestry, Hunt was born near the Kwakwaka'wakw community of Fort Rupert, where his family were employed by the Hudsons Bay Company (HBC) fur trading post. Based in Fort Rupert, marrying into the community, Hunt recorded Kwakwaka'wakw ethnographic information and collaborated with Boas in its translation (Boas 1966). Yet Hunt encountered many barriers to the transmission of traditional knowledge, for many reasons. Social and demographic upheaval in Boas' time allowed untrained people—including some of Hunt's main informants—to make tenuous claims on chiefly titles and unoccupied roles without the benefit of prior training. Through his wife, Hunt was tied to the 'Nak'waxda'xw (Nakoaktok) people, whose language and practices are somewhat distinct from other Kwakwaka'wakw; our recent translation efforts show that some large part of the Boas corpus recorded by Hunt and presented as Kwakwaka'wakw was instead of 'Nak'waxda'xw origin. Hunt also had a sometimes awkward relationship with Kwakwaka'wakw communities by virtue of his family's connection to the HBC, often resulting in a lack of access to traditionally trained clan chiefs and others with specialized cultural knowledge. The Kwakwaka'wakw—especially its clan chiefs and closed "secret societies" that possessed ceremonial property—maintained very strict prohibitions on the public disclosure of specialized or sacred knowledge by specialized practitioners, as well as discussions of the meanings and motives behind traditional ceremony. Thus, lacking open access to the larger corpus of Kwakwaka'wakw philosophy and oral tradition, Hunt and Boas often struggled to translate the Kwak'wala language and many key concepts. They could observe patterns of activity, but had limited access to the organizational principles that put these patterns into motion. These were only some of the challenges the two men faced. Boas, too, famously sought to publish his data in relatively raw form, often presenting translated narratives of Hunt's informants without providing substantive editing, context, or commentary.

Lacking a full and meaningful context for Boas' many accounts of the potlatch, academic audiences could behold only the outwardly apparent mechanics of the potlatch, dazzled by these staggering exchanges of chiefly wealth. To the extent that Boas' data was reliable, its meaning remained largely in the eye of the beholder. In this context, Boas' raw documentation of the "Kwakiutl potlatch" became common academic property, and an epistemic blank slate onto which generations of scholars inscribed their own agendas. Each reinterpreted the potlatch in response to prevailing academic fashions: interpreting the ceremony as ritual warfare (Codere 1950), as a richly symbolic assertion of chiefly status (Drucker and Heizer 1967), as a ritual mediation of social relations through the universal practice of gifting (Mauss 1990), as psychological evidence of underlying paranoias and megalomaniacal tendencies endemic to a Kwakiutl "personality type" (Benedict 1934: 173–221), as status-seeking through the navigation of fundamental unconscious "mental structures" as proposed by Lévi-Strauss (Rosman and Rubel 1971), as a venue for the exchange of Freudian sexual and scatological symbology (Dundes 1979), and so forth. A full treatment of academic interpretations would require a broad, encyclopedic review (though useful starting points are found in sources such as Suttles and Jonaitis (1990) and Drucker and Heizer (1967)). Too often, these competing depictions proved more revealing of the academic allegiances and predilections of the authors than of anything particular regarding Kwakwaka'wakw culture. Moreover, without recourse to the underlying cultural logic as understood by clan chiefs and other officiating knowledge-holders within the potlatch tradition, these writings tended to depict the potlatch and most other Kwakwaka'wakw cultural practices as spectacles, often pathological in intent, serving to advance the most selfish and infantile human instincts. In this regard, the academic discourse is so uniquely longstanding, diverse, and revealing that it provides a valuable cautionary tale for the academic representation of Indigenous cultures worldwide.

Environmental anthropologists also played a role in this discourse. A half century ago, cultural ecologists began to consider whether the potlatch might have ecological influences or consequences. Authors such as Suttles (1960), Vayda (1961), and Piddocke (1965) noted that higher-status clans and chiefs among the Kwakwaka'wakw and their near neighbors tended to also possess the most abundant resources. Resource abundance, they suggested, contributes to the status of clans: such wealth allowed chiefs to enhance their standing relative to other chiefs through displays and redistributions of this wealth in the potlatch. Furthermore, these authors suggested clans might be more successful at holding and expanding their resource holdings by virtue of this enhanced status. In this sense, resource wealth begat the potential for more wealth. These writers suggested that, cumulatively, the arrangement enhanced the resiliency of clans and the Kwakwaka'wakw people as a whole. Piddocke, for example, concluded that

> [I]n aboriginal times the potlatch had a very real pro-survival or subsistence function, serving to counter the effects of varying resource productivity by promoting exchanges of food from those groups enjoying a temporary surplus to those groups suffering a temporary deficit.
>
> *(1965: 244)*

These analyses, however, were tentative at best. And, understandably, were rejected as crude ecological functionalism. In short order, they were eclipsed by a procession of alternative interpretations of the potlatch and of Northwest Coast culture practices (Orans 1975).

In truth, the meaning and context of the potlatch were lost in translation, along with most other aspects of Kwakwaka'wakw culture. The muddled transmission of ethnographic information into the outside world, along with the enduring, almost obsessive academic fixation on interpreting the potlatch, eclipsed some of the most fundamental truths and organizing principles defining Kwakwaka'wakw life. The potlatch has always been important but, we contend, can only be understood as a manifestation of a much larger system of cultural practices and ceremonies. In turn, these practices are guided by an underlying system of values and beliefs permeating all aspects of Kwakwaka'wakw culture. We do not pretend to explain Kwakwaka'wakw cultural values and beliefs in full here, but to offer corrections regarding key points of misinterpretation.

In many ways, reciprocal exchanges, as seen in the potlatch, permeate almost every other traditional Kwakwaka'wakw institution, and have been fundamental to the organization of all social, ceremonial, and economic life. These reciprocal exchanges are guided by a system of ethics and belief asserting the importance of "balance" in all relationships. Aided by an entourage of specialists, clan chiefs traditionally work in diverse arenas to maintain balance between communities through reciprocal exchanges—on the potlatch floor, but in myriad other arenas as well. With the goal of achieving "balance," they actively monitor and correct imbalances through many mechanisms, for example, the giving of material gifts, the repayment of specific material debts, and the offering of ceremonial honors and praise. The exchange or accumulation of wealth is not the ultimate goal. All of these interventions, tangible and intangible, are means to achieve greater ends.

In being attentive to all debts and their meticulous repayment, we build up the status of the clan and its chief. True, those of noble title work to sustain our reputation, to "keep our name good," as the anthropologists recognized, and this is also an important cultural value. But this is no hollow status-seeking, nor is it a selfish zero-sum competition. It is accomplished by giving, by ensuring the well-being of the people, by sustaining human relationships, by maintaining mutually sustaining balances over time. Indeed, these chiefly "names" are enduring. They outlive individual chiefs. The chiefly seat starts at the beginning of remembered human time and is passed down to the right bearer in each living generation. The name holds the person for a generation, just as much as the person holds the name and chiefly identity. In "making our name good," we are therefore working not to raise the status of a single living individual through the potlatch and all other Kwakwaka'wakw cultural practices, but to sustain the seat currently held by that individual over centuries—ensuring that a traditionally trained clan chief takes a long-term view of debt and its repayment, of decorum and reciprocity, and of balances over deep time. We seek to live up to the greatness of our name.

This system of reciprocal obligation creates a sprawling network of interpersonal relationships that bind together people and whole communities according to a set of shared rules. When one chief receives a gift from another chief, for example, some form of repayment is typically due. In any human endeavor, debts might accrue. If repayment is not made, this creates imbalances of many kinds: economic, titular, social, emotional, and otherwise. These imbalances can be unsettling, even damaging, if they cannot be adequately repaid in time. Participants in the potlach, but also other Kwakwaka'wakw ceremonies, engage in a constant inventory of obligations and the systematic rebalancing of relationships. This "systematic inventory and rebalancing" is of such importance that it is a collective enterprise, involving multiple specialists alongside the clan chief. Specially trained "potlatch recorders" (*Qa'qa'stuw'wa* in Kwak'wala) have operated as advisors to clan chiefs, and part of a chief's core entourage, especially before the advent of written language. Recruited in childhood, trained to sharpen their already prodigious memorization skills and fidelity to facts, these individuals track every transmission of property—material and immaterial—within the potlatch and other ritual exchanges. They do this "so that there will be no mistakes," so that all people can maintain appropriate balances between clans, between villages, between chiefs and commoners, between living and future members of the clan, between every imaginable part of human society. Continuously, through the potlatch but also in many other social and ritual contexts, the clan chief and his specialists collaborate to systematically assess their balance in each social domain, on every possible ledger, and to find ways to rebalance relationships in which potentially damaging debts are found. This is not only an economic obligation of the clan chief, but a spiritual obligation rooted in the most fundamental aspects of Kwakwaka'wakw cosmology and belief.

Yet gifts come in many forms. The food one harvests is a gift; even a single fish is understood as a gift from the Creator, from the fish that gave its life for our sustenance, and from all of the other beings dependent on fish for their survival. The weight of this gift is even greater, recognizing that the fish—indeed, any living being consumed for our benefit—is traditionally understood to be alive, sentient, possessing a spiritual identity all its own (Goldman 1975; Walens 1981). Killing is a weighty act, even as it is by necessity an everyday act. To do so lightly, without acknowledging the depth of the gift, is to unsettle a key relationship—a relationship necessary for the very survival of ourselves, our loved ones, and our entire community. Embedded in Kwakwaka'wakw values is an understanding that if we show disrespect, if we upset the balance of our relationships with the fish and other species on which we depend, they are likely to reciprocate in kind. To receive the gift of food requires repayment, then, a meaningful show of respect. Like any gift from a neighboring human community, this precious gift requires an acknowledgment, careful monitoring, and the systematic reciprocation of "interspecific" obligations between human and non-human beings. Our relationship with all near neighbors is "systematically monitored and rebalanced" over time to ensure enduring, stable, and mutually satisfactory relationships.

To be sure, this is a difficult balance to strike when members of one species day in and day out, year after year, kill and eat members of another. This is an awkward foundation for a relationship with any neighbor. For it is not just any member of a species being killed, but a member of that bounded and enduring population of the species that lives in close proximity, as the salmon of a particular stream or the edible plants of a particular valley are consumed by generation after generation of their human neighbors. The great-grandparents of one fed the great-grandparents of the other; if we show proper respect, the great-grandchildren of the present generation, of both species, will continue to honor this arrangement. In this way, all living beings in a clan's territory are bound together in some manner, biologically, but one might also say spiritually and ethically. Interventions are important and, by necessity, are both ceremonial and material. In the systematic review and rebalancing of our debts, non-human species are also included.

Adding to this, in all actions, chiefs are guided by concepts of wealth that hinge not just on resource abundance now, but on the consistency and predictability of that abundance to support all of these relationships over deep time—to sustain future generations of clan chiefs, to uphold their wealth and nobility, and to sustain future generations of the species on which they depend. A clan chief and his advisors are thus obligated to sustain those "natural resources" in their control, building additional motives for conservation atop those already suggested.

Together, it is appropriate to say that such cultural values and practices gave Native communities of coastal British Columbia incentives to sustain natural resource wealth over deep time, to avoid resource overexploitation, and to incrementally intensify natural resources through diverse means (Turner 2014). Moreover, these values and practices are manifested in many ways that are still seen on the land. The traditional ecological teachings of Chief Kwaxistalla suggest the importance of such values in the active management of a number of marine resources, including salmonids, eulachon runs, cultivated "clam gardens," and estuarine root gardens. Long overlooked by academic researchers, these practices demonstrate not only a sophisticated appreciation of nearshore ecology and its potentials. They also illuminate how the overarching conservation values of the Kwakwaka'wakw people, the constant "inventory and rebalancing" within reciprocal relationships, have manifested in specific maritime resource management traditions.

The care traditionally applied to the management of staple fish species helps to illuminate this point. The most important staple of the Northwest Coast, Pacific Salmon (*Oncorhynchus* spp.) have always occupied a preeminent role in Kwakwaka'wakw subsistence, and still play a central role in the diet today. So too, another anadromous species, eulachon (*Thaleichthys pacificus*) has served as a source of staple food and a nutritious oil used widely within Kwakwaka'wakw cuisine. Displays of material and symbolic "respect" have been applied to each in an attempt to sustain a balanced relationship. Chief Adam Dick recalls that ceremonies traditionally mark the first arrival of fish, and thanks are offered throughout the harvest—honoring the fish and the Creator for their sacrifice. So too, under certain circumstances fish bones are ceremonially placed back into the water, or fish transported to replenish other waterways—actions at once undertaken to "gift" the fish, to demonstrate "respect" to many beings, while also having biophysical consequences likely to enhance fish productivity (Thornton, Deur, and Kitka 2015).

Chief Adam Dick also recalls teachings that clan chiefs, aided by shamans and other specialists, traditionally monitored the harvest of these species within the rivers in their jurisdiction. When fishers had caught enough to meet dietary needs, plus a little extra for feasting, gifting, and trade, the chiefs ordered fishing to cease. Factored into the decision were also the scale of that year's fish run, the scale of those of preceding years, and any recent natural events such as landslides that might affect future runs. This was not done simply to avoid the mechanical reality of overharvest. It was also done to show due respect to the fish, the Creator, and all living beings—human and non-human—that rely on the same fish for survival, in addition to future generations of the clan. In these ways and others, the clan chief and his advisors carry out a systematic inventory of the many interrelated obligations—to the fish, to other people, to all the other sentient debtors and creditors of the world—in order to make an informed decision as to when harvests must cease and when other interventions are required. The consequences of disrespect, including overharvest or other adverse impacts, are understood to come back to the harvesters through multiple modalities, requiring attention to each ledger, requiring balance on every ledger.

This attentiveness to fish populations reflects not only their centrality within the diet, culture, and cosmology of Kwakwaka'wakw people, but suggests a shared cultural memory of occasional experiences with fish scarcity. And, as Chief Adam Dick attests, there were sometimes years when the fish, particularly salmon, did not return (*wi'yum'galleese* in Kwak'wala). At these times, clan chiefs and their advisors sometimes made the difficult decision to pack up the entire community and move to clam "gardens" (*lokiwey* "rocks rolled to clear the ground") many kilometers away. As taught to Chief Adam Dick when he was a child, these clam gardens were often constructed by moving stones out of clam beds to form low, wall-like structures in the lower intertidal zone. In addition to clearing obstacles for clam occupation of mudflats, the low walls served to entrain sediment and laterally expand productive clam beds (Deur, Dick, Recalma-Clutesi, and Turner 2015; Lepofsky et al. 2015). Largely undocumented in the anthropological literature until recently, with the ethnographic data provided by Chief Adam Dick, these structures verifiably enhanced the scale, productivity, and predictability of clam bed output within clan territories

(Groesbeck 2014). Yet these were not simply engineered environments, meant to increase food output. They were manifestations of much deeper Kwakwaka'wakw knowledge and values. As with other types of resources, ancestral Kwakwaka'wakw observed what modifications intensified clam output, but their assessments were not simply mechanical. If the clams appreciated a change in their environment, they came back more abundantly. If they were to be consumed in large quantities, respects were required: tending the beds; removing rocks, sticks, and the occasional predatory starfish; improving the quality and extent of shoreline habitable to clams. Clams, famously immobile, were a stable food, and could be critical as a backup resource in lean times. Yet clam gardens were not only a risk-reducing locus of emergency food. They were arguably a "locus of respect," where long-term investment in a particular clam bed and the clams living therein ensures their assistance in a time of need. In maintaining this balance with the clams on which the people depend, a clan chief might maintain the required balances in human domains as well.

The same fundamental logic applies to the traditional management of many other coastal plants and animals. It applies to the estuarine root gardens—*tekilakw* ("places of manufactured soil"), where Kwakwaka'wakw harvesters enhanced culturally preferred estuarine root foods, including Pacific silverweed (*Argentina pacifica*), Springbank clover (*Trifolium worksjioldii*), and occasionally such species as Chocolate lily (*Fritillaria camschatcensis*) and Sea milkwort (*Lysimachia maritima*) (Deur 2005). Individually demarcated root garden plots are traditionally weeded, edible roots transplanted, the soil turned, root-eating predators discouraged, rocks removed and placed lower in the tidal column—sometimes allowing for the lateral expansion of good harvestable ground within the intertidal zone. Here too, Chief Adam Dick's teachings have been key in reconstructing elements of the cultivation process. Controlled experiments that mimic traditional management, as per these teachings, have been demonstrated to measurably enhance food plants in these gardens (Lloyd 2011; Pukonen 2008). We see this too in the harvest of eelgrass (*Zostera marina*), which, when done according to methods prescribed by Chief Adam Dick and his teachers, has been shown to enhance plant output for the benefit of plants and human harvesters alike (Cullis-Suzuki et al. 2015). Many other plant and animal species received similar care, and ongoing documentation and testing show similar outcomes: cultivated crabapple groves and camas patches (*Malus fusca* and *Camassia* spp.), carefully tended ancient red cedar (*Thuja plicata*) trees, and other traditionally managed resources along the shoreline. Again, the goals of these techniques are utilitarian to a point, but manifest a much deeper Kwakwaka'wakw appreciation of the world and its workings. By investing in these traditionally managed habitats, the ancestors gave the species or habitat their support, becoming an ally in struggles with prey species, weeds, and space limitations. By doing so, they recruited them to their own cause, providing the clans with sustenance and risk-reduction, but also success in many other domains, including even potlatches and other feasts where those foods supported the good of the clan and others. Treated with due respect, even a humble root or clam can help "keep our name good."

Returning to the classic debates in environmental anthropology, we see a hint of truth in Piddocke's claims from decades ago. There is no doubt that resource intensification and redistribution practices have given Kwakwaka'wakw communities successful mechanisms for dealing with short-term environmental perturbations: temporary crashes in salmon runs could be offset by an abrupt dietary and geographical shift to clam beds meticulously cultivated for that very purpose, or might be offset by chiefs' careful negotiation of obligations between clans, communities, and individuals. Over time, these practices affected their general fortunes.

Yet counter to the claims of those past academic writers, resource wealth is only a small part of the story. The currency gained in this series of transactions is economic, social, and spiritual, with each "resource" possessing its own intrinsic significance. As embodiments of the Creator's will, the "resources" themselves are deserving of respect, of reciprocal care for their part in our sustenance and our success. As clan chiefs seek to strike balances on every ledger, these resources are factored into deliberations in myriad ways. In turn, these values traditionally place clear limits on resource over-exploitation. To overexploit resources would be reckless and would generate unwelcome debts and possible retribution from a long list of creditors. So too, it would be unthinkable, traditionally, to cut down an entire patch of forest and replant that ground with alien species with whom we have no prior relationship. Introduced agriculture was nonsensical to the Kwakwaka'wakw ancestors who encountered it. But the incremental, respectful enhancement of what is already there, those species with which we have longstanding relationships of mutual benefit—this is at the heart of traditional Kwakwaka'wakw cultivation or *qwakqwala'owkw* (literally "keeping it living"), the term used by the ancestors and shared with us by Chief Adam Dick to encapsulate the traditional management methods described here. Potlatches might aid in food redistribution, but it is the underlying Kwakwaka'wakw values, the ceremonial practices, and the traditional management strategies that converge within the potlatch and *qwakqwala'owkw* that supported the resiliency of the Kwakwaka'wakw people and the natural resources of their homeland over deep time. Perhaps settler societies might embrace parts of this tradition in times to come, systematically assessing and rebalancing debts in multiple domains—to recognize forgotten gifts and unpaid debts on multiple ledgers, and to identify and fix broken things. With our mutual interests and our shared fate in this world, we might all benefit from a tradition of assessment, restoration, and rebalance that the Kwakwaka'wakw have long understood to be a sacred trust.

Ultimately, Indigenous Ecological Knowledge is shaped by the most fundamental values of Indigenous Peoples. These values shape organizing principles and motivations, and determine the many mechanisms through which Indigenous Ecological Knowledge is made manifest in the world (Berkes 1999). To speak of Indigenous Knowledge without the broader context of these values is nonsensical, and can lead scholars and other outsiders to misunderstandings— misunderstandings that do violence to the subject and may even begin to affect the practices and recollections of traditional knowledge within Indigenous communities. Here we provide an antidote. By providing this very cursory discussion of traditional Kwakwaka'wakw values relating to nature and culture, we undermine the contradictory and often wildly speculative views put forth in a century or more of academic discourse. In its place, we offer a more coherent and consistent view of key organizing principles, perhaps allowing academic audiences to glimpse how the potlatch, traditional resource management, and so many other aspects of Kwakwaka'wakw culture fit together into a rich and coherent whole. Past representations of Kwakwaka'wakw more often than not were comically inaccurate. As clearly suggested by the account presented here, past academic mistakes can be overcome by long-term collaborations between academic researchers and specialized knowledge holders from Indigenous communities, such as Chief Adam Dick. This approach provides researchers with access to accurate and nuanced understandings of environmental phenomena, the immediate guidance and correction of trained knowledge-holders, and a clearer understanding of how cultural values might manifest within traditional land and resource management practices. On this academic ledger too, we might yet strike a balance.

Acknowledgments

The authors wish to thank Daisy-Sewid Smith (*Mayanilth*), Kwakwaka'wakw cultural specialist and teacher, for her generous assistance with Kwak'wala translation and spellings.

References

Benedict, Ruth 1934. *Patterns of Culture*. New York: Houghton Mifflin.

Berkes, Firket 1999. *Sacred Ecology: Traditional Ecological Knowledge and Resource Management*. Philadelphia, PA: Taylor & Francis.

Boas, Franz 1921. *Ethnology of the Kwakiutl: Bureau of American Ethnology, 35th Annual Report, 1913–1914*. Washington, DC: Bureau of American Ethnology.

Boas, Franz 1966. *Kwakiutl Ethnology*, ed. H. Codere. Chicago: University of Chicago Press.

Codere, Helen 1950. Fighting with property. *Monographs of the American Ethnological Society* 18. ed. Marian W. Smith. New York: J.J. Augustin.

Cullis-Suzuki, Severn, Sandy Wyllie-Echeverria, Chief Adam Dick, and Nancy J. Turner 2015. Tending the meadows of the sea: A disturbance experiment based on traditional indigenous harvesting of *Zostera marina* L. (*Zosteraceae*) in the southern region of Canada's west Coast. *Aquatic Botany* 127, 26–34.

Curtis, Edward S. 1915. *The North American Indian*, vol. 10: *The Kwakiutl*. Norwood, MA: Plimpton Press.

Deur, Douglas 2005. Tending the garden, making the soil: Northwest Coast estuarine gardens as engineered environments. In Douglas E. Deur and Nancy J. Turner (Eds.), *Keeping It Living: Traditions of Plant Use and Cultivation on the Northwest Coast of North America*. Seattle, WA, and Vancouver: University of Washington and University of British Columbia Press, pp. 296–330.

Deur, Douglas, Chief Adam Dick, Kim Recalma-Clutesi, and Nancy J. Turner 2015. Kwakwaka'wakw "Clam Gardens": Motive and Agency in Traditional Northwest Coast Mariculture. *Human Ecology* 43(1): 201–212.

Deur, Douglas and Nancy J. Turner (Eds.) 2005. *Keeping It Living: Traditions of Plant Use and Cultivation on the Northwest Coast of North America*. Seattle, WA, and Vancouver: University of Washington and University of British Columbia Press.

Drucker, Philip and Robert F. Heizer 1967. *To Make My Name Good: A Re-examination of the Southern Kwakiutl Potlatch*. Berkeley, CA: University of California Press.

Dundes, Alan 1979. Heads or tails: A psychoanalytic study of potlatch. *Journal of Psychological Anthropology* 2(4): 395–424.

Goldman, Irving 1975. *The Mouth of Heaven: An Introduction to Kwakiutl Religious Thought*. New York: John Wiley and Sons.

Groesbeck Amy S., Kirsten Rowell, Dana Lepofsky, and Anne K. Salomon 2014. Ancient clam gardens increased shellfish production: adaptive strategies from the past can inform food security today. *PLoS One* 9(3): e91235.

Lepofsky, Dana, Nicole F. Smith, Nathan Cardinal, John Harper, Mary Morris, et al. 2015. Ancient shellfish mariculture on the Northwest Coast of North America. *American Antiquity* 80(2): 236–259.

Lloyd, T. Abraham 2011. Cultivating the Tekkillakw, the Ethnoecology of Tleksem, Pacific silverweed or cinquefoil (Argentina egedii (Wormsk.) Rydb.; Rosaceae): Lessons from Kwaxsistalla, Clan Chief Adam Dick, of the Qawadiliqella Clan of the Dzawadaenuxw of Kingcome Inlet (Kwakwaka'wakw) . Unpublished MSc thesis. School of Environmental Studies, University of Victoria, Victoria, BC.

Mathewes, Darcy and Nancy J. Turner 2017. Ocean cultures: Northwest Coast ecosystems and indigenous management systems. In P. S. Levin and M. R. Poe (Eds.), *Conservation for the Anthropocene Ocean: Interdisciplinary Science in Support of Nature and People*. Cambridge, MA: Academic Press, pp. 169–199.

Mauss, Marcel 1990. *The Gift: Forms and Functions of Exchange in Archaic Societies*, trans. W. D. Hallis, London: Routledge.

Orans, Martin 1975. Domesticating the functional dragon: An analysis of Piddocke's potlatch. *American Anthropologist* 77(2): 312–328.

Piddocke, Stuart 1965. The potlatch system of the Southern Kwakiutl: A new perspective. *Southwestern Journal of Anthropology* 21(3): 244–264.

Pukonen, Jennifer C. 2008. *The λ'aayas Project: Revitalizing Traditional Nuu-chah-nulth Root Gardens in Ahousaht, British Columbia*. Unpublished M.Sc. thesis. School of Environmental Studies, University of Victoria, Victoria, BC.

Rosman, Abraham and Paula G. Rubel 1971. *Feasting with Mine Enemy: Rank and Exchange Among Northwest Coast Societies*. New York: Columbia University Press.

Suttles, Wayne 1960. Affinal ties, subsistence, and prestige among the Coast Salish. *American Anthropologist* 62: 296–305.

Suttles, Wayne and Aldona Jonaitis 1990. History of research in ethnology. In *Handbook of North American Indians*, vol. 7: Northwest Coast. Washington, DC: Smithsonian Institution, pp. 73–87.

Thornton, Thomas Fox, Douglas Deur, and Herman Kitka, Sr. 2015. Cultivation of salmon and other marine resources on the Northwest Coast of North America. *Human Ecology* 43(2): 189–199.

Turner, Nancy J. 2014. *Ancient Pathways, Ancestral Knowledge: Ethnobotany and Ecological Wisdom of Indigenous Peoples of Northwestern North America*, 2 vols. Montreal: McGill-Queens University Press.

Turner, Nancy, Douglas Deur, and Dana Lepofsky 2013. Plant management systems of British Columbia's First Peoples. *BC Studies* 179: 107–133.

Vayda, Andrew P. 1961. A re-examination of Northwest Coast economic systems. *Transactions of the New York Academy of Sciences* 23: 618–624.

Walens, Stanley D. 1981. *Feasting with Cannibals: An Essay on Kwakiutl Cosmology*. Princeton, NJ: Princeton University Press.

11

CHALLENGES SURROUNDING EDUCATION AND TRANSMISSION OF AINU INDIGENOUS ECOLOGICAL KNOWLEDGE IN JAPAN

Disparate valuations of a people and their IEK

Jeff Gayman

Introduction

The Ainu people are the Indigenous inhabitants of northern Honshu, the island of Hokkaido, southern Sakhalin, and the Kurile Islands, a region richly varied in terms of ecology, yet also influenced significantly by the forces of capitalism as well as colonial incursion from Japan and Russia. Ainu Indigenous Knowledge, grounded in a rich cosmology and moderated in turn by time-honored social practices, could provide valuable tips to the management of regional ecosystems and use of resources therein. However, during the process of modernization, Ainu philosophies and knowledge have been historically under-valued, and their society severely constrained by development of their lands under legislation which neglected or denied traditional Ainu livelihoods. In 1997, Ainu Cultural Promotion legislation was implemented, and, in 2019, additional legislation which could further have bolstered support for the practice of "Ainu culture in its broadest sense" established. Nonetheless, perspectives of what constitutes Ainu traditional knowledge remain unimaginatively piecemeal and folkloristic, and education and cultural transmission initiatives based on them lack a vision of Ainu "culture" informed by Indigenous Environmental Knowledge. Ideally, the 2019 legislation would have provided grounds for Ainu community development based on a revitalization of Ainu traditional IEK. This chapter begins by providing an overview of the diversity of Ainu IK through regional examples from different biospheres, illustrating the range of potential categories of application, then goes on to describe how historical policies have warped, and current conceptions curtailed, Ainu praxis of IK and IEK. The chapter then proceeds to introduce several promising but isolated Hokkaido initiatives in the fields of education, cultural transmission, and locally controlled research which have implications for Indigenous Knowledge usage. Finally, the chapter closes by consideration of how new investigation of some traditional mediums (Ainu oral tradition, IEK of other proximal Indigenous groups, technology) might be used in

DOI: 10.4324/9781315270845-13

a revaluation of possibilities for the upcoming cultural legislation which would open it up to a holistic and practical revitalization of Ainu IK/IEK.

Geographical characteristics

Traditionally the Ainu inhabited the area extending from northern Honshu in present-day Japan, to the southern half of Sakhalin Island, and throughout the Kurile Archipelago to the southern tip of the Kamchatka Peninsula. Much of the region consists of forested hills and mountains surrounded by coastal plains, interspersed with tundra and semi-tundra in higher elevations and the northern stretches of the Kurile Islands. The area is also characterized by being volcanic. Vegetation varies from deciduous/mixed deciduous and coniferous forests to grassland and tundra. Larger terrestrial mammalian inhabitants of the region include deer, bear and fox, while sea otters, whales, and a host of pinnipeds populate the ocean. Climactically the region ranges from humid continental climate zone with Köppen climate classification Dfb (hemiboreal) in Hokkaido to sub-Arctic in the northern Kuriles; for all intents and purposes the relevant environmental typologies vis-à-vis this volume's classification scheme are temperate and boreal forest and coastal.

This chapter adopts as its main focus the IEK/IK of the Hokkaido Ainu. Due to various political and other reasons, almost the entire Ainu population of the Kuriles has been forcibly relocated elsewhere, while the population of Sakhalin has not functioned as a visible ethnic community since the end of World War II, and the Ainu population of northern Honshu has similarly not manifested a visible presence for at least the past 100 years. The chapter therefore deals with these non-Hokkaido-Ainu populations only in terms of future prospects. This chapter covers both forest Ainu and coastal Ainu; nonetheless, due to the author's personal research background, descriptions of non-coastal Ainu IEK have received more emphasis. In any event, it should be noted that in many Ainu communities the distinction between forest and coastal Ainu is a false one; certainly, in terms of restorative justice discussed in the last section of this chapter, such arbitrary distinctions should be avoided.

Ainu IEK and IK

Already by the turn of the twentieth century, anthropologists and other social scientists researching the Ainu were clamoring about "salvage anthropology" to save what vestiges of traditional Ainu culture remained; by the time that a major systematic ethnological/anthropological investigation of Ainu societal customs was organized in the Saru River drainage in the early 1950s, the assumption among researchers was that already no vestiges of Ainu traditional society existed. As will be explained below, legal prohibitions on Ainu traditional fish harvests, and severe restrictions on hunting effectively terminated any Ainu-centered local environmental management. The explanation of Ainu IEK/IK in this chapter is thus based upon data from oral testimony combined with secondary evidence from archeological and historical research. Nevertheless, as will be explained, a significant amount of Ainu IK which could have a bearing on such diverse fields as food studies, pharmacology, geography, and architecture continues to be transmitted to the present day, and it is important to give it due attention for this reason, in addition to the unspoken priority which such knowledge should receive as a fundamental principle of Ainu human rights.

In order to conceptualize Ainu IEK properly, it is first necessary to understand the Ainu residential pattern of a village (*kotan*) located at the center of its own sovereign territory. Forest Ainu were a non-nomadic people who lived in fixed settlements along riverbanks safely out of reach of floods, and who periodically ventured on mid-to-long-term hunting forages during the winter season. Between 5–20 houses (*cise*) clustered together formed a typical Ainu village, which was always strategically located near a potable water source as well as proximate to a river in order to easily harvest spawning fish. Hokkaido Ainu customarily resided in thatched wooden huts (*cise*) built from locally available materials and adapted to local climactic conditions. Dwellings in Sakhalin included the use of logs, possibly a Russian influence, while meanwhile, in the Kuriles, subterranean houses were the Kurile Ainu solution to heavy winds which prevented the construction of structures above ground.

All materials necessary for daily implements, clothing, and foodstuffs in the life of the *kotan* were locally obtained by hunting, fishing and gathering in the *kotan*'s territory, known as an *iwor*, which was centered around the waterway on which the *kotan* was located, bounded upstream and downstream by the river's tributaries, and, between these tributaries by mountain ridges running parallel to the main river. Forest Ainu subsisted on the meat of salmon, trout, deer, and bear, edible mountain plants, and limited grain harvesting. In the case of coastal Ainu, this diet was supplemented with sea mammal meat. Due to the long and often severe winters, preserved foodstuffs provided the only source of nutrition during the winter months, and preservation technologies were highly developed. Importantly, since tools and other daily implements were made from locally obtainable materials, this lifestyle was an inherently sustainable one.

Other than this sustainability, however, little is known about environmental management aspects of Ainu IEK. Since Ainu are a hunting-gathering people, it is assumed that they may have regulated the hunting of deer and bear and fishing for salmon. Since Ainu communal hunting and fishing have been essentially banned for the past 150 years, there is no way of knowing. However, the raising and "Spirit-Sending" of bear cubs whose mother had been killed by hunters hint at the philosophy potentially underlying such management.

In any event, numerous admonitions against profligacy exist in Ainu society to this day. Historical records on inter-tribal covenants and personal testimony from Elders regarding the daily norms affecting sustainability and non-pollution indicate that rules for these matters possessed both a communal as well as an inter-tribal character. In other words, members of neighboring *kotan* were well aware of the boundaries of one another's properties, to the extent that felling a tree even 1 meter beyond the ridge demarcating a neighboring *kotan*'s territory was a criminally punishable offense. Likewise, adult and senior members sustained the purity of water systems by admonishing children and youth never to urinate or defecate in the vicinity of a waterway. This taboo was so strong that one episode recounts an elder who had been hospitalized leaving the hospital to come home when she was faced with the necessity of having to use the hospital toilets which operated using running water.

That the values of moderation underlying these practices of sustainability are encompassed in an Ainu cosmology manifested, *inter alia*, in Ainu oral literature may come as no surprise to anyone studying IEK. The Ainu's relational ontology to their gods/spirits, as with that of many Indigenous Peoples, is expressed in moral tales (*ucaskuma*) of Ainu spirits admonishing humans to treat animals with respect, as in the tale of a fox spirit cautioning hunters to leave some portion of their catch for the local animals, lest the wrath of the gods befall the humans, or alternatively in tales praising such behavior, as in the story of a carrion crow, which together with its comrades, saved the life of a hunter who had made it a habit of sharing his harvest with the foxes and crows. The sine qua non of the Ainu belief in reciprocity between humans and spirits is the Ainu Bear Spirit Sending Ceremony (*iomante*), the largest of all Ainu ceremonies in which the entire community gathers in appreciation to sacrifice a bear in order to send its spirit back to the realm of the spirits, in celebration of the mutually beneficial relations between humans and spirits.

Understanding the twin fundamental Ainu values of modesty plus expression of appreciation/generosity undergirded by this world-view is key to properly understanding Ainu culture vis-à-vis Ainu IEK/IK. The ongoing importance to this day in Ainu communities of dictums such as, "Only take as many (fish) as you need," and "Never harvest more than 2/3s of a patch of wild mountain vegetables/mushrooms," can be comprehended through the philosophy undergirding such stories.

In any event, questions of environmental management aside, the sophistication of Ainu IK is manifest in an extremely high degree of familiarity with the floral, faunal, geological, and climactic characteristics of the local environment surrounding Ainu villages. This knowledge – in a proper ethical context – demands attention for its potential scientific value, in addition to its practical value in maintaining those Ainu cultural practices that have lasted to the current day. To start with an example of floral knowledge, Ainu men were, and in many cases still are, familiar with, among other knowledge of local flora, the types of rot-resistant local trees suitable for use in house frames, flexible yet strong woods appropriate for use in making hunting bows, sturdy trees appropriate for the carving of pestles, trees possessing a solid exterior yet soft interior suitable for gouging out dugout canoes, and easily whittled branches whose wood would curl well for the prayer sticks used in a variety of Ainu ceremonies and rituals.

On a related note, the author personally knows of at least 30 edible mountain vegetables regularly gathered by the Ainu. Some of these plants possess curative properties for common yet serious ailments, such as heart disease, rheumatism, and diabetes (Keira 2003), as well as for such uncommon diseases as bone disintegration. Among plants useful to the Ainu many species, such as silver birch, must only be gathered during certain times of the year, and some plants, which serve as material for crucial household items, such as cattails, must not only be gathered during a certain season (September), but must also be dried outside in the sun at a certain temperature in order to reach a condition at which they can be stored permanently. Needless to say, a thorough knowledge of the characteristics of these trees, plants, and shrubs is necessary in order to be able distinguish them from similar flora, and a keen memory required in order to remember their particular location amidst the vast forest surrounding a *kotan*.

A continuing list of Ainu practices with obvious benefit to the Ainu, and potential benefit to modern society could be additionally provided: traditional Ainu dwellings possess insulating properties which make them warm in winter and cool in summer; the mineral composition of certain volcanically-fed mud pools is ideal for softening tree baste to a pliable state appropriate for tearing into strips later to be wound into thread for use in weaving bags, robes, and sword straps; the gall bladder of a bear contains certain medicinal compounds which are only secreted when the bear becomes excited, which is why the Ainu would agitate a bear before sending its spirit back to the spirit world; the nutritional value of wild mountain vegetables which have been sun-dried is double to multiple times that of the same plants eaten fresh; the Ainu custom of grafting a willow shoot into the ground every time they harvested willow branches – when planted at the edge of bends in a river, they allow for the slow erosion of pools which serve as ideal homes for fish; the understanding that the productivity of the nearshore fisheries is dependent upon the health of the local forests, and so on. In sum, these customs are all based on keen observance of and ingenious adaptation to local circumstances, yet all stand threatened by the processes of modernity, especially as manifested in capitalist economies and the agendas of the nation-state.

The effect of colonialism, development, and assimilation policies

The author's professional background is in Indigenous education. By the standards of Indigenous education as practiced in locations like rural Alaska, Ainu education could be perceived to be doomed before it even has started. A long history of oppression of the Ainu people by the ethnic Japanese, driven at first by values of capitalist expansion, and then tied into the development of the nation-state, has served to thoroughly transform the natural and cultural landscape of the island of Hokkaido, disenfranchising the Ainu from their lands and resources before exploiting and depleting those very same resources, and instilling feelings of inferiority into generations of Ainu which concomitantly functioned to distance them from their identity, heritage language, and culture.

It needs to be emphasized from the outset that the colonial projects of forced disloca- tion of the Ainu (including mass relocations of Sakhalin and Kurile Ainu), annexation, and re-distribution of Ainu lands, prohibition of fishing and severe regulation of hunting all served to force the Ainu into abject poverty and in almost all cases sever them from their traditional ties to the land. Forced dislocation and mandatory congregation of separate villages completely removed the Ainu from their traditional territories and the accrued local knowledge cultivated through centuries of living there, throwing them into a completely alien natural and social environment.

Expropriation of Ainu lands in concert with unilateral top-down legislation implemented by the Japanese state transformed the Ainu into trespassers on their own territory, their former livelihood activities of fishing and hunting now deemed as "poaching," and the mere gathering of firewood punishable as a crime of theft. Bear Spirit Sending Ceremonies were likewise prohibited by law from 1952–2007, and essentially made completely unfeasible thereafter due to the stringent regulations on raising wild animals. Approximately 150 years of assimilationist policy and discriminatory discourse since the start of the Meiji Era (1868~1912) have further functioned to instill feelings of ambivalence into Ainu about their heritage and to distance individuals and communities from their roots, with the result that today only in rare cases and isolated pockets will one find individuals and communities engaged with their Ainu heritage. (Although, as noted below, impassioned cultural practitioners do exist.)

A survey by the Hokkaido Prefectural Government in 2006 put the Ainu population at approximately 24,000 (Hokkaido Prefectural Government, 2006). The author's continuing fieldwork at pan-tribal Ainu events suggests that the actual number of Ainu who normally pass as Japanese in their daily life but will attend cultural events is approximately 1,000, and out-of- these, that perhaps only 100 will identify themselves as Ainu at first meeting. Additionally, the number of individuals actually possessing specialized Ainu IK is likely to be less than 200.

Crucially, the disaggregated nature of Ainu land holdings, in other words, the absence of reservations such as can be found in other countries which mandate collective governance of traditional territories, has been a categorical factor in the weakening of Ainu IEK. Needless to say, profit-driven models of agricultural and forestry development, as well as riverine flood- prevention construction projects have seriously depleted the natural resources, such as the trees, cattails, and rushes necessary for the above-mentioned Ainu traditional cultural activities. In the case of the Ainu people, these particular historical circumstances of land and resource disenfran- chisement, and the prohibition of their traditional livelihood combined with social marginaliza- tion have led to a contemporary situation that warrants special consideration when examining systems of world IEK.

Ainu Cultural Promotion legislation and contemporary revitalization measures

In order to understand the contemporary problems and prospects of Ainu IEK, this section reviews the limited contemporary revitalization measures in place, and evaluates the establishment of the 2019 legislation which fell short of its purported goal of supporting "Ainu cultural praxis in the broad sense."

In response to demands by the Ainu Association of Hokkaido as well as international pressure, in 1997, the Japanese state established the first-ever legislation geared toward Ainu empowerment, the Ainu Cultural Promotion Act (hereafter, CPA). This Act functioned to supplement welfare measures then in place which subsidized low-income Ainu through housing and education loans. The basic policies of the Ainu cultural foundation established as a result of the legislation are:

- Promotion of comprehensive and practical research on the Ainu
- Promotion of the Ainu language
- Promotion of the Ainu culture
- Dissemination of knowledge about Ainu traditions
- Revival of Ainu traditional life style (FRPAC 2018).

Importantly for the purposes of the analysis here, the "culture" being subsidized for revival under the 1997 CPA was limited to such manifestations as material culture, songs, and dances, a point for which it has been repeatedly criticized by Ainu activists. Although the Act contains no provisions for hunting or fishing activities, one element critical to the examination in this chapter was the stipulation of measures for the eventual creation of seven "Restoration Projects for Traditional Ainu Territories" throughout the island of Hokkaido in the highly Ainu-populated regions of Shiraoi, Biratori, Shizunai, Tokachi/Obihiro, Kushiro, Asahikawa, and Sapporo. This initiative has granted Ainu access to public property for the sake of collection of cultural materials, pilot projects in cultivating *Ohyo* elm trees and cattails necessary as materials for Ainu cultural objects, and has provided the means for employment of full-time employees at least in Biratori Town, a hub of forest Ainu located along the Saru River in southern Hokkaido. These resource-use gains, however, provide no income via which Ainu economic autonomy might be furthered, and certainly do not account for the incorporation of Ainu IEK into resource management, both critical points which needed to be covered in the provisions of the legislation described below.

Next, following Japan's vote of support for the ratification of the United Nations Declaration on the Rights of Indigenous Peoples (UNDRIP) in September, 2007, both houses of the Japanese Diet passed a Resolution for the Recognition of the Ainu as Japan's Indigenous People. Based upon this, a Cabinet-level Advisory Council on Ainu Policy (hereafter, Advisory Council) was formed to investigate the possibilities for the implementation of the articles of the UNDRIP in Japan in regard to the Ainu people. In July 2009, the Council issued its Final Report (Advisory Council for Future Ainu Policy 2011), which stated that Japan's situation was unique among world nations, and called for an Ainu policy centered on rights to "culture in the broadest sense," including economic measures, but grounded in pre-existing bureaucratic and legislative frameworks and not containing substantive powers of autonomy nor political rights, until such time as the general populace of Japan was ready to lend its support to the granting of such liberties. The latter objective was to be worked toward through the creation of a National Ainu Museum and Park in Shiraoi for the purposes of (1) information dissemination and consciousness-raising about the Ainu, and (2) high-level international quality research on the Ainu culture, which would serve as a hub for information dissemination and empowerment of Ainu in the provinces (Minzoku Kyosei no Shouchou to Naru Kuukan no Gaiyou 2018). In the same year, 2009, the Culture Bearers Training Initiative at the Ainu Museum at Shiraoi, and the Urespa Project at Sapporo University, a private university, were instigated, thus putting into place the final components of the current configurations for education and cultural transmission.

As reflected in Shiraoi Museum research projects, however, as well as in the content of the curriculum of the latter two projects, the vision of the Ainu being pursued is one restricted to the realms of classical fields of study on the Ainu: ethnology, linguistics, history, archeology, and museum studies. Although fieldwork is a major component of the two latter projects, no mention of Ainu IK, or IEK, has been made in their curriculum statements; in this sense, they remain piecemeal, trapped in the fragmented categories of academic disciplinary hubris.

Although the Advisory Council originally called for a holistic policy approach which would involve the Ministry of Land, Infrastructure and Transport, the Ministry of Justice, and the Ministry of Education, their promises have yet to be actualized. Essentially, the culture of the Ainu envisioned is one firmly under the control of the Japanese state. In 2019, new legislation, the Ainu Policy Promotion Act, was implemented, but the degree of success in revitalizing Ainu IK continues to be hampered by notions therein of Ainu society as being limited merely to "culture." One issue therein which could have potentially huge ramifications for the Ainu people is the granting or denial of Ainu fishing and/or hunting rights.

Praxis of Ainu IK as included in educational and cultural transmission activities

Ainu IK exists only in a truncated fashion, removed for the most part from any traditional Ainu subsistence activities taking place on the land. To date, what activities that have tangentially involved Ainu IK have been formal educational and cultural transmission activities, isolated units or curriculums at local primary schools and high schools, the Urespa Project at Sapporo University, the Culture Bearers Training Initiative at the Ainu Museum at Shiraoi, and the Cultural Impact Environmental Assessment Survey Office in Biratori. Since current Ainu policies do not entail local Ainu resource governance, local education and training initiatives are forced to focus their fieldwork activities on local ceremonies and festivals, or alternatively on special workshops in skills being transmitted at Ainu tourist centers, such as carving.

In such a state of affairs, one initiative which shows promise in drawing nearer to Ainu IEK as it may have been practiced in the past is the Cultural Impact Environmental Assessment Survey Office (hereafter, "Assessment Office") which has been in existence since 2003 and is currently housed in Biratori Town in the museum complex on the east side of Lake Nibutani (Biratori Town 2006). Originally set up by the Hokkaido Development Bureau of the Ministry of Land, Infrastructure and Transport as an act of mitigation to offset the destruction of Ainu cultural sites by the construction of the Biratori Dam, the Office employs approximately 10 members per year and has as its mission the systematic investigation and documentation of local Ainu knowledge threatened with disappearance. Local Ainu and Wajin (ethnic Japanese) residents employed as survey personnel have been engaged in visits to local Ainu Elders to conduct interview surveys on such topics as local flora and fauna and spiritually significant natural sites forecast to be submerged. Working Groups have undertaken such activities as experimental cultivation of edible and medicinal Ainu plants, and research into traditional Ainu cuisine. Meanwhile, the Iwor Initiative housed in the same building employs three full-time workers who essentially are receiving on-the-job training in traditional Ainu skills, such as harvesting materials necessary for cultural activities, carving, and construction of traditional thatched houses. Biratori Town, in conjunction with the Hokkaido Development Bureau, has also piloted projects to re-establish natural habitats for the cultivation of cattails, an essential element in mats used for flooring and decoration in Ainu ceremonies, and a crucial item for Ainu cultural transmission. Additionally, they have hired the local Sikerpe Farm/NPO Cikornai to pilot cultivation of *Ohyo* elm trees, also an important natural material for Ainu traditional clothing. Recently, the first Ainu professional hunter in over a century, Mr. Monbetsu Atsushi,[1] was also hired as a part-time employee. Yet, the Assessment Office is limited in function mainly to recording of traditional knowledge.

Continuing barriers

In the third section of this chapter numerous examples of Ainu IK were introduced. The wealth of potential applications to a holistic pursuit of revitalization of Ainu IK based on restoration of Ainu land and resource rights, should be obvious. However, given the damage that has been wrought upon Ainu traditional lands, as well as the extremely fragmented condition of existing Ainu IK, drastic measures will be necessary.

Numerous barriers exist to Ainu praxis of IEK. Development of Hokkaido according to a capitalist expansionist model has led to a rapid population expansion, the transformation of forest to agricultural land, the decimation of the herring population, changes in the ecology due to monocultural forestry, contamination of wild mountain vegetables due to fertilizer from golf courses, and an explosion in the deer population due in part to extermination of the local wolf species. Ainu people, whose lands and resources have thus been irreparably altered, have had to bear the brunt of this on top of the social discrimination they endure.

Nowadays those cultural bearers who would still attempt to embrace their Ainu heritage have been forced to lead impoverished lifestyles. Research collaborators who have had no choice but to use their knowledge to "poach" in order to support a hand-to-mouth existence report that they have had to share midnight spaces at poaching sites with Japanese criminal gangs who have "stuffed their car trunks full" of illegally harvested salmon. On the other hand, Japanese gathering enthusiasts who seek wild edible mountain plants do not share Ainu values of conservation but rather tend to remove entire plants by their roots, thus obviating any possibility of regeneration. Ainu harvesting rights as manifested in protection of their lands and resources need to be acknowledged.

Ainu research collaborators have informed the author of certain climactic conditions in which fish and shellfish are likely to become beached or stranded in shallow waters, making for easy harvesting and the opportunity for making a considerable subsidiary income; however, sharing this knowledge freely with the public would deprive them of a much-needed source of subsistence and income. In order to capitalize on Ainu IK and make it a source of motivation for up-and-coming Ainu cultural bearers, Ainu subsistence activities based on that IK need to be legalized.

Unfortunately, the calls of Ainu Elders for a restoration of Ainu traditional values have gone unheeded, despite their importance for all contemporary residents of Japan and the islands surrounding its northern borders. For example, Hatakeyama Satoshi, a fisherman from Monbetsu on the Sea of Okhotsk, espousing concern for the environment, has expressed grave concern at the damage which weights used to hold down seine nets are doing to the local ocean floor. Hatakeyama also engaged in a protest against lack of Ainu harvesting rights in September 2019, by harvesting salmon in a river near his house without a permit required by the Hokkaido Government (Uzawa and Gayman, 2020), which would allow the Ainu to control their own economy, in line with their own IEK. Fellow Ainu Elder, poet and activist Ukaji Shizue of Urakawa, has joined Hatakeyama in calling for the restoration of harvesting rights for their Ainu brethren, the majority of whom are living hand-to-mouth, as a means to empowerment that would help the Ainu share their wisdom with the world. The restoration of Ainu fishing rights, reportedly debated by the Advisory Council, would be the bare minimum necessary to reverse the decline in Ainu knowledge. It remains to be seen whether Hatakeyama's protests will have any effect toward changing this situation.

Ultimately, ideological and bureaucratic obstacles to IK revitalization need to be reconsidered. Episodes recounted by Ainu Elders of being turned away from local primary schools upon offering their services as volunteers – in one case, the reason given was that the topic of Ainu spirituality would be a violation of the principle of separation of church and state – are not only a tremendous waste of resources, but also a violation of the Ainu right enshrined in the UNDRIP to education in one's own language and culture. Such accounts reveal a lack of systematization in the current approach – in other words, the know-how of Elders not being matched to praxis within initiatives – and furthermore highlight the serious issue of how Ainu alternative ways of knowing such as premonitions, dreams, and revelations can be taken into account in any cultural revitalization initiative undertaken through public education.

Thus, to paraphrase a question by a fellow researcher, "Can schools save Indigenous Knowledges?," the author's answer as an educational researcher would be a resounding, "No." In fact, given the experiential/situational nature in general of IEK, even promising initiatives like the "Assessment Office" will not be enough to reverse trends of loss of Ainu IK, let alone IEK.

Above all else, considering the already precarious state of Ainu IK, the above episodes should remind us of the adage that "When an Elder dies, a library burns."

Prospects for the future

This chapter has outlined the limited progress by which restorative justice in terms of Ainu IEK/IK has been developing. Colonial processes of development, assimilatory policies, and discrimination have decimated Ainu IEK, and devastated the positive identity ascription which would be essential for an embrace of Ainu culture and associated knowledge. Meanwhile, cultural revitalization projects are currently focused mainly on classical fields of research, without targeting the potential for Ainu community development through their own local knowledge. Although new legislation was implemented in 2019, it failed to include the forecast provisions for Ainu community development through collective economic rights. In this sense, the author suggested that an approach to the development of Ainu society based on a rights-based approach to harvesting which would provide employment opportunities and motivate the Ainu is necessary to overcome current exploitative conditions.

Ultimately, the case of the Ainu provides ample food for thought for considering Indigenous groups whose IK/IEK has been severely damaged over the course of history. In the end, what the author is suggesting is an approach to Ainu policy based on the possibilities of Ainu IEK as a holistic knowledge system which would deploy the still-existing knowledge of Ainu Elders and other cultural bearers in a systematic and imaginative way. On this point, two supplementary mediums, which have not yet been mentioned in this chapter, potentially containing valuable hints for the work of piecing Ainu IEK back together, which, to the author's knowledge, have not been fully mined yet, are Ainu place-names and Ainu oral literature, and in the case in which certain elements of Ainu IK have become extinct, it should be possible to rely upon the knowledge of neighboring Indigenous groups such as the Uilta and Nivkh to supplement the fragments of memory into one whole, as the Makah Indians of Washington State did in the restoration of their whaling traditions.

Curriculum innovations such as the cultural atlas approach (ANKN 2018) being employed in Alaska and elsewhere would be an ideal way to employ modern technology to fuse macro level development approaches to all levels of the curriculum, thus achieving synergy in cultural education as well. This approach would fill in the gaps in coastal Ainu knowledge as well as bridge Ainu separated by the shifting Russo-Japanese border. It is hoped that at the time of reappraisal of the Ainu Policy Promotion Act, scheduled for 2024, that international pressure will have sufficed to convince the Japanese state to include amendments which would allow the Ainu to control their own economy, in line with their own IEK.

Note

1 Note that names in this chapter have been written in the Japanese style; that is, with the family name first.

References

Advisory Council for Future Ainu Policy 2011. *Final Report of the Advisory Council for Future Ainu Policy (Provisional Translation)*. Chief Cabinet Office of Japan. Available at: www.kantei.go.jp/jp/singi/ainu/dai10/siryou1_en.pdf (accessed 19 February 2018).

ANKN (Alaska Native Knowledge Network) n.d. Oral tradition and cultural atlases. Available at: ankn.uaf.edu/NPE/oral.html (accessed 19 February 2018).

Biratori Town 2006. *Ainu Bunka Kankyo Hozen Taisaku Chousa: Sarugawa Chiiki Bunka Hyouka Gyomu Soukatsu Houkokusho* [Survey on the Measures for Preservation of Ainu Cultural Environment: Final Report of the Initiative on Cultural Evaluation of the Saru River Drainage]. Hokkaido, Japan: Biratori Town.

FRPAC (Foundation for the Research and Promotion of Ainu Culture) 2018. Basic policies. Available at: www.frpac.or.jp/web/english/project.html (accessed 19 February 2018).

Hokkaido Prefectural Government, Department of Environment and Lifestyle, Administrative Division, Ainu Affairs Office 2006. To understand the Ainu. Trans. Foundation for Research and Promotion of Ainu Culture. Hokkaido, Japan: Hokkaido Prefectural Government.

Keira, Tomoko. 2003. *Ainu no Shiki: Fuchi no Tutaeru Kokoro* [Four Seasons of the Ainu: The Heart That Fuchi Conveyed]. Tokyo: Ashiya Shoten.

Minzoku Kyosei no Shouchou to Naru Kuukan no Gaiyou [Overview of the Symbolic Space for Ethnic Harmony]. Available at: www.mlit.go.jp/common/000164473.pdf#search=%27%E6%B0%91%E6%97%8F%E5%85%B1%E7%94%9F%E3%81%AE%E8%B1%A1%E5%BE%B4%E3%81%A8%E3%81%AA%E3%82%8B%E7%A9%BA%E9%96%93%27 (accessed 19 February 2018).

Uzawa, K. and Gayman, J. 2020. The Indigenous World 2020. Available at: www.iwgia.org/en/japan/3603-iw-2020-japan.html

12

ENGAGING WITH INDIGENOUS ENVIRONMENTAL KNOWLEDGE IN THE NORTH AMERICAN ARCTIC

Moving from documentation to decisions in environmental governance

Henry P. Huntington

Introduction

Indigenous Environmental Knowledge (IEK) has been documented all around the world (e.g., Johannes 1981; Berkes 1999, other chapters in this volume). We can debate what constitute the best methods in different settings and for different purposes, we can discuss what success means in these efforts, and we can certainly improve our ability to engage with IEK in ways that make sense to IEK holders as well as to academic researchers. Nonetheless, the big challenge now, at least in the Arctic, is to apply IEK to issues beyond the traditional areas and activities of the communities where IEK is generated. Of course, IEK remains relevant to traditional activities and the lives of its holders. In that sense, the application of IEK happens every day, as it has through countless generations (e.g., Nageak 1991). But as newcomers have arrived in the North American Arctic, as resources are developed and travel routes established, as policies and decisions are made that affect the lives of those who live in the Arctic, Indigenous leaders have called for greater attention to IEK and greater use of IEK throughout decision-making and policy-setting processes (e.g., Brooke 1993). Here is where the engagement of outsiders with IEK has largely stalled (e.g., Huntington 2018).

Let us consider a few steps in the policy-making and decision-making process, and how IEK is or can be included at each. It is worth noting that one can ask a similar question about scientific knowledge, and how it is applicable at different stages of making decisions or policies. Not all questions are scientific ones, so even a science-based policy will likely be shaped by other factors as well. An IEK-based policy might similarly take into account other information and perspectives. As one example, whether to open an area to oil drilling is not a scientific question.

DOI: 10.4324/9781315270845-14

Scientific information can help evaluate the likely effects of drilling, the probability of a spill or well blowout, and the effectiveness of actions to prevent or mitigate problems. None of this information answers the question of whether the expected benefits are worth the effects and risks. IEK, which can incorporate judgments and values that are typically excluded from scientific studies and analyses, may speak more directly to the question of whether to drill, but is still rarely the sole source of relevant information, so may play a greater or lesser role at each stage, depending on the circumstances of the specific case.

The engagement of IEK in the steps toward decisions

Observations

IEK is based on keen observations of the world around us. Visitors to the Arctic have long recognized that they have much to learn from local expertise (e.g., Amundsen 1908), and the acknowledged use of IEK in Arctic scientific research has a long history, including recognition of local experts as co-authors (e.g., Irving and Paneak 1954). Much work has been done in the past few decades to document IEK on its own (e.g., Ferguson and Messier 1997) or to include it in scientific studies (e.g., Carmack and Macdonald 2008). This is no doubt a valuable activity and an important recognition of what IEK can offer that is inaccessible by any other means. IEK holders may be able to provide living memory going back many decades, through all seasons of the year, and across large areas of the Arctic. Collective knowledge can extend back much further. Few scientific data sets have such time depth and extensive seasonal and spatial coverage. For these reasons alone, it is worth engaging with IEK. Documenting observations and associated knowledge has been done, as noted earlier, all around the world, and is relatively straightforward. Yet there is much more to be gained, for IEK holders and for others.

Interpretation

Observations mean relatively little without interpretation. Interpretation speaks to patterns and perspective. For example, "We saw bowhead whales here in July" is an observation. Interpretation adds further insight, such as "It is unusual to see bowhead whales here at that time of year," or "The whales that come early are setting a path for the others that follow." Some research has focused on the ways IEK holders interpret their observations. For example, Weatherhead et al. (2010) considered changes in weather patterns in Nunavut, leading to new insights into changes in the persistence of weather conditions. Such insights would not have been made without recognizing the changes in patterns reported by Inuit, which constituted an interpretation of their own observations and knowledge. While making and remembering observations are hardly automatic, interpreting those observations is a greater intellectual contribution, drawing not just on the sensory capabilities of IEK holders but also on their ability to think critically about what they see, hear, feel, taste, and smell. In practice, one way to engage IEK holders in interpretation is by fostering discussion during interviews, participant observations, or similar methods while documenting IEK (e.g., Huntington et al. 2011). This requires that the interviewer knows enough about the subject to recognize salient points and to keep the discussion going on a topic of particular interest or relevance, which is not necessarily the case when simply documenting observations. Interpretation also requires that the IEK holders are engaged in ways that stimulate thinking beyond a recitation of what one has seen.

Analysis

Beyond interpretation comes analysis, which, among other things, involves the evaluation of significance. Hunters recognize the importance of the first animals in many species migrations, as the other animals are likely to follow the leaders (Huntington et al. 2017). In many cases, hunters avoid disturbing the leading animals lest they lead the other animals farther from the hunters. After the first animals have passed, the later ones are more likely to stick to the established route, even if hunters pursue them. Recognizing this behavior is essential to the ways hunters manage their own activities. In managing other activities, for example, offshore oil and gas activity or commercial shipping, the question is how IEK of this kind can and should be applied. How important is it to let the first animals pass undisturbed? What happens if the leaders change course due to human activity? What are the alternatives?

In this step, critical thinking needs to be joined by judgment, to help determine what is worth considering further. Salomon et al. (2007), for example, explored the reasons behind localized declines in an intertidal invertebrate in southern Alaska, engaging IEK holders in analyzing the respective roles of the 1964 earthquake, the arrival of electricity in local villages in the early 1970s (allowing the purchase of freezers and thus the long-term storage of surplus harvests), the return of sea otters following near-extirpation in the nineteenth century, the *Exxon Valdez* oil spill, climate change, and more recent changes in human behavior. These discussions led to further consideration of local management options, based on cultural practices and customs rather than state-run resource management systems (Salomon et al. 2011). Analysis involves yet deeper engagement by IEK holders, in terms both of time and of participation, than do observations and interpretation. IEK holders, of course, routinely analyze their observations and interpretations for their own purposes. Engaging with scientists and resource managers, however, introduces a number of complicating factors. Language barriers can inhibit or prohibit meaningful interaction and exchange of ideas. Power imbalances, including who sets the agenda, scope, and rules governing a meeting or activity can also operate to the detriment of IEK holders (e.g., Nadasdy 2003; Ellis 2005). Additionally, few IEK holders can devote all of their time to preparing for and engaging with scientists and managers, creating a further imbalance of effort (e.g., Huntington et al. 2012). There is also the logistical hurdle that analyses are often done in distant urban, academic, or government centers, rather than in the communities of IEK holders, creating a lopsided travel burden and placing IEK holders in what may be unfamiliar surroundings. Explaining the significance of the whales leading the bowhead migration is so much harder when separated from the geography and culture in which those observations are made and the resulting practices carried out. The challenge is greater still if those to whom the explanation is being given have little first-hand knowledge of the place, the animals, and the people being described.

Assessment

A further step towards decision-making is that of assessment, or consideration of alternatives. In our bowhead whale example, the question might be whether offshore industrial activity would cause greater disturbance to whales and whalers in July or in September. Such questions usually involve speculation, not to mention limited choices, since there is rarely evidence for all of the alternatives and those making the assessment must use their best judgment of what is likely to happen in new circumstances. Again, IEK holders assess their choices regularly, so have considerable experience with the basic idea. The challenge is applying that experience in a very different setting, with many of the same obstacles as are found in analysis. The importance of engaging IEK holders throughout the decision process is recognized by government agencies (e.g., BOEM 2012) and others, but recognition and action are not the same. Among the best examples of involving IEK holders in assessment are the marine mammal and other co-management organizations that have been created in Alaska and Canada, for example, under the auspices of the Alaska Beluga Whale Committee (Fernandez-Gimenez 2006) and the Inuvialuit Final Agreement in the Northwest Territories (e.g., Smith 2001). In these cases, IEK holders are represented on the organizations in question, and staff support is provided to IEK holders as well as government representatives. These organizations require substantial financial and other support to function as intended, which may limit the willingness of others to follow this example fully. In addition, the paradigm in which even the co-management organizations tend to operate is not that of IEK and its holders, but a bureaucratic and scientific one (Nadasdy 2003). Often, explicitly and intentionally or otherwise, this dynamic leads to the presumption that scientific knowledge and associated processes of analysis and interpretation are correct unless proven otherwise (e.g., Scientific and Statistical Committee 2010), and IEK is treated as a supplement to be applied when convenient.

Decisions

Ultimately, one point of all the steps described so far is to make a decision. For example, should certain waters be closed at certain times to protect bowhead whales and those who hunt them? The authority to make such decisions often lies with government agencies. While those agencies may be open to consultation, ideas, and participation of others, ultimately they retain the responsibility for the decision and its consequences, and so cannot let others make the formal decisions without changes to the laws that govern resource management. That said, the effective involvement of IEK holders in the steps leading to a decision, and a clear description of how IEK has been applied to the decision in question can go a long way toward understanding and acceptance of the decision by its opponents, even if they cannot go so far as to support the outcome they deem unfavorable. In the case of co-management bodies, IEK holders may be involved in making recommendations to those with the ultimate decision-making power, but there is still typically a final stage of review in which those recommendations are evaluated against criteria such as best practices, the state of scientific knowledge, and accepted principles for management. How often these criteria are invoked to reject a recommendation varies from case to case, but the terms are aligned with scientific-bureaucratic paradigms rather than with the cultures in which IEK originates. It is thus not surprising that IEK holders are rarely involved in decision-making itself. Bonesteel (2006) noted that greater involvement of IEK in decision-making will require greater numbers of IEK holders in positions of decision-making authority. Even in Greenland, however, where self-rule political authority lies with Inuit, the principle of involving IEK holders remains as much as aspiration as a practice as environmental impact assessment and other aspects of resource management continue to develop there (Hansen and Kørnøv 2010).

Discussion

There are many reasons to engage with IEK and its holders throughout the steps that lead from observations to decisions. As one progresses through these steps, however, the engagement requires increasing levels of intellectual involvement and exchange, at the same time that the logistical and paradigmatic obstacles grow larger. A superficial attempt to pay heed to the notion of IEK is relatively simple. For example, one need only hold a meeting or workshop in a community where IEK holders live, or invite IEK holders to a similar event elsewhere. Documenting IEK observations properly requires the use of rigorous methods and processes for review of draft materials by the IEK holders and their peers to ensure accuracy. Interpretation involves the exchange of ideas, and thus the involvement of subject matter experts in addition to those skilled in documenting IEK. Analysis and assessment include yet more interaction, which can only be done effectively if IEK holders are engaged as equal partners rather than as providers of colorful but ultimately inessential information often dismissed as "anecdotal," or are not involved in the critical thinking required in these steps. Decisions typically are reserved for those to whom such power has been given by law, which limits the degree to which IEK holders can be involved, unless the decision-makers are also IEK holders. Nonetheless, the explicit and acknowledged use of IEK throughout the decision process, if done in the right way, can help increase the legitimacy and acceptance of decisions. By contrast, the exclusion of IEK and its holders can sour even decisions that otherwise might have been popular. In 2015, President Obama withdrew from oil and gas leasing a marine mammal migration corridor along the coast of the Chukchi Sea in northwestern Alaska. Leaders of the predominantly Iñupiaq region had recommended such a move to protect their hunting practices. Nonetheless, the announcement of the withdrawal was made without local consultation, leading to some resentment of what should have generated considerable local support.

A quarter of a century ago, IEK was often seen in Arctic North America as an ethnographic curiosity or a useful contribution to operations but not as an intellectual contribution. Attitudes have changed, thanks to many efforts to document IEK and the openness of many scientists to learning from IEK holders. Taking IEK seriously in research has also meant the commitment of resources for doing so. Greater engagement with IEK and its holders will take more resources, and advancing along the path to decisions takes greater commitment at each step. There are thus financial and practical barriers to further progress, in addition to the need to develop new modes of interaction and new paradigms for incorporating knowledge and its holders from different traditions and cultures than the standard scientific one upon which modern, state-run environmental management is based.

To continue the progress that has been made, leadership is needed in three areas. First, IEK holders and Indigenous leaders should assess the state of IEK engagement today and its progress in recent years, and determine what remains to be done. While the politicization of knowledge can lead to unnecessary divides, ignoring the political dimensions of knowledge ignores the realities that IEK holders live with when attempting to engage in decision-making processes. Being clear about their aspirations can help IEK holders and their allies keep a focus on their end goals (e.g., Gadamus et al. 2015), rather than achieving only partial successes.

Second, all of those involved in the decision-making process should consider what is required in terms of resources and infrastructure to achieve full engagement of IEK and its holders in all steps. Some IEK holders report "research fatigue," from being asked to participate in too many projects, answering too many questions, and too often answering the same questions over and over. At the same time, when IEK holders are compensated for their participation, it is often only for the time devoted to an interview or meeting or workshop, rather than for the time spent acquiring knowledge. Yet ongoing participation, especially in the more intellectually demanding and time-consuming steps beyond observations, requires greater support lest IEK holders lose other employment opportunities or community responsibilities, including pursuing the very species in question. Meeting times, locations, and preparation such as background reading or preliminary analyses should also take into account the ability of IEK holders to participate fully, on their own terms, and in settings in which they are comfortable.

Third, IEK leaders and scholars should consider carefully the ways in which IEK can and cannot contribute to the kinds of decisions typical of environmental governance. IEK encompasses far more than simple observations and understanding of the world around us. It includes ethics and judgments about human behavior and our interactions with the natural world. As such, what may be a seamless continuum of knowledge to IEK holders may connect with the decision-making process in very different ways at different times and steps (e.g., Usher 2000). Furthermore, the purposes for which IEK is generated are not always perfectly aligned with the needs of environmental governance decisions. The same can be said for scientific knowledge, much of which will be irrelevant to any given decision. A better understanding of how IEK can effectively engage with the decision process may include adjusting the decision process itself to better incorporate IEK and better involve its holders on their terms. The Alaska Beluga Whale Committee, a marine mammal co-management organization involving hunters, scientists, and managers, developed regional management plans to reflect specific local hunting practices. It also recognized some reluctance to make rules formal and fixed, at the same time taking a careful and deliberate approach to documenting IEK and using that information in its decisions (Fernandez-Gimenez et al. 2006, 2008). Such practices can help achieve greater success and less frustration than simply hoping that available IEK will be relevant or that available IEK holders will be able to contribute to whatever topic is currently under discussion.

In conclusion, much progress has been made in the North American Arctic in engaging IEK in research and, to some degree, decision-making with regard to natural resource management and other forms of environmental governance. The full promise of doing so has yet to be realized, and much remains to be done. A key piece of further progress will be the assumption by IEK holders of greater leadership roles in shaping what comes next. In practice, this may involve significant organizational and political reform as well as investment, However, like power-sharing in governance, or peer review in science, it is necessary for effective decision-making and management.

References

Amundsen, R. E. G. 1908. *The Northwest Passage*. London: Archibald Constable.

Berkes, F. 1999. *Sacred Ecology: Traditional Ecological Knowledge and Resource Management*. Philadelphia, PA: Taylor & Francis.

BOEM 2012. Special issue on traditional knowledge. *BOEM Ocean Science* 9(2): 1–16.

Bonesteel, S. 2006. Use of traditional Inuit culture in the policies and organization of the Government of Nunavut. In B. Collignon and M. Therrien (Eds.), *Orality in the 21st Century: Inuit Discourse and Practices. Proceedings of the 15th Inuit Studies Conference*. Paris: CERLOM. Available at: www.inuitoralityconference.com/art/Bonesteel.pdf (accessed 19 July 2020).

Brooke, L. F. 1993. *The Participation of Indigenous Peoples and the Application of Their Environmental and Ecological Knowledge in the Arctic Environmental Protection Strategy*, vol. 1. Ottawa: Inuit Circumpolar Conference.

Carmack, E., and R. Macdonald. 2008. Water and ice-related phenomena in the coastal region of the Beaufort Sea: Some parallels between Native experience and western science. *Arctic* 61(3): 265–280.

Ellis, S. C. 2005. Meaningful consideration? A review of traditional knowledge in environmental decision making. *Arctic* 58(1): 66–77.

Ferguson, M. A. D. and F. Messier. 1997. Collection and analysis of traditional ecological knowledge about a population of arctic tundra caribou. *Arctic* 50(1): 17–28.

Fernandez-Gimenez, M. E., J. U. Hays Jr., H. P. Huntington, R. Andrew, and W. Goodwin. 2008. Ambivalence toward formalizing customary resource management norms among Alaska Native beluga whale hunters and Tohono O'odham livestock owners. *Human Organization* 67(2): 137–150.

Fernandez-Gimenez, M. E., H. P. Huntington, and K. J. Frost. 2006. Integration or co-optation? Traditional knowledge and science in the Alaska Beluga Whale Committee. *Environmental Conservation* 33(4): 306–315.

Gadamus, L., J. Raymond-Yakoubian, R. Ashenfelter, A. Ahmasuk, V. Metcalf, and G. Noongwook. 2015. Building an indigenous evidence-base for tribally-led habitat conservation policies. *Journal of Marine Policy* 62: 116–124. http://dx.doi.org/10.1016/j.marpol.2015.09.008.

Hansen, A. M. and L. Kørnøv. 2010. A value-rational view of impact assessment of mega industry in a Greenland planning and policy context. *Impact Assessment and Project Appraisal* 28(2): 135–145.

Huntington, H. P. 2018. Traditional knowledge and resource development. In C. Southcott, F. Abele, D. Natcher, and B. Parlee (Eds.), *Resources and Sustainable Development in the Arctic.* London: Routledge, pp. 234–250.

Huntington, H. P., S. Gearheard, A. Mahoney, and A. K. Salomon. 2011. Integrating traditional and scientific knowledge through collaborative natural science field research: identifying elements for success. *Arctic* 64(4): 437–445.

Huntington, H. P., A. Lynge, J. Stotts, A. Hartsig, L. Porta, and C. Debicki. 2012. Less ice, more talk: The benefits and burdens for Arctic communities of consultations concerning development activities. *Carbon and Climate Law Review* 1: 33–46.

Huntington, H. P., L. T. Quakenbush, and M. Nelson. 2017. Evaluating the effects of climate change on Indigenous marine mammal hunting in northern and western Alaska using traditional knowledge. *Frontiers in Marine Science* 4: 319. doi:10.3389/fmars.2017.00319.

Irving, L. and S. Paneak. 1954. Biological reconnaissance along the Ahlasuruk River east of Howard Pass, Brooks Range, Alaska. *Journal of the Washington Academy of Sciences* 44(7): 201–211.

Johannes, R. E. 1981. *Words from the Lagoon: Fishing and Marine Lore in the Palau District of Micronesia.* Berkeley, CA: University of California Press.

Nageak, J. M. 1991. *An Unusual Whaling Season: An Interview with Waldo Bodfish, Sr.* Fairbanks, AK: University of Alaska, pp. 237–246.

Nadasdy, P. 2003. *Hunters and Bureaucrats: Power, Knowledge, and Aboriginal-State Relations in the Southwest Yuko.* Vancouver, BC: UBC Press.

Salomon, A., H. P. Huntington, and N. Tanape Sr. 2011. *Imam Cimiucia: Our Changing Sea.* Fairbanks, AK: Alaska Sea Grant.

Salomon, A. K., N. M. Tanape Sr., and H. P. Huntington. 2007. Serial depletion of marine invertebrates leads to the decline of a strongly interacting grazer. *Ecological Applications* 17(6): 1752–1770.

Scientific and Statistical Committee 2010. Final report to the North Pacific Fisheries Management Council, February 8–10. Available at: www.npfmc.org/wp-content/PDFdocuments/minutes/SSC210.pdf (accessed 19 July 2020).

Smith, D. 2001. Co-management in the Inuvialuit Settlement Region. In CAFF (Ed.), *Arctic Flora and Fauna.* Helsinki: Edita, pp. 64–65.

Usher, P. J. 2000. Traditional ecological knowledge in environmental assessment and management. *Arctic* 53(2): 183–193.

Weatherhead, E., S. Gearheard, and R. G. Barry. 2010. Changes in weather persistence: Insight from Inuit knowledge. *Global Environmental Change* 20(3): 523–528.

13

TAIGA FOREST REINDEER HERDERS AND HUNTERS, SUBSISTENCE, STEWARDSHIP

Nadezhda Mamontova

Introduction

This chapter provides an account of the Ewenki people's (also Evenki, Evenks) hunting, reindeer herding, and fishing terminologies. The Ewenki are known as the most dispersed Indigenous minority in Siberia with the total population of 38,396. They occupy the huge territory between the Yenisei River in the west and the Okhotsk Sea in the east, with a few communities living outside this area, namely, on the western bank of the Yenisei River (Sym and Turukhansk Ewenki) and on Sakhalin Island (Figure 13.1). Hunting, fishing, and gathering have long been at the heart of the Ewenki subsistence livelihoods and ontologies, but it is through their distinct type of reindeer herding, known as the Ewenki or forest type, that the people have managed to spread their culture and language within such a wide and diverse geographic area, flexibly adjusting to and affecting new landscapes and environmental surroundings. Ewenki traditional ecological knowledge (TEK) is closely related to and reflected in their native language which belongs to the Manchu-Tungusic branch of Altaic language family and is nowadays seriously in danger of disappearing. The importance of language to TEK stems from the fact that every language is a culturally embedded system which is linked to people's experience and practices. Hence, different languages conceptualize the surroundings in different ways (see Mark et al. 2011). This is particularly evident in the case of focal or specialized terminology which is related to the local knowledge of land, vegetation, and animals, their traditional means of classification and management, as well as cultural values and beliefs (see Waddy 1988; Berlin 1992; Ellen 1993; Berkes et al. 2000; Douglas and Turner 2005; Messineo and Cúneo 2011). Due to the vast expanse of the Ewenki's habitat, this part of the vocabulary is very rich and diverse. Although it is not fully comprehensive, this account provides insight into the most commonly used specialized words and categories which can be found in almost all Ewenki dialects as they constitute the backbone of the Ewenki subsistence economy and way of life.

Background: the Ewenki people

Most Ewenki live in the mountainous taiga. However, some also reside in the mixed forest-tundra zone of the Arctic, the steppes in Southern Siberia, and the coasts in the Russian Far East. Hence, they occupy rather diverse environments in terms of ecosystems. Historically,

DOI: 10.4324/9781315270845-15

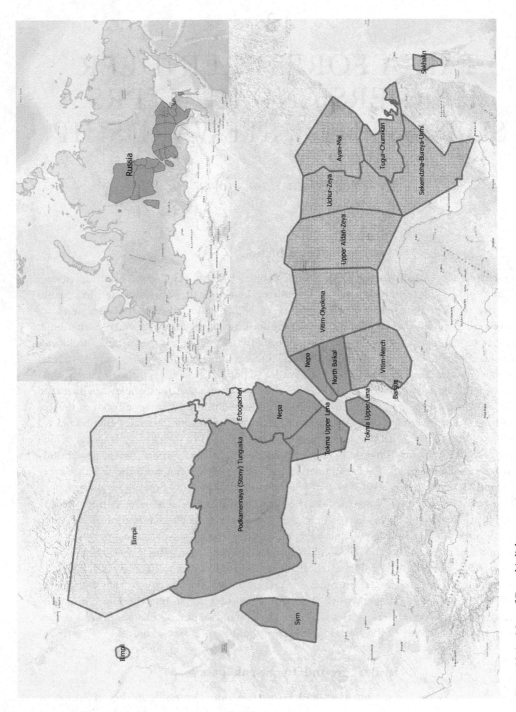

Figure 13.1 Map of Ewenki dialects

the Ewenki subsistence economy was bound to nomadic reindeer herding and hunting, and later linked to the emerging non-herding economy through contacts with traders, miners, and cattlebreeders (see Safonova and Sántha 2016). It is probable that the ancestors of the Ewenki recently came to inhabit Eastern and Central Siberia from the Central Amur (see Pevnov 2012). By the seventeenth century or even earlier, they crossed the Yenisei River and assimilated some local Indigenous groups living in Central and Western Siberia, who adopted the Ewenki language (see Tugolukov 1985; Janhunen 2012). The huge territory occupied by the Ewenki makes their language among the most widely spoken in Eurasia and an interesting case for research on IEK. Although nowadays only a comparatively small percentage of the Ewenki are still involved in the reindeer economy, either in Soviet-style brigades or with private herds, they still spend a lot of time in the forest, hunting, gathering, and visiting ritual, habitation, and other sites. These activities are considered both by the Indigenous people and researchers as culturally significant, helping the Ewenki to preserve mobility in the form of 'walking in the taiga' and conserving their traditional environmental knowledge and livelihoods (see Anderson 2000; Safonova and Sántha 2010, 2011; Davydov 2011; Brandišauskas 2017). The continuation of living on the land has meant that terms related to subsistence and reindeer livelihoods persist despite language shift and other cultural changes. Such retained terms always convey a concept or distinction that is not well defined in the dominant Russian language. In the following sections I will consider some of these terms in the context of IEK.

Materials and methods

Basic Ewenki subsistence terminology has been compiled by Anna Myreeva (2001), a scholar of IEK and an native Ewenki speaker. I build on this work, elaborating especially on terminology related to hunting, reindeer herding, and fishing. IEK terminology is discussed in its cultural context, using ethnographic accounts and my own field data collected over the course of several expeditions to the Ewenki District of Krasnoyarsk Krai, carried out between 2008 and 2014 and among the Tugur-Chumikan Ewenki in Khabarovsk Krai in 2017. Most of the native speakers I interviewed belonged to the Ilimpii speech community of the Northern set of dialects. The residents living in the villages located on the Lower Tunguska River and in the northern part of the district were mainly involved in hunting for sable (*Martes zibellina*) and wild reindeer, gathering, fishing, and small-scale farming. In some villages, for example, Surinda and Econda, reindeer herding is still vital. Language documentation included the pronunciation of dictionary entries which were divided into a number of categories, including 'Flora and Fauna' and 'Traditional Economy'.[1] This data helped to identify which parts of the subsistence vocabulary were well conserved and which were not. Additionally, I recorded and analysed narratives in Ewenki and then translated them into Russian and English. These served as an additional source of IEK.

Ewenki subsistence terminology

Hunting

The huge and varied territory occupied by the Ewenki over time has led to diversification in subsistence portfolios (see Sirina 2012). Nevertheless, it seems that among all traditional activities, hunting has long played a leading role in the people's way of life and their perception of space through the unique practices of movement and dwelling in the mountainous taiga. Traditionally, Ewenki hunted for wild reindeer, moose, goats, cabarga, bears, and some forest birds (the choice of species depending on the local environment). While hunting, some Ewenki groups rode reindeer as animals to travel long distances, a unique practice among the Siberian peoples. This advanced mode of herding enabled people to establish wide social contacts and exchange networks with neighboring groups. However, not all Ewenki communities were involved in reindeer herding, and some shifted their mode of subsistence from hunting to reindeer herding and back, or combined these activities interchangeably (Takakura 2012: 43; cf. also Kolås and Xie 2015). When there were no reindeer due to epidemics, the Ewenki continued hunting on foot or fishing. This is also the reason why some Ewenki groups are known in the literature as 'wandering Tungus' (a pre-revolutionary term designating the Ewenki, Ewen, and Negidal).

Extensive reindeer herding was introduced only in Soviet times when it became collectivized and industrialized, resulting in the over-exploitation of reindeer pastures. In the post-Soviet 1990s, many Ewenki people lost their reindeer and the state infrastructure that supported the system in Siberia basically collapsed. In the wake of this, many Ewenki had no choice but to revitalize their traditional ways of herding and living off the land. They relied upon subsistence knowledge and resource tracking to live off the taiga. Stories about hunters who would walk or ski the whole way through the taiga searching for a remote settlement or a winter dwelling (*zimovye*) are quite popular among the people and celebrate the importance of navigation skills, physical strength, and traditional knowledge.

For the Ewenki, moving or walking is an important practice in its own right, which supports the resilience and continuity of Ewenki culture even when nomadic practices are restricted (see Safonova and Sántha 2010; Lavrillier 2011; Brandišauskas 2017: 126–144). To some extent, this ability also helped the people survive the collapse of the planned economy and dissolution of the Soviet *kolkhoz* system, of collectives, beginning in the 1990s. A hunting family on average used to cover about 300 km on foot and 500 km by reindeer per year (Vasilevich 1963: 306). Nowadays in some regions the people have to cover up to 1,500–2,000 km annually to combine reindeer herding and hunting (Gabyshev and Lavrillier 2017: 42). Contemporary forms of mobility include a variety of vehicles and patterns of movement which continue to exist along with traditional means of transportation (see Mertens 2016).

The Ewenki language possesses sophisticated verbs describing not only different types of movement but also the means by which this movement is performed, for example, *agīkta-mī* 'to wander searching for animal's footprints on the snow (without skis)', *d'avrā-mī* 'to go by boat', *darivul-mī* 'to step aside from the main path for hunting', *hōnta-mī* 'to go without a path', *nulğī-mī* 'to be a nomad', *ūd'a-mī* 'to follow footprints', *mūlēsin-mī* 'to go and bring some water', and so forth (Vasilevich 1940: 172). As one of my Ewenki interlocutors explained, this language appears to be convenient when it is needed to express *what* you are going to do and *how* without spending too much time on formulating the sentence. The precision and economy in Ewenki language are helpful during hunting when people are not supposed to say much aloud for practical and spiritual reasons (the animals will sense you). This rule is still characterized by the circumlocution, the avoidance of particular words referring to hunting animals, tools, and plans.

Movement also requires a good understanding of the local environment and landscape. For example, Lenore Grenoble (2014) highlights the importance of snow for the Ewenki in relation to mobility and subsistence. There are a number of terms to refer to different conditions of snow according to its hardness, thickness, depth (for ice and snow typology in Ewenki, see also Gabyshev and Lavrillier 2017: 243–369). The local knowledge of vegetation and its practical use is also of high importance. The list provided by Myreeva includes more than 300 dialect terms for different types of forests, trees, bushes, plants, grass, flowers, berries, and moss.

We documented approximately 60 terms related to vegetation. This part of vocabulary is significantly diverse across the dialects. In contrast, it seems that there are not so many terms for forest variations in Ewenki. Forest *mōha/mōsa/mōsha, agī,* or *dȳle* 'trees on the slope' is itself a rather ambiguous category. For example, the Ewenki living in the Ewenki District tend to call the forest *dūnne/dūnde* 'earth', 'land', 'space' or even *mutkī dūnnenit* 'our land', in many cases, making no distinction between such terms as forest, taiga, and mountainous taiga. They may even refer to the latter by using the word *ure* 'mountain'. At the same time, there is a tendency toward over-differentiation (in relation to the Russian geographical terminology) to meaningfully reference distinct parts of the forest that are relevant to navigation, mobility, and subsistence. Thus, there are terms describing a dead forest, a forest hit by a hurricane, open parts of the forest covered with different types of vegetation, marshy areas in the forest, and so forth. On the whole, the terminology referring to different parts of the forest, as well as different kinds of trees, their parts, and characteristics, appears to be more developed and highly precise than that for forest types. Of them, the larch tree (*Larix sibirica*) has the largest diversity of terms (Myreeva 2001: 73–74). I have confirmed that some of them, if not all, are still used (*hēkī, epkere, irēkte*). The language also differentiates a number of types of trees according to their shapes and conditions rather than to phylogenetic characteristics as in: *nengīrī* ('inclined tree'), *nelge* ('tree removed from the ground with its roots'), *hēgī/hēgipchā* ('tree struck by lightning'), and so forth (ibid.: 72–73). These categories are important not only for reckoning places but also in assessing their affordances and utility.

Among the most elaborate IEK categories in Ewenki are thickets and burnt forests. Ewenki has a number of terms designating thickets, among which the term *sigikāg/higikāg/heikāg* can be found in most of the dialects. Some Western Ewenki also use the term *hupire/supire* 'thicket'. Ewenki hunters collect rotten wood (*kongnomo mōl* 'black wood') there as it does not crackle and, hence, does not scare animals away. This category is also largely bound to mythology and is associated with spirits. Thus, another word for thicket is *hārgil*, which has a meaning of an evil spirit or the master of the lower world in Ewenki cosmology. In this regard, Alexandra Lavrillier (2011), who works among the Amur Ewenki, highlights the distinction between *bejechi* 'domestic space' and *kanula* 'wild space' that helps people to arrange their nomadic practices and build up relationships with the spirits. Local Ewenki believe that wild animals come from the wild space of the forest where the master of the taiga resides. Some Ewenki I interviewed represent this master in a form of a woman with long hair and in traditional clothes who could come to visit a hunter in his dream. This is a rather typical representation that can be found in many ethnographic accounts. Yet the distinction between *bejechi* and *kanula* fails to take into account the multiple nature and relationality of each *space* in terms of its ethno-ecological and ontological context. Most importantly, the notion of space in its relation to the taiga needs further clarification.

Finally, some places in the taiga can be considered *bugadyl* 'sacred places' or the places inhabited by spirits, which are either not supposed to be visited by people or require a special attitude toward them (Ssorin-Chaikov 2001; Sirina 2008). In the past, *bugady* was a name of a master of a certain hunting area (P-T) or a skin of a sacred animal (Vasilevich 1958: 63; see Table 13.4). Still, the names for *bugady*-places can vary in Ewenki communities. For example, along with *bugadyl* the Ilimpii Ewenki use the term *hulgakīt*, 'sacred place where one should give offerings'. The network of different places and the awareness of patches in the taiga determine the behaviour of a hunter and his patterns of movement.

Burnt places in the forest are another significant and unexplored category in Ewenki reflecting historical-ecological disturbance. While forest fires are considered a problem by modern forest managers (see Lavrillier 2013), burnt places are used for hunting purposes, as fresh bushes, grass, and berries attract moose and reindeer (Vasilevich 1969: 50–51). Some Ewenki have long cultivated these kinds of places through controlled burning. For example, Donatas Brandišauskas reports that the Ewenki-Orochen burn last year's grass to create open fields with fresh grass (2017: 198). Hence, there is a distinction between natural burnt sites and anthropogenic ones. There are a few terms designating natural burnt places in Ewenki, including: *n'adi, nēgdy, nirgal, churgi, jaktan, gōndakta* (Myreeva 2001: 81). Some of them are dialect variations, but others can be found in the same dialect often as distinct terms. Not all distinctions between terms for burnt places have been clarified, but one marked characteristic is the age of the place, i.e. whether the burn is new or old. The other is the presence/absence of fresh grass and bushes (see Table 13.1). The importance of such places is highlighted in the Ewenki ritual festivities called *ikēnipkē* (see below), a set of rites devoted to 'catching' hunting luck. During the festivities, a shaman walks through the area in the forest destroyed by fire, which is called in Ewenki *agijand'akīch* 'burnt path' (Vasilevich 1957: 159). Additionally, in order to obtain hunting luck, the hunters used to create a symbolic model of the landscape which would be attractive to animals. Such models invariably included burnt areas (Anisimov 1949 in Brandišauskas 2017: 85) and which the Ewenki used with symbolic actions to attract prey and thereby increased the hunters' 'luck' (*kute*).

Wayfinding and orientation skills include not only the detailed knowledge of the landscape and vegetation, but also the ability to recognize a variety of signs, both natural and manmade. The latter include *ilkēr* 'marks' or notches on the trees, which served as a means of communication between hunters. Twigs, arrows, or moss placed in the notches were called *sāmelkī*. Pictures the hunters left on the trees were called *on'ovūn*. The vocabulary designating these marks used

Table 13.1 The names for burnt places in Ewenki dialects

Name in Ewenki	English gloss	Dialect
N'adi	Burnt place in the forest with burnt trees	E
Nēgdy	Burnt place	M, Tt, Sak, Urm
Nirgal	Burnt place/forest	Uchr
Churgi	Burnt place	
Jaktan	Old burnt place	P-T, N, E, I
Gōndakta	Burnt place in the forest with fresh grass and bushes	P-T
Bugar	New burnt place in the forest without fresh grass	P-T, N, E, S, Z, Ald
	Old burnt place	N-B
Degdek, degdenne, degdende, degderne	Burnt place	P-T, N, E, Z, Ald, I, Tng, Urm, Uchr
Kurung	Old burn place with fresh grass and bushes	Ald, Uchr, Urm
Hulākāg	Burnt place with fresh aspen-trees	Tmt

Notes:

Ald Aldan dialect; E Erbogachen dialect; I Ilimpii dialect ; M Mai dialect ; N-B Northern Baikal dialect; P-T Podkamenno-Tungus dialect; S Sym dialect; Sak Sakhalin dialect; TmT Tommot dialect; Tng Tungir dialect of the Tungir-Olekma group of dialects; Tt Totta dialect; Uchr Uchur-Zey dialect; Urm Urmi dialect of the Burein-Urmi-Amgun group of dialects

Source: Adapted from Vasilevich (1958) and Myreeva (2001).

to be very elaborate and consisted of both nouns and verbs describing what kind of action was applied (see Vasilevich 1963: 309). Some marking techniques are still in use among hunters today (Simonova 2015). As for natural signs, the sun, the stars, and the moon have been widely recognized as the fundamental axes of spatial and cosmological orientation. According to the position of Ursa Minor (*Heglen* in Ewenki), Ewenki hunters identify time during the polar night. *Heglen* is a name of a mythological Ewenki hunter who was chasing a moose across the sky. It is believed that the track of his skis is still visible there (the Milky Way). The following narrative recorded in the Ewenki District in 2014 describes this characteristic of identifying time still being used:

> *Medveditsa nadan ohikākte. Tyrgāldakin, girkud'avkī tyrgañ d'apkadūn. Tarit-ta hādengāhun, tyrgaldȳvan, gūnne. Tar ugiskī ōvkī, irgindyn nōdāvkī, tyrgāldakin. Amaskī-ka chāila āsir, et hādere. Tar potom ahikākte, tyrgañ d'apkadūn hegdymēmē ohikākte bivkī, umukēn. Tar omolgin. Gūndengkītyn, "Tyrgaldān, tyrgaldān, ohikākte emelden.*
>
> *(recorded by N. Mamontova in Tura, Ewenki District, in 2014)*

[The Bear [Ursa Minor in Russian] has seven stars. When it begins to dawn, she walks on the edge of the day [i.e., at the dawn]. Thus, you know that the day is breaking, [people] say. When it dawns/a new day begins, [the bear] appears at the top, her tail falls [she turns her tail down]. Before [people] didn't have a watch, this way, they found out what time it was. Then a little star, before dawn, a little star shows huge, one [little star]. This is her son [the North Star]. People say, "Dawn breaks, it dawns, the little star appears."]

Finally, hunting activities occupy a special place in the Ewenki traditional calendar and mythology (Vasilevich 1969: 42–53). Based on linguistic data, Alexander Pevnov points out that the "basic and dearest occupation" of the Manchu-Tungusic ancestors was hunting; that is still reflected in their native languages (2012: 37). The traditional calendars of the Ewenki, the Ewen, and the Oroch people are based predominantly on hunting and gathering (Vasilevich 1966: 119). An Ewenki calendar divides a year into four seasons. Each season contains several subdivisions. The length of a year and every season may vary. The names of the months relate to the type of the main activity or specific environmental conditions. For example, *mire* ('shoulder' – when the snow lies on the branches), *turan* ('the time when crows come back'), *sōnkān* ('calving'), *irbe* ('spawning'), *ulunūkīt* ('hunting for squirrels'), and so forth. The names and the order of months significantly vary from one community to another (see Vasilevich 1963; Sirina 2012).

Table 13.2 shows how the months of year are traditionally named in three Ewenki communities: the Sym Ewenki in Western Siberia, the Ilimpii Ewenki in Central Siberia, and the Chumikan Ewenki in Eastern Siberia. Although this division of the year and the names of the months are going out of use, the key activities still are largely performed in accordance with traditional calendars. As a rule, people remember only those names which are still significant for them. Thus, among the surveyed Ilimpii Ewenki the most commonly used month is *higelehe* 'the period when people hunt for sable' (October–November). This term has been known mainly among the Ilimpii Ewenki and has not spread to other groups. The third column of Table 13.2, however, does not contain this month, as the data was collected by Vasilevich among the Ilimpii Ewenki living in the semi-tundra area, who were involved in extensive reindeer herding, unlike the taiga-dwelling Ilimpii, for whom sable hunting is a more salient seasonal time-activity maker.

Table 13.2 The names of the months in three Ewenki communities

No.	Ud (Chumikan) Ewenki	Sym Ewenki	Ilimpii Ewenki
1	*Mire* 'shoulder/period of snow on the branches'	*Giravūn, geraūn*	*Giravūn*
2	*Giravūn* 'period of snow crust'; *Ēktenkirē* 'period when snow falls off the branches'	*Mire*	*Mire*
3	*Shōnkān* 'period of calving'; *Tyglan* 'period when ice breaks up on the rivers'	*Ēktenkirē/ēptankirō*	*Turan*
4	*Sōnkān; Nengne* 'spring'; *Ilagā* 'period of berries, blossom/fresh grass and leaves'	*Turan* 'period when crows arrive'	*Huōnkān*
5	*Ilkun* 'period when berries start to ripen'	*Shōnkān; Nganman* 'period of mosquitos'	*Chukalaha* 'period of grass'
6	*Irkin* 'period when the tissue on reindeer antlers begins to peel off'	*Muchun* 'period when a larch tree turns green'	*Nganman; Gor* 'period when wild geese shed feathers'
7	*Ugun* 'period of shallow snow and frozen water by the riverbanks'	*Irkin; D'ulgalasani* 'period of birch juice'	*Irkin*
8	*Hugdarpi/ugdarpi* 'wrist/period of the best hunt'	*Hunmin* 'period of insects'	*Hirud'an*
9	*Otkī* 'period of great frost'	*Shirūd'an* 'period of pairing of (wild) reindeer'; *Erkin*	*Ulumūkīt* 'period of hunt for squirrels'
10		*Uvun; Sirūd'an*	*Hegdy gildeme* 'period of great frost'
11		*Hugdarpi*	
12		*Otkī*	

Source: Adapted from Vasilevich (1969: 43–44).

The importance of hunting culminated in the above-mentioned festivities *ikēnipkē* described by Vasilevich (1957). She writes that, in the 1930s, the festivities were preserved only among the most Western Ewenki, the Sym group, and lasted eight days. They consisted of a set of rites, including the chase for an imaginary reindeer by a shaman, other community members, and spirits, its killing and sharing of the meat, as well as imitating the behaviour of different animals. The shaman also travelled down and then up the mythological river *engdekīt* 'the place of disappearance' that connects the upper and lower shamanic worlds. During this trip he or she communicated with a number of spirits, both good and potentially harmful. *Ikēnipkē* was meant to celebrate and mediate the spiritual unity between different human-animal spirit actors and to demonstrate all the phases of the hunting calendar. It was crucial for the shaman to know the exact way to the shamanic worlds. Vasilevich (1966) even suggested that previously it might have been the shaman who was responsible for the transmission of geographic knowledge in the community, based on his or her journeying. Either way, IEK and mediation of places of spiritual and economic significance were integral to the festivities. Recently, these festivities have been to some extent revitalized in several Ewenki communities.

Reindeer herding

The distribution of the Ewenki language across the whole of Siberia became largely possible due to the Ewenki type of reindeer herding. Vasilevich distinguished two types of reindeer economy among the Tungus groups which she named as the Ewenki and the Orochen types (Vasilevich 1964). The Ewenki-type, also known as a forest type, is characterized by a small number of reindeer, usually from 10–40 animals in one herd which were used mainly as pack animals. This type was spread in the basin of the Stony Tunguska River, in the upper part of the basin of the Lower Tunguska River and along the tributaries of the Lena River. Some variations of this type could be found among the Tungus inhabiting the tributary of the Angara River, in the area of the Yenisei River and west of it, north of the Lower Tunguska River, in the northern part of the Pribaikalskiy region, between the rivers of Bureya and Amgun, and along the rivers flowing into the Okhotsk Sea. In contrast, the Orochen type is characterized by big herds, from 20–500 reindeer, which were used both as pack and riding animals (ibid.) (Figure 13.2). Therefore, among the proper Ewenki communities, extensive reindeer herding has largely prevailed in the economy of the Orochen.

It is significant that despite the fact that not all Ewenki communities were fully involved in reindeer herding, in mythology the skill of treating reindeer is considered a sign of belonging to humans as opposed to *sulugdy*, a mythological creature. The Ewenki represent *sulugdy* as a being having only half of the body covered with hair. It does not know Ewenki language. These beings have no reindeer and do not know how to treat them properly. When a *sulugdy* turns itself into a human by eating a woman, the Ewenki easily recognize its real nature by the absence of skills for caring for reindeer:

> *Nungan enīnmen huvulven tetygeven tetten, enīngesīn-te ōran. Tar beje, hele, gūndevkī: "Ke, namakaldu, namalgāt!" Namād'avkī, namād'avkī. Taduk ororvō uisilden. Tar ahī irgilduktyn uitchevkī, ororvī. Tar bejengīn dōkēncheren, gūnen: "Eda tug ōsān ahī ere?" ...Taduk: "Kē-tē*

Figure 13.2 Riding a reindeer

anggāt, hulgĭlhilgēt, tar kungakānme tegevkekel!" Nungan gadan-da kungakānme tegekēnem amaskĭ dereteden.

(recorded by N. Mamontova in Tura, Ewenki District, in 2010)

[She put on all the mother's clothing, became like the mother. That man says now, "Well, load up, let's load up [reindeer]!" They're loading, loading up [the reindeer]. Then the woman has begun binding the reindeer together. That woman is tying them together by the tails, the reindeer. The man is eyeing her; he wonders, "Why has this woman become like this?" ... Next, "Come on, let's do that, move on, mount the child!" She picked up the child and sat him with his back to the muzzles [of the reindeer]. That child says, "I cannot sit on the riding deer like this!"].

(recorded and translated by N. Mamontova in Tura, Ewenki District, in 2010)

The terms designating reindeer herding are well developed in the native language. This is a part of vocabulary which is preserved and passed on despite the language shift. Even young reindeer herders who do not speak Ewenki fluently still master a variety of terms for different types of reindeer and the tools associated with the reindeer economy. The utility of retaining Ewenki reindeer vocabulary does not prevent broader language loss, however; instead it preserves certain domains as a kind of professional jargon. Nevertheless, these linguistic domains are resilient based on their utilitarian function (Hunn 1982) (see some common terms in Table 13.3).

The focal vocabulary related to reindeer herding can be divided into at least two sets of terms: (1) nouns designating different types of animals, material objects referring to reindeer herding, and the names for the periods and months which are the most important in reindeer economy; and (2) adjectives characterizing the qualities of reindeer (age, colour, character, etc.) and their position in a sledge (cf. Magga 2006). In addition, there are a number of specific verbs describing movement on reindeer backs (i.e. riding) or by sledge which are not discussed in this chapter.

The Ewenki generally divide reindeer into the following categories: *oron* 'domestic reindeer', *bejun* or *bejngē* 'wild reindeer', and *delmi* 'previously domestic reindeer which became wild'. Additionally, there is a separate term *beyuktēn* with a meaning of a semi-wild and semi-domestic calf that is also used among the Ilimpii Ewenki in colloquial language to describe a person of mixed ethnic background. The classification of domestic reindeer according to

Table 13.3 Classification of reindeer according to age and sex

Category	Male reindeer	Female reindeer
< 1 year old	Engnekēn	Enītkēn
1–2 years old	Avlakān	Gēbd'anī, gēbd'añ, ēpchakān
2 years old	Evkān, chonoko	Sachañ, umnenī
2–3 years old	Iktenē, ektanā, gerbīchēn	Ilivd'anī
3–4 years old	Nēgārkān	Dygĭvd'anī
4 years old	Amarkān, nēgārkān	Gēvnī
≥ 5 years old	D'uptyñ, amarkān, gilduka	
6 years old	Aminan, amginan, mucheñ	
≥ 7 years old	Gemuren	

age and sex is very detailed in Ewenki and well preserved, as this information is important for tracking reproduction and keeping records on the constitution of the herd. Significantly, the latter are usually kept in Ewenki because the language has the vocabulary to do precise, descriptive accounting of the herd. Still, the terminology may slightly vary and change in different Ewenki communities.

The second most important mode of classification is that according to reproduction and the life cycle. The overall term for a female reindeer is *n'amī*. A female reindeer ready to give birth is called *surki*. After the birth of the calf it becomes *entȳ* 'female reindeer with a calf'. If the animal loses its calf, it is referred to *umīrī*. The latter term may also be used to refer to an animal that has not given birth this year and can be milked until the next year. As for male reindeer, the main distinction lies between *sirū* 'bull' and *gilge* 'castrated male reindeer used for transportation'. The general term for a castrated reindeer is *aktakī/akta*. As noted above, the Ewenki use reindeer as a pack animal which is called *inivugdēn* and as a riding/saddled reindeer *ugūchak/ūchak*. Pack animals are further divided according to the item they transport. For example, a reindeer transporting an infant in a cradle is called *un'arūk*, and *nilgu* is an unloaded animal. The terminology for a sledge reindeer is also rather diverse, depending on a position of the reindeer in the sleigh reins. The lead reindeer is called *nōgū*.

The reindeer can be also classified according to size, colour, personality, function, and the shape of the body or its parts. The basic terminology for colour includes the following terms: *cholko* 'white', *kongnomo* 'black', *ugd'ama* 'grey', and *merilē* 'multicoloured'. If any part of the body has a specific colour or shape, this characteristic can be used to name the animal that helps a reindeer herder to distinguish it in a herd. In this sense, antler terminology is one of the most rich and diverse in Ewenki. An animal with antlers is *iechi/gujagin* and without antlers is *kultumek/nultumak/tempuri/tepuli*. A number of horns and their quality are reflected in the terminology as well: *kaltamak* 'reindeer with one horn', *chororīn* 'long, straight and sharp horns', *sarbarīn/laparā* 'branchy horns', and so forth. Some terms for reindeer are occasionally used to describe a person or refer to a bear.

As well as hunting terminology, the reindeer terminology varies from one Ewenki community to another (see Table 13.4). Besides, there are many differences between the two above-mentioned types of reindeer herding in terms of the presence of some terms and their meaning (some Orochen-Ewenki reindeer terminology is discussed in Brandišauskas 2017: 190–192). However, the most basic words appear to be the same across the whole Ewenki dialect continuum, or at least they are comprehensible to native speakers from different groups.

Reindeer terminology differs greatly from terms of differentiation not only among other livestock, for example, cows and horses, but also among wild reindeer. Some wild animals, such as bear and moose, also have detailed classifications in Ewenki, according to their sex, age, size, season, and specific characteristics. Additionally, the terms for male and female animals vary depending on species. Thus, male dogs, wolves, sables, polar foxes are called *mur* (sg.) and female ones are *ukusēn* (I). The classification of bears is more complicated and contains both generic and taboo terms for this animal and parts of its body. For example, a male bear can be called *amākā beje* 'grandfather man' and a female one is *amākā ǝnīn* 'grandfather's mother' (I) highlighting kin relations between Ewenki and bears. Finally, dogs (*nginakir*) are classified according to their hunting abilities. The Ewenki differentiate dogs bred for hunting wild reindeer and moose (*bejūmēn*), squirrel (*ulukīmēn*), sable (*nekēmēn*), bear (*homotȳmān*), and so forth. Ewenki hunters always highlight the importance of partnership between people and dogs as the basis of good hunting. A skilful dog is a source of pride.

Table 13.4 Reindeer terminology in two communities (Northern set of dialects, Tutonchany village and Southern set of dialects, Kislokan village)

English equivalent	Northern (m), Tutoncnany	Northern (f), Tutonchany	Southern (m), Kislokan	Southern (f), Kislokan
Reindeer	Oron	Oron	Oron	Oron
Left reindeer	Kostur	Kostur	Kostur	Kostur
Leading reindeer	Nōgū	Nōgū		Nōvūshik
Wild reindeer	Bejengē	Bejngē	Bagdāku	Bagdāke
Riding reindeer	Ugusak	Ugusak	Uguchak	Uchak
Bull	Hirū	Hirū	Hirū	Hirū
A former domestic reindeer which has become wild	Hokorno 'lost rendeer'	Huktylne oron 'a reindeer which runs away'	Kangal	
A reindeer which is difficult to domesticate	Mekei	Gelūn	Gelūn	
Castrated reindeer	Gilge	Gilge, nārā	Aktapchā	Gilge
Thin reindeer	Epu oron	Epu, n'urkusā 'poor reindeer'	D'alanga	
Reindeer without antlers	Kamuly (Rus. komolyi)	Tapu	Tapu	
Reindeer with big antlers	Ierēgdy	Iektūre		
Old reindeer	Hagdāku	Hagdāku	Hagdāku	
< 1 year old	Avlākān	Engnekēn	Engnekēn	Engnekēn
< 1–2 years old (m)	Avlākān	Avalākān	Avalākān	Avlākān
2–3 years old (m)	Gerbisēn	Gerbisēn	Gilge	Gerbichēn
3–4 years and older (m)	Amīnān	Amīnān	Nōgurkān	Gilge
Female reindeer	N'amisān	N'ami	N'amichān	N'amichān
Barren female reindeer	Vangaj	Vangaj	Umīrī	Vangaj
A female reindeer with calves	Engnemār			
Sacred reindeer				Bugady

Fishing and sea mammal hunting

All Ewenki communities have long combined hunting, reindeer herding, and fishing. Even among some reindeer Ewenki of Central Siberia, fisheries used to determine nomadic routes in summertime. Some of them switched to fishing and sedentary style of life when they lost their reindeer (State Archive of Krasnoyarsk Krai, f. П–1380, l. 2, c. 17, p. 7. Misrepresentation of the Ewenki as primarily reindeer herders resulted in the lack of research on their fishery economy and terminology related to it. Historically, there were a few Ewenki communities which could be chiefly defined as fishing communities. One of these lived in the lake area north and west to the head of the Vilyui River and the other on the coast of the Okhotsk Sea. Unlike other Ewenki groups, their calendar has a few summer months named after fishing activities (Vasilevich 1969: 42).

Currently, mainly the Okhotsk Ewenki are involved in salmon fishery and sea mammal hunting. Their adaptation to the marine environment occurred in the nineteenth century when, having abandoned their reindeer, they migrated from taiga to rivers and bays full of salmon. This largely happened due to the depletion of sable (*Martes zibellina*) in the taiga which the people hunted to pay *yasak* (tribute). Having reached the sea, they came in close contact with the local native groups (Koryak and Nivkh) from whom they are believed to have borrowed most of their fishing techniques (see Turaev 2008).

Apart from that, sturgeon fishing has long been an important activity among the Ewenki of the basin of the Amur River (see Ermolova 1984). The Ewenki used to preserve salmon for winter. They made bags and covers for their tents out of fish skin. Some Ewenki living in proximity to big Russian settlements sold their catches in exchange for cash (Vasilevich 1969: 86). Hence, the terminology for different marine fish species and sea mammals is rich primarily in the Ewenki dialects spoken in the Russian Far East and almost unknown among other groups. There are the following terms for sea mammals in Ewenki: *kuma* 'seal', *kalim* 'walrus', 'whale', *bulege* 'dolphin', and *edar* 'grampus' (Myreeva 2001: 46). The terms for chum and humpback salmon are of common origin in all Tungusic languages (Pevnov 2012: 28). Thus, in Ewenki, chum salmon (*Oncorhynchus keta*) is called *kyata* and humpback salmon (*Oncorhynchus gorbuscha*) is *ukuru/ukurē*. These terms are found only among the easternmost communities. Like the taiga, the sea has its master. Beluga (*Huso huso*) serve it as a messenger. It is believed that the master of the sea covers the fish with a blanket during the spawning season:

> Every year before the fishing season starts, I say, "Oh, this is the blanket for herring [lit. the herring blanket has come], this is the blanket for smelt fish, the pink salmon' blanket, the chum salmon's blanket." I have already lived for many years, and there is the fog every time when the fish comes. As if the master of the sea covers the fish [with the blanket]. This is the blanket, for example, for herring. And then we go there where the herring spawns, and there we catch it.
>
> *(recorded by N. Mamontova in Chumikan, Khabarovsk Krai, 2017)*

As for riverine and lacustrine fish names, the most common fish names in Ewenki are *delī/delbēn* 'taimen' (Hucho taimen/Siberian taimen), *gutkēn* 'pike' (*Esox lucius*), *nekechēn* 'perch' (*Perca fluviatilis*), *kolemtē* 'crucian' (*Carassius auratus*), *nirū* 'grayling' (*Thymallus arcnicus*), and *sēngan* 'burbot' (*Lota lota*) (ibid.: 59–61). Along with birds, some freshwater fish played a very important role in shamanistic rituals. Among the Yenisei Ewenki, *delbēn* 'taimen' was considered to be one of the spiritual helpers of the shaman. From its wooden figure the shaman made a raft to travel along the mythological river (Vasilevich 1969: 219–220). Another important fish is *sēngan/hēngan* 'burbot'. This is one of the most famous characters in folklore which managed to outwit even the cunning fox, a trickster in Ewenki mythology. The importance of burbot is further evidenced by the fact that the thunderbird Kingit with a shamanic power is afraid of its skin. The Western Ewenki, including the Ilimpii Ewenki, believe that Kingit can take a lonely traveller away to its nest. However, if a burbot skin is tied to the gun, it will not touch the person (Nikulshin 1939).

To conclude, it seems that the terminology for fish species and sea mammals is less resistant in comparison with hunting and reindeer breeding terminologies. It is being easily replaced with Russian terms, especially in those Ewenki communities which are not directly involved in fisheries. The Ewenki language has no specific classification of fish species.

Conclusion

This brief account of some Ewenki subsistence terminology shows that this native language is especially rich in specialized terms, which embed important IEK and distinctions, and thus have proved rather stable and resilient, since there is no comparable Russian terminology to replace them. Hence, they continue to be used despite a language shift even among the younger members of the community due to their functional role in environmental cognition, which is linked to classification systems of keystone species. For example, terminologies for such species as reindeer (domestic and wild), moose, dogs, and bears are the most elaborate and precise in the native language. This allows people to easily categorize these animals according to sex, age, size, season, and other useful characteristics. In contrast, the terms for fish species are somewhat less developed and more comparable to Russian classificatory terms, and therefore, less resistant. At a broader geographic level, we find some Ewenki landscape forms and types of vegetation are linked to indigenous ontologies and encode very complex ideas. For example, the basis for the richness of the terms designating burnt places in the forest is their close associations with economic activities (hunting), spirits, and shamanistic rituals.

However, it is important to highlight that although the Ewenki use more or less the same concepts and categories across the whole dialect continuum, specific terms can significantly vary from one dialect to another, or some terms may have slightly different meanings in different dialects. Therefore, some words marked as synonyms in dialect dictionaries may have distinct meanings in different Ewenki communities. For this reason, it is advisable to employ both comparative linguistic and ethnographic approaches to the study of Indigenous Environmental Knowledge among an ethnic group.

While certain Ewenki IEK focal vocabulary appears resilient at least as a professional lexicon, it is not enough to reverse language loss. Perhaps equally significant to IEK and language maintenance is the terminology bound to spiritual aspects of culture, which also relate to the foundations of Ewenki cultural identity, ceremonialism, and their world-view. Over the past decade, we have witnessed the revitalization of some Ewenki festivities, for example, *ikēnipkē*, which evoke both the focal vocabulary and elements of traditional spirituality. This revitalization is developing within a new social context but on the basis of traditional IEK.

Acknowledgments

The linguistic data (dictionary entries) for this research was collected in the project on Endangered Language Documentation led by Dr. Olga Kazakevich among the Ewenki people of the Lower Tunguska River, Ewenki District, Krasnoyarsk Krai, 2008 (grant RSSF 07-04-00332a). The chapter is written in the framework of the project 'Anthropology of Extractivism: Research and Design of Social Changes in the Regions with Resource-Based Economy' (RSCF, grant No. 20-68-46043, PI Professor D.A. Funk).

Key to dialect names

Ald	Aldan dialect
E	Erbogachen dialect
I	Ilimpii dialect
M	Mai dialect
N-B	Northern Baikal dialect
P-T	Podkamenno-Tungus dialect
S	Sym dialect
Sak	Sakhalin dialect
TmT	Tommot dialect
Tng	Tungir dialect of the Tungir-Olekma group of dialects

Tt   Totta dialect
Uchr   Uchur-Zeya dialect
Urm   Urmii dialect of the Burein-Urmi-Amgun group of dialects

References

Anderson, D. G. 2000. *Identity and Ecology in Arctic Siberia: The Number One Reindeer Brigade.* Oxford: Oxford University Press.

Anisimov, A. F. 1949. Predstavleniya ewenkov o shingenah i problema proishozhdeniya religii. *Sbornik Muzeya Antropologii i Etnografii* 12: 160–194.

Berkes, F., Colding, J. and Folke, C. 2000. Rediscovery of traditional ecological knowledge as adaptive management. *Ecological Applications* 10(5): 1251–1262.

Berlin, B. 1992. *Ethnobiological Classification: Principles of Categorization of Plants and Animals in Traditional Societies.* Princeton, NJ: Princeton University Press.

Brandišauskas, D. 2017. *Leaving Footprints in the Taiga: Luck, Spirits and Ambivalence among the Siberian Orochen Reindeer Herders and Hunters.* Oxford: Berghahn,

Davydov, V. 2011. *People on the Move: Development Projects and the Use of Space by Northern Baikal Reindeer Herders, Hunters and Fishermen.* PhD thesis. University of Aberdeen, Aberdeen.

Douglas, D. and Turner, N. J. (eds) 2005. *Keeping in Living: Traditions of Plant Use and Cultivation on the Northwest Coast of North America.* Seattle, WA: University of Washington Press and Vancouver: UBC Press.

Ellen, R. F. 1993. *The Cultural Relations of Classification: An Analysis of Nuaulu Animal Categories from Central Seram.* Cambridge: Cambridge University Press.

Ermolova, N. V. 1984. *Ewenki Priamurya is Sakhalina. Formirovanie i kulturno-istoricheskie svyazi. XVII–nachalo XX vv.* Unpublished PhD thesis. Institute of Ethnography, Russian Academy of Sciences, St. Petersburg.

Gabyshev, S. and Lavrillier, A. 2017. *An Arctic Indigenous Knowledge System of Landscape, Climate, and Human Interactions: Evenki Reindeer Herders and Hunters.* Fürstenberg: Kulturstiftung Sibirien.

Grenoble, L. 2014. On thin ice: Language, culture and environment in the Arctic. *Language Documentation and Description* 9: 14–34.

Hunn, E. 1982. The utilitarian factor in folk biological classification. *American Anthropologist* 84(4): 830–847.

Janhunen, J. 2012. The expansion of Tungusic as an ethnic and linguistic process. In A. L. Malchukov and L. J. Whaley (eds), *Recent Advances in Tungusic Linguistics.* Wiesbaden: Harrassowitz Verlag, pp. 5–16.

Kolås, Å. and Xie, Y. (eds) 2015. *Reclaiming the Forest: The Ewenki Reindeer Herders of Aoluguya.* New York: Berghahn.

Lavrillier, A. 2011. The creation and persistence of cultural landscapes among the Siberian Evenkis: Two conceptions of 'sacred' space. In P. Jordan (ed.). *Landscape and Culture in Northern Eurasia.* Walnut Creek, CA: Left Coast Press, pp. 215–231.

Lavrillier, A. 2013. Climate change among nomadic and settled Tungus of Siberia: continuity and changes in economic and ritual relationships with the natural environment. *Polar Record* 49(250): 260–271.

Magga, O. H. 2006. Diversity in Saami terminology for reindeer, snow, and ice. *International Social Science Journal* 58(187): 25–34.

Mark, D. M, Turk, A. G , Burenhult, N. and Stea, D. (eds) 2011. *Landscape in Language: Transdisciplinary Perspectives.* Philadelphia, PA: John Benjamins Publishing.

Mertens, K. 2016. Patterns of Evenki mobility in Eastern Siberia. *Sibirica* 15(1): 1–40.

Messineo, C. and Cúneo, P. 2011. Ethnobiological classification in two indigenous languages of the gran Chaco Region: Toba (Guaycuruan) and Maká (Mataco-Mataguayan). *Anthropological Linguistics* 53(2): 132–169.

Myreeva, A. N. 2001. *Leksika ewenkiyskogo yazyka. Rastitelnyi i zhivotnyi mir.* Novosibirsk: Nauka.

Nikulshin, N. P. 1939. Unpublished material recorded at the River of Kataramba, Ewenki Autonomous Okrug. Archive of the Krasnoyarsk State Museum of Local Culture, f. 7886/220, folder No. 6 "Northern expedition".

Pevnov, A. M. 2012. The problem of the localization of the Manchu-Tungusic homeland. In A. L. Malchukov and L. J. Whaley (eds), *Turcologica 89: Recent Advances in Tungusic Linguistics.* Wiesbaden: Harrassowitz Verlag, pp. 17–40.

Safonova, T. and Sántha, I. 2010. Walking mind: The pattern that connects Evenki land, companionship and person. In A. Bammé, G. Getzinger, and B. Wieser (eds), *Yearbook 2009 of the Institute for Advanced Studies on Science, Technology and Society*. Vienna: Profil, pp. 311–323.

Safonova, T. and Sántha, I. 2011. Mapping Evenki land: The study of mobility patterns in Eastern Siberia. *Estonian Folklore* 149: 71–96. Available at: www.folklore.ee/folklore/vol49/evenki.pdf

Safonova, T. and Sántha, I. 2016. Evenki hunter-gathering style and cultural contact. In: K. Ikeya and R. K. Hitchcock (eds), *Hunter-Gatherers and Their Neighbours in Asia, Africa and South America*. Osaka: National Museum of Ethnology, pp. 59–79.

Simonova, V. V. 2015. "Imya sobstvennoe" i taezhnyi alyans: ob initsialah na derevyah, istorii podpisi i politike gramotnosti u ewenkov Severnogo Baykala. *Zhurnal sotsiologii i sotsialnoy antropologii* 2(78): 193–208.

Sirina, A. A. 2008. People who feel the land: The ecological ethic of the Evenki and Eveny. *Anthropology & Archeology of Eurasia* 47(3): 9–37.

Sirina, A. A. 2012. *Ewenki i eweny v sovremennom mire. Samosoznanie, prirodopolzovanie, mirovozzrenie.* Moscow: Vostochnaya Literatura.

Ssorin-Chaikov, N. 2001. Evenki shamanistic practices in Soviet present and ethnographic present perfect. *Anthropology of Consciousness* 12(1): 1–18.

Takakura, H. 2012. The shift from herding to hunting among the Siberian Evenki: Indigenous knowledge and subsistence change in Northwestern Yakutia. *Asian Ethnology* 71(1): 31–47.

Tugolukov, V. A. 1985. *Tungusy (ewenki i eweny) Sredney i Zapadnoy Sibiri.* Moscow: Nauka.

Turaev, V. A. 2008. *Dalnevostochnie ewenki: ethnokulturnye i ethnosotsialnye protsessy v 20 veke.* Vladivostok: Dalnauka.

Vasilevich, G. M. 1940. *Ocherk grammatiki ewenkiyskogo (tungusskogo) yazyka.* Narkompros, RSFSR: Leningrad.

Vasilevich, G. M. 1957. Drevnie ohotnichyi i olenevodcheskie obryady ewenkov. In S. P. Tolstov (ed.), *Sbornik muzeya antropologii i etnografii* 17. Leningrad: Rossiyskaya Akademiya nauk, pp. 151–185.

Vasilevich, G. M. 1958. *Ewenkiysko-russkiy slovar.* Moscow: Gosudarstvennoye izdatelstvo inostrannyh i natsionalnyh slovarey.

Vasilevich, G. M. 1963. Drevneyshchie geograficheskie predstavleniya ewenkov i risunki kart. *Izvestiya vsesoyuznogo geograficheskogo obshchestva* 4: 306–319.

Vasilevich, G. M. 1964. Tipy olenevodstva u tungusoyazychnyh narodov (v svyazi s problemoy rasseleniya po Sibiri). In *VII Mezhdunarodniy kongress antropologicheskh i etnologicheskikh nauk, Moskva, avgust 1964.* Moskva: Nauka.

Vasilevich, G. M. 1966. Rol' etnograficheskogo kompleksa ewenkov i ih yazyka v etnogeneze tungusoyazychnyh narodov Nizhnego Priamurya (archival document). Museum of Anthropology and Ethnography named after Peter the Great, Russian Academy of Sciences, fond 22, inventory list 1.

Vasilevich, G. M. 1969. *Ewenki: Istoriko-etnograficheskie ocherki (XVIII – nachalo XX veka).* Leningrad: Nauka.

Waddy, J. A. 1988. *Classification of Plants and Animals from a Groote Eylandt Aboriginal Point of View*, vol. 1. Darwin: Australian National University.

14

TLINGIT ENGAGEMENT WITH SALMON

The philosophy and practice of relational sustainability

Steve J. Langdon

Introduction

The Tlingit are the Indigenous people, who live on the islands and the mainland of the northeastern coast of the Pacific Ocean, from the Bering River in the north to Dixon Entrance in the south. Linguistic and archeological evidence suggest that they were present in the region 5,000–6,000 years ago but it is possible that they arrived earlier as human presence in the region dates to over 10,000 years ago. The five species of Pacific salmon (*onchorynchus sp.*) were the primary source of food on which they relied as typically abundant numbers of anadromous salmon returned to the more than 3,000 streams and rivers in the region where they were born to spawn and die from summer to fall. In the spring, eggs deposited and fertilized in the previous summer and fall hatch out and head to their freshwater homes from the ocean where after several years many would once again travel back to the streams from which they departed. Tlingit harvested and consumed thousands of salmon fresh, processed, and stored even more to provide for their sustenance and to feed guests invited to elaborate feasts and ceremonies over the winter months.

Salmon are likely to have been present at the time of the earliest human arrivals in southeast Alaska, thus Tlingit use of and relationships with salmon have been ongoing for thousands of years. The Tlingit understanding of salmon – their traditional knowledge and the practices derived from that knowledge – are both deeply spiritual and at the same time empirical and practical. The two components of the relationship are complexly interwoven into an existence scape – an envisioning and experiencing system that can be characterized as one of relational sustainability (Langdon 2007). Perceptions of and behaviors toward salmon are channeled by the Salmon Boy story, a mythic charter that establishes a covenant between salmon and humans premised on appropriate behavior by humans that ensures the return of salmon to their natal streams.

Tlingit existence scape and salmon

A common and well-worn anthropological term used to characterize fundamental cultural orientations to existence is worldview. I find this term unsatisfactory for three major reasons.

DOI: 10.4324/9781315270845-16

First, in using the term world, there is a foregrounding of the material "reality" of the planet and in using the term, the Western sensual preference for empirical sight ("seeing is believing") is privileged. Second, the term strongly implicates a coherent structure that is given and lacks a sense of openness to new observations and flexibility to change, develop, and evolve. Finally, the term lacks an evocation of the fundamental emotional nature of human behavior built on deeply grounded neurological synapses that fire often without conscious attention and dictate responses to stimuli and contexts. Human experience from birth builds neurological networks and synaptic connections that create a habitus – a basic set of perceptual, neuronal orientations to observation and interpretation (Bourdieu [1977] 2002). Core elements of the habitus are embedded neurologically through repetitive iteration below the level of attentive consciousness. Examples of this include eye movement, spatial orientation, interpersonal contact, personal spacing, environmental monitoring, and patterns of speech. In order for humans to participate as members of cultural communities, key components of the habitus are shared among the members, allowing for mutual understandings to be the basis for meaningful exchanges and shared actions. Narvaez (2014, 2016) has made the case that moral foundations of human behavior and morality are deeply positioned in our neurobiology and are established at an early age in ways similar to the examples above. She observes that as "biosocial creatures [humans] are co-constructed by their social experiences, especially in early life when built-in maturational schedules of different body/brain systems" are coordinated through specific patterns of interaction and care (Narvaez 2016: 7).

Salmon were and remain iconic for the Tlingit whose existential relations were fused with them through a mythic charter and covenant known as *Aak'wtaatseen* or the Salmon Boy Story (Swanton 1909; Littlefield et al. 2003; Sitka version also recounted in Thornton 2008, 2012)). In interviews conducted in 2002–2003 with Tlingit elders and experts from the communities of Klawock and Hoonah, I asked the question: "What were you taught about salmon as a child?" Almost universally, at some point in their responses, aspects of the Salmon Boy Story would appear (Langdon 2006a). The mythic account recounted with some variation by the Tlingit elders tells the story of how a young Tlingit boy spoke and acted disrespectfully toward a piece of dried fish and subsequently was saved by the Salmon People from drowning. He was then transported to the village of the Salmon People on the ocean bottom offshore where he saw that Salmon were people, like himself under their skins, who lived similarly to how he had lived as a human. He learned that, like humans, Salmon People were sentient, attentive, volitional – they decided to which stream they returned. He was taught many lessons by the Salmon Chief about how salmon were to be treated when they returned to the streams where they had been born and, if treated in those respectful ways by humans, they would continue to return to those streams. After several years, the Salmon Chief announced that it was time to return to their home streams and directed Salmon Boy to get into his canoe and sit in the bow. As they approached the stream where Salmon Boy's parents were in their fish camp, the Chief told Salmon Boy to stand up to see where they were. Because he was a salmon at that moment, standing in the canoe translated into leaping out of the water to observe the location. Having established that they were in the correct location, Salmon Boy proceeded to present himself to his mother who then directed Salmon Boy's father to catch him. When Salmon Boy was landed, his mother noticed that there was a copper necklace around the gills and recognized it as the necklace that had been worn by her son when he disappeared several years ago. The parents showed the salmon to a shaman who told them the fish should be placed on a shelf high in the smokehouse and left overnight. In the morning, the couple arose to find that the necklaced salmon was gone and their son had returned! Salmon Boy then told them of his experiences and taught them all the lessons he had learned from the Salmon Chief about how salmon must be treated so that they would be able to return to the same streams. Those lessons became shared widely among the Tlingit who adopted them and embellished them with additional practices designed to show respect and honor to the Salmon People. In a summation of his understanding of the Salmon Boy story, James Osborne – a member of the Chookaneidí clan of Hoonah – stated: "You have to realize that we treat salmon like we wish to be treated" (Osborne, pers. comm. 2003).

There are many examples of how the Tlingit elaborated on the principles laid down in the Salmon Boy mythic charter/covenant. For example, Clara Peratrovitch's mother's ancestral house was located in Tuxican about 40 miles north of *Lawáak* (Klawock). *T'il hít* or [Raven moiety] Dog Salmon House was the first or lead house of the *L'eenedí* clan and had a prominent position on the beach front at the north end of the village. Visitors approaching the community would have been struck by the exceptional painting that stretched 40 feet across the front of the house (Figure 14.1). The image was of a dog salmon late in life just prior to spawning, displaying the colorful dark green and purple stripes on its side and the huge teeth characteristic of the late life stage. There were two additional elements on the painting that relate understandings about dog salmon in Tlingit social life and existence. On the top of the salmon, the dorsal fin is drawn as a Raven, thereby indicating the moiety of the clan – hence Raven Dog Salmon. At the juncture of the tail and body, an image of a face is apparent that represents information conveyed in the Salmon Boy story that the spirit of the salmon was located in that spot and therefore it should not be carried roughly or grabbed by those processing the fish in that area. More astonishingly, in the *L'eeneidí* Salmon Boy story, upon returning from the village of the Salmon People, Salmon Boy decided to replicate the house front painting that he had observed on the front of the Salmon Chief's house in the offshore village who had taught him how Salmon People must be treated (Garfield and Forrest 1961: 144). The house front painting served several purposes: informing visitors of the social identity of the occupants and displaying knowledge of the Salmon Boy mythic charter/covenant that demonstrated respect for the dog salmon, their

Figure 14.1   Raven Dog Salmon House front painting

crest. There are many other examples of the deeply intertwined existences of Tlingit and salmon that are grounded in the Salmon Boy story. For example, the *L'eeneidí* clan has many personal names relating to their primary crest, Dog Salmon, such as "Last One Up When Ice is on the Leaves", and "Jumping Sideways Many Times Around the Bay."

As a charter and a covenant, the Salmon Boy story portrays the terms of interdependence through which salmon and humans can continue to benefit from the actions of the other party. Salmon return to those locations where humans catch them, treat them right and ensure that they can be reborn, while people can catch, consume and by treating them respectfully, ensure that salmon will return and again make themselves available to humans. Anthropologist Julie Cruikshank has observed that the Tlingit occupied a moral universe inhabited by a community of beings in constant communication and exchange (Thornton 2012). A core component of this philosophic position I have termed "relational sustainability" and the Tlingit lived lives to fulfill their moral responsibilities to maintain the cycle of existence. The Salmon Boy story establishes the fundamental principles for relationships between salmon and humans. One of its underlying themes is that salmon people return to those who act in appropriate ways toward them. This is taken by Tlingit as a point of departure for exploration and elaboration of other ways of displaying respect for salmon as examples later in this chapter demonstrate. Because the Salmon Boy story reveals that salmon are composed of "persons" fundamentally similar to human "persons," the social relationship between human beings provides a template to use to generate new forms of interactions with salmon. This practice of extending appropriate principles of behavior between humans to behaviors between humans and salmon (or other living forms) I term "empathic reflexivity." Empathy constitutes identification of the circumstances of another as similar to one's own while reflexivity implicates the critical evaluation of that relation based on experience.

Two critical features of Tlingit social organization inform these extensions of "empathic reflexivity." First, Tlingit society is ordered by the principle of "obligatory reciprocity" (Langdon 2014). Upon birth, a person becomes a member of the clan and moiety of the mother and participates in social activities in that role. Marital partners must come from a clan in the other ("opposite") moiety and transgressions are regarded as incest subject to severe reprisals, such as banishment and death. At critical junctures (death, first menses, house building), clan members conduct ceremonies and rituals in which they act as hosts, inviting members of their opposite intermarrying clans as guests who observe social claims and receive gifts while in attendance. In turn, guests on one occasion reciprocally become hosts on similar occasions and invite as guests their relatives who previously served as hosts. Without such mutual co-production, Tlingit society could not exist. Hence, "obligatory reciprocity."

Existence is ordered and productive when relationships are in balance. Balance means that appropriate forms of exchange and respect have been exchanged between parties such that both feel content with their circumstances. When things are in disorder and unproductive, it becomes imperative for the parties to seek balance (*Wooch yáx*). This form of thinking resonates in Tlingit understandings about wrongdoings, insults, and warfare. When such conditions exist, participants typically seek to return to order by establishing through formal discussions and negotiations what each party regards as essential to return to balance. A ritual, known as the Deer Ceremony (*Ḵuwaak'aan*), formally recognizes the re-establishment of balance, ordered, productive relations between parties after negotiated adjustments have been completed.

The concept of balance in relations among humans was extended upon occasion to understanding of how ordered and productive existence was arranged in natural settings. Application of the balance concept to the position of salmon is shown later in the chapter.

The intertwining of mythic and practical understandings

Tlingit engagement with salmon is grounded in the intersection of the mythic charter/covenant and practical understandings that arise from careful, empirical observations of salmon behavior. These in turn are encountered through the practice of empathic reflexivity generating additional and in some cases new behaviors by humans. It must be emphasized that careful attention to actual salmon behaviors and the contexts in which they occur are critical to the successful relationship that Tlingit established with salmon. This section provides examples of practical understandings of salmon circumstances.

In the Salmon Boy story, the migratory path of salmon traveling from their ocean dwellings to their stream homes is recounted and their conversations among themselves reported (Thornton 2012: xx). On the west coast of the Prince of Wales Archipelago, near the village of *Lawáak* (Klawock), is a point on the southside of which in the intertidal zone a set of intertidal stone semi-circular fish traps is located, designed according to the principles of "tidal pulse fishing" to capture salmon only on the ebb tide (Langdon 2006b). Large schools of salmon heading for numerous streams in the area pass by this point and these traps allow Tlingit salmon harvesting to begin two weeks earlier and higher quality (more fat) fish to be caught than at the stream mouths (Langdon 1986: 24). In the 1950s, a commercial salmon fish trap operating just south of the site was the highest producing such trap for the nearby cannery. There are several other locations on the west coast of the Prince of Wales Archipelago where the placement of intertidal stone fish traps at points distant from stream mouths indicates Tlingit knowledge of salmon migratory patterns.

As salmon travel through the channels and especially as they approach their home streams, they often leap out of the water. The jumping of the salmon out of the water "to see where they are" is a central element of the story. When they see these performances in front of them, Tlingit across the entirety of their geographic range, give out a cry of greeting. Tlingit elder Joe Hotch describes the Tlingit practice as follows:

> When they're [salmon] jumping, we are supposed to say 'Ey Ho!'; you see a fish Jump, 'Ey Ho,' [then] they know they're being appreciated so they keep jumping. And I guess our people say it so they can know which way it's going. Just keep saying 'Ey Ho', and that's the way they want to be talked to; the fish want to be appreciated.
>
> *(Thornton 2012: 50–51)*

Tlingit knew that salmon had to ascend the streams to reach their homes in order to spawn. The southern Tlingit developed a technology that respected and ensured that salmon would be able to ascend to their spawning grounds on each tide (Langdon 2006b). Tlingit recognized that salmon move up the bays following the incoming tide and enter the streams at high tide when access is easiest. They also observed that many of the salmon approaching the mouth do not ascend but rather back out into the bays where they await another tidal cycle. The region is characterized by a substantial tidal range of 14–18 feet, exposing a large intertidal zone. In that zone, usually at approximately half-tide, the Tlingit constructed semi-circular stone and/or wood stake walls on beaches in immediate proximity to the stream channel that would capture the salmon on the falling tide. Most of the trap designs allowed salmon not taken by the Tlingit to return to the saltwater on the next tidal cycle. Tlingit/Haida elder Christine Edenso of the *L'eineidí* clan explained the traps as follows:

> I've observed in my younger days that ... Tlingits used to trap fish at the mouth of the streams. If you go around today by the mouths of the old creek flats, you will see these

rocks still piled up as they did in the old days. You will be able to see the outline of where they laid a bunch of rocks to form a wall. In that way, when the tide went out, the fish were trapped behind them and they were easier to catch then. They used to catch all the fish they needed as time went on. Some of the creeks were readily adaptable to this kind of fishing, and that was why they caught their fish by this method. The fish would go up to the mouth of the creeks at high tide. They would get behind the wall and would be trapped then the people would gaff them and pull up all the fish they needed right there. That was how they used to catch their fish. When you go along the beach at low tides, you can still see these places where they made these rock walls and traps and they are quite visible. They are the works of the people a long time ago ... You can see these rock enclosures all over Southeastern Alaska on the west coast, in the tidal flats ... and at any place where there was a good number of people ... They used the network of fish traps to corral the fish momentarily while the tide was going out. They used to gather their fish in that way.

(Pulu 1983: 36)

Another example of engagement based on empathic reflexivity can be seen in the placement of carved representations of the Salmon Boy or clan crests on traps that would stick out above the water. When the jumping salmon would approach the stream, they would see these representations that conveyed messages of introduction (who the clan owners of the traps were) and commitment to the understandings and practices laid out in the Salmon Boy story. These carvings can be seen as aesthetic prestations whose beauty is designed to please the observers (Langdon 2007: 267).

A further example of the application of empathic reflexivity can be found in Tlingit behavior toward salmon as they arrive at the stream mouths where the people have constructed the traps to harvest them. The Tlingit greet the salmon with songs and dances while wearing their regalia – this is the same kind of ceremonial greeting that are given by Tlingit hosts when guests arrive for events to which they have been invited. In the songs greeting the salmon, singers urge them to enter their "forts" (ibid.: 265).

As salmon entered and traveled up streams, Tlingit observed their gendered nature as well as their spawning behaviors, i.e. the deposition of eggs in gravel nests in the bottom and the streaming of milt over the eggs. Tlingit recognized that the fertilized eggs would subsequently become fry and then smolts that would travel down the streams and rivers in the spring (Langdon 2006a).

Tlingit applied the concept of balance to stream system recognizing that in addition to themselves, there were others that made use of salmon. While those uses were recognized as rightful, there were parameters within which they could function. A number of different bird species flocked to the streams at the time of salmon spawning. Tlingits observed them feeding on free-floating eggs on the surface but they also noticed that certain birds dove down into the stream and used their beaks to dig up the eggs. Mergansers were considered a particularly troublesome species in this regard (ibid.: 120). In speaking about the presence of birds on the streams during salmon spawning, Huna Tlingit elder Thomas Jack noted that "taking care" of the stream meant the *héen s'aatí* (stream master) had to allow for only "one family" of the relevant bird species to take up residence on his stream. If excess birds appeared, they were harvested and in this way "balance" among the salmon and other species would be maintained (ibid.: 120).

Dolly varden, a small trout, were recognized by the Tlingit as also potentially damaging to salmon reproduction. In the spring, when fry and smolts headed out to the ocean, Dolly varden would voraciously consume the infant salmon. If the *héen s'aatí* determined that the number of

Dolly varden were excessive, special instream traps would be installed to harvest the excess trout and thereby depress their taking of infant salmon (ibid.: 123–125).

Salmon runs are known to fluctuate substantially from year to year. In some cases, major catastrophes such as landslides would severely limit or even destroy salmon runs. Tlingit are known to engage in various types of stock transfers in order to replenish or re-establish runs that have been damaged or destroyed. In one case, Tlingit elder Herman Kitka Sr. undertook a transfer of late spawning chum salmon from one system to his clan system so that there would be a later run, that it would be possible to use after the commercial salmon season had ended (Thornton et al. 2015). Such activities have been referred to as "cultivation" (ibid.). Various methods were used including the release of male and females from a nearby system into the depressed system and the mixing of eggs and milt and placement in nests dug into the stream bed (Langdon 2006a: 134–137).

Traditional knowledge and practice through "streamscaping"

In 2002, I undertook research to document Tlingit knowledge of and relations with salmon (ibid.). The primary methods involved interviews with knowledgeable elders and experts, examination of the ethnographic record, and exploration of oral traditions and transcripts from recorded texts from various Tlingit sources. However, I was of the opinion that place-based knowledge that might be acquired by visiting a salmon stream with a knowledgeable Tlingit elder could conceivably reveal insights unavailable from the interview process. It was important to explore actual on-the-ground relations of Tlingit with salmon through participant observation, if possible, to see what kinds of practices occurred, what their basis was, and what outcomes were anticipated. Through previous research among the Huna Tlingit of the Icy Strait region of northern southeast Alaska, I had become acquainted with and impressed by the knowledge, experiences, and perspectives of two brothers, Patrick and Thomas Mills. They had grown up and spent a great deal of their youth in a village in Excursion Inlet, usually visiting the same streams and using the same resources as their grandparents and clan ancestors who had come before them. I told them of my interest in having them guide me and teach me about the Neva River, its salmon, and their practices at the river. They agreed and we made arrangements for a trip to the river to take place in the fall of 2002. The Neva River is a medium-sized sockeye salmon system located in Excursion Inlet in the northern part of southeastern Alaska, in the traditional territory of the *Xunaa Ḵáawu*. The stream is within the territory of the *Wooshkeetaan*, an Eagle clan, and I was aware that several families from the village of Hoonah with longstanding ties to the stream lived much of the year in close proximity to the stream. The brothers are members of the *T'aḵdeintaan* clan through their mother but both their father and grandfather were members of the *Wooshkeetaan* clan. It is common in Tlingit society for members of two clans to intermarry over many generations, creating sustained ties and shared uses of clan territories.

In 1915, a fish cannery was built near the Neva River in order to process salmon obtained from numerous floating fish traps positioned along the shores of Icy Strait, the major migratory corridor along which salmon, headed for streams on the mainland of northern southeast Alaska, traveled to reach their spawning streams. Local Tlingit of *Xunaa Ḵáawu* and nearby villages began working for the cannery; men harvested and delivered salmon to the cannery and women worked in the cannery processing the salmon. Above the cannery, houses for the Tlingit families were built near the upper waters of the Neva River where they resided during the salmon canning season. The families also had houses in Hoonah and would move between the two sites at various times during the year. The village was in walking distance along a trail to the Neva

River, the traditional source of the sockeye salmon for the *Wooshkeetaan* of this area. The families built processing areas, drying racks, and smokehouses in proximity to their homes so they could conduct traditional salmon harvesting, processing, and storage operations to obtain their winter stores while also working for the cannery.

In order to reach Excursion Inlet, I flew into Hoonah where I met with Thomas Mills and we traveled in his open 16-foot aluminum skiff across a moderately choppy Icy Straits to the cannery dock. We walked up the dock to the Tlingit village where we found his brother Pat in one of the houses waiting for us. He said he would join us shortly so Thomas and I then took the trail to the Neva River. When we reached the river, Thomas said we should go up to the lake out of which the Neva River flowed and visit the tributary stream above the lake where the sockeye spawn.

We hiked up to the lake and around the north shore to the tributary stream. Thomas said that about 20 years ago, the sockeye return had declined significantly. He recalled a remark made by his grandfather that sometimes beavers dam streams preventing sockeyes from reaching their spawning grounds. With this in mind he had gone up the stream and discovered a beaver dam crossing the stream and blocking passage. He then took down the dam, opening up the passage. In his interview, Thomas Mills described the beaver situation and outcomes as follows:

> We noticed that the sockeyes, we weren't getting as much sockeyes as we used to and pretty soon we just looked up, walked up the whole river to find out why. And when we got up by the lake over there, we saw that the beavers blocked out the whole lake where the sockeyes couldn't get into the lake and the bear and wolves and stuff were just having a field day. And some of the sockeyes that couldn't get up into the lake, some of them turned around and went back down the river. But there was a real poor showing … for those years. And pretty soon we started taking those beaver dams apart. And all that, my son is going to be 14 on the 10th of October so I think it was almost 10 years of his life that he helped take beaver dams apart. So now that we dismantled the whole thing there is a lot of fish going up the river again.
>
> *(Langdon 2006a: 124)*

Other Tlingit sources have also described how beaver dams block salmon access to spawning grounds and require monitoring and destruction if necessary.

We then walked down the stream for about a quarter of a mile, past falls and fast-moving waters. We then came up on a section of slower-moving water where I could clearly see salmon moving slowly in close proximity. Their dark bodies were easier to see through the clear water against the bottom, devoid of rocks with a medium-brown coloration. Thomas then went behind the bushes by the trail and retrieved a long-pole with a curved metal hook at the end. I recognized it as a gaff hook. He then moved gingerly toward the stream gradually pushing the gaff out over the water. I noticed that the far end of the wooden part of the implement was stained very dark compared to the lighter color of the rest of the pole. He later explained that the dark color made it more difficult for the salmon to discern in the tannin-colored water of the stream. Thomas then gently stepped out into the stream, placing his left foot on a flattened circular wooden step that had been positioned in the stream. There was another stepping block beyond where he then placed his right foot – also quietly and carefully causing virtually no ripple in the water. These two blocks provided a base for gaffing and allowed Thomas to enter the water gently, thereby not disturbing the fish and causing them to leave the area. Thomas then began to position the hook end of the gaff above the water near the opposite shore. About 10 salmon, relatively dark, circled slowly near the opposite shore. After watching the fish move in this area for about five minutes, Thomas decided to move downstream to another location to do the gaffing.

On the walk down the trail by the rapidly flowing stream, Thomas explained that the successful gaffing technique involved careful placement of the gaff just above the surface of the stream in a location that anticipated the movement of a salmon just below and inside the position of the gaff. When a salmon swam into the gaffing zone, Thomas explosively dropped the head of the gaff and powerfully pulled the gaff head toward him through the stream. The intention was to impale the salmon on the point of the gaff on the top of the fish in the fleshy area immediately behind the head. The rapid and powerful movement of the gaff with the impaled fish toward the nearshore resulted in the salmon being transferred to the shore.

As we descended the stream, we came to a flatter area with numerous medium-sized rocks in the channel and slower-moving water. Soon we arrived at another gaffing site where Thomas noted that the fish were brighter than above, meaning that they were fresher, more recent arrivals from the saltwater. Thomas then went down to the spot along the shore that was clearly visited frequently and positioned himself in an appropriate spot for gaffing.

Just below the new gaffing site, I noticed a semi-circular arc of rocks in the stream – they clearly had been moved and placed in that position. The half-circle was located from the shore to approximately the middle of the stream. Rocks were randomly positioned across the bottom of the stream elsewhere in the area. The area inside the semi-circular area was clear and the gravel on the bottom inside of the arc was clearly evident. Thomas later told me that the rocks had been rearranged into this pattern in order to provide better access for the salmon to the spawning gravels. Soon after, we were joined by Thomas's brother Pat. He noticed me observing the stone arc and said that they had created the area to improve spawning opportunities in the area.

Just above, Thomas began positioning his gaff as the brighter sockeyes moved about the small area. I watched as he then gaffed several fish and pulled them to shore depositing them in the bushes. After the fish was brought ashore, he removed the fish from the gaff and positioned his body such that he was holding the salmon so that its head was pointing upstream. He then used a club to strike the fish on the top of the head – the fish stunned by the blow, immediately stopped its flopping movements. Thomas explained that positioning the fish so that its head pointed upstream ensured that when the spirit left the fish, as it died following the stunning blow, it would head to the land of spirits above that would allow it to travel home and be reborn.

Thomas continued to gaff fish until he had obtained eight. He noted that they always made sure to take more males than females. In this case, I observed that he had taken six males and two females. Pat joined us, looked over the fish and picked up an especially bright male. He then ripped off the jaw of the fish and took a bite out of it. He held the jaw up and smiled at us stating, "There is nothing more tasty than fresh salmon cheeks." On that note, Thomas took a long slender branch from a nearby bush, stripped it of shorter branches and threaded the branch through the gills of the fish. We then continued down the stream toward saltwater with the salmon being carefully guided to minimize bruising.

Soon we came to another flat area of the stream which broadened considerably in this area. It was shallow enough to ford so I moved across the stream. I turned and looked up the stream to an idyllic image. Above, the stream tumbled through boulders and then slowed as it moved into a wide deeper pool with no ripples and little evidence of movement. I stepped up more closely and noted a low-lying dam made of many rocks positioned tightly together. The rocks formed a straight line which caused the pool to form behind it. Water still flowed over the dam in several areas due to the water level behind it. I then looked into the pool and saw that there were no rocks inside of it and the bottom was a gravel mixture. It suddenly struck me that the dam structure was likely constructed by moving the rocks from behind it to form the pool. Just then I heard Pat, who had followed me across the stream and stood behind my right shoulder comment, "We are not the first ones on the river." This was one of those ah ha moments – when a set of observations suddenly crystallize into an insight. I realized then that the Tlingit term – *ish* – did not merely refer to a quiet deep hole in a stream or river where salmon congregated in slowly moving eddying waters but rather was a nuanced and profound concept with multiple meanings and reverberations in the relations between salmon and Tlingit. From that moment of insight, I began an intellectual journey in an effort to create an assemblage of *ish*-related phenomena in Tlingit culture.

Understanding *ish*

The concept of *ish* is an exquisite example of the interpenetration of the empathic reflexivity generated by the Salmon Boy charter/covenant and astute empirical observation of salmon behavior.

Within the mythic charter/covenant as well in the elders' narratives and answers provided during the fieldwork mentioned above, the concept of *ish* appeared both when speaking in Tlingit and when speaking in English. Since the concept was brought into English without translation, it became evident that there was no literal referent in English for "*ish*." The concept was positioned in accounts in various ways but always in relation to aspects of salmon behavior and human practices associated with salmon.

In the first instance, an *ish* is a deep hole in a stream. George Ramos, a Tlingit scholar from Yakutat, provided the following definition of "*ish*":

> This place called an ish on a river or tributary is where the salmon rest; it is a deep pool of clear water and from here they continue their journey. This is called an ish. It is a deep water place. The place where salmon rest.
>
> *(Langdon 2006a: 143)*

Beyond this definition, *ish* has multiple characteristics and utilizations. Tlingit observe that salmon endure great hardships, overcome challenges and threats and demonstrate enormous perseverance and tenacity in their struggle to return to their stream homes. After "taking a drink" of freshwater in the estuaries, their will to ascend the rivers is on direct display and available for Tlingit to perceive. When witnessing salmon battling fast water, navigating rocky barricades and leaping up steep falls, Tlingit see the determination and enormous expenditure of effort that salmon make to reach their homes in the stream and engage in reproductive activity. Tlingit see, as noted by Ramos above, that salmon congregate in the *ishes* where they are able to rest as they drift around in the eddies to regain their strength in order to continue on their journey.

The *ish* is a site of salmon harvest as Tlingit selectively gaffed and speared salmon, taking males and females in a ratio of 3 to 1 (ibid.).

The *ish* is an index site used to determine whether a stream has sufficient salmon. Huna Tlingit elder Sam Hanlon stated that his grandfather always checked the first *ish* in a stream above the beach to see how many salmon were in it. If it was full of salmon, only then would they be allowed to begin harvesting salmon from the stream.

The concept of *ish* could be applied in different contexts. The Chilkat River is a major system draining into Lynn Canal in the northern part of southeast Alaska. The *Jilḵáat Ḵwáan* occupied, used, and owned most of the drainage. The river valley is one of the corridors used and owned by *Jilḵáat* to the interior where they traded for furs with Athabascan groups. The primary village of the *Jilḵáat Ḵwáan* was Klukwan, situated approximately 22 miles from the mouth, on the east

Figure 14.2   The Ish pictured was created by Tlingit trustees by moving rocks in the river to create the low falls thereby establishing preferred habitat for salmon
Source: Steve Langdon

shore of the river. Immediately opposite the village, on the west shore, the Tsirku River entered the Chilkat. Several miles up the Tsirku River, a short channel exits Chilkat Lake, a substantial body of water and the location of the largest sockeye salmon run of the Chilkat River. Chilkat *Kagwaantaan* clan elder Joe Hotch (Figure 14.2) provided the following account concerning how salmon were engaged with on the Chilkat River:

> We had one leader that lived in Klukwan. When he knew it was time for fish to come up river he would tell some of the young men 'Go down, 6-mile, 12-mile, 18-mile and watch.' He controlled it … only one elder controlled the whole river. They were to watch for salmon. If there were salmon there, they would come up and tell the chief. When they [salmon] get to Chilkat Lake, the chief would say 'Gook, Gook go ahead. Go up and get what you need but don't take more than you need. Take enough for yourself and take enough to share with others.' That was always important – to have some to share with others … The chief was *Yeilxáak* of the *Gaanaxteidí*.

The *Ganax'tedi* are a Raven clan whose ancestors founded Klukwan.

This account shows how the use of the *ish* as an index site for the purposes of establishing escapement was scaled up by the *Jilkáat* Tlingit and applied to Chilkat Lake. No salmon were allowed to be harvested in the river or lake until they had arrived at Chilkat Lake. Finally, Chilkat Lake is located up the River Tsirku, a tributary of the Chilkat, but salmon also run up the main channel of the river so that escapement to those streams would also result from restraining harvest until after salmon reached Chilkat Lake.

Figure 14.3   Arch House front painting of *ish* scene

Source: Walter Soboleff Collection, Alaska Historical Library.

The *ish* is also viewed by the Tlingit as a site of existential contemplation where salmon consider their impending passage from life to death. Richard Dauenhauer, a Tlingit speaker and scholar of Tlingit culture, translated the name *Aak'wtaatseen* – the name of the Salmon Boy in the mythic charter/covenant – as "Alive in the Eddy" demonstrating this facet of Tlingit thought.

Another revelation about the significance of *ish* that came into view was the aesthetic representation of the *ish* in a number of Tlingit artistic objects. The most striking of these is a painting on the front of a house in the village of Killisnoo called the "Salmon Dance House" or the "Arch House," that represents an *ish* in the clan territory (Figure 14.3). The image represents a complex translation of the *ish* that begins with the circle in the center of the image. This is the symbolic form chosen by the Tlingit to represent this concept, as elsewhere in the artwork the circle is a minor incidental embellishment. Here it is the center around which all other elements pivot. Around the circle are six sockeye salmon, all with their noses pointed toward the circle, giving a sense of desire and intent to enter the *ish*. This is a comment on the potency or perhaps sacred quality of the space in Tlingit thought. On the outer edges are claws that imply the presence of bears, another species that also shares an interest in the *ish* as a place to obtain salmon. At the bottom oriented toward the *ish* is a bird (perhaps a saw bill, canvasback, or merganser) also present in the real world at the *ish* seeking the eggs that will be released by the female salmon. As discussed above, birds are part of the existential frame of the *ish* but, as discussed earlier, their numbers must be in balance so as not to damage the salmon. Sitting in the background is the arching form that has been interpreted as a mountain, or the arch (from which comes "Arch House") that in the actual geography sits near and above the *ish* on the landscape. It is one of the most profoundly elegant depictions in Tlingit art that presents a multi-layered and textured account of forms and relationships at once mirroring certain realities while privileging elements of significance.

The passage below from revered Tlingit scholar Walter Soboleff demonstrates the position of the *ish* in Tlingit thought:

> There were those who were knowledgeable about all kinds of subjects. This thing named ish - it was almost as if it were human and it was spoken to in that way, this ish. This is how they valued this resource. It was as if their life depended on it so they treated it with respect. Because they got their food from this place is why they would speak to it. There was pride, there was honor (given to the ish) so no one was to say anything foolish about it or to it. If it was said that we could laugh at it, it was not so. We were told not to talk to it in a foolish way but to respect it. This is what the white man calls taboo. When you do this, there is a discipline, a law that will correct you. It will be like it falls on you; this is the way this is. All that is seen around us is said to be alive around us is what it is called. The Tlingít people have known this to be true from time immemorial.
>
> *(Langdon 2006a: 145)*

Protecting the *ish*: attentive agentive practice

In 2012, I was conducting research on traditional knowledge of and practices with members of the *Jilḵáat Ḵwáan and Jilḵáat Ḵwáan* who live in northern southeast Alaska, when I had the opportunity to once again learn about the practice of Tlingit trusteeship over salmon streams. In a background interview with David Light of the *Jilḵáat Ḡaanaxteidí* on the *Jilkoot River* he had been associated with through his father-in-law Austin Hammond, I learned about the Glory Hole, a location in the upper portion of Chilkoot Lake renowned for the concentration of salmon that are found there during spawning season. The Tlingit place name is *Goonk'* translated as Little Spring (Thornton 2012: 58). I sensed that it might be an example of an *ish* so I wanted to visit it. This was later confirmed by *Lukaax̱.ádi* elder and renowned Tlingit artist Nathan Jackson. I inquired of David Light how we might be able to reach the site. I was told that a logging road had been built along the west shore of Chilkoot Lake many years ago, that might be accessible and take us to the area. However, we learned that the road had not been maintained and now was impassable. Light then contacted Tim Ackerman, a young Tlingit whom he first met as a young man at his uncle Austin Hammond's culture camp on the Chilkoot River and whose mother had recorded and published local cultural traditions (Ackerman 1975). Tim had been designated as one of the watchmen for the camp facilities after its closure, due to his deep interest in his Tlingit culture and his heritage. David inquired if Tim had a skiff and whether he would be interested in taking us to the Glory Hole. Tim responded that, indeed, he would be able to take us to the Glory Hole and we arranged to meet at the lake outlet.

The three of us boarded Tim's skiff and he navigated up the west shore of Chilkoot Lake, stopping at culturally significant locations to give us information on them. As we approached the north end, we noticed that the lake was shallowing up and Tim pulled into shore a hundred yards below the outlet to a stream in the northwest corner of the lake. Tim had brought his rifle with him and directed us to follow on the trail behind him as it was possible that, given the time of year when salmon are returning, bears would be in the area.

The trail followed the stream from its outlet in Chilkoot Lake north where it would culminate at the Glory Hole. As we carefully followed Tim, he began an account of a recent trusteeship experience. Nearly a decade ago, he had noticed that sockeye returns to the Chilkoot River had declined. He had worked for the Alaska Department of Fish and Game for several years as a stream technician, had made numerous trips up the main tributary at the head of the lake to make escapement counts, and had visited the smaller stream with the Glory Hole at its head. About 50 yards above the stream outlet, Tim stopped and pointed to the opposite shore where there was a substantial pile of what appeared to be stream debris – sticks, rocks, and dirt. Tim said that was the remains of a beaver dam that he and his friend had taken out about six years previously. He described how as they began their earlier investigation of the sockeye decline, they had first come to this stream in the spring several years back. They discovered a large and compact beaver dam built across the entire stream that had essentially halted the stream flow. Behind the dam, a pond had emerged that was holding the water from the stream. In the pond, Tim saw literally thousands of salmon fry who had begun their outward migration to saltwater but were trapped behind the dam. He and his friend immediately began breaking a hole in the dam and soon had created a passage large enough for water to flow out of the pond. Shortly after they were excited to see that the salmon fry were on their way downstream through the dam and soon there were no fry remaining in the pond. He and his friend decided that they would have to come back and remove the entire dam, as they knew that such a small opening could easily be repaired by the beavers. Eventually they did remove the dam completely, leaving only the remnant end that we had observed earlier.

We then continued on the trail until we arrived at the Glory Hole. Tim explained that the Glory Hole was a pond created by a freshwater spring. Less than 20 yards to the immediate north of the Glory Hole is the end of a landslide that had come down the mountain some time ago. The mountains around Chilkoot Lake are extremely unstable and previous landslides in other nearby locations have played an enormous role in the history of the *Jilkoot Kwáan*. Tim commented that there was a significant amount of copper in that area and infused the upwelling water of the spring that fed the Glory Hole such that the bottom of the pond had the iridescent blue/turquoise hue that derives from copper.

As we acclimated our eyes to the Glory Hole and walked along its shore, we noticed that indeed there were numerous bright red sockeyes moving languidly through the waters (Figure 14.4). We also observed that most of the pond bottom was of a medium-gray hue. Only on the edges did we see areas of the blue/turquoise color. Tim explained that when the beaver dam was in place, the blockage of the stream had led to the deposits of sediments that changed the color of the bottom. The blue/turquoise areas that we saw now were due to the halting of the sedimentation as the stream now running again carried the sediments out. The Glory Hole was gradually returning to its beautiful blue/turquoise hue.

The *Lukaax..ádi* are the clan owners of Chilkoot Lake area including *Goonk'*. One of the clan's most prized *at.oow* (sacred property) is a Chilkat blanket (*Naaxein*) that represents *Goonk'* through the large prominent bold circle at the center of the blanket (Figure 14.6). Inside the circle is a large salmon with small circles standing for eggs that are deposited in *Goonk'*. A wonderful touch in the blanket is the liberal use of turquoise as the color of symbols within the blanket, a representation of the color of *Goonk'*. The blanket is another example of the Tlingit aesthetic representation of place and characteristics that are seen as valued and prized by salmon. It is at once a social statement of ownership and a gift of beauty to its viewers, including salmon. As Thornton (2012: xxi) noted: "Looking at crests from the perspective of place opens up new horizons of meaning."

Figure 14.4   Goonk' or Glory Hole, Chilkoot Lake. Author Steve Langdon and *Jilḵáat* elder David Light stand beside *Goonk'*. At the top of the image the turquoise coloring on the *ish* bottom is present as it returns and at the bottom, sockeye salmon turning red can be seen

Source: Steve Langdon.

Figure 14.5   Tlingit elder Joe Hotch

Figure 14.6   Lukaax̱.ádi Naax̱ein showing *Goonk'* as the circle in the middle

Source: Steve Langdon.

Empirical understandings and pragmatic undertakings

The concept of traditional ecological knowledge has been critiqued on a number of fronts, including that it fails to direct attention to the totality of cultural constructions – the existence scapes – and processes through which people organize their positions in nature. In addition, its emphasis on traditional implies a body of fixed ancient and ecological material so that it attends only to "natural" environmental phenomena. This chapter has demonstrated that in order to approach an understanding of Tlingit relations with salmon, a multiplicity of sites of cultural knowledge and practice must be accessed, and connections among them explored. Multiple methodologies from archeological fieldwork to artistic representation have been required to convey the perspective presented in this chapter. Further, such investigations must not merely codify knowledge but they must also document practices that derive from that knowledge and the linkages between them. In order to provide as accurate an account as possible and fulfill the scholarly obligation for integrity in the representation of materials from other cultures, researchers must explore and be open to a myriad of evidence that can illuminate how people engage with the world.

Tlingit engagement with salmon, in addition to being grounded in a foundational mythic charter/covenant that generates an empathic reflexivity, is nonetheless profoundly pragmatic as it is based on astute empirical observation and contemplation of environmental conditions and salmon behavior. As this chapter has demonstrated, the meanings derived from both braid together continuously stimulating recursive attention to the needs of salmon, thereby ensuring the continued existence of human people and salmon people.

References

Ackerman, M. 1975. *Tlingit Stories*. Anchorage, AK: AMU Press.

Bourdieu, P. [1977] 2002. *Outline of a Theory of Practice*. Cambridge: Cambridge University Press.

Garfield, V. and L. Forrest. 1961. *The Wolf and the Raven: Totem Poles of Southeastern Alaska*. Seattle, WA: University of Washington Press.

Langdon, S. 1986 Traditional Tlingit stone fishing technologies. *Alaska Native News* 4(3): 21–26.

Langdon, S. 2006a. Traditional knowledge and harvesting of salmon by HUNA and HINYAA LINGIT. FIS Final Report 02-104. Anchorage, AK: US Department of Interior, Fish and Wildlife Service, Office of Subsistence Management.

Langdon, S. 2006b. Tidal pulse fishing: selective traditional tlingit salmon fishing techniques on the West Coast of the Prince of Wales Archipelago. In C. Menzies (Ed.), *Traditional Ecological Knowledge and Natural Resource Management*. Lincoln, NE: University of Nebraska Press, pp. 21–46.

Langdon, S. 2007. Sustaining a relationship: Inquiry into a logic of engagement with salmon among the Southern Tlingits. In M. Harkin and D. R. Lewis (Eds.), *Perspectives on the Ecological Indian: Native Americans and the Environment*. Lincoln, NE: University of Nebraska Press, pp. 233–273.

Langdon, S. 2014. *The Native People of Alaska*, 5th edn. Anchorage, AK: Greatland Graphics.

Langdon, S. 2019. Spiritual relations, moral obligations and existential continuity. In D. Narvaez (Ed.), *Indigenous Sustainable Wisdom*. New York: Peter Lang, pp. 153–182.

Littlefield, R., E. Makinene, L. George, N. Dauenhauer, and R. Dauenhauer (Eds.) 2003. Aak'wtatseen, "Shanyaak'utlaax." Told by Deikeenaak'w, Sitka, 1904. Originally transcribed by John R. Swanton, 1904. Transliterated into modern orthography by Roby Littlefield and Ethel Makinen. Alaska Native Knowledge Network.

Narvaez, D. 2014. *Neurobiology and the Development of Human Morality: Evolution, Culture, and Wisdom*. New York: W.W. Norton and Company.

Narvaez, D. 2016. *Embodied Morality: Protectionism, Engagement, and Imagination*. London: Palgrave Pivot.

Pulu, T. (Ed.) 1983. *The Transcribed Tapes of Christine Edenso*. Anchorage, AK: Materials Development Center, University of Alaska.

Swanton, J. (Ed.) 1909. Tlingit myths and texts. *Smithsonian Institution Bureau of American Ethnology Bulletin 39*. Washington, DC: Government Printing Office.

Thornton, T. 2008. *Being and Place among the Tlingit*. Seattle, WA: University of Washington Press/Juneau, AK: Sealaska Heritage Institute.

Thornton, T. (Ed.) 2012. *Haa Léelk'w Has Aaní Saax'ú. Our Grandparents: Names on the Land*. Seattle, WA: University of Washington Press.

Thornton, T., D. Deur, and H. Kitka Sr. 2015. Cultivation of salmon and other marine resources on the Northwest Coast of North America. *Human Ecology* 43: 189–199.

15

MĀTAURANGA AS KNOWLEDGE, PROCESS AND PRACTICE IN AOTEAROA NEW ZEALAND

Priscilla Wehi, Hēmi Whaanga, Krushil Watene and Tammy Steeves

Aotearoa New Zealand (ANZ) is an archipelago of around 600 islands in the South Pacific. Dominated by three large islands, ANZ was the last sizable landmass to be peopled around 800 years ago (Wilmshurst et al. 2008). Positioned in remote Oceania more than 2,000 km from other large islands and archipelagos, ANZ has a unique environment that has evolved in relative isolation for around 80 million years, with no extant native terrestrial mammals other than bats, and a rich seascape that both separates and connects island biota, cultures and peoples. On reaching ANZ, the Polynesian ancestors of Māori had already travelled the vast Pacific Ocean for around 2,000 years, settling on islands along the way. An understanding of Māori environmental philosophies and knowledge requires an understanding of the way that these systems are intricately rooted in, and shaped by, these Pacific pathways (see, for example, Matamua et al. 2013; Tuaupiki 2017).

A focus on relationships in Māori environmental philosophy grounds the significance of *whakapapa* (meaning to place in layers, and indicating connection, lineage and genealogy between people, ecosystems, and all living and non-living things). Just as *whakapapa* can map the navigational lines connecting the peoples of the Pacific, so it can map the connections between Pacific landscapes and oceans, and the vast knowledge embedded within them. Within a *whakapapa* framework, knowledge is mapped in both descent (linear) and kinship (lateral) layers, providing both deep and broad insights, while also acknowledging the entanglement of people, places and practices (Kawharu 2000). The *whakapapa* of kūmara (*Ipomoea batatas*) in ANZ, for instance, details weeds that plague kūmara plantations, as well as optimal soil and growing conditions (Roberts et al. 2004; Wehi and Roa 2020). *Whakapapa* is the foundation of Māori philosophy, encompassing social, cultural, environmental and ecological knowledge, the systems used to generate knowledge, and the basic assumptions that ground it. In ANZ, this is collectively referred to as *mātauranga* and is a philosophical tradition in its own right.

Mātauranga is therefore a system of thought that incorporates Māori concepts, world-view, and their application (see, for example, Berkes 2004). Ecological knowledge has a particularly important role articulating *whakapapa* relationships within this knowledge system, and as such is deeply embedded in tribal histories. People descended from one of the many voyaging vessels

DOI: 10.4324/9781315270845-17

that arrived in ANZ, the Tainui, recount narratives of Whakaotirangi, an early ancestor famous for her horticultural prowess, and Kahupeka, who traversed the Pirongia region to discover medicinal remedies in an attempt to heal herself (Jones and Biggs 1995; Wehi and Roa 2020). Similarly, Hineamaru, the ancestress of Ngāti Hine in the north, is said to have planted kūmara at Paparata, where it thrived and sustained her people (Shortland 2012). Connections between the natural environment and historical narratives are chronicled in many forms of oral and written tradition extant throughout the country, from the *pātere* or chant Te Koko ki Ohiwa from the Bay of Plenty (Black 2014) through to the diaries of customary harvesters who have spent many decades muttonbirding (Clucas et al. 2012). Recognising this to be the case, Māori environmental philosophies are inclusive of and shaped by the oral traditions that connect Pacific journeys, encounters with new landscapes, and the ongoing socio-environmental and political experiences of today (Smith et al. 2016).

Māori environmental philosophies and knowledge are now as critical as ever. More than 40 per cent of New Zealand's birds found nowhere else in the world are extinct, and a third of those remaining are in serious trouble (Cassey 2001; PCE 2017). Other terrestrial taxa such as reptiles are also in decline (Hitchmough et al. 2015). The unique concepts and knowledge systems that make up Māori environmental philosophies are attracting new-found attention as conservationists, policymakers and community groups alike seek to help stem this tide of environmental loss. Māori environmental philosophies encompass both principle and practice. As a practical philosophy, the knowledge itself cannot be separated from the principles and responsibilities that surround its use (Watene 2016a; Hikuroa 2017). Practices such as *manaakitanga* (to care for) and *kaitiakitanga* (trusteeship) are active expressions of customary responsibilities to protect and enhance our relationships across and within our landscapes (Bioethics Report 2019). These concepts are now visible in public concerns for and debate around both land and water (Watene 2016b).

The central place of environmental ethics in *mātauranga* means that Māori have become a voice for change in freshwater management in ANZ. With many rivers and lakes currently polluted and no longer safe to swim (likely c.30–35 per cent of river stretches; Ministry for the Environment and Stats NZ 2019), river and lake health is a long-term issue requiring urgent and sustained attention (Young et al. 2018). *Mātauranga* emphasises the multidimensional value of water, valuing its physical (life-giving) properties as well as its social dimensions, such as the role it plays in facilitating relationships of water bodies with people through customary usage (see, for example, Kitson et al. 2018; Paul-Burke et al. 2018). This approach has led *mātauranga* practitioners to develop cultural indicators that communities can easily apply to measure water health, and that are based on Māori concepts such as *mauri* (life force) (Morgan 2006; Tipa and Teirney 2006; Crow et al. 2018; Hikuroa et al. 2018; Hopkins 2018; Kitson et al. 2018). These cultural health indices assess factors such as site accessibility, the ability to undertake *mahinga kai* activities (customary food gathering), for example, by considering the richness and abundance of species, as well as water quality and land use (Tipa and Teirney 2006).

European settlement and colonisation from the early nineteenth century have added layers of complexity to the intergenerational transmission and expression of *mātauranga* in ANZ. A history of colonisation and land alienation has led to social and environmental inequities between Māori and the dominant Pākehā or European population that remain unresolved (Waitangi Tribunal 2019). Widespread alienation of tribal lands has impacted language, knowledge and practice, and Māori resistance to land alienation and the accompanying cultural destruction is well documented (Mikaere 2011; Kawharu 2018; Mutu 2018). As just one example, letters from Māori in the nineteenth-century Māori newspapers have detailed both impacts and concerns in relation to the Native Lands Act 1862 and 1865 and their amendments (Whaanga and Wehi

2017). Other legislative directives, from the New Zealand Settlements Act 1863 through to the Conservation Act 1987, have made it inherently difficult, if not impossible, for Māori to maintain cultural practices, such as harvesting of traditional food species (Ruru 2004; Wehi and Lord 2017). Despite this powerful historic legacy, contemporary attempts to address these wrongs are now underway. Recent laws such as Te Urewera Act 2014 and Te Awa Tupua (Whanganui River Claims Settlement) 2017 have pioneered co-management solutions that embody partnership with Māori tribal groups, enable customary responsibilities and thus enhance revitalisation and the continuing relevance of *mātauranga*.

Politically, Māori cultural revitalisation has also changed how Māori environmental philosophies are perceived and acted upon in New Zealand ecological and conservation research and policy. Strong calls for Māori sovereignty and an end to inequitable resourcing have led to changes in government funding policies that support *mātauranga*-based research (Wehi et al. 2019a). These calls repeatedly reference the 1840 Tiriti o Waitangi/Treaty of Waitangi which commits to a partnership between Māori chiefs and the British Crown and affirms the rights of Māori. These Treaty principles underlie non-binding rulings by the Waitangi Tribunal, such as a report on the flora and fauna claim WAI 262 (Waitangi Tribunal 2011). In addition, a broad shift by scientists towards the prioritisation of building biocultural partnerships represents a genuine desire to incorporate Māori voices in biodiversity conservation. For example, Cisternas et al. (2019) use native frog conservation as proof of concept to describe a novel framework that enables culturally respectful exchange between researchers, practitioners and relevant communities. This in turn can lead to the co-creation of conservation policies and practices for threatened species that are responsive to diverse knowledges, values and motivations. Collier-Robinson et al. (2019) advocate for the importance of co-developing genomic research for culturally significant species that is responsive to the needs, aspirations and circumstances of local communities, including meaningful consideration of critical issues related to data generation, storage and access, using their research on ANZ freshwater fish and crayfish as an example. Although this shift towards best practice biocultural research is not ubiquitous, it nevertheless represents clear intentions, especially in ecological research (Wehi et al. 2019a). In addition to addressing inequity, these commitments to Tiriti o Waitangi are building the capacity of Māori researchers, to whom research involving Māori environmental philosophies is generally entrusted.

Despite these advances, the relationship between Māori environmental philosophies and what is often termed "western science" is uneasy, with many arguing not only that these are two different knowledge systems (Moller et al. 2009; Wehi et al. 2019a), but also that dominant discourses of ecology continue to erase Indigenous Knowledge (Sangha et al. 2019). *Mātauranga* has certainly been disregarded by *Pākehā* (non-Māori) ecologists and others at some critical junctures in ANZ's history, including during the submission process on proposed legislative changes to Māori customary use of resources in the 1990s (New Zealand Ecological Society 1995), in which the Society submission supported constraints on customary harvest. In contrast, Māori experts challenged the Society's views and the suggested constraints on Māori practices in their own submission (Wright et al. 1995). As such, *mātauranga* "contests mainstream science for recognition, support and implementation" (Lambert et al. 2018). In addition, although Māori philosophy includes scientific knowledge, it is a broad philosophical tradition in its own right, and as such is much deeper and broader than western science in its scope of concerns.

Modern realities

As ecologists begin to explore the intersections between Māori environmental philosophies and conservation policies and practice, a number of challenges arise. In addition to the challenge of working within and across environmental philosophies that begin from different foundational assumptions, a tension exists between romantic conceptualisations of Māori concepts (such as *kaitiakitanga*), and the reality of ANZ's current environmental state with many species in rapid decline. By romanticising Māori concepts, we not only limit the value of these concepts today, we also limit the extent to which these concepts, having evolved under pressures of species decline and scarcity, can speak with experience and insight to contemporary pressures. The arrival of Polynesian ancestors to remote Oceania involved a change from known tropical systems to an unknown temperate environment with many low-reproducing, long-lived species, including large flightless endemic birds, and the concomitant pressure to grow knowledge quickly in an environment that was extremely sensitive to change. The consequent extinctions of vulnerable species in ANZ are mourned in Māori oral tradition (Wehi et al. 2018).

Further ecological degradation after European colonisation in the early nineteenth century exacerbated and amplified such losses (Innes et al. 2010). In practical terms, modern ANZ now has a landscape of extreme human modification, with more than 90 per cent of wetlands destroyed (Cromarty and Scott 1996), a decrease in native forest coverage after human arrival from 85 per cent to 23 per cent (McGlone 1989; LCRIT 2015), and large numbers of introduced mammals such as stoats, rats and possums which pose a major threat to endemic birds and other native species (Fea et al. 2020). In the Pacific, therefore, discussions and acknowledgement of environmental knowledge incorporate a tradition of change, embedding stewardship that both acknowledges lessons learned, and emphasises the prerogative to undertake new as well as old practices that enhance rather than destroy the natural environment.

In 2002, Te Ahukaramu Royal argued that a continual focus on grounding authenticity and tradition in pre-colonial times works to obscure the realities of dynamic growth and development in Māori philosophy (Royal TAC 2002). It is now more widely accepted in ANZ that *mātauranga* is dynamic, and readily incorporates new ideas, methodologies and circumstances. One of the few cultural harvesting traditions that has continued uninterrupted since European arrival is tītī or muttonbird (*Puffinus griseus*) harvesting on islands in the south of the South Island, where young chicks have been harvested and preserved by Rakiura descendants for centuries (Kitson and Moller 2008; Moller et al. 2009; Clucas et al. 2012; Humphries 2014). Many technological advances now expedite this customary harvest, including the use of plastic buckets to store graded chicks rather than woven flax and bark baskets, and use of helicopters to reach some locations. Despite these differences, the harvest continues to rely on an intimate knowledge of biology, weather patterns and other environmental knowledge for its successful harvest. What's more, in order to safeguard this practice for future generations, the Rakiura muttonbirding community partnered with scientists to better understand some of the global migration patterns and demographics affecting the tītī harvest. This resulted in the seminal 14-year partnership "Kia Mau Te Tītī Mo Ake Tonu" between Rakiura and scientists, which also led the way in its ethical approach to examining *mātauranga* and "western science", and public engagement with the research (Moller et al. 2009). In other parts of the country, cultural practice has suffered substantially under the impact of colonisation, but recent partnerships between communities and researchers have resulted in the reclamation and revitalisation of Māori environmental knowledge and reinstatement of some harvesting traditions, including that of oi (*Pterodroma macroptera gouldi*) in Hauraki (Lyver et al. 2008).

Despite these successes, tension remains between state governance groups and iwi (descendant communities) seeking autonomy to reinstate Māori environmental philosophies into ecological management. Inconsistencies include the application of conservation law that prohibits customary harvest in some instances but not others (Lyver et al. 2019). And, for iwi such as Ngātiwai, maintaining viable populations of kiore (*Rattus exulans*), a culturally valued species classed as invasive, is an ongoing issue with the weighing of *kaitiaki* responsibilities for multiple species (Te Iwi o Ngatiwai 2007).

New challenges

Both local and global mobility have led to new challenges, from the extinction of nature to increased risk of biological invasions. Land alienation has led to Māori being a highly urban people, with more than 80 per cent of Māori now living in cities, many outside of their tribal areas (Ryks et al. 2016). Destabilisation of cultural practice has resulted in the loss of accompanying language and knowledge as both individuals and family groups have left trad- itional lands, a story repeated elsewhere in colonised nations. The strong links between linguistic and biocultural diversity (Maffi 2005) mean that investment in supporting customary practice upholds both the Māori language itself, and the cultural and social practices that constitute *mātauranga*. *Mātauranga* and language revitalisation have thus become a concern to many, with solutions being sought via a range of means from ecological and cultural restoration (Taura et al. 2017; Walker et al. 2019) through to governmental policy goals, and interdisciplinary research such as dynamical modelling of language speakers that seeks to maximise language revitalisation (Barrett-Walker et al. 2020).

Walker et al. (2019) describe some of the relational responsibilities and practices that underpin the learning and transmission of *mātauranga*. They show the importance of tribal narratives as a means of imparting *mātauranga*, and thus how connection to place is a vital thread in *kaitiakitanga*. Crucially, they highlight the challenge of participation in cultural practices that support biocultural diversity for those who have ancestral roots in other places, and they discuss opportunities for engagement with *mātauranga* in an urbanised world that is also under pressure from climate change and anthropogenic impacts. Barriers to experiencing nature observed within urban Māori communities appear to be a microcosm of the global extinction of experi- ence with nature (Soga and Gaston 2019; Walker et al. 2019).

The focus on *mātauranga* within urban ecosystems is part of a larger wave of new work responding to the challenge of rebuilding ecosystem knowledge in ANZ. Reihana et al. (2019) developed an app for children in Māori language immersion schools, to re-grow their ecological knowledge and begin the journey of reconnecting to non-human kin. During workshops with school children to explore the implementation of a cultural monitoring framework, they noticed a lack of ecological literacy among the students. Reihana et al. (2019) then created and trialled a game that fused basic ecological knowledge and Māori environ- mental knowledge to offer a culturally and ecologically integrated view of the environment. In doing so, they increased student knowledge acquisition. The use of technology to create tools that support Māori environmental knowledge is thus one effective way to overcome intergenerational barriers to transmission of *mātauranga* and understanding of our natural world. Wehi et al. (2019b) took a complementary approach, and advocate highlighting Māori bird names in biodiversity reporting. Using birds as a case study, they unpacked some of the Māori environmental knowledge within these names to demonstrate observations of seasonal change, ecological relationships and behaviour, as well as regional nuances that connect tribal knowledge to specific landscapes. Understanding the Māori environmental knowledge that sits within such names offers an opportunity to appreciate non-human kin more deeply, a first step in environmental restoration. Wehi et al. urged policymakers to fund initiatives that resonate with communities to re-grow environmental knowledge. Such investment in future generations is not only key for the future vitality of *mātauranga*, and the future of the planet, it is also a central concern of *mātauranga* itself.

Te Tiriti o Waitangi also entails a responsibility for the Crown, represented by government agencies, to act on collective concerns in partnership with Māori. This includes duties that act to prevent or mitigate biosecurity threats. The ongoing response to kauri dieback disease (*Phytophthora agathidicida*) provides one example of the intersection between biosecurity and *mātauranga*. Believed to be a recent invader from Oceania, kauri dieback disease now threatens the health and existence of ancient kauri (*Agathis australis*) trees that dominate some of ANZ's most cherished forests (Scott and Williams 2014). Māori communities have partnered with government agencies in a long-term management programme that attempts to prevent further spread and have outlined a rationale and framework for monitoring kauri dieback disease based entirely on *mātauranga*, that incorporates both ecological variables and community experiences of forests. Within this climate, *mātauranga* is seen not only a contribution to forest management and health but also as resistance to colonisation (Lambert et al. 2018). Despite continuing challenges, there is a clear shift towards the empowerment of, and partnership with, local communities to implement solutions to ecological management issues, and it is likely that Māori environmental knowledge will be a critical contributor to management, including biosecurity threats, in future.

Data sovereignty, biocultural diversity, genomics and other ethical challenges

Many of the above approaches to enable Māori environmental knowledge, and *mātauranga* more broadly, rely on data of some kind. The rapid progress of ICT innovation, its volume, complexity and computing power, has radically changed how we socialise, exchange, access, manage, create, disseminate, display and research Indigenous data and particular forms of *mātauranga* (Whaanga et al. 2017). The ongoing challenges of representation and alienation of Indigenous Peoples, in terms of the digital divide, digital ethics, cultural and intellectual property rights, governance, and the collection, access, use and management of data about Indigenous Peoples, lands and cultures, is an area of growing concern for Indigenous communities (Kukutai and Taylor 2016; Kukutai and Cormack 2019). With vast amounts of new data being produced about Indigenous Peoples as well as the dominance of Western knowledge systems and methodologies underpinning data collection systems (Wilks et al. 2018), the emergence of Indigenous approaches to data access, use and management and sovereignty is a new and vital step (Kukutai and Taylor 2016; Walter and Suina 2019).

Indigenous data sovereignty is premised on the rights of Indigenous Peoples to "determine the means of collection, access, analysis, interpretation, management, dissemination and reuse of data pertaining to the Indigenous peoples from whom it has been derived, or to whom it relates" (Walter 2018). It focuses on

> Indigenous collective rights to data about our peoples, territories, lifeways and natural resources and is supported by Indigenous peoples' inherent rights of self-determination and governance over their peoples, country and resources as described in the United Nations Declaration on the Rights of Indigenous Peoples (UNDRIP).
>
> *(Walter and Suina 2019, pp. 236–237)*

The reclamation of data rights is happening across colonised nations including Australia via the *Maiam nayri Wingara Indigenous Data Sovereignty Collective*, the United States through the *United States Indigenous Data Sovereignty Network,* and in Aotearoa New Zealand through *Te Mana Raraunga* as the Māori Data Sovereignty Network. Established in 2016, *Te Mana Raraunga* advocates for "Māori rights and interests in relation to data, ensuring data for and about Māori is safeguarded and protected, and that data is utilised to advance Māori aspirations for collective and individual wellbeing" (www.temanararaunga.maori.nz; Hudson et al. 2017, p. 67).

In ANZ, ethical guidelines have been developed to manage data that draw on a foundation of Māori philosophies, ethics and practices for biobanking and genetic or genomic research involving human biospecimens (see www.genomics-aotearoa.org.nz/about/maori-and-genomics/guidelines; Beaton et al. 2017; Hudson et al. 2019). Genomics Aotearoa (GA), an alliance of universities and Crown research institutes and associative organisations across ANZ and internationally, has been established to build ANZ's genomics research capability and international connectivity in ways that are better aligned with Māori philosophies (www.genomics-aotearoa.org.nz/). There is growing national and international demand to generate genomic data for biospecimens sourced from taonga (culturally significant) species for biodiversity conservation, ecological and evolutionary genomic research. To address this issue, GA is leading the development of Te Nohonga Kaitiaki Guidelines for Genomic Research with Taonga Species (www.genomics-aotearoa.org.nz/projects/te-nohonga-kaitiaki). Furthermore, the Equity for Indigenous Research and Innovation – Coordinating Hub (ENRICH) is bringing together international scholars with expertise in Indigenous Data Sovereignty, intellectual property law, genome science and data science to develop the Biocultural Label Initiative, including a pilot study in ANZ, which seeks to create transparency around the provenance of, and cultural responsibilities associated with, genomic data (www.enrich-hub.org/bc-labels). In the meantime, there is growing recognition that researcher quests to generate genomic resources for culturally significant species – including reference genomes and population genomic data (see Collier-Robinson et al. 2019, Figure 1) – may push aside consideration of community priorities for a species, ignoring the cultural significance of that species for the community. Further, publishing genomic resources with no cultural context can result in no benefit for the species itself or those with *kaitiakitanga* responsibilities in relation to it. Instead, publishing these resources can encourage a competitive culture that is not mindful of negative impacts to either local communities or researchers engaged in genuine biocultural partnerships. Indeed, to build capability in genomics, especially among Indigenous Peoples, a growing number of researchers are protesting against a research culture where research groups view biota as a shopping basket of interesting species they can exploit, continuing a legacy of colonialism (see also the Nagoya Protocol on Access to Genetic Resources and the Fair and Equitable Sharing of Benefits Arising from their Utilization to the Convention on Biological Diversity [www.cbd.int/abs/]). These issues also exist in the context of medicinal plants.

Data are also the backbone of museum collections, embedded within the objects that are protected within these institutions, but that are often orphaned from their traditional owners and contexts. Scientists – including genomic scientists – are increasingly exploring museum collections and herbaria to explore past biodiversity patterns and distributions, the structure of ecological communities, geographical origins for artefacts and more. As yet, little of this research links adequately to cultural knowledge or Indigenous relationships (Wehi et al. 2012). Nonetheless, museum collections and herbaria have a great deal to offer future research that integrates *mātauranga*. Herbaria specimens were a core contributor to data used to model the spatial distribution of two culturally significant plants, one used for weaving (kuta; *Eleocharis sphacelata*) and one medicinal plant (kumarahou; *Pomaderris kumaraho*). Bond et al. (2019) examined a range of future climate projections to determine how climate change might affect weavers' and medicinal practitioners' access to these signature plants. Altered plant distributions will have ramifications for gift-giving practices between neighbours. The research team of weavers, medicinal users and scientists returned the findings to both tribal and weaving groups. If we are to harness the potential of museum collections to reveal future insights and address cultural and biodiversity loss, we need to decolonise museum processes and structures to create space for collaborations that value the potential of *mātauranga* (Salick et al. 2014; Palavi et al. 2018).

Unpacking Māori environmental knowledge to reveal insights for the future

The future of ANZ's biodiversity is intimately linked to the health of Māori environmental knowledge, the ability to work within different philosophical traditions, and a willingness to work at the intersections of philosophy, cultural practice and science. We require innovative and sustainable solutions that acknowledge not only the inter-relationships of social and environmental goals, as highlighted in the Sustainable Development Goals, but the centrality of culture in framing these solutions also (Watene and Yap 2015; Dawes 2019; Yap and Watene 2019). It is critical for regional and government policy to address social justice, including inequalities such as housing, health and employment that are intertwined with the ability of tribal members to live and thrive within rural homelands, simultaneously strengthening relationships to the environment as kin, if *kaitiakitanga* and Māori environmental knowledge are to thrive (Harawira 2015).

Smith et al. (2016) argue both for the recognition of expert knowledge keepers in communities, and a shift away from some of the false binaries that separate science and Indigenous Knowledge. It is clear that we need diverse researchers and other experts who can work respectfully and collaboratively across disciplines, in ways that address power imbalances between knowledge-holders. Indeed, we need *much* more of Albert Marshall's "two-eyed seeing", first mooted within the Canadian health system (Bartlett 2012), as researchers attempt to address critical issues in landscape and ecological management and conservation, incorporating insights from Māori environmental knowledge, practices and value systems.

The pairing of Māori environmental knowledge with a range of other scientific tools, models and analyses is a trend that will help provide beneficial indicators of population and ecosystem health, that will in turn feed the growth and continuance of *mātauranga*. Our best endeavours are required to support *mātauranga* in the future. For practitioners, access is a prerequisite to practice; locking up the environment, and locking up collections in museums, act to alienate Māori and reduce the capacity to respond to both cultural prerogatives and national challenges. Scientists who work with Māori environmental knowledge or *mātauranga* more broadly are part of a shifting inter-disciplinary landscape of solution building at the national and the international scales, where considerations of intellectual property rights, ethical research partnerships, data sovereignty and community empowerment all contribute to best outcomes.

Acknowledgements

We thank colleagues on the Predator Free Bioethics Panel for discussions on social ethics. PMW was funded by a Rutherford Discovery Fellowship from the Royal Society of New Zealand (14-LCR-001). KW was funded by a Rutherford Discovery Fellowship from the Royal Society of New Zealand (18-MAU-01), Marsden Funding (MAU-16-03) and the National Science Challenge.

References

Barrett-Walker, T., Plank, M. J., Ka'ai-Mahuta, R., Hikuroa, D., and James, A. 2020. Kia kaua te reo e rite ki te moa, ka ngaro: Do not let the language suffer the same fate as the moa. *bioRxiv January*: 817148.

Bartlett, C., Marshall, M., and Marshall, A. 2012. Two-eyed seeing and other lessons learned within a co-learning journey of bringing together Indigenous and mainstream knowledges and ways of knowing. *Journal of Environmental Studies* 2: 331–340.

Bioethics Report 2019. Predator Free New Zealand: Social, cultural, and ethical challenges. BioHeritage Challenge. Available at: bioheritage.nz/.../04/2019-MAY-Bioethics-Report.pdf

Beaton, A., Hudson, M., Milne, M., Port, R.V., Russell, K., et al. 2017. Engaging Māori in biobanking and genomic research: A model for biobanks to guide culturally informed governance, operational, and community engagement activities. *Genetics in Medicine* 19: 345–351.

Berkes, F. 2004. *Sacred Ecology*. London: Routledge.

Black, T. 2014. Te Koko ki Ohiwa (the surge at Ohiwa). In T. Black (ed.), *Enhancing Mātauranga Māori and Global Indigenous Knowledge*. Wellington, New Zealand Qualifications Authority, pp. 12–28.

Bond, M. O., Anderson, B. J., Henare, T. H. A. and Wehi, P. M. 2019. Effects of climatically shifting species distributions on biocultural relationships. *People and Nature* 1(1): 87–102.

Cassey, P. 2001. Determining variation in the success of New Zealand land birds. *Global Ecology and Biogeography* 10: 161–172.

Cisternas, J., Wehi, P. M., Haupokia, N., Hughes, F., Hughes, M., et al. 2019. Get together, work together, write together. *New Zealand Journal of Ecology* 43: 3392.

Clucas, R., Moller, H. and Bragg, C. 2012. Rakiura Māori muttonbirding diaries: monitoring trends in tītī (*Puffinus griseus*) abundance in New Zealand. *New Zealand Journal of Zoology* 39: 37–41.

Collier-Robinson, L., Rayne, A., Rupene, M., Thoms, C., and Steeves, T. 2019. Embedding indigenous principles in genomic research of culturally significant species. *New Zealand Journal of Ecology* 43: 3389.

Cromarty, P. and Scott, D. A. 1996. *A Directory of Wetlands in New Zealand*. Wellington, New Zealand: Department of Conservation.

Crow, S. K., Tipa, G. T., Booker, D. J., and Nelson, K. D. 2018. Relationships between Maori values and streamflow: Tools for incorporating cultural values into freshwater management decisions. *New Zealand Journal of Marine and Freshwater Research* 52: 626–642.

Dawes, J. H. 2019. Are the Sustainable Development Goals self-consistent and mutually achievable? *Sustainable Development* 28(1): 101–117. DOI:10.1002/sd.1975

Fea, N., Linklater, W. and Hartley, S. 2020. Responses of New Zealand forest birds to management of introduced mammals. *Conservation Biology*: https://doi.org/10.1111/cobi.13456.

Harawira, H. 2015. *Kia hikoi ake ahau i ngā tapuwae ō tōku tipuna, ō Teepa*. MA thesis, Te Whare Wānanga o Awanui-ā-Rangi, Whakatane.

Hikuroa, D. 2017. Mātauranga Māori: The ūkaipō of knowledge in New Zealand. *Journal of the Royal Society of New Zealand* 47: 5–10.

Hikuroa, D., Clark, J., Olsen, A. and Camp, E. 2018. Severed at the head: Towards revitalising the mauri of Te Awa o te Atua. *New Zealand Journal of Marine and Freshwater Research* 52: 643–656.

Hitchmough, R., Barr, B., Lettink, M., Monks, J., Reardon, J., et al. 2015. *Conservation status of New Zealand reptiles*. New Zealand Threat Classification Series 17. Wellington, New Zealand: Department of Conservation.

Hopkins, A. 2018. Classifying the mauri of wai in the Matahuru Awa in North Waikato. *New Zealand Journal of Marine and Freshwater Research* 52: 657–665.

Hudson, M., Anderson, T., Dewes, T. K., Temara, P., Whaanga, H. and Roa, T. 2017. He Matapihi ki te Mana Raraunga: Conceptualising Big Data through a Māori lens. In H. Whaanga, T. T. Keegan and T. Apperley (eds), *He Whare Hangarau Māori: Language, Culture and Technology*. Hamilton: Te Pua Wānanga ki te Ao, Te Whare Wānanga o Waikato.

Hudson, M., Mead, A. T. P., Chagné, D., Roskruge, N., Morrison, S. et al. 2019. Indigenous perspectives and gene editing in Aotearoa New Zealand. *Frontiers in Bioengineering and Biotechnology* 7: 1–9.

Humphries, G. R. W. 2014. *Using long term harvest records of sooty shearwaters (Titi; Puffinus griseus) to predict shifts in the Southern Oscillation*. PhD thesis. University of Otago, Dunedin.

Innes, J., Kelly, D., Overton, J., and Gillies, C. 2010. Predation and other factors currently limiting New Zealand forest birds. *New Zealand Journal of Ecology* 34: 86–114.

Jones, P. T. H. and Biggs, B. 1995. *Te Iwi o Tainui*. Auckland, New Zealand: Auckland University Press.

Kawharu, M. 2000. Kaitiakitanga: A Maori anthropological perspective of the Maori socioenvironmental ethic of resource management. *Journal of the Polynesian Society* 110(4): 349–70.

Kawharu, M. 2018. The 'Unsettledness' of treaty claims settlements. *The Round Table* 107(4): 483–492.

Kitson, J. C., Cain, A. M., Johnstone, M. N. T. H., Anglem, R., Davis, J., et al. 2018. Murihiku Cultural Water Classification System: Enduring partnerships between people, disciplines and knowledge systems. *New Zealand Journal of Marine and Freshwater Research* 52: 511–525.

Kitson, J. C. and Moller, H. 2008. Looking after your ground: Resource management practice by Rakiura Maori titi harvesters. *Papers and Proceedings of the Royal Society of Tasmania* 142: 161–176.

Kukutai, T. and Cormack, D. 2019. Mana motuhake ā-rarauanga: Datafication and social science research in Aotearoa. *Kōtuitui: New Zealand Journal of Social Sciences Online* 14(2): 201–208.

Kukutai, T. and Taylor, J. (eds) 2016. *Indigenous Data Sovereignty: Toward an Agenda*. Canberra: ANU Press.

Lambert, S., Waipara, N., Black, A., Mark-Shadbolt, M., and Wood, W. 2018. Indigenous biosecurity: Māori responses to kauri dieback and myrtle rust in Aotearoa New Zealand. In J. Urquhart, M. Marzano and C. Potter (eds), *The Human Dimensions of Forest and Tree Health*. Basingstoke: Palgrave Macmillan. pp. 109–137.

LCRIT (Landcare Research Informatics Team) 2015. LCDB v4.1. Land Cover Database version 4.1, Mainland New Zealand. Available at: https://lris.scinfo.org.nz/layer/48423-lcdb-v41-land-cover-database-version-41-mainland-new-zealand/

Lyver, P. O., Davis, J., Ngamane, L., Anderson, L., and Clarkin, P. 2008. Hauraki Maori matauranga for the conservation and harvest of titi, *Pterodroma macroptera gouldi*. *Papers and Proceedings of the Royal Society of Tasmania* 142(1): 149–159.

Lyver, P. O., Ruru, J., Scott, N., Tylianakis, J. M., Arnold, J., et al. 2019. Building biocultural approaches into Aotearoa–New Zealand's conservation future. *Journal of the Royal Society of New Zealand* 49(3): 394–411.

Maffi, L. 2005. Linguistic, cultural, and biological diversity. *Annual Review of Anthropology* 34: 599–617.

Matamua, R., Harris, P., and Kerr, H. 2013. Māori navigation. In G. Christie (ed.), *New Zealand Astronomical Society Yearbook 2013* Auckland: Stardome Observatory Planetarium, pp. 28–34.

McGlone, M. S. 1989. Postglacial history of New Zealand wetlands and implications for their conservation. *New Zealand Journal of Ecology* 33(1): 1–23.

Mikaere, A. 2011. *Colonising Myths: Māori Realities*. Wellington: Huia.

Ministry of the Environment and Stats NZ 2019. Report. Available at: www.mfe.govt.nz/sites/default/files/media/Environmental%20reporting/environment-aotearoa-2019.pdf.

Moller, H., Fletcher, D., Johnson, P. N., Bell, B. D., Flack, D., et al. 2009. Changes in sooty shearwater (*Puffinus griseus*) abundance and harvesting on the Rakiura Titi islands. *New Zealand Journal of Zoology* 36: 325–341.

Morgan, T. K. K. B. 2006. Decision-support tools and the indigenous paradigm. *Proceedings of the Institution of Civil Engineers: Engineering Sustainability* 159: 169–177. DOI:10.1680/ensu.2006.159.4.169.

Mutu, M. 2018. Behind the smoke and mirrors of the Treaty of Waitangi claims settlement process in New Zealand: No prospect for justice and reconciliation for Māori without constitutional transformation. *Journal of Global Ethics* 14(2): 208–221.

New Zealand Ecological Society 1995. Maori customary use of native birds, plants and other traditional materials. *New Zealand Journal of Ecology* 1: 77–82.

Palavi, V., Railton, N., and Waitai, S. 2018. Collaborative Kaitiakitanga-new Joint pathways in guardianship. *Biodiversity Information Science and Standards* 2: e26954. DOI:10.3897/biss.2.26954.

Parliamentary Commissioner for the Environment (PCE) 2017. *Taonga of an Island Nation: Saving New Zealand's Birds*. Wellington: Parliamentary Commissioner for the Environment.

Paul-Burke, K., Burke J., Te Ūpokorehe Resource Management Team, Bluett, C., and Senior, T. 2018. Using Māori knowledge to assist understandings and management of shellfish populations in Ōhiwa harbour, Aotearoa New Zealand. *New Zealand Journal of Marine and Freshwater Research* 52(4): 542–556.

Reihana, K,. Taura, Y. and Harcourt, N. 2019. He tohu o te wā–Hangarau pūtaiao. Signs of our times: Fusing technology with environmental sciences. *New Zealand Journal of Ecology* 43(3): 3382.

Roberts, M., Haami, B., Benton, R., Satterfield, T., Finucane, M. L., et al. 2004. Whakapapa as a Māori mental construct: Some implications for the debate over genetic modification of organisms. *The Contemporary Pacific* 16: 1–28.

Royal TAC 2002. Indigenous worldviews: A comparative study. Report for Ngati Kikopiri Te Wananga-o-Raukawa, Te Puni Kokiri. Fulbright, New Zealand: Winston Churchill Memorial Trust.

Ruru, J. 2004. Managing our treasured home: The conservation estate and the principles of the Treaty of Waitangi. *New Zealand Journal of Environmental Law* 8: 243–266.

Ryks, J., Pearson, A. L., and Waa, A. 2016. Mapping urban Māori: A population-based study of Māori heterogeneity. *New Zealand Geographer* 72: 28–40.

Salick, J., Konchar, K., and Nesbitt, M. 2014. Biocultural collections: Needs, ethics and goals. In J. Salick, K. Konchar and M. Nesbitt (eds), *Curating Bio-Cultural Collections: A Handbook*. Kew, London: Royal Botanic Gardens Kew Publishing.

Sangha, K. K., Maynard, S., Pearson, J., Dobriyal, P., Badola, R., and Ainul Hussain, S. 2019. Recognising the role of local and indigenous communities in managing natural resources for the greater public benefit: Case studies from Asia and Oceania region. *Ecosystem Services* 39: 1–12.

Scott, P. and Williams, N. 2014. Phytophthora diseases in New Zealand forests. *New Zealand Journal of Forestry* 59(2): 15.

Shortland T. 2012. Taumarere, the River of Chiefs Taumarere, te Awa o nga Rangatira. Available at: www.epa.govt.nz/assets/FileAPI/hsno-ar/APP201365/3aa4b7fee4/APP201365-SUBMISSION102614-Nga-Tirairaka-o-Ngati-Hine.pdf (accessed 12 January 2020).

Smith, L., Maxwell, T. K. K., Puke, H. and Temara, P. 2016. Indigenous knowledge, methodology and mayhem: What is the role of methodology in producing indigenous insights? A discussion from mātauranga Māori. *Knowledge Cultures* 4(3): 131–156.

Soga, M. and Gaston, K. J. 2019. Extinction of experience: The loss of human–nature interactions. *Frontiers in Ecology and the Environment* 14(2): 94–101.

Taura, Y., Van Schravendijk-Goodman, C., and Clarkson, B. (eds) 2017. Te reo o te repo. The voice of the wetland: connections, understandings and learnings for the restoration of our wetlands. Hamilton, New Zealand: Manaaki Whenua – Landcare Research and Waikato Raupatu River Trust.

Te Iwi o Ngatiwai 2007. Ngatiwai Environmental policy document. Available at: www.wdc.govt.nz/PlansPoliciesandBylaws/Plans/DistrictPlan/Documents/Iwi-Management-Plan-Te-Iwi-o-Ngatiwai-Iwi-Environmental-Policy-Document-2007.pdf (accessed 12 January 2020).

Tipa, G. and Teirney, L. 2006. *A Cultural Health Index for Streams and Waterways: A Tool for Nationwide Use.* Wellington, New Zealand: Ministry for the Environment.

Tuaupiki, J. 2017. *E kore e ngaro, he takere waka nui: Te mātauranga whakatere waka me ōna take nunui.* PhD thesis, University of Waikato, Hamilton, New Zealand.

Waitangi Tribunal 2011. Ko Aotearoa tēnei: A report into claims concerning New Zealand law and policy affecting Māori culture and identity. Available at: https://forms.justice.govt.nz/search/Documents/WT/wt_DOC_68356416/KoAotearoaTeneiTT2Vol1W.pdf (accessed 12 January 2020).

Waitangi Tribunal 2019. Hauora: Report on Stage One of the Health Services and Outcomes Kaupapa Inquiry. Available at: https://forms.justice.govt.nz/search/Documents/WT/wt_DOC_152801817/Hauora%20W.pdf (accessed 12 January 2020).

Walker, E. T., Wehi, P. M., Nelson, N. J., Beggs, J. R. and Whaanga, H. 2019. Kaitiakitanga, place and the urban restoration agenda. *New Zealand Journal of Ecology* 43(3): 3381.

Walter, M. 2018. The voice of Indigenous data. Beyond the markers of disadvantage. *Griffith Review* 60. Available at: www.griffithreview.com/articles/voice-indigenous-data-beyond-disadvantage/

Walter, M. and Suina, M. 2019. Indigenous data, indigenous methodologies and indigenous data sovereignty. *International Journal of Social Research Methodology* 22(3): 233–243.

Watene, K. 2016a. Valuing nature: Māori philosophy and the capability approach. *Oxford Development Studies* 44(3): 287–296.

Watene, K. P. M. 2016b. Land and water. In N. Leggat (ed.), *Journal of Urgent Writing 2016.* Wellington, New Zealand: Massey University Press, pp. 180–189.

Watene, K. and Yap, M. 2015. Culture and sustainable development: Indigenous contributions. *Journal of Global Ethics* 11(1): 51–55.

Wehi, P. M., Beggs, J. R., and McAllister, T. G. 2019a. Ka mua, ka muri: The inclusion of mātauranga in New Zealand ecology. *New Zealand Journal of Ecology* 43(3): 1–8.

Wehi, P. M., Carter, L., Harawira, T. W., Fitzgerald, G., Lloyd, K., et al. 2019b. Enhancing awareness and adoption of cultural values through use of Māori bird names in science communication and environmental reporting. *New Zealand Journal of Ecology* 43(3): 3387.

Wehi, P. M., Cox, M. P., Roa, T., and Whaanga, H. 2018. Human perceptions of megafaunal extinction events revealed by linguistic analysis of indigenous oral traditions. *Human Ecology* 46: 461–470.

Wehi, P. M. and Lord, J. M. 2017. Importance of including cultural practices in ecological restoration. *Conservation Biology* 31:1109–1118.

Wehi, P. M. and Roa, T. 2020. Reciprocal relationships: Identity, tradition and food in the Kīngitanga poukai. He Manaakitanga: O Te Tuakiri, O Te Tikanga Me Te Kai Ki Te Poukai O Te Kīngitanga. In

C. Loos (ed.), *Interdisciplinary Perspectives on Participation and Food Justice*. Fayetteville, AK: University of Arkansas Press.

Wehi, P. M., Whaanga, H. and Trewick, S. A. 2012. Artefacts, biology and bias in museum collection research. *Molecular Ecology* 21(13): 3103–3109.

Whaanga, H., Simmonds, N. and Keegan, T. T. 2017. Iwi, institutes, societies and community led initiatives. In H. Whaanga, T. T. Keegan and T. Apperley (eds), *He Whare Hangarau Māori: Language, Culture & Technology*. Hamilton, New Zealand: Te Pua Wānanga ki te Ao, Te Whare Wānanga o Waikato.

Whaanga, H. and Wehi, P. 2017. Rāhui and conservation? Māori voices in the nineteenth century niupepa Māori. *Journal of the Royal Society of New Zealand* 47:100–106.

Wilks, J.. Kennedy, G., Drew, N. and Wilson, K. 2018. Indigenous data sovereignty in higher education. *Australian Universities' Review* 60(2): 4–14.

Wilmshurst, J. M., Anderson, A. J., Higham, T. F. and Worthy, T. H. 2008. Dating the late prehistoric dispersal of Polynesians to New Zealand using the commensal Pacific rat. *Proceedings of the National Academy of Sciences* 105: 7676–7680.

Wright, S. D., Nugent, G. and Parata, H. G. 1995. Customary management of indigenous species: A Maori perspective. *New Zealand Journal of Ecology* 19(1): 83–86.

Yap, M. L. M. and Watene, K 2019. The Sustainable Development Goals (SDGs) and indigenous peoples: Another missed opportunity? *Journal of Human Development and Capabilities* 20(4): 1–17.

Young, R., Wagenhoff, A., Holmes, R., Newton, M. and Clapcott, J. 2018. What is a healthy river? Prepared for Cawthron Foundation. Cawthron Report No. 3035.

PART III

Applications of IEK for adaptation, conservation, and coexistence

16

INTEGRATING AMAZIGH CULTURAL PRACTICES IN MOROCCAN HIGH ATLAS BIODIVERSITY CONSERVATION

Irene Teixidor-Toneu, Gary Martin, Soufiane M'sou and Ugo D'Ambrosio

Local ecological knowledge and biocultural conservation in Morocco

Characteristic ecosystems in the Mediterranean have co-evolved with people, producing varied cultural landscapes that require active human management to sustain biodiversity (Blondel, 2006; Bugalho et al., 2011; Thompson, 2005). Biodiversity richness in the Mediterranean is often linked to ecological spatial heterogeneity, shaped by diverse climatic and geographical conditions as well as traditional agro-ecological practices, historical processes and other elements linked to local livelihoods (Atauri and de Lucio, 2001; Thompson, 2005). The idea that certain culture-specific practices may have a positive impact on biodiversity conservation is based on observed sustainable use and intimate knowledge of biodiversity by peoples who rely on natural resources for their livelihoods (Gadgil et al., 1993; Xu et al., 2009). This implicit understanding of interrelated ecological processes guides practices of environmental management. The contribution of local knowledge to the sustainable use of natural resources is widely acknowledged and has stimulated the development of less centralized models for biodiversity conservation (Berkes, 2003; Berkes et al., 2000; Fernandez-Gimenez, 2000; Jarvis and Hodgkin, 2000; Oba et al., 2000; Soleri and Smith, 1999; Wilkes, 1991). Socio-environmental and biocultural approaches are not new, yet have been inadequately represented in conservation programmes (Wehi and Lord 2017) despite the growing amount of research pointing to the relevance of cultural practices to biodiversity conservation in Africa and beyond (Kideghesho 2009; Asante et al., 2017). Conservation managers and policy-makers still struggle to implement a socio-environmental approach despite the well-documented and increasingly acknowledged role of Indigenous ecological knowledge and traditional practices in sustaining resource use, conserving biodiversity and contributing to ecosystem resilience. This is especially true for North Africa, where predominantly top-down conservation and development policies have given little voice to local communities (Montanari and Bergh, 2014) and where human and environmental factors have been only slightly integrated into management plans and policies. This partly results from the lack of effective transdisciplinary action research (Dorward, 2014; Ostrom 2009), the focus on biodiversity studies as compared to ethnobiological ones, and the

DOI: 10.4324/9781315270845-19

absence of engaged collaboration between various actors and stakeholders (Berkes, 2007, 2011; Ostrom, 2009).

Local ecological knowledge and culture-specific resource management have been a central focus of contemporary academic inquiry in Morocco, including studies on the argan agroforestry system (El Harousse et al., 2012; Simenel et al., 2009), Jebala cultural landscapes (Aumeeruddy et al., 2017), oasis agroecosystems and the *agdal* sylvo-pastoral resource management system (see, for example, Auclair and Alifriqui, 2012). *Agdals*, of particular importance in the High Atlas, are systems "of seasonal prohibitions that limit access to one or more agro-sylvo-pastoral resources in order to allow them to recover from direct or indirect human pressure during their most critical period of growth" (Domínguez et al., 2012, p. 278). Only pastoral and forest *agdals* have been studied in depth (see for example, Auclair et al., 2012; Auclair and Alifriqui, 2012). Outside the *agdal* system, Baumann (2009) shows that transhumant grazing practices in southern Morocco are adapted to maintain the resilience of pasturelands and to maximize forage plant production. Southern Moroccan herdsmen adapt their mobility according to the amount of rainfall and the specific buffering capacity of each pasture type. This sustains both the ability of the vegetation to build up a surplus of standing biomass and maintains the vitality of perennial plants even under intensive seasonal grazing (ibid.). Other aspects of local livelihoods may have an impact on maintaining local habitats, ecological processes and biodiversity, including home gardens (Teixidor-Toneu, 2015), terraced agriculture (Mediterranean Consortium for Nature and Culture, 2013) and sacred groves (Deil et al., 2005).

The Moroccan High Atlas Mountains have a particularly high biological diversity (Médail and Quézel, 1997) within the Mediterranean biodiversity hotspot (Olson and Dinerstein, 1998) accompanied by rich historical, cultural and linguistic heterogeneity (Taibi et al., 2015; Haut-Commissariat au Plan, 2014). This mountain range stretches along 560 kilometres through central Morocco, separating the heartland of Moroccan economic activity to the north from the Saharan influence to the south. Our fieldwork in the High Atlas focused on describing cultural practices that impact environmental conservation in the rural communes of Ait M'hamed and Imegdale, using a community-based participatory approach. As part of our research, we developed an operational definition of "cultural practices of conservation" based on the definition proposed by the Mediterranean Consortium for Nature & Culture (2013): "all actions carried out by local peoples that foster and maintain biodiversity, sustainable land management and viable use of water". Based on the seminal *kosmos-corpus-praxis* ethnoecological framework proposed by Toledo (1991) and the knowledge-practice-belief concept described by Berkes et al. (2000), our definition aims to contextualize cultural practices according to the local cosmology or belief system and ground them by taking into account the traditional ecological knowledge necessary for decision-making.

Identifying and describing cultural practices of conservation

With a multicultural, multilingual and interdisciplinary team of eight community, national and international researchers, we documented local ecological knowledge and cultural practices of conservation from May 2016 to October 2017. Research questions and documentation methods were co-designed by the team members. We developed the research process complementary stages of design, revision of methodology, data collection and analysis. In order to effectively co-document cultural practices of conservation with community members, we discussed the term "cultural practices of conservation" with research team members and adopted an operational definition to reflect both local and institutional conservation perspectives (COMBIOSERVE,

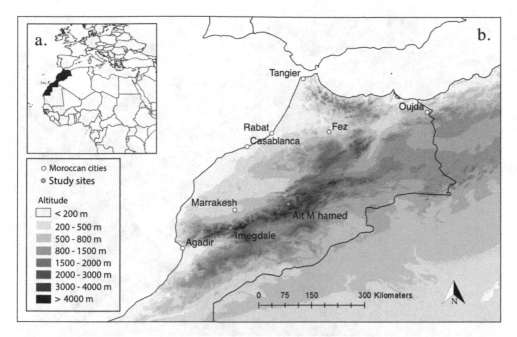

Figure 16.1 Geographical situation of the two research localities, Ait M'hamed and Imegdale

2015). We took care to conceptualize cultural practices of conservation in a way that could be easily understood by the local community.

In our interviews, the term referred to "what is done by the community to, with or in the environment" as well as how, how much, when, where and by whom. Since the local environment is differentiated by various levels of ownership, practices were documented in reference to private, communal and inter-communal land. Data were collected with local inhabitants primarily through structured interviews and questionnaires conducted in Tashelhit or Tamazight, two variants of the Amazigh language. Toward the end of the field research, complementary information essential to contextualize and elaborate responses from structured interviews was collected through participant observation and informal open-ended interviews. In total, 19 men and 21 women were interviewed from 7 localities in Imegdale, and 28 men and 15 women from 10 localities in Ait M'hamed (Figures 16.1 and 16.2).

Ait M'hamed (31.87 N, 6.51 W) and Imegdale (31.12 N, 8.14 W) are approximately 235 km apart from each other along the SW-NE High Atlas axis. The climate is arid Mediterranean with an annual rainfall of around 300 mm and an average temperature of 28°C. A hot and dry summer contrasts with a cold and wet winter. Imegdale is located approximately 75 km south of Marrakech. With altitudes ranging from 900 to 2,500 m, the commune has an area of approximately 278 km² with a population of 5,537 people in 1,156 households dispersed in 28 small villages (*douars*) (Haut-Commissariat au Plan, 2014). The commune has a highly diverse vegetation comprising Mediterranean forests and shrublands as well as montane grasslands and scrublands (Olson et al., 2001). Ait M'hamed lies approximately 180 km east of Marrakech. The commune has an approximate area of 560 km² and altitudes ranging from 950 to 2,600 m with a population of 23,696 inhabitants in 3,493 households in 45 small villages (*douars*) (ibid.). Natural subhumid Mediterranean vegetation is found in Ait M'hamed at lower elevations with evergreen oaks (*Quercus ilex* subsp. *ballota*), patches of *Juniperus phoenicea, J. thurifera* and *Pinus*

Figure 16.2 Community researcher interviewing an elder in Ait M'hamed

Source: © Inanc Tekguc.

halepensis and, at higher elevations, scrublands featuring spiny xerophytes (Emberger, 1939), degraded in various degrees.

Most of Imegdale's inhabitants are Ishelhin (sing., Ashelhi) and speak Tashelhit, while in Ait M'hamed people are Imazighen (sing., Amazigh) and speak Central High Atlas Tamazight. In Ait M'hamed, half of the men are also fluent in Moroccan Arabic (Darija) whereas in Imegdale just over a tenth of the men are fluent in Darija (Haut-Commissariat au Plan, 2014). However, most men have basic communication skills in this language, which is not the case for women. Imegdale inhabitants do not claim identity as any particular tribe, whereas in Ait M'hamed residents are predominantly part of the Ait Messat tribe mixed with members of the neighbouring Ait Atta, Ait Bouguemez, Hansala, Ait Bouzid, Ait Attab, Ait Abbas, Soukhmane and Anetifa tribes. The inhabitants of both Imegdale and Ait M'hamed base their livelihoods on subsistence agriculture complemented by other economic activities. Most households rear livestock, mainly cows, sheep and goats, and cultivate various cereal, vegetable and tree crops. Apples, carob, walnuts and other nuts and fruits, marginal crops such as orris root (*Iris germanica*) and livestock are sold in local markets. Migration of men (and to a lesser extent women) to urban areas is an important source of revenue, contributing to the local economy yet affecting local livelihoods.

Cultural practices of conservation in the Moroccan High Atlas

The terms biodiversity and cultural diversity have no close translation in Amazigh languages and "cultural practices of conservation" do not constitute a cultural domain in rural Morocco. Thus, the notion of conservation of diversity is not easily understood and requires an open dialogue with local community members. The inhabitants of the High Atlas identify local practices and traditions that maintain their environment and are aware of change in their cultural landscapes and traditions. Community members identify concrete actions, often named, that they carry out to maintain the environment as they know it. Practices identified as important by scholars, such as sacred groves or agricultural terraces (Deil et al., 2005; Mediterranean Consortium for Nature and Culture, 2013), may be so obvious to community members that they are taken for granted and not distinguished as distinct practices.

Practices that use, manage or maintain natural resources exist in a context of shared values, attitudes and ecological knowledge and result in specific livelihoods. With changing socio-economic and environmental local conditions, some practices have been abandoned and others are adapted to new contexts. Over 20 named cultural practices of conservation were described by informants during this research; Table 16.1 provides a summary of the most salient ones. Practices have been organized using an etic approach in broad, interconnected domains.

The cultural practices of conservation presented in Table 16.1 are all interrelated directly or indirectly and represent key elements of a more complex agro-ethnoecological system. Enclosure practices, for example, shape and delimit the landscape while also serving as a soil management strategy; walls made with stones (*aderass*) or branches (*afrague*) protect fields and home gardens from grazing and erosion. Fences made with branches are renovated approximately once a year, and old wood is used as fuel. Fields can also have stone lines that stabilize the soil (*imarine*) and single trees can be protected by individual barriers around the trunk (*astour*) mostly made by stones, but also branches (Figure 16.3). Enclosures made with branches can also retain livestock (*tafergant*). In turn, agricultural practices improve soil quality by the addition of manure (*amazer*) or the removal of stones (*taoudia*). Agro-forestry practices include harvesting techniques for fruits and cereals (*azzwui* and *tawala n anrar*, respectively), cutting wood (*oboy n okchoud*) and building infrastructure for stocking agricultural production (*ighrem*), as well as coordinated livestock management strategies to ensure the protection of agricultural production (*azzayin*). Practices managing livestock grazing patterns include seasonal transhumance to areas formally protected by customary law (*agdal*) (Figure 16.4) or free of such restrictions (*laêzib*) and rotational grazing by the movement of branch enclosures (*assemgonou*).

Water is a key resource for both agriculture and pastoralism. Irrigation water is distributed in an egalitarian way among families and across cultivated areas during drought months (*tawala n waman*) and the canals are carefully maintained (*arras n targa*). Many of the plants growing in this managed landscape have a high cultural value as they are used in traditional recipes (*isenwi abdeldi*) or as medicines. Traditional dishes are prepared during celebrations that express local values (*lemarouf*) or mark key moments in the agricultural calendar (*asseft*).

Community work is a key aspect of many of these practices. *Tiwizi* is the voluntary pooling of effort to conduct activities that benefit the whole community. Labour, just as limited and valuable as water, is organized collaboratively and in turns (*tawala*) to ensure agricultural production.

These practices impact High Atlas biodiversity in three different ways. Some shape the landscape and maintain specific topographic features and biodiversity patterns by delimiting cultivation and grazing areas and managing their water supply (*aderass, afrague, imarine, astour, tafergant, assemgonou, laêzib, azzayin, agdal*). Others contribute to the rich knowledge and use of the local flora (*isenwi abdeldi, oboy n okchoud*). Yet others facilitate the embodiment of local values that regulate interactions among people and between them and the local environment (*tiwizi, asseft, lemarouf*).

Although documented as separate entities, these agro-ecological practices cannot be understood in isolation from each other. They all contribute to High Atlas biodiversity distribution patterns and to local livelihoods. For example, some of the practices named by informants are complementary and achieve multiple aims at the same time. Building stone walls creates

Table 16.1 Cultural practices of conservation

Agriculture and forest use	
Amazer	Using natural manure to increase the quality of the cultivated soil
Azzayin	Restricting grazing non-irrigated agroforestry areas temporarily to protect fields and trees
Azzwui	Manual fruit harvesting from trees by hitting the branches with a long stick
Ighrem	Silos to store grain (mostly barley and wheat) as well as other agricultural produce
Oboy n okchoud	Harvesting wood to provide construction and fuel materials
Taoudia	Act of clearing stones from the soil to increase the quality of cultivation
Tawala n anrar	Taking turns to thresh cereals by collaborative work (*tiwizi*, see below)
Ceremonies and celebrations	
Asseft	Celebration linked to the agricultural calendar, to indicate the start of the cereal harvest season
Lemarouf	Gathering linked to charity or prayer, to celebrate a festivity or commemorate a saint or dead person
Cooperation between community members	
Tiwizi	Cooperative, collaborative and solidarity work to more effectively conduct tasks that require significant effort
Enclosures and soil management	
Aderass	Enclosing wall made with rocks to protect fields and home gardens from animal grazing and erosion
Afrague	Enclosing wall made with branches to protect fields, home gardens and single trees from animals and soil erosion
Imarine	Lines of rocks and small terraces to reduce erosion and clear fields of rocks
Astour	Enclosing wall made with rocks to protect trees and home gardens. They provide shade and increase humidity.
Tafergant	Enclosure made with branches to keep animals grazing in a delimited space
Food and cooking	
Isenwi abdeldi	Preparation of special recipes for specific celebrations or seasonally according to the availability of local wild edible plants
Pastoralism and transhumance	
Assemgonou	Spatial rotation of *tafergant* (animal enclosure) to fertilize the soil
Agdal	Intercommunal resource management within a defined territory in which there is a temporary restriction on the use of specific biological resources, maximizing their availability in critical periods of need
Laêzib	Seasonal transhumance of (mostly) goat herds to access best forage
Water management	
Arras n targa	Cleaning and repairing irrigation canals to maintain them and ensure efficient irrigation
Tawala n waman	Taking turns to irrigate fields during the drought season in order to allow different households the right of use of a spring to have equitable access to water

important landscape features, clears stones from the soil, enhancing its arability, and are used to manage the mobility of livestock. Animal resources, such as manure, increase agricultural production and plants in turn are used for fodder and forage. Above all, water supply is necessary for all activities and celebrations and recipes are the foremost expression of local cultural and biological diversity. This material and immaterial connection between practices is manifested not only in the impact of the practices, but also in their temporal and spatial organization.

Figure 16.3 *Astour* enclosures provide suitable growing conditions for trees and protection from grazing

Source: © Irene Teixidor-Toneu.

Figure 16.4 A diversity of cultural practices of conservation regulates the environmental impact of grazing, an important livelihood activity

Source: © Inanc Tekguc.

Cultural landscapes in the High Atlas are structured by various levels of appropriation (Auclair et al., 2012), from areas and resources privately owned at a household level to those shared among families of one village (*douar*), among villages of the same rural commune (*jamaa karawiya*) and among people of different rural communes. The inhabitants of the High Atlas categorize their interaction with the environment according to these different levels, following principles of land management and ownership. In some cases, a single practice occurs at multiple levels: at a household level on private land, at village level, by collaboration of all families in the village, or through coordination of different villages. This is often the case for water management (*tawala n wamen*), as well as harvest restriction and practices (*azzayin* and *azzwui*). These communal practices can bring together people in neighbouring villages that share land or water resources, even if they do not necessarily belong to the same rural commune, which is a governmental administrative unit. Only the practice of *agdal* is in most cases intercommunal, requiring the coordination of people from more than one rural commune and often from different tribal origins. Some of these landscapes are not free from conflict between different tribal groups or between the local community and the national governmental institutions.

Most practices occur at specific times of the year. Following the four seasons – *taguerst* (winter), *tadrar* (spring), *tambdout* or *sif* (summer) and *tamanzouyt* or *elkhrif* (autumn) – these are generally linked to plant growth cycle and are dependent on one another. In winter (December to February), snow in the high mountains keeps the *agdals* and alpine pastures unused. While the last snow can fall in April, May is already hot and cereals are harvested. The extreme heat in the summer (July to mid-September) triggers the ripening of fruits and nuts and *tawala n wamen* (distribution of water in turns) is used to distribute the limited water after *arras n targa* (cleaning of irrigation canals). Fencing linked to *taoudia* (clearing the soil from stones) is carried out "during the season of working the land" (winter) and also after the cereal harvest (end of spring).

Land ownership and management also shape spatial ecological diversity by enhancing agricultural and farming production in specific ways. Fields and grazing areas are distributed along altitudinal gradients, from the humid and irrigated valley bottoms, with seasonal or year-round streams, to dry mountain slopes. Terraced agriculture combined with an intricate irrigation system comprising *aghbalous* (springs), *tergew* (pl. of *targa*, irrigation canals) and *tafrouts* (water reservoirs), all contribute to shaping the landscape. Water is distributed from springs to irrigated fields through a dendritic system of canals in which reservoirs are strategically integrated. Vegetables can be grown in the margins of cereal and tree fields or separately in small adjacent fields. Planted tree cover and dry cereal cultivation can continue beyond the immediate areas around the village, in non-irrigated fields where mostly nut trees are planted. These trees can also be grown alongside mountain streams in privately owned and managed irrigated areas, referred to as *tagdalt* (lit. "little *agdal*"), that have an understory of weedy plants (*touga*) harvested as fodder at specific regulated times of the year.

Moreover, enclosing is key in creating High Atlas vegetation mosaics in cultural landscapes. Enclosures contribute to terracing and the separation of fields of different owners and of areas where different types of agricultural activities are conducted. They delimit home gardens, sacred groves and livestock passages across cultivated areas from home enclosures to pasturelands. In Ait M'hamed, *astour* is used around fruit trees in gardens or to protect saplings of juniper (*Juniperus* spp.), ash (*Fraxinus dimorpha*) and oak (*Quercus ilex*) in non-irrigated lands. Herds are taken to graze beyond the cultivated areas toward mountain sites following seasonal transhumance, closer to villages in wintertime (when there is snow in the higher mountains) and upland in summertime, within or outside *agdal* areas.

Beyond particularities in space and time, these practices take place in a social context where labour and responsibility are highly gendered. Women and men have specific roles in each aspect of the agro-ecological system and practices are carried out by only one or the other sex, with very few exceptions. Decisions on specific dates for all major agricultural activities, including water management, celebrations linked to the agricultural calendar and appointment of guardians for *azzayin*, are taken on Fridays in the mosque primarily by men. Male representatives from different communities gather once a year to decide on the closing and opening dates of the *agdal*, and appoint one or more guardians. Women rarely participate in decision-making regarding the agro-ecological calendar, unless they are the heads of the household (typically widows).

In contrast, women are in charge of keeping the household running: they are responsible for cooking, family health and domestic animals such as cows, chickens and rabbits. They are primarily in charge of food preparation for celebrations, or during both men's and women's *tiwizi* work. Non-daily routine tasks include weaving rugs, weeding, harvesting cereals, shelling nuts and collecting specific fodder plants in remote areas (e.g., *auri, Stipa tenacissima*). All these tasks are carried out in *tiwizi*. While men are in charge of the cultivation of cereals, vegetables and fruit and nut trees, women contribute by caring for young animals of the herds, harvesting cereals, carrying crops to rooftops for drying and shelling nuts, among other tasks. Men take care of ploughing, building *aderas* and *imarine*, planting and threshing cereal, harvesting fruits and nuts, growing vegetables, shepherding (including transhumance to areas seasonally restricted or open year-round), maintaining the irrigation infrastructure and doing most of the construction work. Both men and women build *astour* (protective walls around saplings and small trees) as needed, and may engage in commercial medicinal and aromatic plant collection.

In this layered management system, regular decisions are taken for the temporal coordination and implementation of many practices. Community decision-making bodies set rules for the use of biodiversity and water resources, including agreement on restrictions and sanctions. Of all customary laws relating to the management of the local environment, *agdal* grazing practices have received most scholarly attention (Auclair and Alifriqui, 2012). This is likely due to the size of the land involved, its economic importance and the social complexity of its sustainable management (Auclair et al., 2011; Auclair and Alifriqui, 2012; Domínguez et al., 2012). However, other agroecological practices at smaller scales are also regulated by customary law and involve restrictions on the use of resources (importantly, *laêzib, tawala n wamen* and *azzayin*, see Table 16.1 for translations). These can be just as economically important as pastoral and forestry *agdals* and also have a significant impact on High Atlas biodiversity. At the village and household level, private family lands and those shared among a small number of households follow access rules comparable to those of the *agdal*, though simpler in implementation. *Azzayin* (temporal grazing restriction on cultivated areas) matches the description of *agdal* as a practice of temporary grazing restrictions on the use of specific resources within a delimited territory (Auclair et al., 2012). In Imegdale, although *laêzib* (transhumance) mostly involves grazing in common unregulated lands, access to some areas is restricted. The distribution of water by *tawala n wamen* is cautiously controlled in space and time. None of these practices is emically identified as *agdal*. This gradient from private to inter-community owned and managed lands has been overlooked by scholars (ibid.).

The role of local knowledge and practice in conservation of North African biocultural landscapes

Our research documented cultural practices in the High Atlas Mountains that have an impact on maintaining cultural and biological diversity beyond the well-known pastoral *agdals*. The natural and social environments in the High Atlas are rapidly changing, and local cosmologies and livelihoods are shifting to adapt to a globalized world (Domínguez, 2017). Many of the practices we documented are disappearing or drastically changing. The complex and long-standing systems of management of landscapes in the High Atlas by Amazigh communities should be understood and reinvigorated with a socio-environmentally resilient, economically viable and enriching approach to land use. Studies such as the one presented here can serve as tools for local development programmes on a regional scale. Fortunately, in recent years, governmental and especially non-governmental institutions are beginning to include in their strategic mission, vision and values the need to engage actively with local communities to better understand the role of traditional practices in the conservation of natural resources and strengthen best practices and knowledge (CEPF, 2017; McIvor et al., 2008). This transition, although still being relatively under-recognized on the ground (Wehi and Lord, 2017) will be key to ensure long-term sustainability and governance in bioculturally rich regions such as the Maghreb. Understanding the praxes and associated knowledge that guide people's decision-making in the twenty-first-century context of changing economies and environments is essential to devising appropriate biodiversity conservation strategies.

Despite the fact that neither the conservation of biodiversity *per se* nor the maintenance of ecosystem services are concepts that exist in the world-view and language of the inhabitants of the High Atlas, the term "cultural practices of conservation" can be instrumental in developing conservation initiatives locally. The term as developed and used in this research mixes perspectives from both the social and natural sciences and reconciles etic and emic world-views. As knowledge and practice are transformed by globalization, cultural practices of conservation can become reservoirs of knowledge and know-how similar to the *lieux de mémoire* characterized by Nora as material, symbolic and functional enclaves for traditional livelihoods in a modernizing world (1989). Rather than just documenting Indigenous or local ecological knowledge to render it accessible to international audiences, the concept of 'cultural practices of conservation' can help fuel efforts to actively maintain activities that would no longer occur spontaneously.

The study of local ecological knowledge and of biocultural diversity conservation requires an integrated research approach in order to understand the various *kosmos-corpus-praxis* and knowledge-practice-belief complexes in any geographical area (Berkes et al., 2000; Toledo, 1991). This information can later be combined with results from other disciplines, for which integrative research frameworks across the natural and social sciences are key (e.g., Binder et al., 2013; McGinnis and Ostrom, 2014; Ostrom, 2009). Given that cultural knowledge and practice are place-dependent (Basso, 1996; Maffi and Woodley, 2010), biodiversity conservation and sustainable development can be facilitated by strengthening local cultural practices of conservation that integrate the natural and social elements of cultural landscapes.

Sustainable development can only be achieved through a dialectical approach that fully integrates the human and non-human elements that comprise different cultural landscapes. Complex long-standing systems, such as Amazigh landscape management in the High Atlas, should be understood and reinvigorated through participatory ethnoecological research and other empirical studies to enable a socio-environmentally resilient, economically viable and enriching approach to sustainable land use.

Acknowledgments

This research was carried out collaboratively with community researchers Touda Atyah, Hamid Ait Baskad, Fadma Ait Iligh, Adel Merzough and Said Oughzif. We would also like to express our gratitude to the rest of the Global Diversity Foundation (GDF) and Moroccan Biodiversity and Livelihoods Association teams, especially Emily Caruso, Mohamed El Haouzi, and Hassan Rankou, and intern Giada Bellia. The study was supported by a grant to GDF from the MAVA Foundation for a project on "Integrated approach to plant conservation in the Moroccan High Atlas", and contributed to projects on "Mobilising useful plant conservation to enhance Atlas mountain community livelihoods" (24-010) funded by the UK Darwin Initiative and "Cultural Landscape Management in the Moroccan High Atlas" (17005) funded by the MAVA Foundation. We would like to acknowledge the collaboration of local and regional authorities as well as all the people who made this research possible, especially the interviewees from Imegdale and Ait M'hamed rural communes who shared their time and knowledge.

References

Asante, E. A., Ababio, S. and Boadu, K. B. (2017). The use of indigenous cultural practices by the Ashantis for the conservation of forests in Ghana. *SAGE Open January-March* 2017: 1–7. https://doi.org/10.1177/2158244016687611.

Atauri, J. A. and de Lucio, J. V. (2001). The role of landscape structure in species richness distribution of birds, amphibians, reptiles and lepidopterans in Mediterranean landscapes. *Landscape Ecology* 16: 147–159.

Auclair, L. and Alifriqui, M. (2012). *Agdal: Patrimoine socio-écologique de l'Atlas marocain.* Rabat, Morocco: IRCAM-IRD Editions.

Auclair, L., Baudot, P., Genin, D., Romagny, B., and Simenel, R. (2011). Patrimony for resilience: Evidence from the forest *agdal* in the Moroccan High Atlas Mountains. *Ecology and Society* 16(4): 1–24.

Aumeeruddy-Thomas, Y., Caubet, D., Hmimsa, Y. and Vicente, A. 2017. Les sociétés jbala et la nature: Parlers et relations à autrui dans le Rif, nord du Maroc. *Revue d'ethnoécologie* Suppl. 1: 1–18.

Baumann, G. (2009). *How to assess rangeland condition in semiarid ecosystems? The indicative value of vegetation in the High Atlas Mountains, Morocco.* PhD dissertation. Mathematisch-Naturwissenschaftlichen Fakultät, Universität zu Köln, Cologne, Germany:

Basso, K. H. (1996). *Wisdom sits in places: Landscape and language among the Western Apache.* Albuquerque, NM: University of New Mexico Press.

Berkes, F. (2003). Rethinking community-based conservation. *Conservation Biology* 18(3): 621–630.

Berkes, F. (2007). Community-based conservation in a globalized world. *PNAS* 104(39): 15188–15193.

Berkes, F. (2011). Implementing ecosystem-based management: Evolution or revolution? *Fish and Fisheries* 13(4): 465–476. DOI:10.1111/j.1467-2979.2011.00452.x.

Berkes, F., Colding, J. and Folke, C. (2000). Rediscovery of traditional ecological knowledge as adaptive management. *Ecological Applications* 10: 1251–1262.

Binder, C. R., Hinkel, J., Bots, P. W. G. and Pahl-Wostl, C. (2013). Comparison of frameworks for analyzing social-ecological systems. *Ecology and Society* 18(4): 26.

Blondel, J. (2006). The 'design' of Mediterranean landscapes: A millennial story of humans and ecological systems during the historic period. *Human Ecology* 35: 713–729.

Bugalho, M. N., Caldeira, M. C., Pereira J. S., Aronson, J., and Pausa, J. G. (2011). Mediterranean cork oak savannas require human use to sustain biodiversity and ecosystem services. *Frontiers in Ecology and the Environment* 9: 278–286.

CEPF (Critical Ecosystem Partnership Fund) (2017). *Ecosystem profile: Mediterranean Basin biodiversity hotspot. Extended technical summary.* Arlington, VA: CEPF.

COMBIOSERVE. (2015). Co-enquiry and participatory research for community conservation: Methods manual. Available at: www.global-diversity.org/wp-content/uploads/2016/01/Methods-Manual-Co-enquiry-and-Participatory-Research-GDF-2015.pdf

Deil, U., Culmsee, H. and Berriane, M. (2005). Sacred groves in Morocco: A society's conservation of nature for spiritual reasons. *Silva Carelica* 49: 185–201.

Domínguez, P. (2017). Political ecology of shifting cosmologies and epistemologies among Berber agro-sylvo-pastoralists in a globalizing world. *Journal for the Study of Religion, Nature & Culture* 11(2): 227–248.

Domínguez, P., Bourbouze, A., Demay, S., Genin, D. and Kosoy, N. (2012). Diverse ecological, economic and socio-cultural values of a traditional common natural resource management system in the Moroccan High Atlas: The Ait Ikiss Tagdalts. *Environmental Values* 21: 277–296.

Dorward, A. R. (2014). Livelisystems: A conceptual framework integrating social, ecosystem, development, and evolutionary theory. *Ecology and Society* 19(2): 44. DOI:10.5751/ES-06494-190244.

El Harousse, L., Aziz, L., Bellefontaine, R. and El Amrani, M. (2012). Le savoir écologique de deux populations habitant l'arganeraie (Essaouira). *Sécheresse* 23: 67–77.

Emberger, L. (1939). *Aperçu général de la végétation du Maroc.* Rabat: Société des sciences naturelles et physiques du Maroc. Mem. H.S.

Fernandez-Gimenez, M. E. (2000). The role of Mongolian nomadic pastoralists' ecological knowledge in rangeland management. *Ecological Applications* 10: 1318–1326.

Gadgil, M., Berkes, F. and Folke, C. (1993). Indigenous knowledge for biodiversity conservation. *Ambio* 22(2/3):151–156.

Haut-Commissariat au Plan. (2014). Recensement général de la population et de l'habitat. Available at: www.hcp.ma/ (accessed 20 July 2017).

Jarvis, D. and Hodgkin, T. (2000). Farmer decision making and genetic diversity: linking multidisciplinary research to implementation on-farm. In S. B. Brush (ed.), *Genes in the Field: On-Farm Conservation of Crop Diversity.* Rome: International Development Research Centre, International Plant Genetics Resource Institute, pp. 261–78.

Kideghesho, J. R. (2009). The potentials of traditional African cultural practices in mitigating overexploitation of wildlife species and habitat loss: Experience of Tanzania. *International Journal of Biodiversity Science & Management* 5(2): 83–94. DOI:10.1080/17451590903065579.

Mediterranean Consortium for Nature and Culture (2013). A rapid assessment of cultural conservation practices in the Mediterranean. Available at: http://medconsortium.org/wp-content/uploads/RapidAssessmentReport_web_1.pdf (accessed 15 May 2016).

Maffi, L. and Woodley, E. (2010). *Biocultural Diversity Conservation: A Global Sourcebook.* London: Earthscan.

McGinnis, M. D. and Ostrom, E. (2014). Social-ecological system framework: Initial changes and continuing challenges. *Ecology and Society* 19(2):30. DOI:10.5751/ES-06387-190230.

Médail, F. and Quézel, P. (1997). Hot-spots analysis for conservation of plant biodiversity in the Mediterranean basin. *Annals of the Missouri Botanical Garden* 84: 112–127.

McIvor, A., Fincke, A. and Oviedo, G. (2008). Bio-cultural diversity and indigenous peoples' journey. Report from the 4th IUCN World Conservation Congress Forum, 6–9 October 2008, Barcelona, Spain.

Montanari, B. and Bergh, S. I. (2014). The challenges of 'participatory' development in a semi-authoritarian context: The case of an essential oil distillation project in the High Atlas Mountains of Morocco. *Journal of North African Studies.* DOI:10.1080/13629387.2013.878247.

Nora, P. (1989). Between memory and history: Les lieux de mémoire. *Representations* 26: 7–24.

Oba, G., Stenseth, N. C. and Lusigi, W. J. (2000). New perspectives on sustainable grazing management in arid zones of sub-Saharan Africa. *BioScience* 50: 35–51.

Olson, D. and Dinerstein, E. (1998). The global 200: A representation approach to conserving the Earth's most biologically valuable ecoregions. *Conservation Biology* 12(3): 502–515.

Olson, D. M., Dinerstein, E., Wikramanayake, E. D., Burgess, N. D., Powell, G. V. N., et al. (2001). Terrestrial ecoregions of the world: A new map of life on Earth. *BioScience* 51: 933–938.

Ostrom, E. 2009. A general framework for analyzing sustainability of social-ecological systems. *Science* 325: 419–422.

Simenel, R., Michon, G., Auclair, L., Romagny, B., Thomas, Y. and Guyon, M. (2009). L'argan: l'huile qui cache la forêt domestique: de la valorisation du produit à la naturalisation de l'ecosystème. *Autrepart* 50: 51–74.

Soleri, D. and Smith, S. E. (1999). Conserving folk crop varieties: Different agricultures, different goals. In V. D. Nazarea (ed.), *Ethnoecology: Situated Knowledge/Located Lives.* Tucson, AZ, University of Arizona Press, pp. 133–154.

Taibi, A. N., El Khalki, Y. and El Hannani, M. (2015). *Atlas régional de la région du Tadla-Azilal.* Angers: Université d'Angers.

Teixidor-Toneu, I. (2015). Isafarn nudrar: Flowerpots help preserve biocultural diversity in the High Atlas, Morocco. *Langscape* 4(2): 53–55.

Thompson, J. D. (2005). *Plant Evolution in the Mediterranean.* Oxford: Oxford University Press.

Toledo,V. M. (1991). *El juego de la supervivencia: un manual para la investigación etnoecológica en Latinoamérica.* Santiago de Chile: Consorcio Latinoamericano de Agroecología y Desarrollo (CLADES).

Wehi, P. M. and Lord, J. M. (2017). Importance of including cultural practices in ecological restoration. *Conservation Biology* 31: 1109–1118. DOI:10.1111/cobi.12915.

Wilkes, G. (1991). In situ conservation of agricultural systems. In M. Oldfield and J. Alcorn (eds), *Biodiversity, Culture, Conservation and Ecodevelopment.* Boulder, CO: Westview Press, pp. 86–101.

Xu, J., Lebel, L. and Sturgeon, J. (2009). Functional links between biodiversity, livelihoods, and culture in a Hani swidden landscape in southwest China. *Ecology and Society* 14(2): 20. Available at: www.ecologyandsociety.org/vol14/iss2/art20/.

17

SACRED GROVES OF SIERRA LEONE

Preserving Indigenous Environmental Knowledge

Alison A. Ormsby

What are sacred groves?

Sacred groves are small forested areas that have local cultural and spiritual significance. These forests, found in numerous countries around the world, are conserved for various reasons, including traditional rituals and ceremonies, burial grounds, and ecosystem services, such as watershed value. Groves are community-managed forests that are often protected through local taboos or prescriptions on resource extraction activities (Colding and Folke 2001; Virtanen 2002; Tengö et al. 2007). They are a type of biocultural conservation (Gavin et al. 2015) and are an example of an Indigenous and Community Conserved Area (ICCA), as opposed to a governmental national park (Robson and Berkes 2013). Knowing what cultural practices and traditional knowledge are associated with the groves can help support conservation of the groves, and in many cases, these traditional systems of management are the reasons the groves still exist.

Research has been conducted on sacred forests in several countries in Africa, for example, on the groves of Ghana (Fargey 1992; Ormsby and Edelman, 2010; Ormsby 2012), Ethiopia's church forests (Wassie et al. 2005; Aerts et al. 2006; Wassie et al. 2010; Cardelus et al. 2013), and the sacred Mijikenda Kaya forests in Kenya (Nyamweru and Kimaru 2008; Nyamweru et al. 2008; Metcalfe et al. 2009). However, the only published research on the sacred groves of Sierra Leone was conducted in the Moyamba District of southern Sierra Leone with groves associated with the Mende cultural group (Lebbie and Guries 1995, 2008; Lebbie and Freudenberger 1996), and by Martin Martin et al. (2011) in the Bombali District of northern Sierra Leone. There is a lack of published research about the groves associated with the Temne ethnic group in Sierra Leone. In addition, there is a need for studies that do not solely focus on the ecological aspects of groves. Continued conservation of sacred forests and associated local knowledge is crucial to the future of biodiversity found in the groves.

The cultural significance of sacred groves

In Sierra Leone, sacred groves, also referred to as sacred bush, are critical to community structures and ceremonies. This cultural management helps support conservation of groves.

DOI: 10.4324/9781315270845-20

Sacred groves are areas in Sierra Leone where communities have established initiation rites, cultural rituals, and burial sites. They are believed to be the resting places of ancestors (Kaindaneh and Rigby 2010). These sacred groves are managed by traditional authorities and vary in use depending on the associated secret society. A secret society is a social group that requires initiation for membership, and for access to associated groves. In Sierra Leone, community politics and practices are heavily influenced by male and female societies. Secret societies within the Temne culture include the following: the chiefs' society; the male society known as Poro society; and the women's Bundo (also spelled Bundu) society. The societies perform initiation rituals within the sacred groves involving circumcision and youth training for men and women, separately, during which they learn life skills and forest-related knowledge, such as use of medicinal plants (Dorjahn 1961; Bledsoe 1984; Shaw 1985; Dorkenoo 1994; Sharif 1997; Mgbako et al. 2010). These societies in Sierra Leone have the power to control the natural resources within the sacred groves, forming a community-based local management system. The secret societies serve as the social mechanisms for transmitting Indigenous Environmental Knowledge, providing a way to retain and transfer traditional knowledge. The societies provide the framework for appropriate behavior regarding the groves, and enact sanctions or punishments if grove rules are broken. Indigenous Knowledge associated with the groves includes ethnobotanical knowledge of medicinal and useful plants, social institutions that manage resource use, and traditional knowledge education (Berkes et al. 2000; Berkes 2008).

Multiple taboos are part of the management of sacred groves in central Sierra Leone. These local rules are enforced by informal institutions – traditional authorities such as the men's Poro society and the women's Bundo society. As is typical of Indigenous Environmental Knowledge systems, management is "carried out using rules that are locally crafted and socially enforced by the users themselves" (Berkes et al. 2000: 1259). For example, no person of the opposite sex can enter the male or female groves. Rules also prohibit non-society members from entering any of the three groves, male (Poro), female (Bundo), or the chiefs' grove.

The historical context

The status, management, and possible future of the sacred groves of Sierra Leone must be considered within the context of the political and economic history of the country. Several factors have impacted Sierra Leone's biodiversity, including an 11-year civil war, long-term habitat fragmentation, and resource extraction (Squire 2001; Lindsell et al. 2011). Sierra Leone had a population of nearly 6 million people in 2010; the GDP per capita was US$775 in 2014, and life expectancy was estimated to be only 51 years of age for men and women in 2015 (United Nations Statistics Division 2017). The average years of schooling in Sierra Leone was only 3.3 years in 2010 and rose to 8.6 in 2015 (UNDP 2013, 2015). There are six major ethnic groups in Sierra Leone: Temne, Mende, Limba, Fulani, Kono, and Krio (The Diagram Group 2000). Most Sierra Leoneans are Muslim (60 percent), 10 percent are Christian, and 30 percent practice Indigenous beliefs (ibid.). The country is known for its numerous natural resources such as diamonds, bauxite, titanium, gold, and chromite. These resources played a part in driving the war in Sierra Leone.

In recent years, Sierra Leone gained media attention for its civil war that lasted from 1991 to 2002 (Bah 2011). During the war, approximately 50,000 people were killed (Bellows and Miguel 2009; Millar 2011) and 2 million were displaced (Akinsulure-Smith and Smith 2012). Sacred groves were used as hiding places by fleeing residents as well as base camps for the rebels (members of the Revolutionary United Front, RUF).

Sierra Leone has 48 forest reserves and conservation areas, including two national parks and one wildlife sanctuary (Brown and Crawford 2012). Unlike these formal protected areas, sacred groves do not have legal government protection. In light of the pressures from resource extraction, war, and population growth, it is impressive that any sacred groves still remain in Sierra Leone today. It is a testament to the grove managers, the continued strength (and associated fear) of cultural traditions, and the sustained value of Indigenous Environmental Knowledge that these groves have persisted.

In most cases there has been long-term historical management of each grove by its associated society, with management rules and practices continued by new initiates. The groves therefore serve as "knowledge carriers" for the Indigenous Environmental Knowledge associated with the society that uses and protects each grove, and represent a place for "intergenerational transmission of knowledge" (Berkes et al. 2000: 1253).

Researching Sierra Leone's sacred groves

I conducted research in 2012, with a focus on understanding the traditional management and status of groves in the Tonkolili District in central Sierra Leone, which had not previously been studied. Resident interviews focused on the topics of grove history, protection, resource use, changes over time, and the impacts of the war on the groves.

A semi-structured interview questionnaire with 19 questions was used, which was administered orally in the Temne language with the assistance of a translator. Data were collected through qualitative ethnographic interviews with 99 residents – 55 men and 44 women – in 16 communities. In addition, 51 sacred groves were observed. Most communities have (or had) one to three separate groves – one each for the chiefs, the male society, and the female society. It is not possible to provide the specific locations or place names of groves studied due to the secrecy of the groves. Consistent with the trend across Sierra Leone, the average level of education of those surveyed was 4.7 years; 70 percent identified as Muslim and 30 percent as Christian.

I followed a similar research process in each community studied. Upon arrival at a study site, the chief was visited to obtain permission to conduct interviews and visit the grove(s) with the chief and/or a member of one of the societies. However, due to the secrecy of the groves and associated activities, it was not possible to go into the groves; only initiated members of the society associated with each grove are allowed to enter. Afterward, resident interviews were conducted with both men and women of varying ages and occupations in the community, with the assistance of a translator. An attempt was made to interview key informants, such as secret society members, in each community. In addition, residents who represented a variety of occupations and ages were sought for a diversity of perspectives.

The interview results were analyzed using summary statistics for quantitative data and thematic coding for qualitative data and open-ended questions to identify patterns in the responses. The responses are aggregated from the 99 interviewees and presented in the results section. A similar method was used for sacred forest research in Ghana (Ormsby 2012) and in India (Ormsby 2013).

Sacred groves can usually be identified as small patches of forest surrounded by other land use types, such as farmland or villages. The groves observed in the Tonkolili District were estimated to be 0.5–7 acres (0.2–3 hectares) in size. The sacred groves were easily identified by typically having one or more very distinct and visible large cotton trees (*Ceiba pentandra*) within a small forested area (Figure 17.1), as also observed by Martin Martin (2011) in the Bombali District. The age of the groves is unknown. Although residents were asked about the grove age, there is

Figure 17.1 Cotton tree (*Ceiba pentandra*) in a sacred grove

Source: Photo credit: A. Ormsby.

no oral history documenting age, and taking soil cores or tree core samples would require secret society permission.

Results: Indigenous Environmental Knowledge and sacred groves

Interview results were analyzed and coded by theme. Results are presented in terms of grove protection, taboos, resource use, the impacts of war, and future vision for the groves.

Residents were interviewed about community opinions and values of the groves. When asked, "What is the purpose of protection of the grove?," 75 percent of residents' answers involved matters related to sacredness, including: the secret society, secrecy, ceremonies, sacred items, and/or culture. As one male resident who is a member of the Poro society said about the sacred grove, "It is our culture. We have to respect our culture." A resident in a different community said the grove is protected "because the place is a society place, a secret place." A female interviewee described, "It is an inherited cultural tradition from the ancestors, so we have to maintain it." In a third location, one resident explained that, "when the head of the society dies, they are buried in the bush." When asked the open-ended question, "What would you say is the most important value of the grove?," 53 percent of interviewees' responses related to secret societies, ceremonies, and culture. For example, one man said that the value is "our culture, becoming a man" and another said "it [the grove] was inherited" and that older Poro members taught about herbal medicine there. Forest animals and medicine were named as important values of the grove by 13 percent of respondents.

Taboos, which are local cultural rules, also play a role in the protection of sacred groves. Everyone in each society knows the rules and knows not to go near the sacred grove, to stay away from the area, especially at initiation times (Karim 2017, pers. comm.). When asked, "What actions are not allowed in the sacred grove(s)?" with respect to the rules of the groves, respondents identified a number of prohibited actions, ranging from terms of access to forbidden activities (Figure 17.2). As one male resident said about the grove, "People can only go in for ceremonial business."

As seen in Figure 17.2, cultural taboos play a role in grove resource management, because the groves are off-limits to any people entering (except secret society members during initiation ceremonies), and resource extraction is limited, including hunting, tree cutting, and clearing for farming.

Members of the society affiliated with each grove are typically allowed access to medicinal plants. In an open-ended question asking if they had used anything from the sacred grove, 45 percent of residents responded that they had used resources from the grove, predominantly herbs for medicine. Residents gave examples of herbs taken from local groves that are used for

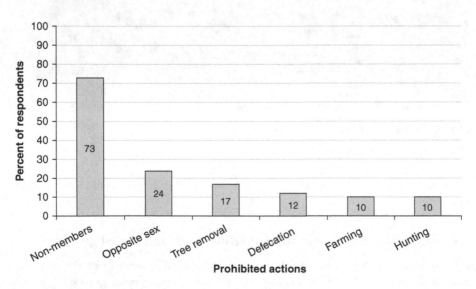

Figure 17.2 Responses to the interview question "What actions are not allowed in the sacred grove(s)?"

numerous ailments, ranging from dysentery to malaria. One resident reported that his local grove has "medicine that can't be found elsewhere."

Residents also spoke of the role of groves during the recent war. It is clear that during this period, the local grove taboos were not upheld and people had to overcome their fear of breaking rules because residents were fleeing for their lives. As one resident said about the local sacred grove (secret bush), "People are afraid to enter, but it is a good place to hide from the rebels – they are scared of the secret bush." When asked the open-ended question, "How did the war affect the sacred grove?," 49 percent of respondents reported that residents hid in the groves, 20 percent of respondents said that rebels hid in the groves, and 20 percent of those interviewed said that rebels destroyed ceremonial items stored in the groves. In addition, some residents reported that rebels had set fire to the sacred grove or cleared part of the grove in order to prevent it from serving as a hiding place for residents during the war.

When asked about the creation of the groves, 50 percent of residents reported that the ancestors had created the groves. Indicating the interest in continuing the tradition of sacred groves, when asked the open-ended question, "What do you think should happen to the grove in the future?," the most common responses were to expand and preserve the grove (Figure 17.3). Interviewees could give more than one response. As one female resident said about the future of her local grove, "Preserve it, it is part of history." In addition, one resident said, he "doesn't want the grove to disappear. The community gets their health from it, from the trees, and trees protect the houses from wind." Regarding use of the groves for initiation ceremonies, one male resident discussed how initiation ceremonies (circumcision) are being criticized, saying, "I do not think culture should be condemned." He explained that if it is, then there will be no performance of the initiation and it is possible that they will lose the forest. Other respondents alluded to modifying culture in order to address human rights issues relating to circumcision.

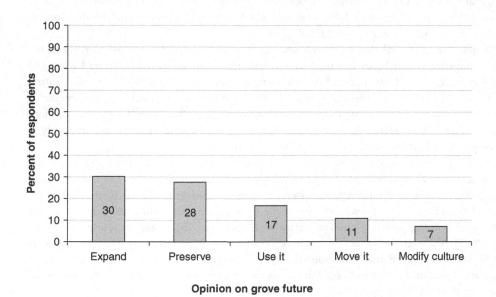

Figure 17.3 Residents' opinions about the future of sacred groves

Discussion

Results indicated that interviewed residents believe the greatest purpose and value of the groves is their sacredness. This study found that sacred groves in the Tonkolili District of Sierra Leone were protected out of respect for the wishes of ancestors, as well as for ritual use during the training and initiation period for each secret society. Similarly, in their study of the role of sacred groves of Sierra Leone particularly relating to the war, Kaindaneh and Rigby (2010: 247) identified groves as "the place where people can connect with their collective past as represented by their ancestors." In contrast, research on sacred forests in Ghana and India found that gods are believed to reside within the sacred groves, which are conserved out of the respect for these god(s) (Dorm-Adzobu et al. 1991; Ormsby and Bhagwat 2010; Ormsby 2012). It is important that the sacredness and Indigenous Environmental Knowledge associated with the groves in Sierra Leone are continuously upheld and respected. Many interviewees in this study explained how the groves were affected during the war, a time when sacred taboos were temporarily ignored. Based on their research on this topic, Kaindaneh and Rigby (2010) explained that through recent community rituals and ceremonies, society members worked to restore the sacredness of the groves in Sierra Leone that had been destroyed or affected by the rebels during the war. In this way, the secret societies help maintain the resilience of the groves and reaffirm the associated cultural traditions and Indigenous Environmental Knowledge.

Results from this study showed that residents felt that the most important grove taboos related to the secret societies and limitations on entry. Other taboo activities identified by interviewees included: tree removal, defecation, and hunting or farming within the groves. Lebbie and Freudenberger (1996) also found that hunting and extraction of timber from sacred groves were prohibited in the groves of southern Sierra Leone. Similar natural resource management taboos have been reported in other regions of the world. Colding and Folke (2001) reviewed the use of resource and habitat taboos in multiple sites around the world, ranging from marine reserves to sacred groves. They identified six major categories of taboos, including temporal taboos and specific species taboos. Most of the sacred grove rules reported in Sierra Leone fall into the category of habitat taboos, which relate to "access to and use of resources from particular habitats in space and time" (ibid.: 590).

Despite the taboos relating to natural resource use within sacred forests, 45 percent of the sample population stated they had used natural resources from the groves in Sierra Leone. The majority of these residents specified the use of medicinal plants. Similarly, Lebbie and Guries (1995) found that over 75 medicinal plant remedies were used by traditional folk medicine practitioners of the Kpaa Mende residents who obtained plants from sacred groves in southern Sierra Leone. They reported that some plants that were once common are now found only in sacred groves, where they can only be used with permission.

Several factors contribute to the resilience of sacred groves in Sierra Leone. In particular, the continued strength of secret societies and their use of the groves, as well as the fear associated with going into the groves, keep people from entering the groves or harming them. Despite the unusual circumstances of the war and use of the groves during that time, the groves still exist, showing the strong power of cultural norms and "social fencing" using society rules (Tengö et al. 2007). Although the groves do not have an actual fence around them, the boundaries are usually very clear, as the groves are bordered by farmland or communities, and these boundary edges are maintained by fears of sanctions, punishment, or general harm that may befall anyone who improperly enters a grove.

Although government conservation areas in Sierra Leone are important to forest protection, they are only partially effective. Sacred groves are key community-based conservation areas that add to the overall stability of forest protection within the country, despite past government instability. If the taboos associated with the groves are upheld, these forest areas may be sustained in the future. As Virtanen (2002: 229) notes on the sacred groves in Mozambique, the groves rely on "supernatural sanctions for their protection ... backed up by human sanctions administered by selected members of the respective community." In Sierra Leone, this is the role of the secret societies as the informal institutions and customary authorities that uphold rules associated with the groves. As in the sacred groves of Mozambique (ibid.), the sacred groves of Sierra Leone are the sites where important rituals are held.

This research in the Tonkolili District confirms that informal institutions, cultural traditions, and rituals remain the most significant elements in continued protection of the sacred groves. However, one male resident interviewed expressed concern regarding the initiation rituals associated with the groves, wondering if the initiations were stopped due to human rights concerns, whether conservation efforts of sacred groves would continue. As he said about the sacred groves, "Certain laws should be modified, like human rights issues."

Residents may be able to continue cultural traditions and conservation efforts while respecting human rights and conserving the biodiversity within the groves, but this will take time. Regarding sacred groves in Madagascar, Tengö et al. (2007: 690) asserted that "a social-ecological system approach" is needed for grove management and conservation. Sacred groves around the world can serve as habitat islands, providing cultural and ecological services (Virtanen 2002). Even with ongoing natural resource pressures in Sierra Leone, it is possible that the long-held tradition of sacred groves and their associated Indigenous Environmental Knowledge will continue to be conserved and respected, particularly since these groves were able to survive a war. It is a testament to the strength of Indigenous Environmental Knowledge and the associated belief systems and social norms that, in most cases, the sacred groves survived the 11-year war – reflecting both cultural and ecosystem resilience. Further research is needed to explore the links between cultural values, local knowledge, and biodiversity conservation, as well as the impacts of civil unrest on systems of biocultural conservation.

Acknowledgments

This research was conducted with Sierra Voss, and I am grateful for her field assistance. Thanks to our host communities and residents who participated in our research. We owe a special debt of gratitude to K. Bokum. We are also grateful for support from the Eckerd College Ford Scholar Program, Explorers Club Youth Activity Fund, and the Garden Club of America's Elizabeth Gardner Norweb Summer Environmental Studies Scholarship.

References

Aerts, R., Koen, V. O, Mitiku, H., Martin, H., Jozef, D., and Bart, M. (2006) Species composition and diversity of small afromontane forest fragments in Northern Ethiopia. *Plant Ecology* 187: 127–142.

Akinsulure-Smith, A. M. and Smith, E. H. (2012) Evolution of family policies in post-conflict Sierra Leone. *Journal of Child & Family Studies* 21: 4–13.

Bah, A. B. (2011) State decay and civil war: A discourse on power in Sierra Leone. *Critical Sociology* 37(2): 199–216.

Bellows, J. and Miguel, E. (2009) War and local collective action in Sierra Leone. *Journal of Public Economics* 93: 1144–1157.

Berkes, F. (2008) *Sacred Ecology*, 2nd edn. New York: Routledge,

Berkes, F., Colding, J. and Folke, C. (2000). Rediscovery of traditional ecological knowledge as adaptive management. *Ecological Applications* 10(5): 1251–1262.

Bledsoe, C. (1984) The political use of Sande ideology and symbolism. *American Ethnologist* 11: 455–472.

Brashares, J., Arcease, S., Peter, S., Moses, K., and Coppolillo, P. (2004) Bushmeat hunting, wildlife declines, and fish supply in West Africa. *Science* 306: 1180–1183.

Brown, O. and Crawford, A. (2012) Conservation and peacebuilding in Sierra Leone. The International Institute for Sustainable Development. Available at: www.iisd.org/pdf/2012/iisd_conservation_in_Sierra_Leone.pdf (accessed 11 November 2012).

Cardelus, C. L., Peter, S., Alemayehu W. E., Carrie, L. W., Peter, K., Eliza, K., and Izabela, O. (2013) A preliminary assessment of Ethiopian sacred grove status at the landscape and ecosystem scales. *Diversity* 5: 320–334.

Colding, J. and Folke, C. (2001) Social taboos: "Invisible" systems of local resource management and biological conservation. *Ecological Applications* 11: 584–600.

Diagram Group (2000) *Encyclopedia of African Peoples*. New York: Facts on File, Inc.

Dorjahn, V. R. (1961) The initiation of Temne Poro officials. *Man* 61, 36–40.

Dorkenoo, E. (1994) *Cutting the Rose: Female Genital Mutilation, the Practice and its Prevention.* London: Minority Rights Publication.

Dorm-Adzobu, C., Ampadu-Agyei, O. and Veit, P. G. (1991) *Religious Beliefs and Environmental Protection: The Malshegu Sacred Grove in Northern Ghana.* Washington, DC: World Resources Institute.

Fargey, P. J. (1992). Boabeng-Fiema Monkey Sanctuary: An example of traditional conservation in Ghana. *Oryx* 26(3): 151–156.

Gavin, M. C., Joe, M., Aroha, M., Fikret, B., John R. S., Debora, P., and Ruifei, T. (2015) Defining biocultural approaches to conservation. *Trends in Ecology & Evolution* 30(3): 140–145.

Kaindaneh, S. and Rigby, A. (2010) Promoting co-existence through sacred places in Sierra Leone. *Peace Review* 22: 244–249.

Lebbie, A. R. and Freudenberger, M. S. (1996) Sacred groves in Africa: Forest patches in transition. In J. Schelhas and R. Greenberg (Eds.), *Forest Patches in Tropical Landscapes.* Washington, DC: Island Press, pp. 300–321.

Lebbie, A. R. and Guries, R. P. (1995) Ethnobotanical value and conservation of sacred groves of the Kpaa Mende in Sierra Leone. *Economic Botany* 49: 297–308.

Lebbie, A. R. and Guries, R. P. (2008) The role of sacred groves in biodiversity conservation in Sierra Leone. In M. Sheridan and C. Nyamweru (Eds.), *African Sacred Groves: Ecological Dynamics and Social Change.* Oxford: James Currey Ltd, pp. 42–61.

Lindsell, J., Klop, E. and Siaka, A. M. (2011) The impact of civil war on forest wildlife in West Africa: Mammals in Gola Forest, Sierra Leone. *Fauna & Flora International* 45 : 69–77.

Martin Martin, A., Martinez de Anguita, P., Vicente Perez, J., and Lanzana, J. (2011) The role of secret societies in the conservation of sacred forests in Sierra Leone. *Bois et Forêts des Tropiques* 310(4): 43–55.

Metcalfe, K., Ffrench-Constant, R.F., and Gordon, I. (2009) Sacred sites as hotspots for biodiversity: The Three Sisters Cave complex in Coastal Kenya. *Oryx* 44(1): 118–123.

Mgbako, C., Saxena, M., Cave, A., Farjad, N., and Shin, H. (2010) Penetrating the silence in Sierra Leone: A blueprint for the eradication of female genital mutilation. *Harvard Human Rights Journal* 23, 1–28.

Millar, G. (2011) Local evaluations of justice through truth telling in Sierra Leone: Postwar needs and transitional justice. *Human Rights Review* 12: 515–535.

Nyamweru, C. and Kimaru, E. (2008) The contribution of ecotourism to the conservation of natural sacred Sites: A case study from coastal Kenya. *Journal for the Study of Religion, Nature and Culture* 2: 327–350.

Nyamweru, C., Staline, K., Mohammed, P., and John A. C. The Kaya Forests of Coastal Kenya: 'Remnant patches' or dynamic entities? In M. J. Sheridan and C. Nyamweru (Eds.), *African Sacred Groves: Ecological Dynamics and Social Change.* Athens, OH: Ohio University Press, pp. 62–86.

Ormsby, A. (2012) Cultural and conservation values of sacred forests in Ghana. In G. Pungetti, G. Oviedo, and D. Hooke (Eds.), *Sacred Species and Sites: Advances in Biocultural Conservation.* Cambridge: Cambridge University Press, pp. 335–350.

Ormsby, A. (2013) Analysis of local attitudes toward sacred groves of Meghalaya and Karnataka, India. *Conservation and Society* 11: 187–197.

Ormsby, A. A. and Bhagwat, S. A. (2010) Sacred forests of India: A strong tradition of community-based natural resource management. *Environmental Conservation* 37: 320–326.

Ormsby, A. and Edelman, C. (2010) Tafi Atome Monkey Sanctuary, Ghana: Community-based ecotourism at a sacred site. In B. Verschuuren and R. Wild (Eds.), *Sacred Natural Sites: Conserving Nature and Culture.* London: Earthscan, pp. 233–243.

Robson, J. P. and Berkes, F. (2013) Sacred nature and community conserved areas. In S. Pilgrim and J. Pretty (Eds.), *Nature and Culture: Rebuilding Lost Connections*. New York: Routledge, pp. 197–216.

Sharif, K. (1997) Female genital mutilation: What does the new federal law really mean? *Fordham Urban Law Journal* 24: 409–426.

Shaw, R. (1985) Gender and the structuring of reality in Temne divination: An interactive study. *Africa* 55: 286–303.

Squire, C. (2001) Sierra Leone's biodiversity and the civil war: a case study prepared for the Biodiversity Support Program. Washington, DC: Biodiversity Support Program.

Tengö, M., Kristin, J., Fanambinantsoa, R., Jakob, L., Jean-Aimé, A., Jean-Aimé, R., and Thomas, E. (2007) Taboos and forest governance: Informal protection of hot spot dry forest in Southern Madagascar. *Ambio* 36: 683–691.

Tiwari, B. K., Barik, S. K., and Tripathi, R. S. (1998) Biodiversity value, status, and strategies for conservation of sacred groves of Meghalaya, India. *Ecosystem Health* 4: 20–32.

UNDP (United Nations Development Program) 2013. The rise of the south: Human progress in a diverse world. Available at: http://hdr.undp.org/en/2013-report. (accessed 25 January 2016).

UNDP (United Nations Development Program) 2015. Work for human development. Available at: http://hdr.undp.org/sites/default/files/2015_human_development_report_1.pdf (accessed 25 January 2016).

United Nations Statistics Division (2017) Sierra Leone country profile. Available at http://data.un.org/CountryProfile.aspx?crName=sierra%20leone (accessed 26 January 2017).

Virtanen, P. (2002) The role of customary institutions in the conservation of biodiversity: Sacred forests in Mozambique. *Environmental Values* 11: 227–241.

Wassie, A., Demel, T., and Neil, P. (2005) Church forests in North Gonder Administrative Zone, Northern Ethiopia. *Forests, Trees, and Livelihoods* 15: 349–373.

Wassie, A., Frank J. S., and Frans B. (2010) Species and structural diversity of Church Forests in a fragmented Ethiopian highland landscape. *Journal of Vegetation Science* 21: 938–948.

18

THE ROLE OF BIODIVERSITY IN THE MAINTENANCE OF ECOSYSTEM SERVICES IN HUMAN-DOMINATED LANDSCAPES

Evidence from the Terai Plains of Nepal

Jessica P. R. Thorn, Thomas F. Thornton, Ariella Helfgott and Kathy J. Willis

Introduction

In the age of the Anthropocene (Mace 2005; Steffen 2007), the predominant paradigm of social and economic development remains largely oblivious to the risk of human-induced environmental disasters at local, continental to planetary scales (Stern 2007), reduced well-being that humans derive from ecosystem services (Guo 2010), and accelerated rates of biodiversity loss (Chapin 2000). A lack of appreciation of humans' dependence on ecosystem services is one of many factors contributing to the complex disruption of the biosphere (IPBES 2019). Despite a rapidly accumulating evidence base quantifying ecosystem services, uncertainties remain about the role of biodiversity in the maintenance of ecosystem services in variable human-dominated landscapes, humans' contributions to biodiversity maintenance and the extent to which such services contribute to social-ecological resilience (Fahrig 1997; Elmqvist et al. 2003; Acharya 2006; Guo 2010; Thorn, Thornton, and Helfgott 2015).

There is strong evidence to suggest agricultural expansion is a major global driver behind loss of ecosystem functioning and services, may adversely affect the regulatory capacities of climate system, hydrological, nitrogen and phosphorus cycle (Rockström 2009), produce lower yields (Plumecocq et al. 2018), and increase the likelihood of systemic threshold changes or tipping points in the near future (Hossain et al. 2016). Approximately 60 per cent of the ecosystem services examined in the Millennium Ecosystem Assessment are being degraded or unsustainably used (MEA 2005). However, a growing body of literature suggests agro-ecosystems in remote, marginal environments are often home to Indigenous farming communities and may support some of the highest overall biodiversity levels of any agricultural system – levels which rival those of conservation planning for three key reasons (Kremen 2015). First, around homesteads and farm boundaries, farmers

DOI: 10.4324/9781315270845-21

often maintain multiple layer systems of trees, herbs, climbers, grasses and herbs (Kumar and Nair 2004; Clarke 2014) – including not only cultivated crops, but also native and close wild relative species often considered more resilient than modern cultivars (Seto et al. 2012; IIED 2015). Such plants have been grown or tended for millennia (Olango 2014), directly contributing to farmers' livelihoods, food and nutritional security (Howard 2009; Dansi 2012). Second, small and medium-sized farms typically have higher variation in community abundance as compared to monoculture croplands, because farmers tend to cultivate a diversity of crops within and across seasons to spread the risk (Lubbe 2011). Third, agricultural landscapes are typically configured as multifunctional and relational mosaics, crops being situated according to both their individual utility and complementarity with other biota (Grieg-Gran 2011; Haenke 2014; Ford 2015). Such landscape heterogeneity and connectivity between croplands and native vegetation can encourage the recolonisation of disturbed habitats, and counterbalance degraded ecosystem function (Fahrig 1997; Elmqvist 2003).

Yet surprisingly little is known about wild and tended plant species' community composition in and around farms. (Here the term 'tended' is used to infer a lower level of cultivation of 'wild species', given the domestic origin of plants, and the fact that what constitutes cultivated or uncultivated is difficult to determine. For example farmers may manage the habitat of useful plants, support growth and reproduction through trimming, protect plants from animals, wind or water damage, or monitor harvesting levels.) Moreover, precise information of local ethnobotanical knowledge, including how communities mediate ecosystem services delivery at the landscape and community level to benefit from genetic resources, remains limited (Fahrig 1997; Elmqvist et al. 2003; MEA 2005). Understanding local agricultural traditions and preferences of plants is important because processes of planting, extraction and domestication of plant populations influence the community structure, rate of species turnover (Clarke 2014) and genetic make-up (Acharya 2006). At the same time, local knowledge and practice remain the foundation for any local response (Boissière 2013), and are often the only interventions to reduce risk (Thorn et al. 2015; Thorn 2019). An integrated social-ecological assessment of ecosystem services in relation to changing cultural knowledge, practices and ecosystem composition and function is needed to understand and appropriately respond to the multiplicity of values inherent in diverse, cultivated landscapes.

Situated in the Terai Plains of Nepal, the aim of this chapter is to explore the role of biodiversity in maintaining ecosystem services by investigating the composition and use of wild and tended plant material found in and around rice production landscapes. Specifically, the analysis is based on the following objectives: first, to survey wild and unmanaged plant abundance and diversity; second, to capture local ethnobotanical knowledge, including use, source and administration of plants, and determine if knowledge of use differs according to caste, gender and age; and, third, to identify what factors incentivise the maintenance of biodiversity in and around farms. As human pressures on forested and non-agricultural lands continue to escalate, conservation planners and land managers are seeking approaches to quantify the numerous economic and non-monetary benefits that ecosystems provide. Understanding the degree of overlap of agricultural lands that provide important ecosystem services, biodiversity and support livelihoods, can reveal hotspots for conservation and build common ground for new partnerships, with implications for broader areas of adaptive planning.

Methodology

Description of the study area

The study was situated in Central and Western Nepal, where 78 per cent of the workforce depend on subsistence agriculture (World Bank 2015). The Terai Plains of Nepal (hereafter

Terai) stretches 1360 km² across the south of the country, produces 68 per cent of Nepal's agricultural output (Malla 2008) and 38.2 per cent of Nepal's GDP (Manandhar 2010).

Inhabitants are predominantly the Tharu – the largest ethnic Indigenous minority in Nepal – comprising of over 2,000 subdivisions. Historically, the Tharu were semi-nomadic, practising short fallow-shifting rice cultivation with livestock, but today are increasingly sedentary. During the resettlement programme, the cultural-demographic profile shifted from small pockets of Tharu to a mixture of Brahmin, Chettri, other castes, and Indian migrants (Regmi 1994). This amalgamation, in turn, has led to homogenisation of culture and knowledge, resulting in a loss of ethnobotanical knowledge, identities, and agricultural practices which sustain Tharu livelihoods. Changing land rights, urbanisation, and new income streams have also contributed to these shifts (Guneratne et al. 2002, 2014).

While numerous studies have considered ethnobotanical knowledge in Nepal, most have focused on the ecological structure or specific ecosystem services (Mohri 2013), particularly plants with pharmacological value (Manandhar 1998; Bhattarai 2006; Kunwar, Shreshtra, and Bussman 2010), or cultural keystone species (Garibaldi and Turner 2004), neglecting a comprehensive set of ecosystem goods and services values at the landscape scale. Although a few scholars have studied the traditional knowledge systems of the Tharu, many regions in the Terai remain understudied and no studies have considered how smallholders mediate non-agricultural plant species community composition in and around farms.

Field sampling

Field sampling was conducted in the Indo-Mayalan Tropical Dry Zone of the Terai in 2012. Sampling was carried out in four landscapes, 22 village district committees (VDCs) and 40 wards (Figure 18.1): (1) four VDCs in Madi Valley, Chitwan district (N27°28.305' E084°17.244', 204masl); (2) six VDCs in Rupandehi district (N27°35.414' E083°31.180', 138masl); (3) six VDCs surrounding Gohari, Dang district (N27°50.783' E082°30.068', 256masl) (referred to hereafter as Dang); and (4) six VDCs in the Deukhuri Valley, Dang district (N28°03.086' E082°18.712', 597masl) (Deukhuri).

Ethnobotanical data collection

Standardised sampling procedures were used to collect specimens (Bridson 1998) involving transects walks in home gardens, farms, and the surrounding landscape within 250 m of homesteads (one sample/species/farm). To identify scientific and English names of species, the nomenclature followed was that employed by Press, Shreshtra, and Sutton (2000). To verify uses, we referred to previous studies. All wild and tended plant material was photographed for further reference. Unidentified species were identified and deposited in the National Herbarium and Plant Laboratories Godawari, Lalitpur in Kathmandu. We were limited to collecting predefined 'key' parts of the plant (e.g. leaf, fruit, stem), rather than the entire plant. Yield was not recorded.

Sampling informants

Ethnobotanical data were gathered from 180 informants using semi-structured interviews, questionnaires, focus group discussions, participant observation, and field observations. As far as possible, the sample was randomly stratified across age (25–67 years), gender (72.5 per cent male, 27.5 per cent female), and caste (n = 10). 82.5 per cent of the study population own land through inheritance, 12.5 per cent through procurement, and 5 per cent through

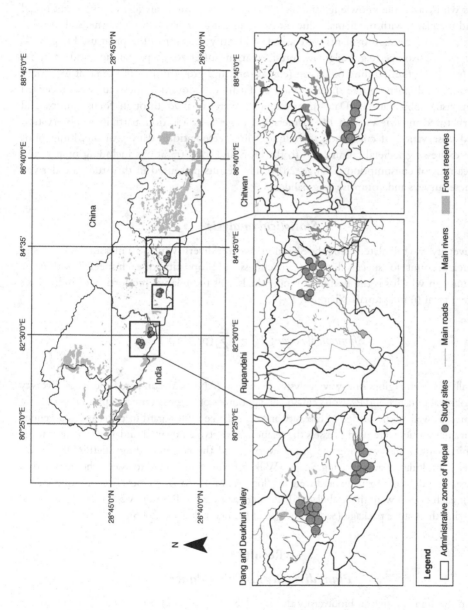

Figure 18.1 Map of study area in the Central and Western zones of the Terai Plains of Nepal (n = 40 villages). Ten villages sampled were distributed across the landscape (watershed catchment), where farmers cultivate rice on 5.78 ± 2.33 ha in the monsoon season (May–September) in terraced landscapes

government provision. Most agricultural land relies on rain-fed agriculture, while 26.5 per cent of cultivated land is irrigated. Forty farmers managing the farms surveyed were asked to free-list vernacular names in Nepali (N) or Tharu (Th), rather than using predefined categories to reduce researcher bias (Bernard and Gravlee 2014), and describe the use-value of wild and tended plant material (e.g. medicine for humans/livestock, fodder, fuel, building material, biocides, food additives, fertilisers, or cultural, religious, aesthetic, ornamental, and ritual purposes). Community members, rather than specialist practitioners, were interviewed to assess wildly available knowledge (Brush 1996). Uses were then categorised into eight broad uses, and for plants with medicinal value into 61 ailment groupings, using biomedical terminology. For each species, informants were asked to identify the part of the plant used (e.g. bark, root), the plant's source (e.g. hedgerow, forest, riparian buffer zone), preparation and administration (e.g. decoction), timing of harvest (e.g. season), and growth form of the plant (e.g. grass, tree, shrub). Multipurpose tree species found on farms and around homesteads were recorded through visual observations. Data inconsistencies were verified through focus groups and semi-structured interviews with 140 farmers and experts. In addition, participant observation involved observation of cultivation techniques, ritual celebrations, daily worship, Indigenous folklore expressing societal cultural ties to the crop, and food traditions including preparation and occasions of consumption (Olango 2014). The results presented in this study are derived from these surveys and comprise original data.

Statistical analysis

Plant diversity was calculated using the Shannon–Weaver diversity index: a measure of biodiversity which accounts for species dominance (richness and proportion of each species) within the community, in which s is the number of individuals, p_i is the relative proportion of individuals belonging total (i) individuals (Shannon 1949).

$$Shannon\ Weaver: H' = -\sum_{i=1}^{s} (p_i^0 * \ln p_i)$$

Across all and each region, one-way analysis of variance (ANOVA) compared means of diversity and absolute abundance and Pearson's chi-squared goodness of fit tests compared proportional plant abundance, as well as knowledge of diversity and abundance with regard to caste, age and gender. Multivariate statistics were used to assess the relationship between plant abundance and caste, using hierarchical cluster analysis using the group average, and the corresponding SIMPROF test for non-metric multidimensional scaling. Shapiro Wilk tests were performed to assess whether the data met the assumptions of normality and were log/log^{10} transformed where necessary. Unique species found in each region were then tabulated. Data were analysed in R Studio V.3.1.1. (R Studio Team 2020), using the Lattice package (Sarkar 2008), and Primer-E (Clarke 2006).

Results

Plant diversity and abundance

Farms in the Terai are rich in biodiversity that provide essential goods that sustain livelihoods and, correspondingly, land managers maintain a rich ethnobotanical knowledge. Overall, 390 vascular plant specimens were collected and identified as belonging to 76 distinct plant species from 49 phylogenetic families. Individual farms have in between 1 and 27 useful plants (i.e., important and

Table 18.1 Comparison across climatic regions of plant species diversity (H'), abundance, and relative constitution of number of plant species, according to use categories

	Chitwan (Wettest)		Rupandehi		Deukhuri		Dang (Driest)		All farms
Species diversity	3 ± 0.26		3.22 ± 0.17		3.14 ± 0.14		2.99 ± 0.08		3.09 ± 0.08
Species abundance	9.4 ± 1.66		12.1 ± 2.06		9.6 ± 1.06		7.9 ± 0.69		9.75 ± 0.74
	Total sp.	Unique sp.	Total sp.	Unique sp.	Total sp.	Unique sp.	Total sp.	Unique sp.	All sp.
All uses	38	5	49	11	44	6	31	5	76
Fuel	11	1	10	0	12	1	15	3	17
Fodder	7	1	9	1	11	2	8	2	17
Food	11	2	15	2	13	2	12	3	25
Timber	15	3	11	1	14	2	13	1	20
Soil	8	1	6	0	9	2	4	0	11
Medicine	29	2	38	8	37	4	24	4	56
Spiritual	15	2	21	5	12	2	8	1	27
Pesticide	4	0	8	1	6	0	4	0	8

Note:
Table 18.1 presents (a) diversity, (b) abundance, and (c) the results of our analysis of the relative constitution of number of plant species, according to use categories. Values show the mean ± SE. Unique plant species were found in each climatic region, and the highest number of unique species was found in Rupandehi. Unique species refers to species found only in one climatic region. (sp.: species).

commonly used). In the entire study area, species diversity (H') is 3.09±0.09 (mean±SE), species abundance is 9.75±0.74 and the average number of plants/farm is 9.75±4.71 (Table 18.1). Of the 76 plant species collected, 56 are used for medicine, 27 for rituals, 25 for food, 20 for timber, 17 for fuel, 17 for fodder, 11 for soil enhancement, and 8 for pesticides. Most plants (73.3 per cent) are used for multiple purposes: 29.3 per cent are used for two, 22.7 per cent for three, 12 per cent for four, 6.7 per cent for five, and 2.7 per cent for six purposes. Eight species are considered to have disservices (e.g. invasive weeds). The most dominant plant families are *Euphorbiaceae* (5 families), *Fabaceae* (4), *Moraceae* (4), *Anacardiaceae*, *Lamiaceae*, and *Rutaceae* (3). The most common species are *Shorea robusta* (Sal tree) (6.7 per cent), *Dalbergia sisoo* (Indian rosewood) (6.4 per cent), *Azadirachta indica* (Mugwort) (6.4 per cent), *Melia azedarach* (Persian lilac) (5.1 per cent), *Leucaena leucocephala* (Leucaena) (4.4 per cent), *Ficus religiosa* (Banaya tree) (4.4 per cent), *Dendrocalamus strictus* (Bamboo) (4.4 per cent), *Ocimum tenuiflorum* (Holy basil) (4.4 per cent), *Mangifera indica* (Mango) (3.9 per cent), and *Jatropha curcas* (Physic nut) (3.1 per cent) (Storrs and Storrs 1998; Singh, Kumar, and Twearu 2012; IUCN 2004). (See Thorn et al. 2020 for full details of all the species identified.)

Source of biological material

Biological material is collected from farm boundaries, around homesteads, home gardens, or uncultivated patches, such as wetlands, small woodlands or riverbanks. Farmed areas are typically adjacent to homesteads, around which smallholders maintain wind and shade barriers, nurseries, fruit orchards, ornamentals, spices, vegetables, zero-grazing pens for raising domestic animals, and multi-storied crops (e.g. grasses, herbs, shrubs, trees) (Figure 18.2). Contour hedgerows

Figure 18.2 Adapted schematic view illustrating a typical multi-layered system of trees, herbs, climbers, grasses and herbs, in a heterogeneous, integrated cropping system in the Terai: (a) road, (b) pathway, (c) rice paddy fields (d) lentil and soya bean grown along boundaries, (e) tube wells, or slurry processing for biogas, (f) vegetable garden (e.g. bottle gourd, cucumber, tomato, beans, okra, sesame), spices (e.g. ginger, turmeric) and cosmetics (e.g. aloe vera) with mulched patches and ridges/bunds for water efficiency, (g) buffalo, goat or pig pen and fuel wood storage, (h) cluster of trees alongside boundary for windbreaks (e.g. *Dendroclamus strictus*), fuel wood and timber (e.g. *Dalbergia sisoo, Shorea robusta, Melia azedarach*), fruit (e.g. *Psidium guajava*), fodder (e.g. *Azadirachta indica, Albizia lebbeck*), religious value (e.g. *Aegle marmelos*), or shade (e.g. *Mangifera indica*), (i) house roof made of reed thatch, covered with creepers and gourds for aesthetic value, insulation, and food

along terrace ridges (e.g. lentils, soya bean) are more common than intercropped hedgerows – and are used to mitigate soil erosion on moderately sloping land, conserve soil nutrients and limit competition with crops for water and sunlight (Haenke et al. 2014). Fuel wood is typically sourced from trees around homes (in 41 per cent of cases), from community forests and national parks (71 per cent) or bought from traders (19.4 per cent). In 25 per cent of cases, fuel wood comes from two or three sources, but in some cases material cultivated around homes is sufficient.

Processing and administration

Of the 76 species recorded in the study area, the most commonly used growth forms are trees (51 per cent of species), herbs (24 per cent) and shrubs (16 per cent). Ten parts of the plant are used – most commonly the leaf (23 per cent), fruit or stem (14 per cent), flower (10 per cent) or bark (9 per cent). Other parts used are the root, flower, bark, seed, latex, shoot, and resin but rarely the entire plant (3 per cent). Generally, fresh plant parts are collected and used immediately. Alternatively, plants are stored in the shade or dry places in their original form, powdered or used as an ash. Plants are consumed directly, roasted, juiced or pickled, or applied externally using the paste of leaves or milky latex. Administration of most medicinal plants is via decoction (mashing, and boiling the plant in water to extract oils, volatile organic compounds, and other chemical substances), although dermatological ailments are usually treated topically. For example the leaves of *Azadirachta indica* (Margosa tree) are used to wash the skin to treat scabies (Storrs and Storrs 1998; IUCN 2004; Singh et al. 2012).

Administration of biological material varies across seasons. For example *Asparagus racemosus* (Asparagus) is used to prepare alcohol in August, and the fruit of *Paris polyphylla* (Herb paris) is used for worship in mid-April. Although most Community Forestry User Groups officially restrict the harvest of fuel wood between December–February and during festivals in November (e.g. *Daishan, Tihar*), in 48 per cent of cases, fuel wood is collected throughout the year. Administration of biological material furthermore varies according to the day of the week. For example on Tuesdays and Thursdays, women practise a ritual which involves grinding and eating the root of *Mimosa pudica* (Touch-me-not plant), or chewing the stem of *Calotropis gigantea* (Crown flower), to promote the well-being of their husbands. Other species are regularly ingested, such as *Aegle marmelos* (Bengal quince), the leaves and fruit pulp used as an offering to Lord Shiva (IUCN 2004; Madhu, Phoboo, and Jha 2010; Singh et al. 2012; Government of Nepal 2014).

Knowledge of plant use

Given the Nepalese are a culturally diverse population, we set out to assess whether knowledge of use and maintenance of species on farms differed according to caste, being sensitive to social norms of disclosure and using reported definitions. The caste system is a traditional classification system of 36 hereditary groups of hierarchical social classes, which are defined through a combination of elements of birthright, ethnicity and financial acumen. This may determine one's education, income, occupation, and social standing (Pigg 1992). Although social classification is socially constructed and evolving, broadly defined, following the *Chaturvarnashram* model, there are four social classes: Brahmin, Kshatriya, Vaishya and Sudra. Of the ten castes represented, SIMPROF tests identified four statistically different caste groups in similarity of knowledge: Chettri, Brahmin and Tharu have the most diverse (H') knowledge of plant uses, and have a 40 per cent similarity in knowledge of species. Dura and Gurung have a 30 per cent similarity in knowledge, as do Chaudhury, Teli, Dalit and Magar, while Sanyasi have a unique knowledge base endogenous to the region (Figure 18.3). No significant relationships between absolute ($F_{(9.30)} = 0.87, p = 0.56$) or proportional abundance ($x^2_{(167)} = 155.65, p = 0.1079$) of plant species and caste were found. All three analyses indicated Brahmin, Tharu and Chhetri generally have the most diverse knowledge, but Sanyasi, Magar and Chaudhury had higher scores when controlling for the number of respondents representing each caste in the sample (Figure 18.3). Males reported more plants (10±0.87) than women (9.09±1.51), however, the difference in the knowledge of plants between genders was not significant ($x^2(15) = 15.45, p = 0.4197$). No significant differences in species diversity according to age ($x^2(330)=356.78, p=0.1489$) were observed.

Reasons for use

Farmers adjust local management to augment certain services. Use depends on the availability of nearby resources and alternatives or supplements (e.g. synthetic building material, electricity infrastructure), affordability, available travel or collection time, effectiveness of use (e.g. for medicinal purposes), and appropriateness based on traditional customs, spiritual beliefs and livelihood strategies. Plants are generally used for domestic purposes in the household economy, rather than for commercial sale.

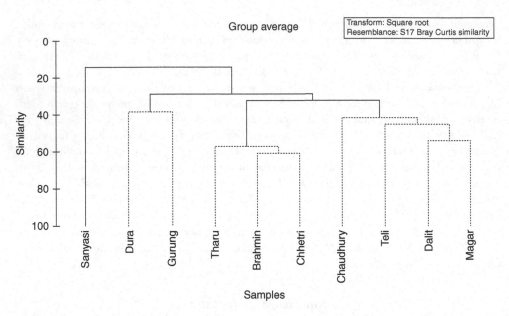

Figure 18.3 Cluster analysis of caste and plant species abundance. It shows Chettri, Brahmin and Tharu have the most diverse (H') knowledge of plant uses, and have a 40 per cent similarity in knowledge of species. Dura and Gurung have a 30 per cent similarity in knowledge, as do Chaudhury, Teli, Dalit and Magar, while Sanyasi have a unique knowledge base. Black lines indicate relationships that are significantly supported, while dashed lines indicate no significant difference was detected

Multipurpose woodlots

Multipurpose woodlots (known as '*Bagaincha*' (N), or '*Fulbari*' (Th)) provide fuel, fodder, timber, food, fertilisers, pesticides and control erosion. In total, 33 multipurpose tree species found on farms and around homesteads were recorded through visual observation. The average number of trees grown around farms is 6.03±4.35, ranging from 1–26. Ten tree species are used for fodder, six of which are found around homesteads (i.e. *Dalbergia sisoo* (Indian rosewood) (62.5 per cent), *Melia azedarach* (Persian lilac) (50 per cent), *Leucaena eucocephala* (Leucaena) (42.5 per cent), *Garuga pinnata* (Garuga) (12.5 per cent), *Artocarpus heterophyllus* (Jackfruit), and *Artemisia indica* (Mugwort)). Fifteen tree species are used for food (mostly fruit), ten of which are found around homesteads (e.g. *Mangifera indica* (Mango), *Syzygium cumini* (Black plum), *Psidium guajava* (Guava) and *Phyllanthus emblica* (Indian gooseberry)). Three tree species are used to improve soil nutrient levels and prevent erosion, two of which are found around homesteads (i.e. *Leucaena leucocephala* (Leucaena), *Albizia lebbeck* (Black siris)). Four tree species are used for pesticides, all of which are found around homesteads (i.e. *Melia azedarach* (Persian lilac), *Azadirachta indica* (Margosa tree), *Artocarpus heterophyllus* (Jackfruit), *Artemisia indica* (Mugwort)). Fourteen tree species are used for fuel wood, nine of which are found around homesteads. Fifteen tree species are used for timber, 11 of which are found around homesteads (Storrs and Storrs 1998; Kerkhoff 2003; IUCN 2004: Singh et al. 2012; Singh et al. 2014).

Energy use

The top ten species used for fuel wood are *Shorea robusta* (used by 65 per cent of households), *Dalbergia sisoo* (Indian rosewood) (62.5 per cent), *Melia azedarach* (Persian lilac) (50 per cent),

Dendrocalamus strictus (Bamboo), *Leucaena leucocephala* (Leucaena) (42.5 per cent), *Mangifera indica* (Mango), *Jatropha curcas* (Physic nut), *Psidium guajava* (Guava), *Phyllanthus emblica* (Indian gooseberry) and *Garuga pinnata* (Garuga) (Storrs and Storrs 1998; Kerkhoff 2003; IUCN 2004; Zheng and Dicke 2008; Singh et al. 2012). Some species of wood have cultural significance, used especially during festivals or for cremation (see below). Wood is the main source of household energy (97 per cent), as at the time of this study, electrical connections are very limited and power outage are common in many of the study sites. Biogas (i.e. ox, cow or human) (70 per cent), liquefied petroleum gas (22 per cent), solar (8 per cent), crop residue (e.g. maize husks (8 per cent)), or kerosene, candles or battery lamps for lighting are also used.

Medicinal use

The collection of medicinal and aromatic plants is vital to Tharu human and animal healthcare, in combination with modern remedies. Plant medicine is preventative, curative, soporific or stimulatory. Respondents reported 61 ailments that are treated with medicinal plants, including gastro-intestinal, dermatological, cardio-vascular, urogenital, respiratory, skeleto-muscular, mental disorders, or dental, eye, ear, nose, throat, birthing or lactation issues. Medicinal plants are most commonly used for fever ($n = 9$), cough ($n = 9$) and cold ($n = 8$), and are often mixed in combination. For example for constipation and a distended stomach, the root of *Mimosa pudica* (Touch-me-not plant) (Government of Nepal 2014) is ground and mixed with the root of *Achyranthes bidentata* (Hill chaff flower) (IUCN 2004), and the leaves of *Pogostemon benghalensis* (Bengal pogostemon) (Dangol 2008), *Psidium guajava* (Guava) (Kerkhoff 2003) and *Artemisia indica* (Mugwort) (Singh et al. 2012). Many remedies are influenced by *Ayurveda* used in India, and the *Bhaidya* system used in far-western Nepal (Kunwar, Shrestha, and Bussmann 2010) (Table 18.2).

Cultural use

Twenty-seven plants have cultural, spiritual or religious significance and are associated with cultivation practices. Plants are used in symbolic rituals, offerings, religious occasions, marriage ceremonies, fasting, or acts of purification. For example various flowers are used for worship in homes and temples as a ritualistic expression of reverence or adoration for deities, present daily offerings (e.g. *Jacaranda mimosifoliam* (Jacaranda) (IUCN 2004), *Lantana camara* (Lantana) (Ghosh 2012), *Michelia champaca* (Champak) (Storrs and Storrs 1998) and *Nyctanthes arbor-trisis* (Night-flowering jasmine) (Government of Nepal 2014)), or to make garlands for celebrations and welcome visitors (e.g. *Pogostemon benghalensis* (Bengal pogostemon) (Dangol 2008)). Women bathe in the leaves of *Achyranthes bidentata* (Hill chaff flower) during the festival of *Teej* to welcome the monsoon and as an act of purification (Singh et al. 2012). Traditionally, female Newaris (a caste) at the age of six months perform a symbolic marriage ceremony with *Aegle marmelos* (Bengal quince), and the size and morphology of the fruit is used to predict the character of the child's future husband (IUCN 2004). Similarly, Hindu women fast on Mondays and offer the leaf of *Cynodon dactylon* (Dog tooth's grass) (Singh et al. 2012) to Lord Ganesh for the well-being of one's husband. On religious celebrations the leaves of *Butea monosperma* (Flame of the forest) and *Shorea robusta* (Sal) (Storrs and Storrs 1998) are pieced together or used singly to make a leaf-plate on which to serve meals ('*patravali*' (N)). In many villages, the leaves of *Musa paradisiaca* (Banana) are used as a surface to cut communal meat shared with every village member for religious occasions (Singh et al. 2012). Various trees considered sacred in Buddhist, Jain and Hindu traditions are planted along roadsides, in public areas, villages and temples. For

Table 18.2 Ailment categories of medicinal plants identified during interviews

Ailment category	Biomedical term	Species used	No uses	No taxa
Gastro-intestinal illness	Stomach pain	*Mangifera indica, Artemisia indica, Vitex negundo, Acorus calamus, Curcuma angustifolia*	11	5
	Dysentery	*Musa paradisiaca, Cannabis sativa, Shorea robusta, Phyllanthus emblica, Aorus calamus*		5
	Intestinal worms	*Melia azedarach*		1
	Diarrhoea	*Psidium guajava, Shorea robusta, Syzygium cumini*		3
	Indigestion	*Artemisia indica, Syzygium cumini*		2
	Vomiting	*Psidium guajava*		1
	Nausea	*Melia azedarach, Mentha arvensis*		2
	Gastritis	*Azadirachta indica*		1
	Constipation	*Syzygium cumini, Phyllanthus emblica, Acorus calamus, Pueraria tuberosa, Mimosa pudica*		5
	Stomach tumours	*Phyllanthus emblica*		1
	Distended stomach	*Mimosa pudica*		1
Fever	Fever	*Mangifera indica, Ocimum tenuiflorum, Cinnamomum tamala, Centella asiatica, Acorus calamus, Paris polyphylla, Cuscuta reflexa, Pogostemon benghalensis, Zanthoxylum armatum*	3	9
	Typhoid	*Centella asiatica*		1
	Malaria	*Crateva unilocularis, Butea monosperma*		2
Dermatological disorders	Scabies	*Melia azedarach, Azadirachta indica*	8	2
	Cut	*Ficus religiosa*		1
	Skin allergies	*Melanochyla caesia*		1
	Scorpion/snake bites	*Melanochyla caesia*		1
	Burns	*Aloe vera*		1
	Styptic	*Pogostemon benghalensis*		1
	Wounds	*Euphorbia hirta*		1
	Itching	*Colocasia affinis*		1
Cardiovascular/ blood	Blood purifier	*Melia azedarach, Centella asiatica*	3	2
	Blood pressure	*Azadirachta indica, Michelia champaca*		2
	Jaundice	*Aloe vera, Cuscuta reflexa*		2
Ear, nose and throat	Throat	*Psidium guajava, Acorus calamus, Zingiber officinale*	4	3
	Salivation	*Sapindus mukorossi*		1
	Bronchitis	*Syzygium cumini*		1
	Sinus infection	*Pogostemon benghalensis*		1
Urogenital problems	Urinary tract infections	*Bombas ceiba, Centella asiatica*	5	2
	Diuretic	*Bombas ceiba, Butea monosperma, Nyctanthes arbor-trisis*		3
	High uric acid	*Azadirachta indica*		1
	Bladder stones	*Crateva unilocularis*		1
	Kidney stones	*Crateva unilocularis*		1

Table 18.2 Cont.

Ailment category	Biomedical term	Species used	No uses	No taxa
Respiratory diseases	Cough	*Terminalia bellirica, Azadirachta indica, Ficus religiosa, Ocimum tenuiflorum, Crateva unilocularis, Acorus calamus, Zanthoxylum armatum, Nyctanthes arbor-trisis, Zingiber officinale*	4	9
	Common cold	*Terminalia bellirica, Acorus calamus, Curcuma angustifolia, Ageratina adenophorum syn. Eupatorium odoratum, Euphorbia royleana, Pogostemon benghalensis, Zanthoxylum armatum, Cymbopogon flexuosus*		8
	Asthma	*Cannabis sativa, Nyctanthes arbor trisis*		2
	Pneumonia	*Azadirachta indica, Centella asiatica*		2
Skeleto-muscular pain and swelling	Swelling	*Crateva unilocularis*	9	1
	Arthritis	*Azadirachta indica*		1
	Muscular pain	*Carica papaya, Ficus benghalensis*		2
	Neck pain	*Mentha arvensis*		1
	Headache	*Cannabis sativa, Syzygium cumini, Crateva unilocularis, Mentha arvensis, Paris polyphylla*		4
	Sprains	*Ricinus communis, Calotropis gigantea*		2
	Back pain	*Phyllanthus emblic, Mentha arvensis, Themeda Triandra*		3
	Inflammation	*Albizia lebbeck*		1
	Joint pain	*Mentha arvensis, Themeda Triandra, Calotropis gigantea*		3
	Rheumatic pain	*Ricinus communis, Crateva unilocularis, Crateva unilocularis, Cuscuta reflexa*		4
Dental and eye disorders	Toothache	*Zanthoxylum armatum*	2	1
	Blindness	*Terminalia chebula*		1
Other	Epilepsy	*Sapindus mukorossi*	11	1
	Dizziness	*Crateva unilocularis*		1
	Stimulant	*Carica papaya, Pogostemon benghalensi, Cymbopogon flexuosus*		3
	Diabetes	*Ficus benghalensis, Syzygium cumini, Michelia champaca*		3
	Tonic	*Litchi chinensis, Crateva unilocularis*		2
	Immune booster	*Ocimum tenuiflorum, Pueraria tuberosa, Asparagus racemosus*		3
	Hypertension	*Acorus calamus*		1
	Sedative	*Acorus calamus, Pogostemon benghalensis*		2
	Tumours	*Terminalia chebula*		1
	Lactation	*Asparagus racemosus*		1
	Tonic after delivery	*Asparagus racemosus*		1

Note:

Respondents free-listed medicinal uses, which were then categorised into the listed medical terms. Many taxa are used for more than one ailment category (n = 76 species). These categories used were based on previous literature from the Terai (Kunwar et al. 2010), rather than that specified by informants.

example *Neolamarckia cadamba* (Kadamba) (RBG Kew 2015) is associated with a tree deity called '*Kadambariyamman*' and according to folklore, the sacred couple of Shiva and Parvati gave birth to a child under the tree. During a harvest festival on the eleventh moon day in the month of *Bhadra* (August/September), a twig is brought and worshipped in the courtyard of the house.

Three species of indigenous fig possess high religious value (Subedi et al. 1998; Kunwar and Bussmann 2006): *Ficus religiosa* (Pipal), *Ficus benghalensis* (Bar), and *Ficus racemosa* (Dumri) (IUCN 2004) – and are often found in the centre of villages next each other or at shrines. In Buddhism, Buddha attained enlightenment underneath the Pipal tree (Chisholm 1991). *Ocimum tenuiflorum* (or *Ocimum sanctum L.*, *Tulsi* or Holy basil) (Singh et al. 2012) is planted outside the homes of Hindus, often in masonry structures to indicate religious inclination of a family. The offering of its leaves is mandatory in the daily ritualistic worship of Lord Vishnu. The plant also has diverse healing properties and is used as an essential oil (Pattanayak, Das and Panda 2010).

Veterinary use

For veterinary use, plant material is used for livestock medicine, fodder, shelter, and fences to deter wild animals. Livestock medicinal plants include *Cannibis sativa* (Cannabis), used for diarrhoea in ox, cattle and buffalo, *Dendrocalamus strictus* (Bamboo) (Storrs and Storrs 1998), used to treat abscesses in goats, and *Litchi chinensis* (Litchi) (Kerkhoff 2003), used to treat animal bites. Common plants used for fences and pens include *Jatropha curcas* and *Dendrocalamus strictus* (Storrs and Storrs 1998). Plants are used to deter wild animals from raiding crops, such as *Euphorbia royleana* (Cactus spurge) (IUCN 2004) planted as a thorny fence to repel largest Asian antelope, the blue bull (*Boselaphus tragocamelus*), or *Ageratina adenophorum* (Sticky snakeroot) (USDA 2015) is planted around houses to deter snakes. Sixteen fodder species are collected on a daily basis.

Timber and building material use

Shorea robusta (Sal) is the most common species used for timber (used in 65 per cent of cases), followed by *Dalbergia sisoo* (Indian rosewood) (62.5 per cent), *Melia azedarach* (Persian lilac) (50 per cent) and *Mangifera indica* (Mango) (32.5 per cent). Timber products include both low-grade (e.g. *Neolamarckia cadamba* (Kadamba)) and hard woods (e.g. *Shorea robusta* (Sal)), used for constructing housing, livestock sheds, furniture, ladders, stairs, doors, and farm equipment. Eight types of grasses, herbs and shrubs are similarly used for constructing houses. Around homes, plants are grown for windbreaks (e.g. *Dendrocalamus strictus* (Bamboo)), fences (e.g. *Euphorbia royleana* (Cactus spurge)) or shade (e.g. *Mangifera indica* (Mango)). Plants are used for ornamental purposes along roadsides (e.g. *Duranta erecta* (Golden dewdrop)), and homes (e.g. *Dendrocalamus strictus* (Bamboo)) (Storrs and Storrs 1998; IUCN 2004).

Household use

Soap is made from the latex of *Carica papaya* (Paw paw), or nuts and seeds of *Sapindus mukorossi* (Soap nut), also used to weave baskets and mats. *Aloe vera* (Aloe vera) is used as a face cream and *Euphorbia hirta* (Asthma plant) is used to wash hair. *Bombax ceiba* (Silk cotton tree) is used to make cotton for mattresses and *Themeda Triandra* is used as a clothing dye. Brooms are made of *Thysanolaena maxima* (Broom grass). Pots used for cultivating yoghurt, called '*taki*' (N), are made of *Artocarpus heterophyllus* (Jackfruit). The seed of *Jatropha curcas* (Physic nut) is burnt for fuel for transport and lighting, and the stem is used as a toothbrush. Plants are also used to make musical instruments, fishing baskets, toys, jewellery and containers to carry wild harvested goods (Storrs and Storrs 1998; Shankar, Lama and Bawa 2001; Kerkhoff 2003; IUCN 2004; Singh et al. 2012; Government of Nepal 2014.

Discussion

Factors that incentivise the maintenance of biodiversity in and around farms

Results offer evidence to support emerging claims that as agriculture intensifies and expands, farmers may increasingly play an important role in conservation beyond protected areas (Ehrlich and Ehrlich 2014). Communities have a comprehensive understanding of the structure and function of the interconnected human-environmental systems in which they live, allowing them to secure necessities from ecosystem services (Ingold 2000; Government of Nepal 2015). Using local plant material is cost-effective, time-tested, situation-specific, practical, and flexible. Much of this agricultural knowledge and practice can be integrated into scientific knowledge, while modifying or transforming existing norms and behaviours to deal with emerging stressors (Gurung 1994). For example traditional multi-cropping systems are maintained for various subsistence priorities, such as providing timber, fuel, medicine, organic plants and pesticides. Yet, land managers' incentives for their conservation are not purely economic. Farmers maintain plants to conserve soil, prevent erosion, fix nitrogen, and decompose organic matter. Trees, shrubs and herbs support important ecosystem functioning, such as photosynthesis, evapotranspiration, watershed regulation, carbon sequestration, and protect against crop raiding by animals (Eilu et al. 2003; Del Angel-Perez and Mendoza 2004). Plants are critical sources of fodder for livestock, which in turn, provide ploughing power for cultivation, milk, meat, and contribute to soil productivity by generating manure (Acharya 2006). Plants are cultivated in rotation plots, fallows, forests, home gardens and are multistory cropped to efficiently use space (Ingold 2000; Salas 2005). Plants furthermore have aesthetic and ornamental value and are interlinked with tradition, religious and cultural heritage (UNEP 2006; Boelee, Chiramba and Khaka 2011). Farmers are aware of both services provided by biodiversity on farms, as well as 'disservices'. For example *Lantana camara* (Lantana) (Ethnobotanical Society of Nepal 2015), the flower which is used as an ornamental and offering, is an introduced plant that forms dense thickets, reduces farmland productivity, prevents growth of new trees, and is toxic to livestock (Bhagwat et al. 2012). *Oxalis corniculata* (Creeping soral) (IUCN 2004) is used for food, but is considered an invasive species that occupies fallow land. *Calotropis gigantea* (Crown flower) (Singh et al. 2012) has milky latex that is massaged into muscles and joints to relieve sprains, but can cause blindness. *Kalanchoe piñata* (Life plant) (Mandal, Panda and Rana 2013), is used for compost and timber, but is also poisonous to livestock. As such, farmers' practices are geared towards augmenting such services, and reducing disservices from and related to wild and tended plant material in and around farms. This results in a restructuring of the agro-ecosystems in terms of diversity and abundance of wild plant species (Ango et al. 2014). This is what can be termed 'servicing ecosystems' (Comberti et al. 2015).

Wild edible plant use

Wild edible plants have noteworthy roles and contributions in Nepalese diets and food security, particularly for many Indigenous, rural, ethnic, and marginalized people in Nepal (Kunwar et al. 2003; Dangol et al. 2017). Wild edible plants are uncultivated plants found in the wild, with nutritious value for fulfilling dietary requirements. A number of agricultural crops, and their wild relatives and edible plants enrich the species and diversity of Nepa (Dangol et al. 2017). Twenty-five unique species are used as wild edible plants for food across the study area.

Vegetables are obtained from some plants, such as *Crateva unilocularis* (Garlic pear) which is high in iron, *Justicia adhatoda* (Malabar nut), and *Artocarpus heterophyllus* (Jackfruit). Others are used for pickle, such as *Mentha arvensis* (Peppermint), or *Terminalia chebula* (Yellow myrobalan). Plant's edible seeds are either consumed after boiling or roasting, such as *Cannabis sativa* (Bhang) and *Artocarpus heterophyllus* (Jackfruit). Some plants are rich in nectar in the flower, such as *Hibiscus rosasinensis* (Chinese hibiscus). Some leaves are edible, such as *Oxalis corniculate* (Creeping sorrel). Others are used for fermenting substrates or alcohol, such as *Syzygium cumini* (Indian blackberry), or to make preserves when boiled with sugar, such as *Terminalia chebula* (Yellow myrobalan). Plants are also used for spices and herbs."

Medicinal plant use

Perhaps unsurprisingly, our analysis reveals that a large proportion of plants with a documented use are used as medicines: from dermatological to gastrological ailments. Despite the reported increased number of healthcare clinics in the last 15 years, most respondents have a high reliance on, and prefer to use, plant-based remedies, rather than going to hospitals. Eighty per cent of respondents consult herbalists or folk healers, often in combination with health workers (97 per cent), visiting health clinics (80 per cent), or hospitals (40 per cent). This finding is supported by previous studies in other regions in Nepal, which suggest two-thirds of the rural population rely on traditional herbal medicine (Kunwar et al. 2010; Singh et al. 2012). Reasons for this include cultural acceptance, a long history of traditional medicinal use; the affordability of traditional remedies compared to modern healthcare; and limited alternatives (i.e. in Nepal, at the time of this study the patient ratio is 1:20,000 for medical professionals (e.g. nurses, doctors) versus 1:100 for traditional healers (Gillam 1989; WRI 2005)).

Ethnoecological knowledge associated with the maintenance of biodiversity on farms

In this study, the average number of useful plants of 8.64±1.23 reported by informants of 18–39 years of age was low as compared to 10.35±0.93 reported by informants of ≥40 years. This suggests potential loss of traditional ethnobotanical knowledge in the younger generation as studies show that most botanical knowledge is acquired by young adulthood (Hunn 2002). Farmers explain this trend as due to a declining transfer of local knowledge, practice and beliefs, which have accumulated over generations of living in particular environments, handed down through cultural transmission (Berkes 1999), including apprenticeships (Kunwar et al. 2010). Young farmers thus have fewer elderly mentors, and agricultural livelihoods are increasingly discouraged in favour of new forms of service employment. Farmers associate this trend with observed declines or disappearance in the last ten years of formerly tended plants, such as *Pueraria tuberosa* (Indian kudzu) (IUCN 2004), a climber, the root and fruit of which is used to relieve constipation and boost immunity; or *Kalanchoe piñata* (Life plant) (Mandal, Panda and Rana 2013), used for compost. This finding is consistent with previous findings showing youngsters of both Tharu and migrant societies are less aware of plant use in the Terai, than previous generations (Singh et al. 2012). On the other hand, the higher number of plants reported by informants of ≥40 years of age corroborates the long-standing local belief that elders possess more botanical knowledge than other segments of society.

Ethnobotanical knowledge and practice within any culture vary according to geographic origin, residence, religion, age, gender, and ethnicity (Pfeiffer and Butz 2005). Nevertheless, some explanations for differences in knowledge of plant use could be that: (1) Brahmin and Chettri castes tend towards more secure land tenure and higher education, leading to a more diverse knowledge of plant use; (2) the Tharu and Dura castes have a long-standing heritage of subsistence agriculture in the Terai, leading to a greater understanding of plant use; and (3) the unique knowledge of the Sanyasi stems from Hindu *Ayurvedic* tradition (Gunaratne 2002). However, the fact that social groups are geographically close, exposed to similar environments and able to exchange knowledge readily could act as compounding factors (Saslis-Lagoudakis et al. 2014).

Building the evidence base

The continued contribution of biodiversity to the maintenance of agro-ecosystem services will depend on how farmers gear management towards augmenting particular services, the extent to which local knowledge is preserved, and this will vary across types of crops, altitudes, seasons and cultural needs (Poudel and Kotani 2012). In each context, evaluations of the optimal species mix or ecosystem type, under multiple possible futures in particular locales are needed along with approximations of substitutability. Future longitudinal research, across seasons or years, could consider how the knowledge and use of high-value species are changing, such as species with high nutritional benefits and opportunities for commercial viability (Storrs and Storrs 1998; IUCN 2004), such as *Azadirachta indica* (Margosa tree) or *Artocarpus heterophyllus* (Jackfruit) .Agricultural and agroforestry landscapes represent a largely unexplored source of pharmacological and phytochemical studies of new bioactive compounds for treating illnesses (Nair 2013; Wynberg and van Niekerk 2014). If it is sensitive to local knowledge ownership regimes, and threats of 'biopiracy' (Shiva 1999), documentation may safeguard the sovereignty of traditional knowledge and protect knowledge from being misappropriated in the form of patents by pharmaceutical companies on non-original innovations (Twilley 2015). Further in-depth anthropological analysis could help to validate and explain differences in knowledge of plant use, to further assess the knowledge of particular groups (e.g. Sanyasi), or to consider the influence of other factors (e.g. income, landholding size) on biodiversity, ecosystem services, or services to ecosystems at the landscape and regional scales.

Implications for broader areas of adaptive planning, policy and management

Our evidence indicates a high prevalence of autochthonous, low-cost, small-scale ethno-botanical practices that enhance social-ecological resilience to global environmental change. Mobilizing and harnessing this knowledge could build common ground for new partnerships between conservation planners and land managers. Among other co-benefits, managing wild and tended plants can do the following:

- prevent erosion with erratic and intense rainfall, e.g. *Leucaena leucocephala* (Leucaena);
- retain soil moisture during droughts by growing groundcover, e.g. *Ocimum tenuiflorum* (Sacred basil) and enhance soil fertility, e.g. *Albizia lebbeck* (Black siris);
- serve as windbreaks, e.g. *Dendrocalamus strictus* (or various species of bamboo) or *Justicia adhatoda* (Malabar nut);
- boost immunity among food-insecure communities where malnutrition is high, e.g. *Pueraria tuberosa* (Indian Kudzu);
- treat human fever and malaria, e.g., *Centella asiatica* (Indian pennywort), *Crateva unilocularis* (Garlic pear), *Butea monosperma* (Flame of the forest));
- treat livestock fever and colds, e.g. *Paris polyphylla* (Herb paris), *Terminalia bellirica* (Yellow myrobalan);

- be used as pesticides, e.g. *Azadirachta indica* (Margosa tree) and *Melia azedarach* (Persian lilac);
- support alternative livelihoods, e.g. *Bombax ceiba* (Silk cotton tree) grown for cotton; or
- have multiple uses, e.g. *Eupatorium odoratum* (Crofton weed)) used for fodder, pest control, medical and religious purposes (Storrs and Storrs 1998; IUCN 2004; Bhattarai, Chaudhary and Taylor 2006; Madhu et al. 2010; Singh et al. 2012; Government of Nepal 2014).

Translating best practices from other developing countries could provide useful insights for Nepalese farmers to harness ecosystem benefits from non-agricultural plants. There is significant potential to reinvest revenue generated from genetic resources in conservation, such as cosmetics, botanical medicines, and pharmaceuticals. Safeguards against Big Pharma expropriation could be strengthened by updating national records of the status and distribution range of non-charismatic protected and threatened species according to IUCN lists (e.g. volumes of Flora of Nepal), or establishing a Traditional Knowledge Digital Library of Nepal, following the example of India (Government of India 2016). There is a need to up-scale local seed warehouses of useful plants that can withstand dry years, saturation, or invasive populations, as well as increase public spending in purchasing native seeds (CBD 2016), and explicitly incorporate the utilization of material on farms in climate adaptation plans (Regmi and Paudyal 2009). Tightening legislative controls on access to genetic resources could be supported by becoming a signatory of the Nagoya Protocol on Access to Genetic Resources and Benefit Sharing – a legally binding instrument which aims to monitor, support long-term use, and ensure equitable access for Indigenous and future generations (Prip and Rosendal 2015; RBG Kew 2016). Such local and national initiatives further tie into the ongoing thematic assessment of the sustainable use of wild species of the International Science-Policy Panel on Biodiversity and Ecosystem Services.

Conclusion

This research provides an original contribution to a growing discourse articulating the role that plant biodiversity plays in maintaining ecosystem services and humans' contributions to biodiversity maintenance. Local knowledge, beliefs, and practices, which have accumulated over generations, support high diversity of non-agricultural plant communities in the Terai Plains of Nepal: a total of 391 vascular plant specimens belong to 76 distinct plant species from 49 phylogenetic families. This high level of plant agrobiodiversity provides a rich source of ecosystem services that contributes to the social, cultural, environmental, and economic enrichment of Nepalese rice farming communities. Farmers' ethnobotanical knowledge-belief-practice complexes (Berkes 1999), which differ by the caste and age group, assist communities to face emerging risks by enhancing the adaptive capacity of both ecosystems and the livelihood resources they provide (Government of Nepal 2015). However, there appears to be a declining transmission of ethnobotanical knowledge to the younger generation, and farmers associate this trend with declines or loss of formerly tended plants. Results can serve as baseline data to initiate further research, be used to plan for a range of potential development trajectories, and be used to conserve valuable, but disappearing, traditional knowledge and practices.

References

Acharya, K. P. 2006. Linking trees on farms with biodiversity conservation in subsistence farming systems in Nepal. *Biodiversity and Conservation* 15: 631–646.

Ango, T. G., L. Borjeson, F. Senbeta and K. Hylander 2014. Balancing ecosystem services and disservices: smallholder farmers' use and management of forest and trees in an agricultural landscape in southwestern Ethiopia. *Ecology and Society* 19(30).

Aziz, M. A. 2007. Role of microorganisms in litter decomposition of Salix spp. MSc, Sher-e-Kashmir University of Agricultural Sciences and Technology of Kashmir.

Berkes, F. 1999. *Sacred Ecology*. London: Routledge.

Bernard, H. R. and Gravlee, C. C. 2014. *Handbook of Methods in Cultural Anthropology*. Lanham, MD: Rowman and Littlefield.

Bhagwat, S. A., B. E. Thekaekara, T. F. Thornton and K. J. Willis. 2012. A battle lost? Report on two centuries of invasion and management of Lantana camara L. in Australia, India and South Africa. *PLoS One* 7: 32407.

Bhattarai, S., R. P. Chaudhary and R. S. Taylor. 2006. Ethnomedicinal plants used by the people of Manang district, Central Nepal. *Journal of Ethnobiology and Ethnomedicine* 2: 41.

Boelee, E., T. Chiramba and E. Khaka. 2011. An ecosystem services approach to water and food security. In International Water Management Institute and United Nations Environment Programme (ed.) *An Ecosystem Services Approach to Water and Food Security*. Colombo: International Water Management Institute.

Boissiere, M., B. Locatelli, D. Sheil, M. Padmanaba and E. Sadjudin. 2013. Local perceptions of climate variability and change in tropical forests of Papua, Indonesia. *Ecology and Society* 18.

Bridson, D. M. and L. Forman. 1998. *The Herbarium Handbook*. London: Royal Botanic Gardens Kew.

Brush, S. B. 1996. Is common heritage outmoded? In S. B. Brush and D. Stabinsky, (eds), *Valuing Local Knowledge: Indigenous People and Intellectual Property Rights*. Washington, DC: Island Press.

CBD (Convention of Biological Diversity) 2016. Nepal overview. Secretariat of the Convention of Biological Diversity. Available at: www.cbd.int/countries/profile/default.shtml?country=np%20-%20measure (accessed 12 September 2020).

Chapin, F. S., E. S. Zaveleta, V. T. Eviner, R. L. Naylor, P.M. Vitousek, S. et al. 2000. Consequences of changing biotic diversity. *Nature* 405: 234–242.

Chisholm, H. 1991. *Encyclopædia Britannica* 4. Cambridge: Cambridge University Press.

Clarke, K. R. and R. N. Gorley 2006. *PRIMER Marine Biology v6: User Manual/Tutorial*. Plymouth.

Clarke, L. W., L. Li, G. D. Jenerette and Z. Yu. 2014. Drivers of plant biodiversity and ecosystem service production in home gardens across the Beijing Municipality of China. *Urban Ecosystems* 17: 741–760.

Comberti, C., T. F. Thornton, V. Wyllie de Echeverria and T. Patterson. 2015. Ecosystem services or services to ecosystems? Valuing cultivation and reciprocal relationships between humans and ecosystems. *Global Environmental Change* 34: 247–262.

Dangol, D. R. 2005. *Dictionary of Forest and Common Land Plants of Western Chitwan*. Chitwan: Institute for Social and Environmental Research-Nepal.

Dangol, D. R., K. L. Maharjan, S. K. Maharjan, and A. K. Acharya. 2017. Wild edible plants of Nepal. In: B. K. Joshi, H. B. KC, and A. K. Acharya (eds), *Conservation and Utilization of Agricultural Plant Genetic Resources in Nepal*. Kathmandu: Proceedings of 2nd National Workshop, 22–23 May 2017 Dhulikhel; NAGRC, FDD, DoA and MoAD.

Dansi, V. R., P. Azokpota, H. Yedomonhan, P. Assogpa, A. Adjatin, et al. 2012. Diversity of the neglected and underutilized crop species of importance in Benin. *The Scientific World Journal*. https://doi.org/10.1100/2012/932947

Del Angel-Perez, A. L. and M. A. Mendoza. 2004. Totonac homegardens and natural resources in Veracruz, Mexico. *Agriculture and Human Values* 21: 329–346.

EFlora of India 2015. Flowers of India. Available at: www.flowersofindia.net/ (accessed 12 September 2020).

Ehrlich, P. R. and A. H. Ehrlich. 2014. Can a collapse of global civilization be avoided? *Proceedings of the Royal Society B*, 280.

Eilu, G., J. Obua, J. K. Tumuhairwe and C. Nkwineb. 2003. Traditional farming and plant species diversity in agricultural landscapes of south-western Uganda. *Agriculture, Ecosystems and Environment* 99: 125–134.

Elmqvist, T., C. Folke, M. Nystrom, G. Peterson, J. Bengtsson, B. Walker, and J. Norberg. 2003. Response diversity, ecosystem change, and resilience. *Frontiers in Ecology and the Environment* 1: 488–494.

Ethnobotanical Society of Nepal 2015. Database. Kathmandu, Nepal: Central Department of Botany, Tribhuvan University. Available at: www.eson.org.np/ (accessed 12 September 2020).

Fahrig, L. 1997. Relative effects of habitat loss and fragmentation on population extinction. *The Journal of Wildlife Management* 61(3): 603–610.

Ford, A. and R. Nigh. 2015. *The Maya Forest Garden: Eight Millennia of Sustainable Cultivation of the Tropical Woodlands.* Altamira, CA: Left Coast Press.

Garibaldi, A. and N. Turner. 2004. Cultural keystone species: Implications for ecological conservation and restoration. *Ecology and Society* 9(1).

Ghosh, D. 2012. Anti-diabetic and anti-oxidative potencies study of ethyl acetate fraction of hydromethanolic (40:60) extract of seed of *Eugenia jambolana Linn* and its chromatographic purification. *Journal of Pharmacy Research* 5: 696–703.

Gillam, S. 1989. The traditional healer as village health worker. *Journal of Institute of Medicine* 11: 67–76.

Government of India 2016. Traditional knowledge digital library. Available at: www.tkdl.res.in/tkdl/langdefault/common/Home.asp?GL=Eng (accessed 12 September 2020).

Government of Nepal 2002. Nepal census of agriculture 2001/2. National Planning Commission Secretariat. Available at: www.cbs.gov.np (accessed 12 September 2020).

Government of Nepal 2014. Plant resources. Kathmandu: Bulletin of Department of Plant Resources Government of Nepal.

Government of Nepal 2015. Indigenous and local knowledge and practices for climate resilience in Nepal: Mainstreaming climate change risk management in development. Kathmandu, Nepal: Ministry of Science, Technology and Environment Government of Nepal.

Grieg-Gran, M. and B. Gemmill-Herren. 2011. *Handbook for Participatory Socio-Economic Evaluation of Pollinator-Friendly Practices.* Rome: Food and Agricultural Organization.

Gunaratne, A. 2002. *Many Tongues, One People: The Making of Tharu Identity in Nepal.* Ithaca, NY: Cornell University Press.

Gunaratne, A. M. T. A., C. V. S. Gunatilleke, I. A. U. N. Gunatilleke, H. M. S. P. Madawala and Burslem, D. F. R. P. 2014. Overcoming ecological barriers to tropical lower montane forest succession on anthropogenic grasslands: Synthesis and future prospects. *Forest Ecology and Management* 329: 340–350.

Guo, Z., L. Zhang and Y. Li. 2010. Increased dependence of humans on ecosystem services and biodiversity. *PLoS ONE* 5: 1–8.

Gurung, B. 1994. *A cultural approach to natural resource management: A case study from Eastern Nepal. Summary report of FAO regional expert consultation on non-wood forest products: social, economic and cultural dimensions.* Rome: Food and Agricultural Organization.

Haenke, S., A. Kovacs-Hostyanszki, J. Freund, P. Batary, B. Jauker, T. Tscharntke and A. Holzschuh. 2014. Landscape configuration of crops and hedgerows drives local syrphid fly abundance. *Journal of Applied Ecology* 51(2): 505–513.

Hossain, M. S., J. A. Dearing, M. M. Rahman and M. Salehin. 2016. Recent changes in ecosystem services and human well-being in the Bangladesh coastal zone. *Regional Environmental Change* 16: 429–443.

Howard, P. 2009. Human adaptation to biodiversity change: Facing the challenges of global governance without science? Paper presented at Human Dimensions of Global Environmental Change, Earth Systems Governance – People, Places, and the Planet. Amsterdam.

Hunn, E. S. 2002. Evidence for the precocious acquisition of plant knowledge by Zapotec children. *Ethnobiology and Biocultural Diversity* 604: 13–31.

Ingold, T. 2000. *Perception of the Environment: Essays on Livelihood, Dwelling and Skill.* London: Routledge.

IIED (International Institute for Environment and Development) 2015. Smallholder innovation for resilience: Promoting resilient farming systems and local economies. Available at: www.biocultural.iied.org/smallholder-innovation-resilience-sifor (accessed 12 September 2020).

IPBES. 2019. *Summary for Policymakers of the Global Assessment Report on Biodiversity and Ecosystem Services of the Intergovernmental Science–Policy Platform on Biodiversity and Ecosystem Services.* In: S. Díaz, E. S. Brondizio, H. T. Ngo, M. Guèze, J. Agard, A. Arneth, P. Balvanera, K. A. Brauman, S. Butchart, K. Chan, L. A. Garibaldi, K. Ichii, J. Liu, S. M. Subramanian, G. F. Midgley, P. Miloslavich, Z. Molnár, D. Obura, A. Pfaff, S. Polasky, A. Purvis, J. Razzaque, B. Reyers, R. Chowdhury, Y. J. Shin, I. J. Visseren-Hamakers, K. J. Willis, and C. N. Zayas (eds). Germany: International Science-Policy Panel on Biodiversity and Ecosystem Services (IPBES) Secretariat.

IUCN (International Union for the Conservation of Nature Nepal) 2004. *National Register of Medicinal and Aromatic Plants.* Kathmandu: The World Conservation Union.

Jones, H. P., D. G. Hole and E. S. Zavalet. 2012. Harnessing nature to help people adapt to climate change. *Nature Climate Change* 2: 504–509.

Kerkhoff, E. E. 2003. Sustainable sloping lands and watershed management. Paper presented at International Centre for Integrated Mountain Development (ICIMOD) conference.

Kremen, C. 2015. Reframing the land-sparing/land-sharing debate for biodiversity conservation. *Annals of the New York Academy of Sciences* 1355: 52–76.

Kumar, B.M. and P. K. R. Nair. 2004. The enigma of tropical homegardens. *Agroforestry Systems* 61: 135–152.

Kunwar, R. M. and R. W. Bussmann. 2006. *Ficus* (Fig) species in Nepal: A review of diversity and indigenous uses. *Lyonia* 11: 85–97.

Kunwar, R. M. and N. P. S. Duwadee. 2003. Ethnobotanical notes on flora of Khaptad National Park, far western Nepal. *Himalayan Journal of Science* 1(1): 25–30.

Kunwar, R. M., K. P. Shrestha and R. W. Bussmann. 2010. Traditional herbal medicine in Far-west Nepal: A pharmacological appraisal. *Journal of Ethnobiology and Ethnomedicine* 6: 1–18.

Lubbe, C. S., S. J. Siebert and S. S. Cilliers. 2011. Floristic analysis of domestic gardens in the Tlokwe city municipality, South Africa. *Bothalia* 41: 351–361.

Luitel, D. R., M. B. Rokaya, B. Timsina and Z. Munzbergova. 2014. Medicinal plants used by the Tamang community in the Makawanpur district of central Nepal. *Journal of Ethnobiology and Ethnomedicine* 10: 1–11.

Mace, G., H. Masundire and J. Baillie. 2005. Biodiversity. I, H.R.S. Hassan and N. J. Ash (eds), *Ecosystems and Human Wellbeing: Current State and Trends.* Washington, DC: Island Press.

Madhu, K. C., S. Phoboo and P. K. Jha. 2010. Ecological study of *Paris Ployphylla* Sm. *Ecoprint* 17.

Malla, G. 2008. Climate change and its impact on Nepalese agriculture: Review paper. *The Journal of Agriculture and Environment* 9: 62–71.

Manandhar, N. P. 1998. Native phytotherapy among the Raute tribe of Dadeldhura district, Far-west Nepal. *Journal of Ethnopharmacology* 60: 199–206.

Manandhar, S., D. S. Vogt, S. R. Perret and F. Kazama. 2010. Adapting cropping systems to climate change in Nepal: A cross-regional study of farmers' perception and practices. *Regional Environmental Change* 11: 335–348.

Mandal, D., A. K. Panda and M. Rana. 2013. Medicinal plants used in folk medicinal practice available in rich biodiversity of Sikkim. *Environment and Ecology* 31: 1445–1449.

MEA (Millennium Ecosystem Assessment) 2005. *Ecosystems and Human Well-Being: Synthesis.* Washington, DC: MEA.

Mohri, H., S. Lahoti, O. Saito, A. Mahalingam, N. Gunatilleke, et al. 2013. Assessment of ecosystem services in homegarden systems in Indonesia, Sri Lanka, and Vietnam. *Ecosystem Services* 5: 124–136.

Nair, K. P. P. 2013. *The Agronomy and Economy of Turmeric and Ginger: The Invaluable Medicinal Spice Crops.* London: Elsevier.

Olango, T. M., B. Tesfaye, M. Catellani and M. E. Pe. 2014. Indigenous knowledge, use and on-farm management of enset (*Ensete ventricosum* (Welw.) Cheesman) diversity in Wolaita, Southern Ethiopia. *Journal of Ethnobiology and Ethnomedicine* 10: 1–18.

Pattanayak, P, B. P., D. Das and S. K. Panda. 2010. *Ocimum sanctum Linn.* A reservoir plant for therapeutic applications: An overview. *Pharmacognosy Reviews* 4: 95–105.

Pfeiffer, J. M. and R. J. Butz. 2005. Assessing cultural and ecological variation in ethnobiological research: The importance of gender. *Journal of Ethnobiology* 25: 240–278.

Pigg, S. L. 1992. Inventing social categories through place: Social representations and development in Nepal. *Comparative Studies in Society and History* 34: 491–513.

Plumecocq, G., T. Debril, M. Duru, M.-B. Magrini, J. Sarthou and O. Therond. 2018. The plurality of values in sustainable agriculture models: Diverse lock-in and coevolution patterns. *Ecology and Society* 23(1): 21.

Pomeroy-Stevens, A., M. B. Shrestha, M. Biradavolu, K. Hachhethu, R. Houston, et al. 2016. Prioritizing and funding Nepal's Multisector Nutrition Plan. *Food and Nutrition Bulletin* 37(4S): S151–S169.

Poudel, S. and K. Kotani. 2012. Climatic impacts on crop yield and its variability in Nepal: Do they vary across seasons and altitudes? *Climatic Change* 116: 327–355.

Press, J. R., K. K. Shrestha and D. A. Sutton. 2000. *Annotated checklist of the flowering plants of Nepal.* Kathmandu: Natural History Museum of London and Central Department of Botany, Tribhuvan University.

Prip, C. and K. Rosendal. 2015. Access to genetic resources and benefit-sharing from their use (ABS): State of implementation and research gaps. FNI Report 5/2015. Amsterdam: PBL Netherlands Environmental Assessment Agency.

RStudio Team (2020). RStudio: Integrated Development for R. Boston, MA: RStudio, PBC.

RBG (Royal Botanical Gardens Kew) 2015. World checklist of selected plant families. Facilitated by the Royal Botanic Gardens, Kew. London: Kew Royal Botanic Gardens. Available at: http://apps.kew.org/wcsp/ (accessed 12 September 2020).

RBG (Royal Botanical Gardens Kew) 2016. *The State of the World's Plants Report – 2016.* London: Kew Royal Botanic Gardens.

Regmi, B. and A. Paudyal. 2009. *Climate change and agrobiodiversity in Nepal: Opportunities to include agrobiodiversity maintenance to support Nepal's National Adaptation Programme of Action (NAPA).* Edited by P. Bordoni. Kathmandu: LI-BIRD.

Regmi, R. R. 1994. Deforestation and rural society in the Nepalese Terai. *Occasional Papers in Sociology and Anthropology* 4: 72–89.

Rockstrom, J., W. Steffen, K. Noone, Å. Persson, F. S. Chapin III et al. 2009. Planetary boundaries: Exploring the safe operating space for humanity. *Ecology and Society* 14: 32.

Ross, I. A. 2003. *Medicinal Plants of the World.* Champaign, IL: Humana Press.

Salas, M. 2005. Seed songs: Reflections on swidden agriculture agrobiodiversity and food sovereignty. *Indigenous Affairs* 2: 15–22.

Sarkar, D. 2008. *Lattice: Multivariate Data Visualisation.* New York: Springer.

Saslis-Lagoudakis, C. H., J. A. Hawkins, S. J. Greenhill, C. A. Pendry, M. F. Watson, et al. 2014. The evolution of traditional knowledge: Environment shapes medicinal plant use in Nepal. *Proceedings of the Royal Society of London B: Biological Sciences* 281 (1780).

Seto, K. C., B. Guneralp and L. R. Hutyra. 2012. Global forecasts of urban expansion to 2030 and direct impacts on biodiversity and carbon pools. *Proceedings of the National Academy of Sciences* 109: 16083–16088.

Shankar, U., S. D. Lama and K. S. Bawa. 2001. Ecology and economics of domestication of non-timber forest products: An illustration of broomgrass in Darjeeling Himalaya. *Journal of Tropical Forest Science* 13: 171–191.

Shannon, C. and W. Weaver. 1949. *The Mathematical Theory of Communication.* Urbana, IL: University of Illinois Press.

Shiva, V. 1999. *Biopiracy: The Plunder of Nature and Knowledge.* Boston: South End Press.

Singh, A. G., A. Kumar and D. D. Twearu. 2012. An ethnobotanical survey of medicinal plants used in Terai forest of western Nepal. *Journal of Ethnobiology and Ethnomedicine* 8: 19.

Singh, V. K., B. S. Dwivedi, K. N. Tiwari, K. Majumdar, M. Rani, et al. 2014. Optimizing nutrient management strategies for rice–wheat system in the Indo-Gangetic Plains of India and adjacent region for higher productivity, nutrient use efficiency and profits. *Field Crops Research* 164: 30–44.

Steffen, W., P. J. Crutzen and J. R. McNeill. 2007. The Anthropocene: Are humans now overwhelming the great forces of Nature? *Ambio* 36: 614–621.

Stern, N. 2007. *The Economics of Climate Change: The Stern Review.* Cambridge: Cambridge University Press.

Storrs, A. and J. Storrs. 1998. *Trees and Shrubs of Nepal and the Himalayas.* New Dehli: Books Faith India.

Subedi, B. P., L. D. Chintamani and D. A. Messeschmidt. 1998. *Tree and land reuse in the Eastern Terai, Nepal: A case study from the Siraha and Saptari districts, Nepal.* Community Forestry Case Study Series 9 ed. Rome.

Thorn, J. P. R. 2019. Adaptation "from below" to species distribution, habitat and climate-driven changes in agro-ecosystems in the Terai Plains of Nepal. *AMBIO*: 1–16. https://doi.org/10.1007/s13280-019-01202-0.

Thorn, J. P. R., T. F. Thornton and A. Helfgott. 2015. Autonomous adaptation to global environmental change in peri-urban settlements: Evidence of a growing culture of innovation and revitalisation in Mathare Valley Slums, Nairobi. *Global Environmental Change* 31: 121–131.

Thorn, J. P. R., T. F. Thornton, A. Helfgott, and K. J. Willis. 2020. Indigenous uses of wild and tended plant biodiversity maintain ecosystem services in agricultural landscapes of the Terai Plains of Nepal. *Journal of Ethnobiology and Ethnomedicine* 16(33): 1–25. https://doi.org/10.1186/s13002-020-00382-4.

Twilley, N. 2015. Who owns the patent on nutmeg? *The New Yorker,* 26 October.

UNEP (United Nations Environment Programme). 2006. Guidelines for the rapid assessment of inland, coastal and marine wetland biodiversity. Ninth Meeting of the Conference of the Parties to the Convention on Wetlands (Ramsar, Iran, 1971): Wetlands and water: supporting life, sustaining livelihoods. Kampala: UNEP.

USDA (United States Department of Agriculture) 2015. Plants database. Natural Resources Conservation Service (NRCS). Available at: https://plants.usda.gov/java/ (accessed 12 September 2020).

World Bank. 2015. *Climate Risk and Adaptation Country Profiles.* Washington, DC: World Bank.

WRI (World Resources Institute). 2005. *World Resources 2005. The Wealth of the Poor: Managing Ecosystems to Fight Poverty.* New York: WRI.

Wynberg, R. and S. Laird. 2015. *Bioscience at a crossroads: Implementing the NP in a time of scientific, technological and industry changes.* New York: Secretariat of the Convention of Biological Diversity.

Wynberg, R. and J. Van Niekerk. 2014. Global ambitions and local realities: Achieving equity and sustainability in two high-value natural product trade chains. *Forests, Trees and Livelihoods* 23: 19–35.

Zheng, S. and M. Dicke. 2008. Ecological genomics of plant-insect interactions: From gene to community. *American Society of Plant Biologists* 146(3): 812–817.

19

CREATING COEXISTENCE

Traditional knowledge and institutions as a foundation for Maasai-wildlife coexistence in southern Kenya

Guy Western and Samantha Russell

Introduction

For the last several thousand years, Kenya's southern rangelands have been occupied by milk pastoralists, the most recent of whom are the Maasai, who are believed to have arrived in the seventeenth century (Hanotte et al., 2002). Over the last century, Kenya Maasai are now largely confined to an area of 40,000 km² commonly referred to as southern Maasailand. This area spans from Maasai Mara National Reserve to the West and the Amboseli region to the East, along the international border between Kenya and Tanzania. Across the border is also Maasai territory, with the border simply a political barrier rather than practical one. Across this Kenyan "Maasailand", the terrain spans the Great Rift Valley and thus encompasses a diversity of habitats, ranging from the highland forests of the Loita Hills, to the dry shores of the alkaline Lake Magadi and Lake Natron and is home to one of the richest vertebrate assemblages in Africa (MEWNR and RDA, 2015).

Our case study is based in Kenya's southern Maasailand, on the group ranches of Olkiramatian and Shompole. A group ranch is a jointly owned freehold land title given to the customary occupants of communal lands (Kimani and Pickard, 1998). The occupants of Olkiramatian and Shompole number roughly 20,000 people (Bedelian, 2019) in an area of approximately 1,000 km² with the two ranches used as a single management area. The area is at an altitude of 600–700 m and has high temperatures ranging from 18°C at night to 45°C during the day (SORALO, unpublished data) with erratic rainfall that is scattered across the area, averaging 400–600 mm yr⁻¹ (SORALO, unpublished data; Bedelian, 2019) A perennial river, the Ewaso Ngiro, runs through the study area, along with some small streams from the Nguruman Escarpment, these representing the only permanent water sources available in the area. The Ewaso Ngiro river flows through the Shompole swamp before ending up in Lake Natron.

Despite the semi-arid climate, combinations of topography, river and spring discharge through the area create a mosaic of habitats and seasonal pasture fluxes, which in turn support a high density of migratory grazing herds of wildlife and livestock and resident browsing species (Russell, Tyrrell and Western, 2018). Here, the Maasai still sustain seasonal movements and traditional grazing practices alongside a large population of wildlife, with a full complement of herbivore and carnivore species, except for black and white rhinoceros (Schuette, Creel and

DOI: 10.4324/9781315270845-22

Christianson, 2013). The ecology and seasonal dynamics mirror other pastoral and livestock systems in East Africa that have now largely been fragmented and ecologically uncoupled by land use changes (Hobbs et al., 2008).

In Shompole and Olkiramatian, the traditional seasonal livestock movements and herding practices are planned and governed by local committees, which primarily dictate where settlement and grazing are allowed in a given season. The wet season grazing areas in both group ranches are termed "livestock zones". The dry season grazing areas have been retained as "grass banks" for livestock and since 2000 have been established as wildlife conservancies used for ecotourism. Seasonally, the communities of both group ranches move their livestock between the livestock rearing zone in the wet season and the grass bank in the dry season when regional grass biomass and quality decline (Russell et al., 2018).

The principal aim of this case study is to demonstrate how social capital present in traditional Maasai societies and traditional ecological knowledge (TEK) facilitate coexistence between pastoralists and wildlife. More specifically, we integrate and expand on key findings from our research in Olkiramatian and Shompole to provide novel insights into how the Maasai living in the region are able to coexist with lions.

Defining social capital

Definitions of *social capital* vary widely. It is most often thought to be relationships which enable a community to work collaboratively to achieve a common goal (Putnam, 1993; Bowles and Gintis, 2002). Social capital is comprised principally of *generalized trust, reciprocity*, and *social networks* which exist between and within communities (Coffé and Geys, 2007). Fukuyama considers generalized trust to exist when "a community shares a set of moral values in such a way as to create regular expectations of regular and honest behavior" (1995 :. 153). Reciprocity has been proposed as an evolutionary adaption within human society that contributes to resilience to catastrophic environmental shocks. It is the willingness of an individual to cooperate with other members of society even when it is not economically rational or motivated by self-interest (Gintis, 2000). Social networks are the result of linkages or ties to other individuals in society. Galvin (2008) distinguishes between bonding ties and bridging ties. Bonding ties exist within families, extended kinship networks and between friends, while bridging ties link outside an individual's immediate network and give access to resources.

Defining traditional ecological knowledge (TEK)

Ecological knowledge, Indigenous Knowledge and traditional ecological knowledge (TEK) are often used interchangeably and do not adhere to a singular strict definition. Based on Berkes (1999), we consider TEK to be knowledge acquired through experience, adaptive learning, and intergenerational transmission about the natural history of an ecosystem and its inhabitants. TEK has been proposed as a mechanism in the context of conservation and ecosystem management that facilitates adaptive management and allows communities to respond to environmental variability. Similarly, the existence of resource and habitat taboos (RHT) as part of TEK can serve to mitigate and manage human–wildlife interactions (Berkes et al., 2000; Colding and Folke, 2001).

Why live with wildlife?

In order to understand the mechanisms that create coexistence between Maasai and wildlife, one must also explore the individual and societal willingness to live with wildlife and tolerate the risks and damage associated with doing so. Site-specific case studies conducted by Western et al., submitted show, unlike in other parts of Africa, Maasai living in Olkiramatian and Shompole expressed a predominant desire to coexist with lions. This desire has historically been driven by a culture of living with wildlife. Within the last decade, community-based conservation initiatives have fostered personal benefit from conservation and created perceived communal ownership of wildlife, which reinforce the notion of coexistence.

In Shompole and Olkiramatian, a culture of coexistence with wildlife exists and is influenced by societal beliefs relating to wildlife. Wildlife are seen as an integral part of the ecosystem, as important to consider in overall management of their environment as water, grass, herd productivity, their families and the land itself. They have one word which encompasses their overall management of every aspect of well-being: *erematare*, which takes into account their social networks, their environment, their livestock, their families and their resources.

As explained by a Maasai elder who stated: "If I see somewhere without wildlife, I ask myself what is wrong with this place that no wild animals live here". Analogously, cultural belief that tigers contributed to a healthy ecosystem shaped positive perceptions of tigers in Chitwan, Nepal (Carter and Allendorf, 2016).

Conservation benefits in Olkiramatian and Shompole accrued at the individual and community levels (Western et al., submitted), and illustrate how community-based conservation (CBC) can complement rather than undermine existing traditional ecological knowledge and social capital. As in other parts of Kenya, CBC initiatives in Olkiramatian and Shompole have not been constrained by government regulations making them inherently flexible and have not resulted in outside interests usurping control of natural resources. The Olkiramatian and Shompole community conservation areas are based on existing TEK and land-use practices and are managed by pre-existing governance committees. Thus, the creation of conservation areas reinforced existing social capital and provided an economic incentive to maintain traditional management practices. Hence, this results in community-based conservation practices that are adapted to local cultural contexts and based on TEK, setting them apart from the type of rigid solutions advocated for by development organizations and increasingly in the conservation community (Roe, 1993; Western et al., 2019).

Elsewhere in the world, connecting community-based conservation with local traditional institutions and cultural beliefs has also greatly increased the effectiveness of conservation. For instance, in India's Western Ghats, Ormsby and Bhagwat (2010) found that sacred forests proved an effective model for community-based natural resources management (CBNRM) and local resource protection. Together these examples highlight local variation in cultures and the need for a case-by-case conservation approach that is sensitive to varying traditions.

Conversely, the failure of many CBC/CBNRM projects around the world to prevent land conversion and environmental degradation demonstrates that financial incentives from conservation and wildlife cannot always generate substantial revenue or offset the benefits of natural resource exploitation (Calfucura, 2018). As in Shompole and Olkiramatian, there is now an increasing awareness that CBC/CBNRM initiatives should supplement pre-existing tangible and non-tangible benefits, rather than replacing them.

Creating coexistence

Coexistence in Maasailand is created and maintained at multiple scales by the interplay between TEK and social capital existing at the family and household level as well as within the broader Maasai community.

The individual ability to adapt

Transhumance pastoralism present in Maasailand creates the individual ability to adapt to changing environmental and socio-political conditions. This individual adaption in turn allows for coexistence by maintaining an individual's access to seasonal pasture and water resources within an open, resource-abundant (pasture and water), and heterogeneous landscape (BurnSilver et al., 2008; Butt, 2010). Spatial heterogeneity of pasture in the Olkiramatian and Shompole ecosystem allows herders to avoid areas of high lion occupancy for a large portion of the year, creating temporal and spatial separation between lions and livestock within the ecosystem.

Traditional husbandry practices in Maasailand have shaped the ecology of these pastoral rangelands. The Maasai have long believed that landscape heterogeneity is created by coexistence between livestock and wildlife as epitomized by a Maasai saying, "Elephants make grassland; cattle make woodlands". Ecological studies have similarly shown wildlife-livestock interactions are mutually beneficial; areas with cattle grazing have more bush and eventually trees while those used by elephants have more grass because the elephants destroy the bushes and trees (Odadi et al., 2011; Kimuyu et al., 2017).

These grazing dynamics combined with land-use and grazing management in the Olkiramatian and Shompole ecosystem concentrate zebra and wildebeest in conservation areas during periods of highest human presence (Tyrrell et al., 2017). Lions are hence able to avoid areas of human habitation while still maintaining access to key resources, such as prey, counteracting the effect of increased human and livestock presence within the conservation areas. Consequently, the frequency of livestock depredation does not increase when livestock are present within the conservation areas compared to when they are not (Western et al., in press).

Resource tracking and livestock husbandry

The members of Olkiramatian and Shompole communities possess a breadth and depth of traditional ecological knowledge that has helped them select areas of highest resource abundance while simultaneously reducing predation risk through resource tracking and active guarding. The process of resource tracking known as *Ele'enore* enables herders to track changes in pasture, water, and carnivore presence and adapt to changing environmental, social, and political contexts. Herders claimed they were able to identify where large carnivores were likely to be on a given day and understood seasonal changes in occupancy. Ecological knowledge was acquired from experiential learning while collective knowledge came from other community members, as stated by an elder in the Shompole community:

> Even as an old man I know where lions, leopards and all manner of wildlife can be found. When my livestock go to pasture, I go ahead to survey the area to ensure that there is nothing that will bring them harm. If I find lion tracks, I show them to my children in order to teach them the areas that are dangerous for livestock.

Similarly, pastoralists in India used traditional ecological knowledge to track the movement of tigers and wolfs surrounding Panna Tiger reserves (Kolipaka et al., 2015).

The accuracy of traditional ecological knowledge to predict wildlife movements has been questioned by some and yet TEK often closely conforms to scientific studies (Service et al., 2014; Frans and Augé, 2016). This case study could not evaluate the ability of *Ele'enore* to identify areas of high lion occupancy but information provided by local herders frequently assisted research teams to locate lions. Goldman (2007) also advocates integrating Maasai ecological knowledge into conservation science and rangeland management, as is the case in Olkiramatian and Shompole, where settlement and grazing management in conservation areas demonstrates how collective resource tracking and utilization can lead to ecologically sensible rangeland management and year-round access to pasture (Russell et al., 2018). In turn, ecologically sensible land-use and rangeland management allowed lions access to areas of high prey abundance even during times of increased human activity, thereby promoting coexistence (Schuette et al., 2013).

Maasai appear to have developed and continue to adopt cultural practices that allow for coexistence with large carnivores (Kissui, 2008; Goldman et al., 2010). Cattle are highly valued in Maasai society; accordingly, herders traditionally emphasized protection of livestock, active guarding and persecution of problem carnivores (Hollis, 1905; Spencer, 1988). Corralling of livestock in *bomas* reduces predation at night while active herding deters daytime attacks (Kolowski and Holekamp, 2006; Woodroffe et al., 2007; Lichtenfeld et al., 2015). Lion hunts may have helped maintain coexistence by making lions afraid of people while at the same time making people respect lions (Goldman et al., 2010). Herders in Olkiramatian and Shompole were capable of disrupting lion attacks on livestock and frequently did so, thereby reducing the incidence of livestock mortality at pasture. Maasai in Northern Tanzania also perceive adult herders to substantially reduce conflict at pasture (Mkonyi et al., 2017). Conversely, evidence shows that poor livestock husbandry increases the likelihood of depredation (Kuiper et al., 2015; Loveridge et al., 2017).

In Olkiramatian and Shompole, wealthier families choose to send their children to school and employed herders for the pastoral care to their livestock. Like others in the Maasai steppe (Mkonyi et al., 2017), cattle owners in Shompole and Olkiramatian who couldn't pay for herders often combined their herd with stock associates or extended family who employed herders. Those isolated from community and family reciprocity were less able to provide active protection and consequently more likely to be vulnerable to livestock predation. It is not surprising that elders in Olkiramatian expressed concern over the erosion of cultural husbandry practices exaggerated by sending children to school which they felt was leading to loss of pastoral and ecological knowledge and increases the prevalence of livestock depredation, as explained by a herder in Maasailand:

> Before the people in my area used to herd their cows well and guard them so they did not get killed. Carnivores rarely came to attack our livestock because they knew they would be chased and killed. Now people do not herd their cows well and animals are no longer scared of people.
>
> *(translated from Maa)*

Elsewhere in the world, traditional husbandry practices and habitat taboos influence livestock predation. For example, in the Panna Tiger reserve, India, traditional taboos simultaneously prevented and facilitated livestock predation. On one hand, they prevented conflict by dictating that livestock should not be left out to pasture at night and religious beliefs also prohibited the killing of carnivores. On the other hand, however, the practice of disposing of dead cattle on the village periphery attracted predators to settlements, thereby leading to greater conflict and livestock predation (Kolipaka et al., 2015). In Romania, traditional methods of livestock management, namely nightly corralling of livestock and active guarding of livestock with shepherds and dogs, reduced sheep depredation by bears (Dorresteijn et al., 2014).

Communal mechanisms for coexistence

Among pastoral communities, societies' adaption to ecological variability (including changes in wildlife presence) is facilitated by formal and informal institutions that govern communal property rights and access to pooled resources (Fratkin et al., 2004). In the Olkiramatian and Shompole communities, traditional ecological knowledge and social capital further reduced vulnerability and increased resilience to environmental, economic and socio-political perturbations. These same mechanisms enabled coexistence and adaption to the variable presence of lions within the ecosystem. Reciprocity and cooperation emerge as central themes that allow the Shompole and Olkiramatian communities to coexist with wildlife.

Reciprocity and risk reduction

Social capital forms the backbone of traditional Maasai society and is reinforced through *reciprocity*, enabling individuals and communities to reduce vulnerability to environmental change and livestock predation. Reciprocity serves as a societal safety net and acts as an informal insurance scheme (Spear and Waller, 1993; Galvin, 2008). Three types of reciprocity exist in Maasai society: (1) balanced reciprocity; (2) delayed reciprocity; and (3) generalized reciprocity (Potkanski, 1994).

Balanced reciprocity usually takes the form of direct exchange of livestock and operates in similar ways to traditional barter systems. Both parties benefit from a mutually agreed exchange of goods or services (Spear and Waller, 1993). High-grade Sahiwal bulls were frequently exchanged for several lower-quality Zebu animals by livestock owners in Olkiramatian and Shompole, allowing one beneficiary to grow his herd and the other to increase the genetic potential of his breeding herd.

Delayed reciprocity is still common in Maasailand and is exemplified by bridging ties known as the "stock friendships" or "associates". In these, age-mates gift livestock to each other in order to reinforce friendships (Spencer, 1988; Galvin, 2008). Stock friendships dissipate the effects of livestock losses by drought, disease and predation, and provide a mechanism by which herd owners can replenish their herds following livestock losses (Rutten, 1992). Bonding ties and delayed reciprocity in Maasai society expand the communities' ability to adapt, thus reducing a household's vulnerability to livestock predation and increasing the community's resilience to drought. During the droughts of 2012, 2013, 2014, and 2017, community members from Olkiramatian and Shompole moved to Oldonyo Lengai in Northern Tanzania in search of pasture. Their ability to do so depended on existing bridging ties and reciprocal arrangements with Northern Tanzanian Maasai communities. Migration helped cattle herds to survive the drought and reduced the likelihood of human-lion conflict during times of resource scarcity when lion attacks on livestock are likely to be most dangerous.

Generalized reciprocity allowed poor individuals in Maasai society to be financially supported by members of their extended family or clan (Potkanski, 1994). Donations of livestock or money from the wider community were commonly collected through *Harambees* (formal gatherings to solicit donations) and were used by members of the Shompole and Olkiramatian to pay for school fees, medical bills, or to replenish herds following substantial losses.

Maasai are not alone in using delayed and generalized reciprocity. These are important components of many traditional societies, shown to reduce the vulnerability of communities living adjacent to protected areas (Naughton-Treves et al., 2003; Dickman et al., 2011).

Coexistence through cooperation

Cooperation is also important in traditional societies and is enabled by social capital. Surrounding Corbet Tiger Reserve in India, villages with higher social capital were more likely to work collaboratively to deal with the human-tiger conflict (Rastogi et al., 2014). Pastoralism, like many traditional livelihoods, requires strong community cohesion and cooperation (Bekure, 1991; Niamir-Fuller and Turner, 1999). Grazing and settlement management in Olkiramatian and Shompole are rooted in inter- and intra-community collaboration among livestock owners who are willing to collaborate in communal management. This gives them greater access to resources and the ability to adapt to changing environmental conditions, including carnivore presence (Galaty, 1980; Butt et al., 2009). The willingness of Shompole and Olkiramatian communities to adhere to group ranch management and planning at the ecosystem-scale maintains mobility and open space within pastoral rangelands which facilitated coexistence (Russell et al., 2018). Parallel examples of communal management through collaborative processes are present in rancher-led initiatives in the United States of America, such as the Malpai Group in Arizona, which reconstituted individual land parcels into communally managed rangeland (Keough and Blahna, 2006; Curtin and Western, 2008).

By contrast, individualism is promoted by land subdivision with the subsequent loss of social capital. Maasai group ranches in Central Kajiado have moved ahead on land subdivision, leading to limited access, landscape heterogeneity and the concurrent drop in livestock production (Rutten, 1992; BurnSilver et al., 2004).

The traditional system of zoning of settlement into permanent areas and temporary areas gives seasonal access to pasture but also decreases the amount of time livestock spend in areas of high lion presence. Seasonal closure of the conservancies to grazing helps ensure abundant year-round forage and also maintains zebra abundance in areas of high lion presence (Russell et al., 2018). Furthermore, lion presence is seldom a consideration in decisions about land-use planning or even in grazing regulations, however, both types of management further enabled coexistence (Western et al., in press).

Losing coexistence: a culture in transition

Cultures are rarely static. They adapt and change. Maasai culture is well known across the globe for its dedication to tradition. However, it is a culture in transition, facing changes brought about from both internal and external pressures. At the individual level, challenges of religion and formal education, both recognized as positive forces, however, bring new values and changing mindsets to the younger generations in particular. Changes from a communal mindset to individually centred thinking is creeping in (referred to as the "we to me" effect with individual needs now being catered for before the interests of the community at large. Land subdivision and land sales can result from this mindset, creating challenges for the future of the rangelands that the Maasai and their livestock ultimately depend upon. Government policies rarely favour mobile communities, further encouraging sedenterization. International boundaries further exacerbate the problem of mobility for pastoral people, often artificially dividing communities and ecosystems.

Securing land under new forms of tenure, including wildlife conservancies, can combine both new values for the community from assets such as wildlife, with traditional *erematare* practices still able to play a role through increased security of land tenure, critical for cultural perpetuity.

References

Bedelian, C., Moiko, S. & Said, M.Y. (2019) *Harnessing opportunities for climate-resilient economic development in the semi-arid lands: The Kenya southern rangelands beef value chain*. Nairobi: Kenya Markets Trust. London: Overseas Development Institute.

Bekure, S. (1991) *Maasai Herding: An Analysis of the Livestock Production System of Maasai Pastoralists in Eastern Kajiado District*. Kenya: ILRI.

Berkes, F. (1999) *Sacred Ecology: TEK and Resource Management*. London: Routledge.

Berkes, F., Colding, J. and Folke, C. (2000) Rediscovery of traditional ecological knowledge as adaptive management. *Ecological Applications* 10, 1251–1262.

Bowles, S. and Gintis, H. (2002) Social capital and community governance. *The Economic Journal* 112, F419–F436.

BurnSilver, S. B., Boone, R. B. and Galvin, K. A. (2004) Linking pastoralists to a heterogeneous landscape. In J. Fox, R. R. Rindfuss, S. J. Walsh and V. Mishra (eds), *People and the Environment: Approaches for Linking Household and Community Surveys to Remote Sensing and GIS*. Boston: Springer US, pp. 173–199.

BurnSilver, S. B., Worden, J. and Boone, R. B. (2008) Processes of fragmentation in the Amboseli ecosystem, southern Kajiado District, Kenya. In K. A. Galvin, R. S. Reid, R. H. J. Behnke, and N. T. Hobbs (eds), *Fragmentation in Semi-Arid and Arid Landscapes*. Springer, ebook. pp. 225–253.

Butt, B. (2010) Seasonal space-time dynamics of cattle behavior and mobility among Maasai pastoralists in semi-arid Kenya. *Journal of Arid Environments* 74, 403–413.

Butt, B., Shortridge, A. and WinklerPrins, A. M. (2009) Pastoral herd management, drought coping strategies, and cattle mobility in southern Kenya. *Annals of the Association of American Geographers* 99, 309–334.

Calfucura, E. (2018) Governance, land and distribution: A discussion on the political economy of community-based conservation. *Ecological Economics* 145, 18–26.

Carter, N. H. and Allendorf, T. D. (2016) Gendered perceptions of tigers in Chitwan National Park, Nepal. *Biological Conservation* 202, 69–77.

Coffé, H. and Geys, B. (2007) Toward an empirical characterization of bridging and bonding social capital. *Nonprofit and Voluntary Sector Quarterly* 36, 121–139.

Colding, J. and Folke, C. (2001) Social taboos: "invisible" systems of local resource management and biological conservation. *Ecological Applications* 11, 584–600.

Curtin, C. and Western, D. (2008) Grasslands, people, and conservation: Over-the-horizon learning exchanges between African and American pastoralists. *Conservation Biology* 22, 870–877.

Dickman, A. J., Macdonald, E. A. and Macdonald, D. W. (2011) A review of financial instruments to pay for predator conservation and encourage human–carnivore coexistence. *Proceedings of the National Academy of Sciences* 108, 13937–13944.

Dorresteijn, I., Hanspach, J., Kecskés, A., Latková, H., Mezey, Z., et al. (2014) Human-carnivore coexistence in a traditional rural landscape. *Landscape Ecology* 29, 1145–1155.

Frans, V. F. and Augé, A. A. (2016) Use of local ecological knowledge to investigate endangered baleen whale recovery in the Falkland Islands. *Biological Conservation* 202, 127–137.

Fratkin, E., Roth, E. A. and Nathan, M. A. (2004) Pastoral sedentarization and its effects on children's diet, health, and growth among Rendille of Northern Kenya. *Human Ecology* 32, 531–559.

Fukuyama, F. (1995) *Trust: The Social Virtues and the Creation of Prosperity*. New York: Free Pres.

Galaty, J. (1980) The Maasai group-ranch: Politics and development in an African pastoral society. In P. C. Salzman (ed.), *When Nomads Settle*. New York: Praeger, pp. 157–172.

Galvin, K.A. (2008) Responses of pastoralists to land fragmentation: Social capital, connectivity, and resilience. In K. A. Galvin, R. S. Reid, R. Behnke Jr and N. T. Hobbs (eds), *Fragmentation in Semi-Arid and Arid Landscapes*. Boston: Springer US, ebook, pp. 369–389.

Gintis, H. (2000) Strong reciprocity and human sociality. *Journal of Theoretical Biology* 206, 169–179.

Goldman, M. (2007) Tracking wildebeest, locating knowledge: Maasai and conservation biology understandings of wildebeest behavior in Northern Tanzania. *Environment and Planning D: Society and Space* 25, 307–331.

Goldman, M. J., Roque De Pinho, J. and Perry, J. (2010) Maintaining complex relations with large cats: Maasai and lions in Kenya and Tanzania. *Human Dimensions of Wildlife* 15, 332–346.

Hanotte, O., Bradley, D. G., Ochieng, J. W., Verjee, Y., Hill, E. W. and Rege, J. E. O. (2002) African pastoralism: Genetic imprints of origins and migrations. *Science* 296, 336–339.

Hollis, A. C. (1905) *The Masai: Their Language and Folklore*. Oxford: Clarendon Press.

Keough, H. L. and Blahna, D. J. (2006) Achieving integrative, collaborative ecosystem management. *Conservation Biology* 20, 1373–1382.

Kimani, K. and Pickard, J. (1998) Recent trends and implications of group ranch sub-division and fragmentation in Kajiado District, Kenya. *Geographical Journal,* 25, 202–213.

Kimuyu, D. M., Veblen, K. E., Riginos, C., Chira, R. M., Githaiga, J. M. and Young, T. P. (2017) Influence of cattle on browsing and grazing wildlife varies with rainfall and presence of megaherbivores. *Ecological Applications* 27, 786–798.

Kissui, B. (2008) Livestock predation by lions, leopards, spotted hyenas, and their vulnerability to retaliatory killing in the Maasai steppe, Tanzania. *Animal Conservation* 11, 422–432.

Kolipaka, S., Persoon, G., De Iongh, H. and Srivastava, D. (2015) The influence of people's practices and beliefs on conservation: A case study on human-carnivore relationships from the multiple use buffer zone of the Panna Tiger Reserve, India. *Journal of Human Ecology* 52, 192–207.

Kolowski, J. and Holekamp, K. (2006) Spatial, temporal, and physical characteristics of livestock depredations by large carnivores along a Kenyan reserve border. *Biological Conservation* 128, 529–541.

Kuiper, T. R., Loveridge, A. J., Parker, D. M., Johnson, P J., Hunt, J. E., et al. (2015) Seasonal herding practices influence predation on domestic stock by African lions along a protected area boundary. *Biological Conservation* 191, 546–554.

Lichtenfeld, L. L., Trout, C. and Kisimir, E. L. (2015) Evidence-based conservation: predator-proof bomas protect livestock and lions. *Biodiversity and Conservation* 24, 483–491.

Loveridge, A. J., Kuiper, T., Parry, R. H., Sibanda, L., Hunt, J. H., et al. (2017) Bells, bomas and beefsteak: Complex patterns of human-predator conflict at the wildlife-agropastoral interface in Zimbabwe. *Peer Journal* 5, e2898.

Mkonyi, F. J., Estes, A. B., Msuha, M. J., Lichtenfeld, L. L. and Durant, S. M. (2017) Fortified bomas and vigilant herding are perceived to reduce livestock depredation by large carnivores in the Tarangire-Simanjiro ecosystem, Tanzania. *Human Ecology* 45, 513–523.

Naughton-Treves, L., Grossberg, R. and Treves, A. (2003) Paying for tolerance: Rural citizens' attitudes toward wolf depredation and compensation. *Conservation Biology* 17, 1500–1511.

Niamir-Fuller, M. and Turner, M. D. (1999) A review of recent literature on pastoralism and transhumance in Africa. In M. Niamir-Fuller (ed.), *Managing Mobility in African Rangelands: The Legitimization of Transhumance.* New York: Norton, pp. 18–46.

Odadi, W. O., Karachi, M. K., Abdulrazak, S. A. and Young, T. P. (2011) African wild ungulates compete with or facilitate cattle depending on season. *Science* 333, 1753–1755.

Ormsby, A. A. and Bhagwat, S. A. (2010) Sacred forests of India: A strong tradition of community-based natural resource management. *Environmental Conservation* 37, 320–326.

Potkanski, T. (1994) *Property Concepts, Herding Patterns and Management of Natural Resources Among the Ngorongoro and Salei Maasai of Tanzania.* London: IIED.

Putnam, R. D. (1993) The prosperous community. *The American Prospect* 4, 35–42.

Rastogi, A., Thapliyal, S. and Hickey, G. M. (2014) Community action and tiger conservation: Assessing the role of social capital. *Society and Natural Resources* 27, 1271–1287.

Roe, E. M. (1993) Development narratives, or making the best of blueprint development. *World Development* 19(4), 287–300.

Russell, S., Tyrrell, P. and Western, D. (2018) Seasonal interactions of pastoralists and wildlife in relation to pasture in an African savanna ecosystem. *Journal of Arid Environments* 154, 70–81.

Rutten, M. M. E. M. (1992) Selling wealth to buy poverty: the process of the individualization of landownership among the Maasai pastoralists of Kajiado district, Kenya, 1890–1990. Thesis. Leiden University.

Schuette, P., Creel, S. and Christianson, D. (2013) Coexistence of African lions, livestock, and people in a landscape with variable human land use and seasonal movements. *Biological Conservation* 157, 148–154.

Service, C. N., Adams, M. S., Artelle, K. A., Paquet, P., Grant, L. V. and Darimont, C. T. (2014) Indigenous knowledge and science unite to reveal spatial and temporal dimensions of distributional shift in wildlife of conservation concern. *PLoS One* 9.

Spear, T. and Waller, R. (1993) *Being Maasai: Ethnicity and Identity in East Africa.* Athens, OH: Ohio University Press.

Spencer, P. (1988) *The Maasai of Matapato: A Study of Rituals of Rebellion.* Manchester: Manchester University Press.

Tyrrell, P., Russell, S. and Western, D. (2017) Seasonal movements of wildlife and livestock in a heterogenous pastoral landscape: Implications for coexistence and community based conservation. *Global Ecology and Conservation* 12, 59–72.

Western, D., Wright, R. M. and Strum, S. (eds) (1994) *Natural Connections: Perspectives in Community-based Conservation*. Washington, DC. Island Press.

Western, G., Macdonald, D., Loveridge, A. & Dickman, A. (2019) Creating Landscapes of Coexistence: Do Conservation Interventions Promote Tolerance of Lions in Human- dominated Landscapes? *Conservation and Society,* 17, 204–2017.

Western, David, et al. "Conservation from the inside-out: Winning space and a place for wildlife in working landscapes." *People and Nature* 2.2 (2020): 279–291.

Woodroffe, R., Frank, L. G., Lindsey, P. A., Ole Ranah, S. M. and Romanach, S. (2007) Livestock husbandry as a tool for carnivore conservation in Africa's community rangelands: A case-control study. *Biodiversity and Conservation* 16, 1245–1260.

20

CULTURAL KEYSTONE SPECIES AS INDICATORS OF CLIMATIC CHANGES

Victoria Wyllie de Echeverria

Theoretical concepts

One of the most important issues facing the conservation and management of ecosystems is how to quantify the links between biotic and abiotic factors in ecosystem health, and it is impractical to do this without proxy measures. To date, ecologists and conservationists have used several concepts to classify species and illustrate their importance to people, their role in an ecological context, and their relevance in measuring biodiversity. The purpose of creating these proxies is to provide a tool for prioritizing and assessing conservation actions by focusing on one or more key species which identify the broader status of ecosystems and additional species that need to be conserved. In addition, it is vitally important to incorporate the Indigenous Knowledge of species and ecosystem linkages into this framework of conservation and management, as they have a long-standing understanding of their surrounding ecosystem which will contribute ecological knowledge to our understanding of these proxy species.

Keystone species

The earliest of these concepts was that of Ecological Keystone Species (EKS), coined by Paine (1969), who developed this concept by studying how the presence of sea stars (*Pisaster ochraceus*) preserved ecosystem balance and increased biodiversity. Keystone species are those which are considered to have an impact which is much larger than their abundance (Power et al. 1996), and therefore the interactions and population fluctuations of other species are likely to be moderated by keystone species in different communities. Similar to the keystone stones used at the top of doorways to hold the framework of stones in place, keystone species structure their environment and are critical to the integrity of their ecosystems. An important example of a keystone species in the Pacific Northwest is the sea otter (*Enhydra lutris*) (Estes and Palmisano 1974; Mills et al. 1993), which balance sea urchin and kelp abundance.

Umbrella species

A second concept, umbrella species, introduced by Wilcox (1984), suggested that a "target species" should be chosen for conservation programmes wherein the "minimum area requirements"

DOI: 10.4324/9781315270845-23

were of a similar size as the community in which the species lived, which would then in turn provide a "protective umbrella" for the rest of the species included in that community. Also, in order for this target species to be chosen and warrant protection, it would often have the characteristics of a "large body size ... high trophic level ... high metabolic requirements ... patchy distributions ... and species dependent on successional, rare or unpredictable habitats and resources" (Wilcox 1984). Thus, by focusing on the protection of this one target species which has many ecological interactions but whose home range is large, the surrounding species would be protected. Some examples in the Pacific Northwest include the northern spotted owl (*Strix occidentalis caurina*) and the coho salmon (*Oncorhynchus kisutch*) (Simberloff 1998; Branton and Richardson 2014).

Flagship species

In more recent years, the concept of a flagship species has surfaced, which is defined as "popular charismatic species that serve as symbols and rallying points to stimulate conservation awareness and action" (Heywood 1995). Species in this category are often appropriated as icons for conservation campaigns such as those spearheaded by the World Wide Fund for Nature. They are picked because people easily identify with them as proxies for their environments: they are well known, local, cute, majestic and/or fierce, and the hope is to protect whole ecosystems which are being threatened by issues such as deforestation, pollution and climate change, by protecting these species. While many common flagship species are tropical, one well-known example in the Arctic region is polar bears (*Ursus maritimus*) (Peacock et al. 2011).

Indicator species

A fourth species classification concept for conservation is that of an indicator species. While the concept of an indicator species has been discussed in the literature for many years (Landres et al. 1988), Siddig et al. (2016) provide a succinct definition: "Indicator species (IS) are used to monitor environmental changes, assess the efficacy of management and provide warning signals for impending ecological shifts". Chosen species have been used to "indicate" risks within a wide range of contexts, such as levels of pollutants and toxins in a system (Hilty and Merenlender 2000; Parmer et al. 2016), categorizing the local environmental conditions and ecosystem assemblages/biodiversity (Klinka *et al.* 1989; Hilty and Merenlender 2000), observing abundance and fluctuations of resources/taxa/endemic species (ibid.), and monitor ecosystem health, and environmental and ecological stressors and changes, such as changes influenced by climate change (ibid.; Siddig et al. 2016).

One confusing aspect of the term "indicator" is that it can also be used to refer to abiotic features, or community assemblages, for example air and water quality, rainfall amounts, temperature, and access to shellfish beaches (both for harvesting, and when closed due to toxicity) (indicators relevant to this research, see Wong and Rylko 2014). Other issues of applying this term is similar to issues with the other three concepts discussed above (keystone, umbrella and flagship), for example little consensus as to what these species actually indicate, uncertainty surrounding selection criteria, unsuitable taxa selected, unclear definitions, and varied methodologies (Kremen 1992; Simberloff 1998; Hilty and Merenlender 2000; Siddig et al. 2016). Hilty and Merenlender (2000) suggest that the criteria for selecting indicator species are often too "conceptual", making it hard for managers to appropriately apply these concepts. However, the USDA Forest Service seems to have understood the concept well enough to mandate a programme of Management Indicator Species (MIS), which includes a host of species of interest in different protected regions of the United States, for them to monitor to see how management affects these species and the communities they live in.

Advantages and disadvantages

There are advantages and disadvantages in using each of the above conservation concepts and their single species focus to sharply condense conservation research and biodiversity concerns. One advantage is that, regardless of which category a species in classed in, it is likely that the protection of this one species will lead to the preservation of other species in their environments, which is the main aim of conservation campaigns (Simberloff 1998). The largest advantage for all four single species concepts is that by providing this focus on a species which appeals to scientists, local people and/or an international community, conservation is rendered more comprehensible and manageable. However, despite these advantages, Simberloff argues that there have been conflicts identified during the application of these concepts and the species chosen, particularly with regards to umbrella and flagship species. These conflicts include:

- contradictions in management needs between species;
- protection of species is overly expensive;
- management and protection of the chosen species may not confer protection of other species in the ecosystem;
- confusion or disagreement over how species should be classed under the various iconic categories.

Also, there are issues as to what happens when a focal species becomes extinct and is lost, or if an ecosystem does not have any key endangered species to fill the role of a focal species (ibid.).

While Hilty and Merenlender acknowledge that it is unlikely that a selected taxa or species would fit all criteria, they suggest that a group of species or a taxon should be grouped together to display a comprehensive picture, which would then fulfil several criteria. They delineated the following criteria for species/taxa to qualify as viable indicators:

- Knowing baseline information, such as "biology, taxonomy and tolerance" and how humans change their characteristics.
- Having limited biogeographical ranges, but occur in many places (this will increase the number of studies, and also cut out anomalies from migration patterns).
- Being specific in their niche requirements and life history features, as well as having low variability in their genetics and ecology, so that changes in the species/taxa more accurately provide early warnings of ecological changes, rather than the species changing in response to a smaller ecological change.

Also, Hilty and Merenlender urge researchers to take into account how easy it is to measure and identify species/taxa and ecological changes, and the cost of the research. They suggest the following:

- species/taxa should be selected that respond to ecological changes easily and early, while resisting erratic variability;

- being more niche-specific, so they cannot alter their resources easily to limit the impacts of changes and thus accurately represent changes in resource abundance;
- selecting species that are endangered or represent more than one political or social agenda.

Ecosystem conservation

While it can be seen from the above four terms that there are advantages to a single species conservation concept, it is also helpful to expand our view to the entire ecosystem, and it has been suggested that groups or organisms should be used – instead of single species, which I briefly examined in the indicator criteria discussed above (Kremen 1992; Hilty and Merenlender 2000). Thus, another way to look at species conservation is to look at the ecosystem as a whole as it would allow a comprehensive ecosystem-wide view to preserving biodiversity, which has been thought to be beneficial to all processes and species (Daily et al. 1997; Simberloff 1998; Ingram et al. 2012), and might be better than a single species focus (Simberloff, 1998). However, some specific drawbacks to the ecosystem management concept are: (1) the boundaries of an ecosystem are ill defined;(2) "ecosystem healthiness" is a fluctuating concept; and (3) reconciling how an ecosystem's processes and functions can be preserved regardless of certain species being present (thus no need for a focus on protecting individual umbrella and flagship species). However, overall, this concept appears to succeed at combining people and their environment as a comprehensive unit, partially because this idea of using ecosystem management is also a prevalent aspect (Constanza et al. 1997) which can be seen in the recent paradigm of ecosystem services (ES), a framework for valuing what ecosystems provide for people, including provisioning, regulating, supporting and cultural services (ibid.; Daily 1997). By recognizing and monetizing these values, it is argued, humans can more effectively conserve biodiversity and structure sustainable use relationships to ecosystems.

Traditional Ecological Knowledge and ecosystem services

The inclusion of humans as an aspect of conservation challenges the original tenets of the American conservation movement, which focused on the preservation of the environment by excluding people to protect high value lands as wilderness (Library of Congress, 2002). Traditionally, the role of humans in the ecosystem was seen by conservationists and preservationists to be completely removed from the system to protect it, and that systems should be preserved solely for biodiversity and not for resource needs. Partially this was due to early conservationists and preservationists not realizing the often productive impact Indigenous and local people have had on landscape and biodiversity composition over centuries. In fact, there is no such thing as untouched wilderness in the Earth's habitable environments (Cronon 1996; Balee 1998). People have helped shaped the land nearly everywhere and thus need to be considered more inclusively in environmental management.

Going beyond monetizing the environment for humans' needs and quantifying what people can obtain and use sustainably, a new literature has begun to consider how humans provide "services to ecosystems", particularly through the reciprocal relationships inherent in many Indigenous Peoples' interactions with nature (Comberti et al. 2015). This conceptualization opens up space to integrate ecosystem management and single species conservation concepts.

To further elucidate how people interact with their environment, particularly in the arena of TEK, a concept was developed simultaneously in two unique environments. Cristancho and

Vining (2004) developed the term "Culturally Defined Keystone Species" from their research in the Amazon, and the same year Garibaldi and Turner coined the term Cultural Keystone Species (CKS), which they defined as a "culturally salient species that shape[s] in a major way the cultural identity of a people, as reflected in the fundamental roles these species have in diet, materials, medicine, and/or spiritual practices" (2004: 4), based on their research in Canada. This concept, in both instances, holds that there are not only pivotal species in ecological landscapes, but also in cultural landscapes. Garibaldi and Turner further explored how to incorporate CKS into environmental conservation and restoration, and due to regional proximity, this chapter discusses their research more.

Bringing it all together

As we have seen, looking at entire ecosystems and linking interactions within them is incredibly complicated. Thus, it is useful to start with a tighter focus. Using species that fulfil both indicator and keystone functions strikes a pragmatic balance between a whole ecosystem focus and one based on a single species.

To underscore the choice of using "indicators" and keystone species in a conservation framework for TEK, I turn to Borrett et al. (2014), who examined the ranking of different ecological concepts in 1986 to compare how frequently they still appeared in the literature in 2012. In addition, they added an additional 13 terms which did not appear in 1986 but were deemed to appear frequently in 2012. Of the terms I have looked at in this case study, only indicator species and keystone species appeared in this list, with indicator species originally ranked 29th in 1986, and moving up to a ranking of 14th by 2012, and keystone species which was originally ranked 46[th,] but moved down to a ranking of 51th by 2012. Neither umbrella nor flagship terms were seen in the ranked list in either year. Keystone species likely moved down in ranking because all but one of the additionally added terms in 2012 were above a 2012 ranking of 48. This illustrates how important, in particular, the indicator concept is. In addition, in a literature review by Siddig et al. (2016), it was shown that the number of papers published referring to indicator species had increased from 8 papers in 2001 to 149 in 2014, showing a massive jump in the use of this concept and its applicability. Additionally, because keystone species have already been situated in a biocultural framework, I feel that this concept would best be used to carry on into my framework of culturally important species which can detect changes in climate and the environment, which will affect harvesting frequency and the durability of the resource, and the cultural process into the future.

Using indicator species and keystone species concepts would focus our attention on single species for management, but also situate the chosen species to illustrate the interconnections within an entire ecosystem, and thus allow us to have a dual view of a single species focus within a broader ecosystems context. While indicator species have not frequently been used in a climate change context (Siddig et al. 2016), in this review, I argue that we need a way of monitoring the effects of a changing climate on the ecosystem into the future, and use it to link to species of importance, and species as indicators and keystones are applicable for this purpose. Hence my concept of Cultural Keystone Indicator Species (CKIS) was developed, which may be defined as: critical species of both cultural importance and perceptual salience in relation to environmental change (Wyllie de Echeverria and Thornton, 2019). To illustrate how this works in practice, this chapter looks at this concept in the Pacific Northwest, and uses two case studies: Sitka black-tailed deer (*Odocoileus hermionus sitkensis*); and salmonberry (*Rubus spectabilis*) and blueberries (*Vaccinium spp.*) together (as they occupy similar cultural and ecological roles, and are being impacted by climate change in similar ways), to investigate how CKIS can reflect climatic influences in an ecosystem. These case study examples are both indicator and keystone species in their ecosystems and are culturally important, so local people pay attention to their population fluctuations, and manage them through traditional ecological knowledge, leading them to be climate change indicators.

Case studies

In the following two case studies, I illustrate how local Indigenous Knowledge can be incorporated into conservation and management frameworks to better understand how certain species can be used as proxies to indicate climatic changes. For the two case studies, I discuss the role of each species, or in one case, complex of species, as both an indicator and a keystone species ecologically and culturally, and then link this to the interview data describing how each case alters patterns of behaviour and population to indicate climate change, as noted by the Indigenous People in Southeast Alaska and Northern British Columbia. The case study method is important, because it can be used in a variety of contexts; once a species has been identified as culturally and ecologically important as a keystone and indicator, and its links to climate change established, it can then be researched more in depth to provide context as to how this example species can ultimately be monitored and related to ecosystem conservation as a CKIS for climate change. Figure 20.1 shows a map of the region showing the locations of the home communities of the interview participants.

Sitka black-tailed deer (Odocoileus hermionus sitkensis)

Deer can be deemed to be both an indicator and keystone species. As an indicator species, since Sitka black-tailed deer move throughout the year between a mix of habitats, their presence and movement around the landscape can be used to indicate the quality of old growth forests, open younger forests and forest edges. Because deer also eat a varied diet, and need forage with certain characteristics, such as digestibility, their presence can also indicate the floral diversity (Hanley 1996; Lee and Rudd 2003; Schoen and Kirchhoff 2007). They are considered indicator species due to the following broad characteristics: (1) their biological features are well studied; (2) they have large home ranges which they migrate throughout in a seasonal pattern; (3) throughout the year they make use of different habitat types, including varied food sources and canopy cover options; and (4) they are valued by people, primarily as a food source, which makes them "socially relevant" (Hanley 1996). Deer also particularly indicate when undisturbed habitat is lost (Lee and Rudd 2003), as this ties in with their forage and movement requirements. They are considered to be a keystone species because their presence has a great effect on the landscape. When they are removed from the system, they can alter the floral architecture and diversity, the processing of minerals such as nitrogen, and the local soil make-up (Cobb 2014). Next, I discuss how these features show that deer can be considered to be a CKIS for climate change indicators in the Pacific Northwest region.

Indigenous participants interviewed to inform this case have noticed that the way Sitka black-tailed deer modify their seasonal migration habitat use signifies changing weather patterns, and the abiotic factor most closely linked to their distribution was snow. It was noted that since temperatures are warming overall, this will lead to smaller amounts of snow falling, and will facilitate deer migration higher into the surrounding hills and mountains, thus making them less accessible to hunters. Alternatively, if there is a snowfall which is deeper and heavier, the deer will migrate closer to areas of human presence. The browse that is available will also be

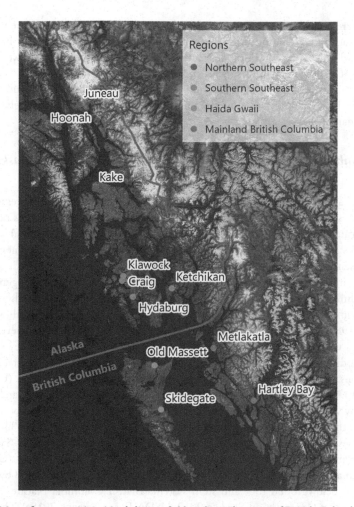

Figure 20.1 Map of communities visited during fieldwork on the coast of British Columbia, Canada, and Alaska, United States of America, and showing the four subregions linked to the climatic and geographical zones

Source: Image created by Conrad Zorn, University of Oxford.

impacted by snow levels which will in turn affect how the deer population shrinks or expands. Warming temperatures not only affect the snow levels, but also affect meat quality, the timing of when they put on their winter fat, and higher frequencies of disease (e.g. ticks). Because the effect of these abiotic factors on varying population levels are not well understood, there may be a disconnect between government regulations on hunting seasons, and times of the year when deer are actually available, which also impacts the ability of local people to harvest this traditional resource.

In summary, a case can be made for deer to be a climate change indicator due to their signposting of snow levels and distribution both in the current year, and into the following years, based on their location and population size, for how they are responding to this abiotic factor. As deer are heavily impacted by snowfall amounts, current snow depth levels determine where the deer are distributed. If deer are present in low elevations, it is likely there is snow on the mountains, forcing the deer lower, and if the deer are not present in low elevations, there is not as much snow higher up, and thus deer can be tracked throughout the year in relation to changing snow levels. Several interviewees noted that different communities would self-manage hunting, restricting harvests if deer populations were low. These observations of TEK concerning species movement and how Indigenous People manage populations in reaction to the observed fluctuations in response to climatic changes are important for local people and government bodies working collaboratively to develop appropriate management regimes.

Salmonberry (Rubus spectabilis) and blueberry (Vaccinium alaskense, V. ovalifolium, V. caespitosum, V. uliginosum)

Salmonberries and the several species of blueberries that occur in the study region are both indicator species and keystone species. First, the soil and moisture characteristics of an area can often be indicated by the plant assemblages that are living there (Halverson et al. 1986; Klinka et al. 1989). Because salmonberry and blueberry have slightly differing requirements ecologically, their presence is used to indicate different soil regimes. Salmonberries indicate locations that have wet and moist conditions and areas that are particularly rich in nitrogen (Halverson et al. 1986; Klinka et al. 1989), whereas blueberries indicate locations that are also wet, but the opposite of salmonberries for nitrogen, indicating acidic and nitrogen poor soils, such as bogs (Klinka et al. 1989; Hilty 2015; USDA Forest Service 2018b). Additionally, blueberries can indicate both land use changes and the presence of older ecosystems, because they often appear shortly after a disturbance occurs due to their characteristic of being able to establish easily in such conditions, but they are also a common species in old growth ecosystems (USDA Forest Service 2018b). These two groups of plants are good to use as indicators for several reasons: (1) they are closely watched (harvested) by people; (2) they are common enough to be seen in many places; and (3) they are influenced by weather patterns. Both salmonberries and blueberries are keystone species for the role they play as forage and habitat for various animals. Salmonberries provide cover and nesting sites (e.g. red squirrels, mice, black bears and beavers (USDA Forest Service 2018a) and stabilize soil in eroded or disturbed sites as the soil is bound by their quick growing, dense belowground root and stem system (British Columbia Nature 2002; USDA National Resources Conservation Service 2012; USDA Forest Service 2018a). They are also a vital food sources for a wide range of animals, both with the leaves and twigs (e.g. deer, mountain goats, moose, rabbits, porcupine, beaver), fruit (e.g. grouse, songbirds, American robins, raccoons, squirrels, foxes, mice and rodents (primarily the seeds), black bears, and grizzly bears (USDA National Resources Conservation Service 2012; USDA Forest Service 2018a), and nectar (e.g. bees, butterflies, other insects, and hummingbirds (USDA National Resources Conservation Service 2012; USDA Forest Service 2018a). Blueberries play a very similar keystone role to salmonberries, but because they are taller, it is thought that more taller animal species take cover in blueberries (USDA Forest Service 2018b) than salmonberries, and they similarly stabilize soil in eroded or disturbed sites, but in this case due to their characteristics of growing quickly and easily from seed, cuttings, or damaged parent plants, and thus they spread (or are planted) rapidly into cleared areas, additionally providing an early food sources (USDA Forest Service 2018b, 2018c). Their leaves are a vital food source for many animals (deer, mountain goats, elk, and the larval stages of Lepidoptera (USDA Forest Service 2018b, 2018c; www.nhm.ac.uk), fruit (e.g. grouse, ptarmigan, pheasants, songbirds, raccoons, squirrels, foxes, black bears, and grizzly bears [USDA Forest Service 2018b, 2018c]), and nectar (e.g. long- and short-tongued bees [Hilty

2015]). The two case study examples are also linked, because blueberries are a favourite food of black-tailed deer (*Odocoileus hermionus*) in Western Washington (USDA Forest Service 2018b). In addition to its food value to animals, blueberry harvesting contributes to the economy in the region through both commercial and wild harvesting (www.adfg.alaska.gov/sb/CSIS/). In the following paragraph, I discuss how these features show that salmonberries and blueberries can be considered to be a CKIS complex to indicate climate change in the Pacific Northwest region.

Participants interviewed to inform this case revealed that both salmonberries and blueberries were very important to them for food, and that particularly blueberries were also economically important. As a climate change indicator, both berries have had various growth stages affected by the weather. Earlier and warmer springs have been noted as initiating an earlier bud time than normal, leading to either this early bud growth being killed off by a later frost, or a disjunction with available pollinators, both of which in turn decreases the fruit set rates. If flowers manage to have fruit set, there are then further challenges with regards to the quality and quantity of the berries, which can vary greatly depending on the weather conditions in both summer and autumn. For example wet conditions result in berries either become saturated with water and thus not producing the necessary sugars to ripen fully, or they rot before they ripen, and dry conditions cause the berries to desiccate and dry out before they ripen. People also reported that years with increased rain have led to higher rates of insects in the berries, which is not a desirable trait for the local harvesters, or the blueberry economic industry.

In summary, from these observations, it can be seen that the late winter, spring, summer, and autumn weather conditions can be tracked, including the effect of changing climatic conditions throughout the growing season, by observing the phenology, quality, and quantity of these berry resources. While Indigenous Peoples do not self-manage berries in response to population fluctuations in the same way that they did for deer, they did adapt their usage patterns based on availability. Several interview participants described how in low yield years, people would use store-bought berries for feasts and other important events, and of course low yield years will impact the economic commercial harvests. Thus, linking the knowledge of Indigenous Peoples in respect to berry fluctuations will be important in the co-management between local people and governmental bodies when considering the management of these resources.

Conclusion

In conclusion, the TEK case studies examined in this chapter not only add to the limited literature on climate change indicators (Siddig et al. 2016), but also present evidence for how Indigenous Peoples' knowledge of the landscape and environment can be incorporated into management decisions, and lead to more effective co-management schemes between the government and other stakeholders and local people. Effective, collaborative co-management regimes are vital to advance modern conservation frameworks in relation to climatic change and local Indigenous needs. As mentioned in the theoretical concepts above, it is often hard to conceptualize how conservation and management can work within a ecosystem conservation framework, but this chapter presents the case that conservation researchers are better able to understand conservation at an ecosystem level, and management will be more comprehensive, when the knowledge of Indigenous Peoples and their relations to the landscape are consulted and included. CKIS provide a focused lens through which to view ecosystem conservation and change in an integrated ecological and cultural context.

References

Balée, W. 1998. *Advances in Historical Ecology.* New York: Columbia University Press.

Borrett, S. R., J. Moody and A. Edelmann. 2014. The rise of network ecology: Maps of the topic diversity and scientific collaboration. *Ecological Modelling* 293: 111–127. http://dx.doi.org/10.1016/j.ecolmodel.2014.02.019.

Branton, M. A. and J. S. Richardson. 2014. A test of the umbrella species approach in restored floodplain ponds. *Journal of Applied Ecology* 51: 776–785.

British Columbia Nature. 2002. Appendix 2, Native plant species. Available at: www.bcnature.ca/wp-content/uploads/ 2015/03/Appendix2.pdf (accessed 1 September 2018).

Cobb, M. A. 2014. Using mark-recapture distance sampling to estimate sitka black- tailed deer densities in non-forested habitats of Kodiak Island, Alaska. Refuge Report no. 2014.5. Kodiak, AK: Kodiak National Wildlife Refuge, U.S. Fish and Wildlife Service.

Comberti, C., T. F. Thornton, V. Wyllie de Echeverria and T. Patterson. 2015. Ecosystem services or services to ecosystems? Valuing cultivation and reciprocal relationships between humans and ecosystems. *Global Environmental Change* 34: 247–262. :http://dx.doi.org/10.1016/j.gloenvcha.2015.07.007.

Costanza, R., R. d'Arge, R. de Groot, S. Farber, M. Grasso, et al. 1997. The value of the world's ecosystem services and natural capital. *Nature* 387: 253–260.

Cristancho S. and J. Vining. 2004. Culturally defined keystone species. *Human Ecology Review* 11(2): 153–164.

Cronon, W. 1996. The trouble with wilderness; or, getting back to the wrong nature. *Environmental History* 1(1): 7–28.

Daily, G. C. 1997. *Nature's Services: Societal Dependence on Natural Ecosystems.* Washington, DC: Island Press.

Daily, G. C., S. Alexander, P. R. Ehrlich, L. Goulder, J. Lubchenco, P. A. et al. 1997. Ecosystem services: Benefits supplied to human societies by natural ecosystems. *Issues in Ecology* 2. Available at: www.esa.org/esa/wp-content/uploads/2013/03/issue2.pdf

Estes, J. A. and J. F. Palmisano. 1974. Sea otters: Their role in structuring nearshore communities. *Science* 185(4156): 1058–1060.

Garibaldi, A. and N. Turner. 2004. Cultural keystone species: Implications for ecological conservation and restoration. *Ecology and Society* 9(3). Available at: www.ecologyandsociety.org/vol9/iss3/art1

Halverson, N. M., R. D. Lesher, R. H. McClure, Jr., J. Riley, C. Topik, et al. 1986. *Major indicator shrubs and herbs in national forests of Western Oregon and Southwestern Washington.* USDA Forest Service. R6-TM-229–1986. Available at: http://agris.fao.org/agris-search/search.do?recordID=US8860347.

Hanley, T. A. 1996. Potential role of deer (*cervidae*) as ecological indicators of forest management. *Forest Ecology and Management* 88: 199–204. https://doi.org/10.1016/S0378-1127(96)03803-0.

Heywood, V. H. (ed.) 1995. *Global Biodiversity Assessment.* Cambridge: Cambridge University Press.

Hilty, J. (ed.) 2015. Insect visitors of Illinois wildflowers, version (09/2015). Available at: www.illinoiswildflowers.info/ (accessed 1 September 2018).

Hilty, J. and A. Merenlender. 2000. Faunal indicator taxa selection for monitoring ecosystem health. *Biological Conservation* 92: 185–197.

Ingram, J. C., K. H. Redford and J. E. M. Watson. 2012. Applying ecosystem services Approaches for biodiversity conservation: benefits and challenges. *S.A.P.I.E.N.S.* 5(1).

Klinka, K., V. J. Krajina, A. Ceska and A. M. Scagel. 1989. *Indicator plants of coastal British Columbia.* Vancouver, BC: UBC Press.

Kremen, C. 1992. Assessing the indicator properties of species assemblages for natural areas monitoring. *Ecological Applications* 2(2): 203–217.

Landres, P. B., J. Verner and J. W. Thomas. 1988. Ecological uses of vertebrate indicator species: A critique. *Conservation Biology* 2(4): 316–329.

Lee, N. and H. Rudd. 2003. Conserving biodiversity in Greater Vancouver: Indicator species and habitat quality. Surrey, Canada: Douglas College Institute of Urban Ecology, for The Environmental Stewardship Division, Ministry of Water, Land and Air Protection – Lower Mainland Region,. Available at http://citeseerx.ist.psu.edu/viewdoc/download?doi=10.1.1.122.3418&rep=rep1&type=pdf.

Library of Congress. 2002. The evolution of the conservation movement: Chronology of selected events in the development of the American Conservation Movement, c.1850–1920. Available at: http://lcweb2.loc.gov/ammem/amrvhtml/conshome.html (accessed 9 February 2018).

Mills, L. S., M. E. Soulé and D. F. Doak. 1993. The keystone-species concept in ecology and conservation. *BioScience* 43(4): 219–224.

Paine, R. T. 1969. A note on trophic complexity and community stability. *American Naturalist* 103: 91–93.

Parmer, T. K., D. Rawtani and Y. K. Agrawal. 2016. Bioindicators: The natural indicator of environmental pollution. *Frontiers in Life Science* 9(2): 110–118.

Peacock, E., A. E. Derocher, G. W. Thiemann and I. Stirling. 2011. Conservation and management of Canada's polar bears (*Ursus maritimus*) in a changing Arctic. *Canadian Journal of Zoology* 89: 371–385.

Power, M. E., D. Tilman, J. A. Estes, B. A. Menge, W. J. Bond, et al. 1996. Challenges in the quest for keystones: Identifying keystone species is difficult – but essential to understanding how loss of species will affect ecosystems. *BioScience* 46(8): 609–620.

Schoen, J. W. and M. Kirchhoff. 2007. Sitka black-tailed deer (*Odocoileus hermonius sitkensis*). In J. W. Schoen and E. Dovichin (eds), *The Coastal Forests and Mountains Ecoregion of Southeastern Alaska and the Tongass National Forest: A Conservation Assessment and Resource Synthesis*. Anchorage, AK: Audubon Alaska and The Nature Conservancy. Available at: www.conservationgateway.org/ConservationByGeography/NorthAmerica/UnitedStates/alaska/seak/era/cfm/Pages/default.aspx (accessed 4 November 2018).

Siddig, A. A. H., A. M. Ellison, A. Ochs, C. Villar-Leeman, and M. K. Lau. 2016. How do ecologists select and use indicator species to monitor ecological change? Insights from 14 years of publication in *Ecological Indicators*. *Ecological Indicators* 60: 223–230. DOI:10.1016/j.ecolind.2015.06.036.

Simberloff, D. 1998. Flagships, umbrellas, and keystones: Is single-species management passé in the landscape era? *Biological Conservation* 83(3): 247–257.

USDA Forest Service. 2018a. Fire effects information sheet for *Rubus spectabilis*. Available at: www.fs.fed.us/database/feis/ plants/shrub/rubspe/all.html (accessed 1 September 2018).

USDA Forest Service. 2018b. Fire effects information sheet for *Vaccinium ovalifolium*. Available at: www.fs.fed.us/database/feis/plants/shrub/vacovl/all.html (accessed 1 September 2018).

USDA Forest Service. 2018c. Fire effects information sheet for *Vaccinium alaskaense*. Available at: www.fs.fed.us/database/feis/plants/shrub/vacala/all.html (accessed 1 September 2018).

USDA Natural Resources Conservation Service. 2012. Plant guide for *Rubus spectabilis*. Available at: www.nrcs.usda.gov/Internet/FSE_PLANTMATERIALS/publications/orpmcpg10952.pdf (accessed 1 September 2018).

Wilcox, B. A. 1984. In situ conservation of genetic resources: Determinants of minimum area requirements. In J. A. McNeely and K. R. Miller (eds), *National Parks, Conservation and Development: Proceedings of the World Congress on National Parks*. Washington, DC: Smithsonian Institution Press, pp. 18–30.

Wong, C. and M. Rylko. 2014. Health of the Salish Sea as measured using transboundary ecosystem indicators. *Aquatic Ecosystem Health and Management* 17(4): 463–471.

Wyllie de Echeverria, V. R. and T. F. Thornton. 2019. Using traditional ecological knowledge to understand and adapt to climate and biodiversity change on the Pacific Coast of North America. *Ambio* 48(12): 1447–1469.

21

LIVING WITH ELEPHANTS

Indigenous world-views

Tarshish Thekaekara

Living with animals is perhaps going to be the key future challenge for conservation globally. Across much of the developed world, numerous "rewilding" projects are under way to bring back a suite of locally extinct animals or other suitable replacements (Donlan 2005;Vera 2009). While most of these projects are experimental and restricted within fenced-off, human-free regions, some large and potentially dangerous mammals, including wolves and bears, are making significant comebacks with populations increasing and ranges expanding across Europe, North America and Japan (Boitani 2003; Chapron et al. 2014; Saito et al. 2016), putting them into direct contact with people. People and wildlife already live in close proximity and at very high densities across much of South and South-East Asia, though with seemingly rising levels of "conflict" with each other. Africa is perhaps only now beginning to see a significant human population expansion (World Bank 2014), but is already experiencing significant and increasing human-wildlife conflict (HWC) (Weladji and Tchamba 2003; Madden 2008), which could potentially become worse in the future.

Key insights into living more peacefully with animals could be gleaned from Indigenous cultural perspectives, which have often negotiated these relationships sustainably in particular locales for centuries, if not millennia (Nelson 1995). Anthropologists have highlighted this for some time; hunter-gatherer communities' relationships with nature, for example, may be completely at odds with the dominant Judaeo-Christian world-view wherein "man" is at the pinnacle of creation with all of nature at his disposal, for, as the Bible asserts, "the Lord God sent him [Man] forth from the Garden of Eden, to till the ground ..." (Genesis 3:23; see also White 1967). It is assumed that modern, developed societies are able to dominate nature, while traditional, so-called "undeveloped" societies are at the mercy of nature (Manfredo and Dayer 2004). But Ingold (2000) disputes this view, and argues that hunter-gatherers' relationship with nature is more one of trust, with no separation between people and their environment. Nurit Bird-David, who had done most of her fieldwork with the Kattunayakans in the Nilgiri hill of South India, writes that they "look to the forest as they do on a mother or father. For them, it is not something 'out there' that responds mechanically or passively, but like a parent, it provides food unconditionally to its children" (Bird-David 1990: 190). Such a view is also held by the Batek Negritos people of Malaysia (Endicott 1979), the Mbuti Pygmies of the Congo (Turnbull 1965) and many others groups with a kin-centric ecological view of nature (Salmón 2000). Concepts of animism, where animals, non-living beings (stone,

DOI: 10.4324/9781315270845-24

the sun, ancestors) and natural phenomena (thunder or wind) are all thought to be "other-than-human persons", as described by (Hallowell 1960), have a direct impact on how people perceive and respond to nature and wild animals. A "collaborative reciprocity" is expected between human and non-human persons, as when hunting, "the animals gave themselves to the hunter in response to the hunter's respectful treatment of them as non-human persons" (Fienup-Riordan 1995: 50).

Thekaekara and Thornton (2016) have argued that there is a significant difference in how various ethnic communities – with varying modes of subsistence, socio-economic status and belief systems – interact with elephants, with the Indigenous people better equipped to deal with HWC than other migrant communities. In this chapter, based on ethnographic fieldwork in Southern India, I focus on the Kattunayakan and Bettakurumba peoples' relationships with, and strategies of living alongside, Asian elephants. I start with a range of grounded descriptions of people's perceptions and interactions with elephants, and then attempt to describe the key elements of their "alternate world-view" that allow them to live with elephants relatively peacefully.

Kattunayakans

We have no problem with these elephants. We know them, and they know us. Every year we do pooja for 'Aane devaru'[1] and ask them not to disturb our village. They listen to us. They don't come and trouble us here even though there are lots of jack fruit trees, but all the other people in this whole area have lot of problems with elephants.

This quote is from a discussion with a group of Kattunayakans, traditionally a hunter-gatherer tribe, living in Therpakolly, a village just south of the Mudumalai Tiger Reserve in the Nilgiri Hills, South India. The village is relatively well known among the other villages in the area for its rather unique approach to dealing with elephant problems.

Kattunayakans are the most forest-dependent of all the communities in the region, as is described by their name: Kattu (forest) Nayakans (rulers). They have stayed away from most of the "development" schemes run by both government and NGOs in the region, and are largely landless, with most of their villages located on the forest fringe. Working occasionally as wage labourers for both the forest department and local landowners, they also still routinely collect wild food and forest produce for consumption and sale.

They have been the focus of much of Nurit Bird-David's work (1990, 1992, 1996, 1999, 2006), continued by her student Daniel Naveh (Naveh and Bird-David 2013, 2014), and key anthropological perspectives on Indigenous communities' "alternative world-views" and epistemologies are arguably based on this ethnography of the Kattunayakans. Given that they have been so extensively documented, I include significant ethnographic descriptions from Bird-David and Naveh's work along with my own field data and analysis.

Kattunayakans' understanding of an elephant as a non-human person is well articulated:

Nayaka described some elephants as 'devaru'. They did not apply this word to all the elephants .. because of their assumed, shared, inert 'elephantness'. Rather, Nayaka used the word for specific elephants, in particular situations ... characterized by immediacy not just in the physical sense of close distance, but in a social-phenomenological one.

(Naveh and Bird-David 2014: 60)

This is further elaborated with examples; an elephant that carefully walks between houses without damaging them and being respectful towards people, or one which you can "look straight into his eyes" and "communicate with non-verbally" is "*aana-devaru*", but an elephant that breaks houses, behaves unpredictably, or where there is no mutual engagement, is just an ordinary "*aana*" (Bird-David and Naveh 2008).

There is also an understanding of elephants having "idiosyncratic" personalities, and much of their behaviour is attributed to this as described by Bird-David and Naveh (ibid.: 65):

> There are good budi (olle budi) elephants and bad budi elephants. When we walk in the forest, if there is an elephant with good budi, the elephant makes noise to make us know he is there. If there is bad budi elephant, the elephant is not making any sound, just wait silently. When you get near, this elephant attacks.

There is an understanding of elephant emotions as well, which allows them to be more accepting of elephants killing people in accidental encounters, which is a growing problem in India with about 500 people killed every year across the country in such "human elephant conflict". Bird-David and Naveh (ibid.) describe an incident when an elephant killed a person in the village, but still the others in the village did not agree to help the forest department capture the elephant and take it to Mudumalai, since the reason it killed a person was that it was angry and upset that the forest department had previously captured the same elephant's partner/companion elephant.

Kattunayakans often talk to elephants, particularly the "*devaru*" elephants that they relate to, as non-human persons. Bird David and Naveh observe:

> One October night in 2003, elephants entered KK [the village]; they trampled one of the huts, walked through the wetland paddies, and started to eat banana plants. While doing so, they also emitted loud bellows that were heard all over the village. One man went to about eight meters from where the elephants were standing, a distance that, should the need have arisen, would still have enabled him to run away. From there he approached the elephants boldly. In a typical blaming tone he said: 'Seri [in this sense 'OK'], if you want to eat, you silently eat and go. We have children here!'
> The elephants, then, stopped bellowing, and a few minutes later went away, out of the village.
>
> *(ibid.: 63)*

> When a Nayaka finds himself in front of an elephant, he prefers to stand still and, as calmly as possible, to address the elephant in a persuasive tone of voice (characterized both by the tone and by the substance):
> 'I am not coming to disturb you, or to do any harm to you.'
> The most frequently used rhetoric in such cases stresses what is common to both sides of the encounter:
> 'You are living in the forest, I am also living in the forest; you come to eat here, I am coming to take roots (fruits, firewood, etc.) ... I am not coming to do any harm to you.'
>
> *(ibid.: 63–64)*

Through these descriptions, it is evident that the Kattunayakans are well adapted to living with elephants, partly on account of their non-agricultural and non-confrontational mode of subsistence, but also on account of deeper cultural beliefs and values. Their understanding of elephants as other-than-human persons is arguably the most relevant; they relate to particular individuals rather than the species as a whole, and believe they are able to communicate with and maintain good relations with these individuals. Some particular elephants are not attributed personhood, and this allows for the accommodation of the occasional breakdown in the Kattunayakan-elephant relationship.

Bettakurumbas

While the Kattunayakans have been written about extensively, there is almost no contemporary literature on the Bettakurumbas. In much of the early writing, all the Kurumbas were grouped together, making precise description of this group a challenge. Thurston and Rangachari (1909) suggest that the Bettakurumbas originally lived on a mountain range called the Vollagamalai in Karnataka, but in their own oral history they prefer to think of themselves as always being forest people. While the Kattunayakans have shied away from development, the Bettakurumbas have been more ambivalent: almost all the children are enrolled in schools; they routinely access government schemes and public distribution system; and they are more integrated into "mainstream society", while also retaining their links with the forest. They believe their exposure to the outside world began centuries ago; narratives around capturing and taming wild elephants are vibrant in their stories, where they insist that Maharajas depended on them for "*keddah*"[2] operations, with British and Indian forest departments continuing this tradition. This facility with elephants is also mentioned in some of the early literature:

> The Betta Kurumbas are, I am told, excellent elephant mahauts (drivers), and very useful at keddah (elephant-catching) operations.
>
> *(Thurston and Rangachari 1909: 162)*

> I have heard of a clever Kurumba, who caught an elephant by growing pumpkins and vegetable marrow, for which elephants have a partiality, over a pit on the outskirts of his field.
>
> *(ibid.: 163)*

Even today, almost all the mahouts managing the captive elephants in the Mudumalai Tiger Reserve are Bettakurumbas, and they also are employed as guards and watchers and as guides for tourists and researchers entering wildlife areas.

In their handling of tame elephants, the Bettakurumbas are unique in that they are one of the few groups who do not use the "*ankush*" or the bull hook (a pointed metal hook) that is widely used to manage and control captive elephants. They sometimes carry a small stick, but communicate with the elephant mostly by moving their toes behind the elephant's ears. Each of the captive elephants is attached to one mahout and "*cavady*" (assistant) for most of their lives. With the mahout and *cavady* invariably being related, each elephant is in some sense a part of a human extended family, with a strong bond between the elephant and its mahout family. There is a well-known story of Bhama, one of the elephants in the camp rescuing her mahout Bomman from a leopard attack, and "after driving the predator away, she carried the unconscious mahout with her trunk through a distance of around three kms to the safety of the camp".[3]

An excerpt from a discussion with some elderly mahouts in 2009 brings out a version of elephant capture rather different from the *keddah* operations:

> In the old days there was no fuss like there is now to capture elephants; hundreds of people and shooting the elephants with sleeping medicine and all that.
>
> On the correct day, the elders in the village will do all the required poojas for the spirit. Then some selected men will go into the forests, to a particular area that the spirits tell us where to find the elephants. When they see the herd, they go up to them and ask some elephants to come and join us to work for the Kings. Some particular elephants would separate out from the herd and give themselves up to be caught. On their own they would come out and enter the kraal for training.

The idea of Bettakurumbas being able to communicate and cooperate with wild elephants is also not new, and finds mention in the 1908 *Gazetteer* of the Nilgiris: "Stories are told of how they can summon wild elephants at will" (Francis 1908: 156). This indicates animistic ideas of elephants as "other-than-human persons" capable of mutual respect and cooperation.

In summary, Bettakurumbas are slightly more removed from the forests than the Kattunayakans, and have embraced "modernity" to a larger extent, interacting less with wild elephants on a daily basis. They have limited challenges in living with elephants on account of their non-agricultural mode of subsistence and also their alternative world-views. But elephants, particularly captive ones, are more central to their culture, with Bettakurumbas considered the "elephant experts" by most of the other communities and the state forest department, and again ideas of elephants being other-than-human persons capable of mutual respect and reciprocity remain. They have a clear understanding of elephant personalities, and also relate to and interact with elephants as other-than-human persons.

Fostering better human-elephant coexistence

Given the urgent need to better integrate human and elephants needs in an increasingly crowded world, deeper understanding of what makes some communities more tolerant than others is imperative. I argue that there are two main cultural-ecological threads that allow for tolerance and a more peaceful sharing of space: (1) elephant ontologies, or what the community thinks an elephant is; and (2) their modes of subsistence and the varying agricultural crops types.

Elephant ontologies

First, concerning the characterisation of elephants, or the varied elephant ontologies. How are they conceived and how are their interactions with people explained? I have argued that there are four broad conceptualisations that emerge, where people understand elephants as: (1) other-than-human persons; (2) gods; (3) victims; and (4) wild/unpredictable animals.

First is the Indigenous idea of other-than-human persons, where some individual elephants are accorded some form of person-hood, capable of mutual respect, communication and even relationships with humans, that is clearly prevalent among the Kattunayakans and Bettakurumbas, as I have described above. This conceptualisation of elephants allows for accepting varying behaviour in elephants based on individuality, personality and agency. Elephants are expected to behave in accordance with human values and morality, and elephants that have been wronged are expected to be angry or sad and to behave unpredictably (where even killing of a person is not seen as unusual). Correlatively, aberrant individuals who behave badly with no provocation are liable for punishment. This understanding of elephants is perhaps the most conducive for a peaceful sharing of space.

Another conceptualisation of the elephant is that of Ganesha or Ganapati, one of the best known and most worshipped deities of the Hindu pantheon. Ganesha's fame is prevalent across India; attributing divine status to elephants automatically confers a certain reverence and tolerance. Accordingly, negative encounters between people and elephants may be rationalised in terms of divine retribution, and there is a certain acceptance of elephant attacks as instances of this. While this disposition appears to be ideally suited for tolerance and a sharing of space, such divine reverence does not allow for individuality in elephants. Continuous exposure to violence from elephants leads to a complete breakdown in the human-elephant relationship, since there is no room to blame or attempt to punish the elephants for wrongdoing, and elephants can then quickly become "demons". This duality exists in Hindu mythology; Gajasura is the "elephant demon", and Gajasurasamhara, an avatar of Shiva, is the "slayer of the elephant demon", who appears in Pallava and Chola art and iconography from over a thousand years ago, portrayed dancing on an elephant's head (Peterson 1991).

The third is the idea of elephants as victims, with humans expanding into and destroying their habitat, and forcing elephants into contact with people. With this conceptualisation, there is again limited scope to accommodate individuality, personality or agency in elephants. The underlying assumption is that elephants are passive victims not in control of their circumstances, who interact with people only because they have been forced to do so. This idea is arguably the basis of the global narrative around "fortress conservation" (Brockington 2002), protecting animals from humans, but, ironically, it is not shared by most of the communities living with elephants. While there has been a significant reduction of natural cover over the last century with immigration and growing human population into the Southern Indian region, elephants are also expanding their range over the last decade (Thekaekara 2019).

Finally, is the idea of elephants as wild and unpredictable animals. This stems from a very anthropocentric view of the world, arguably rooted in the Judaeo-Christian ideology, and Lynn White (1967) and others have argued that this ideology is perhaps the root of the current ecological crisis. While this understanding accommodates all the varying elephant behaviour that is experienced by the humans who share space with them, it does not allow elephants (or any elements of nature) and humans to be ontological equals, and there is no moral obligation to behave well or live well with animals, and killing elephants is acceptable.

It is evident that Indigenous communities in South Asia ascribe to multiple conceptualisations of the elephant. While all of these different ideas around "what is an elephant" are important, from the point of sharing space, the most relevant is perhaps in the hunter-gatherers' other-than-human ontology of elephants, that allows for significant mutual accommodation and variation in the behaviour of both elephants and people.

Modes of subsistence or agricultural crop types

Another important factor that mediates human-elephant interaction is the type of land use humans engage in, and this is very relevant in shared spaces. HWC has traditionally been understood from a "competition over space and resources" (Treves and Karanth 2003) approach, but, as we have seen, this is minimised by hunter-gatherer communities that do not engage in any large-scale agricultural activities that could result in competition or confrontation over crops. Interestingly, as hunter-gatherer communities in the Nilgiris have started engaging in agriculture, they have taken to planting tea and coffee over the last few decades, partly as a means of proving their possessory rights over the land they occupied, but also perhaps as a strategy of avoiding negative interactions with elephants. They almost never planted bananas, even though that crop is more remunerative than tea or coffee. When queried about why they did not grow bananas, the answer from a Kattunayakan was "because elephants will eat it of course". They have also been critical of other communities planting bananas and the increased risk it poses in attracting elephants to the human settlements (Thekaekara 2019). In this respect, Kattunayakans are still engaging elephants as other-than-human counterparts, and respect their agency, and specifically their tastes, wants and needs as relatives of the forests.

In summary, I argue that that are three underlying drivers of people's tolerance to elephants and the ability to share space more peacefully: (1) elephant ontologies and the very conceptualisation of what is an elephant; (2) the (non-competitive and non-confrontational) mode of subsistence, including the kind of crops people choose to gather or grow; and (3) a shared history of living together. All three of these factors vary significantly between the different Indigenous and local communities, but "tolerance" does not vary linearly with each of them. That is, communities who plant conflicting crops are sometimes more tolerant than others who do not engage in agriculture, or communities who have had a longer exposure to elephants are sometimes less tolerant than those with a shorter exposure to elephants. Yet, from a management perspective, some generalisations are required, and given the monolithic understanding of "humans" in policy around HEC, these three factors are arguably a reasonable way of heuristically understanding the propensity for people to be able to share space with elephants.

Conclusion

Indigenous world-views are clearly relevant to allow people and animals to share space more peacefully, but none of the policies relating to human-elephant interactions even begin to recognise that there is considerable variation in how people in the landscape understand elephants. There is an underlying assumption that all people are impacted by elephants in the same way. Factoring cultural diversity into policy is a significant challenge: labelling entire groups of people with certain tags of tolerance or intolerance has very serious shortcomings – it does not allow for the individual variation that always exists, or account for temporality and how individuals change over time. Nevertheless, there is perhaps room for some broad ideas that could feed meaningfully into policy.

Hunter-gatherers' relationships with particular other-than-human elephants is very useful in allowing them to live with elephants more peacefully. Given the "remarkable consistency of animism across the world" among hunter-gatherer communities (Praet 2013: 341), it is perhaps safe to assume that this world-view is common to a majority of forest-based people who share space with animals. The Kattunayakans' understanding of "idiosyncratic personalities" and behaviour among elephants that Naveh and Bird-David (2008) describe is very similar to what modern ethologists have discovered through careful elephant behavioural studies (Srinivasaiah et al. 2012). Perhaps linked to this is that people who have been living with elephants for some time also seem to have a nuanced idea of personality and culture in elephants, where they distinguish between, for example, "good" and "friendly" elephants and "bad" or "rowdy" elephants. This is not the same as hunter-gatherers' ontologically being equal to elephants as other-than-human persons, but it does nevertheless allow people to cope with negative interactions with elephants as individual "persons" in a more tolerant way, and thus allows for a more peaceful sharing of space.

Notes

1 While literally translated as "elephant god", the phrase is more nuanced in the Kattunayakan context, relating to their animistic relationship with elephants and other "non-human persons", rather than the better-known Hindu "Ganesha", the elephant deity.
2 A method of capturing elephants where an entire herd is driven into a specially constructed stockade or "*keddah*", followed by mahouts entering the *keddah* on tame elephants and lassoing and separating out the elephants for individual training.
3 See www.thehindu.com/2000/01/23/stories/13231087.htm

References

Bird-David, Nurit. 1990. The giving environment: Another perspective on the economic system of gatherer-hunters. *Current Anthropology* 31(2): 189–196.
Bird-David, Nurit. 1992. 'Beyond the hunting and gathering mode of subsistence': culture-sensitive observations on the Nayaka and other modern hunter-gatherers. *Man* 27(1): 19–44.
Bird-David, Nurit. 1996. Puja or sharing with the gods?: On ritualized possession among Nayaka of South India. *The Eastern Anthropologist* 49(3–4): 259–276.
Bird-David, Nurit. 1999. "Animism" revisited: Personhood, environment, and relational epistemology 1. *Current Anthropology* 40(S1): S67–S91.
Bird-David, Nurit. 2006. Animistic epistemology: Why do some hunter-gatherers not depict animals? *Ethnos* 71(1): 33–50.
Bird-David, Nurit, and Daniel Naveh. 2008. Relational epistemology, immediacy, and conservation: or, what do the Nayaka try to conserve? *Journal for the Study of Religion, Nature & Culture* 2(1): 55–73.
Boitani, L. David. 2003. Wolf conservation and recovery. In L. David Mech and Luigi Boitani (eds), *Wolves: Behavior, Ecology, and Conservation*. Chicago: University of Chicago Press, pp. 317–340.
Brockington, Dan. 2002. *Fortress Conservation: The Preservation of the Mkomazi Game Reserve, Tanzania*. Bloomington, IN: Indiana University Press.
Chapron, Guillaume, Kaczensky, Petra, Linnell, John D. C., von Arx, Manuela, Huber, Djuro et al. 2014. Recovery of large carnivores in Europe's modern human-dominated landscapes. *Science* 346(6216): 1517–1519.
Donlan, Josh. 2005. Re-wilding North America. *Nature* 436(7053): 913–914.
Endicott, K.M. 1979. *Batek Negrito Religion: The World-View and Rituals of a Hunting and Gathering People of Peninsular Malaysia*. Oxford: Clarendon Press.
Fienup-Riordan, Ann. 1995. *Boundaries and Passages: Rule and Ritual in Yup'ik Eskimo Oral Tradition*. New edn. Norman, OK: University of Oklahoma Press.
Francis, Walter. 1908. *Madras District Gazetteers: The Nilgiris*. Superintendent, Government Press.
Hallowell, Alfred. 1960. Ojibwa ontology, behavior, and world view. In S. Diamond (ed.), *Culture in History: Essays in Honor of Paul Radin*. New York: Columbia University Press, pp. 19–52.
Ingold, Tim. 2000. From trust to domination: An alternative history of human-animal relations. In Tim Ingold, *The Perception of the Environment: Essays on Livelihood, Dwelling and Skill*. London: Routledge, pp. 61–76.
Madden, Francine M. 2008. The growing conflict between humans and wildlife: Law and policy as contributing and mitigating factors. *Journal of International Wildlife Law & Policy* 11(2–3): 189–206. https://doi.org/10.1080/13880290802470281.
Manfredo, Michael J. and Dayer, Ashley A. 2004. Concepts for exploring the social aspects of human–wildlife conflict in a global context. *Human Dimensions of Wildlife* 9(4): 1–20. https://doi.org/10.1080/10871200490505765.
Naveh, Daniel and Bird-David, Nurit. 2013. On animisms, conservation, and immediacy. In Graham Harvey (ed.), *A Handbook on Contemporary Animism*. Durham: Acumen Publishing, pp. 27–37.
Naveh, Danny, and Nurit Bird-David. 2014. How persons become things: economic and epistemological changes among Nayaka hunter-gatherers. *Journal of the Royal Anthropological Institute* 20(1): 74–92.
Nelson, Richard. 1995. Searching for the lost arrow: Physical and spiritual ecology. In Stephen Kellert (ed.), *The Biophilia Hypothesis*. Washington, DC: Island Press.

Peterson, Indira Viswanathan. 1991. *Poems to Śiva: The Hymns of the Tamil Saints*. Delhi: Motilal Banarsidass Publisher.

Praet, Istvan. 2013. The positional quality of life and death: A theory of human–animal relations in animism. *Anthrozoös* 26(3): 341–355.

Saito, Masayuki U., Momose, Hiroshi, Inoue, Satoshi, Kurashima, Osamu and Matsuda, Hiroyuki. 2016. Range-expanding wildlife: Modelling the distribution of large mammals in Japan, with management implications. *International Journal of Geographical Information Science* 30(1): 20–35.

Salmón, Enrique. 2000. Kincentric ecology: Indigenous perceptions of the human–nature relationship. *Ecological Applications* 10(5): 1327–1332.

Srinivasaiah, Nishant M., Vijay D. Anand, Srinivas Vaidyanathan, and Anindya Sinha. 2012. Usual populations, unusual individuals: Insights into the behavior and management of Asian elephants in fragmented landscapes. *PloS One* 7(8): e42571.

Thekaekara, Tarshish. 2019. *Living with Elephants, Living with People: Understanding the Complexities of Human-Elephant Interactions in the Nilgiris, South India*. The Open University.

Thekaekara, Tarshish and Thomas F. Thornton. 2016. Ethnic diversity and human-elephant conflict in South India. In Piers Locke and Jane Buckingham (eds), *Conflict, Negotiation, and Coexistence: Rethinking Human-Elephant Relations in South Asia*. Oxford: Oxford University Press.

Thurston, Edgar, and Rangachari, K. 1909. Castes and tribes of South India. *Madras Government Press* 5.

Treves, Adrian, and Ullas Karanth, K. 2003. Human-carnivore conflict and perspectives on carnivore management worldwide. *Conservation Biology* 17(6): 1491–1499.

Turnbull, C. M. 1965. *The Mbuti Pygmies: An Ethnographic Survey*. Anthropological Papers of the American Museum of Natural History, 50, Pt. 3. New York: American Museum of Natural History.

Vera, Frans W. M. 2009. Large-scale nature development: The Oostvaardersplassen. *British Wildlife* 20(5): 28–36.

Weladji, Robert B. and Tchamba, Martin N. 2003. Conflict between people and protected areas within the Bénoué Wildlife Conservation Area, North Cameroon. *Oryx* 37(01). https://doi.org/10.1017/S0030605303000140.

White, Lynn 1967. The historical roots of our environmental crisis. *Science* 155(10.3): 1967.

World Bank. 2014. Africa's urban population growth: Trends and projections. Open data. Available at: http://blogs.worldbank.org/opendata/africa-s-urban-population-growth-trends-and-projections.

22

DO DRAGONS PREVENT DEFORESTATION?

The Gambia's sacred forests

Ashley Massey Marks, Joshua B. Fisher and Shonil A. Bhagwat

Introduction

One in four people live in agricultural villages, which cover 7.7 million km² of the Earth's terrestrial surface (Ellis and Ramankutty, 2008). Urbanization and increased exports of agricultural products are linked to deforestation in the humid tropics (DeFries et al., 2010). Drivers of forest clearance at the rural household level vary by region, and include availability of male labor, amount of forest cover, market access, and asset holdings (Babigumira et al., 2014). Destruction of forest habitat due to changing land use has precipitated an unparalleled rate of global biodiversity loss, altering ecosystem functioning and the provision of ecosystem services (Cardinale, 2012).

However, in the face of intense pressure of agricultural conversion, patches of forest persist in agricultural landscapes. Custodians conserve these forest patches due to their religious and spiritual values, cosmologies, or world-views. For example, a forest patch may be conserved due to the presence of medicinal plants or cosmological beliefs, for example, spirits believed to inhabit a geographic feature such as a tree, spring, or hilltop.

While performing biodiversity research in the Gambia, West Africa, local community members warned us to avoid certain parts of the forest due to the presence of *ninkananka*, a Mandinka word they translated into English as "dragons" (Figures 22.1a and 22.1b).

They told us that if someone sees the *ninkananka*, he or she will die on the spot or soon after. Belief in the *ninkananka* spans ethnic groups across West Africa, including Senegal, the Gambia, Guinea, and Sierra Leone (see the film, *Adventure Ninka Nanka: Samba Lives in Africa* by Ax, 2000). In 2006, a team of "dragon hunters" went on an expedition to the Gambia, seeking proof of the *ninkananka*'s existence, yet they determined it was likely to be "local folklore" (BBC News, 2006). We asked, does the belief in dragons prevent deforestation (or other forest uses) in forests where they are believed to reside?

The study area

Primary forest covers only 0.2 percent of the Gambia today (FAO Statistics Division, 2015). Mangroves have declined by 47 percent in the Gambia in 28 years since 2012. The population density of the Gambia is one of the highest in Sub-Saharan Africa; nearly 70 percent of people live in rural areas and depend on forest resources for fuel wood and building materials. Overgrazing

DOI: 10.4324/9781315270845-25

Figure 22.1a A patch of forest in a cleared agricultural landscape in Jiffarong, Kiang West, the Gambia, conserved by belief in a *ninkananka*. *Ninkananka* are believed to inhabit patches of forest across West Africa

Figure 22.1b A drawing of a *ninkananka* by Adama M. Saidy, grade six at Dumbuto Lower Basic School, Kiang West, the Gambia

by livestock, annual bushfires, and the exporting of timber also threaten Gambian forests (ibid.). Formal forest management in the Gambia includes protected areas serving as wildlife sanctuaries (37,400 ha), nationally gazetted forest parks (77,000 ha), community forest reserves (300,000 ha), and coastal mangrove forests (66,000 ha) (Bojang, 2001). Gazetted in 1987, Kiang West National Park is the Gambia's largest and most biodiverse wildlife reserve, its, 11,000 ha comprise half of the forest cover of Kiang West District. Most of all other land is owned by the state but in practice is treated as open access or managed by local institutions including village chiefs and elders. The 104,957 ha study area in Kiang West, Lower River Region of the Gambia (13.38 N–15.92 W) lies at an ecoregional boundary between Guinean forest-savanna mosaic and West Sudanian savanna, with Guinean mangroves buffering the Gambia River (Figure 22.2) (Table 22.1).

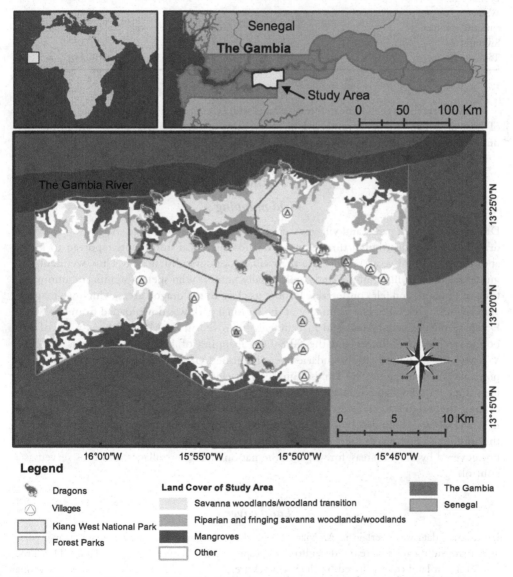

Figure 22.2 Map of study area (104,957 ha) including 15 dragon areas, 3 forest parks, Kiang West National Park and surrounding areas

Table 22.1 Land cover of study area: area (hectares) and percentage area

Land cover of study area	Hectares	Percentage
Savanna woodlands/woodland transition	24785.91	47.23
Mangroves	6264.54	11.94
Riparian and fringing savanna woodlands and woodlands	6216.03	11.84
Current agriculture	5677.38	10.82
Shrub and tree savanna	5102.46	9.72
Herbaceous steppes	1914.21	3.65
Grass savanna	1123.47	2.14
Barren flats	980.46	1.87
Water	345.42	0.66
Savanna woodlands	39.33	0.07
Swamp rice	29.25	0.06
Total	52478.46	100.00

Only 2 percent of the Guinean forest-savanna mosaic ecoregion is protected and 6.7 percent of West Sudanian savanna is protected; however, most of these protected areas are in practice under-patrolled "paper parks" (World Wildlife Fund, 2007, 2008).

Methods
Field mapping

Dragon areas, forest parks, and villages were field mapped via GPS June–August 2009. While dragon areas are distributed throughout the landscape, local informants reported to know about only one or two areas closest to their villages or agricultural areas. Thus, we identified guides from each village in the study area, usually hunters who spent a significant amount of time in the bush and felt comfortable leading us close to the dragon areas, which we mapped via Garmin eTrek GPS. The boundary of the national park was downloaded from the World Database on Protected Areas and inaccuracies improved in ArcMap v.9.1 via a list of 320 boundary post coordinates and the GPS field mapping of 115 boundary posts (UNEP-WCMC, 2012). Forest park boundaries were identified via GPS field mapping of boundary posts and referencing a 1984 Landsat image with clearly defined fire breaks. Villages were identified via GPS, Landsat imagery, and Google Earth. A study area of 104,957 ha incorporating the dragon areas, forest parks, national park, and villages was drawn as a polygon in ArcMap v.9.3, bounded by the Gambia River to the north, a mangrove-lined *bolong* to the south, and Senegal to the east. Surrounding areas comprised the parts of the study area not covered by dragon areas, forest parks, the national park, or villages, and thus served as a control.

Land cover

Land cover data was overlaid in ArcMap v.9.3 and the dragon areas, forest parks, national park, and surrounding areas were divided into their respective land covers (Tyldum, 1982). The three predominant land covers shared by all area types were selected for analysis: mangroves, savanna

Table 22.2 Size, number of pixels, and number of clusters for area types in each land cover category (mangroves, savanna woodlands/woodland transition, and riparian and fringing savanna woodlands/woodlands)

Land cover	Type	Area (hectares)	Number of pixels (30m x 30m)	Number of clusters
Mangroves	Dragon areas	7.74	86	3
	National park	1,760.58	19,562	8
	Surrounding areas	4,509.90	50,110	16
Savanna woodlands/woodland transition	Dragon areas	80.73	897	9
	Forest parks	2,468.70	27,430	3
	National park	10,953.00	121,700	12
	Surrounding areas	49,759.65	552,885	20
Riparian and fringing savanna woodlands/woodlands	Dragon areas	14.94	166	7
	Forest parks	273.06	3,034	8
	National park	2,289.33	25,437	7
	Surrounding areas	3,600.27	40,003	22

woodlands/woodland transition, and riparian and fringing savanna woodlands/woodlands. Contiguous pixels of a land cover served as the units of analysis (n) within each area type: dragon areas, forest parks, national park, and surrounding areas (Table 22.2).

Pre-processing satellite imagery

Landsat Thematic Mapper (TM) 5 and Landsat 7 satellite imagery were obtained from the United States Geological Survey for 1984, 1988, 1991, 1999, 2002, 2006 and 2009 (United States Geological Survey, 2012). Scene years are 3–4 years apart, except for a seven-year gap from 1992–1998 due to satellite sensor malfunction. To maintain seasonality across images and reduce climatic variation, scene dates were chosen for the end of the rainy season, when vegetation is green and bushfires are not present (26 Oct. 1984, 30 Oct. 1988, 15 Nov. 1991, 29 Nov. 1999, 5 Nov. 2002, 24 Nov 2006, 16 Nov 2009). All scenes (WGS 84, utm-28n, ellipsoidal projection) are characterized by 0 percent cloud cover. Landsat 7 scenes (1999 and 2002) were calibrated so as to be comparable with the Landsat 5 scenes using empirically derived values in Idrisi v.15.0 (Vogelmann et al., 2001). Bands from all scenes were atmospherically corrected using the Cos(t) model in Idrisi v.15.0 where the Dn value was assigned as the darkest pixel value observed in the Gambia River in each scene (Chavez Jr, 1988). All scenes had a resolution of 30 meters, except 1988, which had a resolution of 28.5 meters and was resampled to 30 meters via Idrisi v.15.0; all seven scenes were then accurately geocoded.

The normalized difference vegetation index (NDVI) is a spectral transformation of two reflectance bands from remote sensing imagery used to identify and quantify the greenness of vegetated areas. NDVI has been widely employed as an indicator of change in vegetation cover over time, including in Sub-Saharan Africa (Lambin and Ehrlich, 1997). We compare the NDVI of dragon areas, formally protected areas, and unprotected surrounding areas within each year as opposed to the inter-annual variations in NDVI[1] which geophysical scientists attribute to regional climate signals (Jarlan et al., 2005). NDVI was calculated:

$$\frac{near - infrared - red}{near - infrared + red}$$

where near-infrared is band 4 (0.76–0.90 μm) and red is band 3 (0.63–0.69 μm).

Statistics

The mean NDVI of contiguous land cover pixels (n) within each area type – dragon areas, forest parks, national park, and surrounding areas – was calculated in Idrisi Andes v.15.0 for all image years (1984, 1988, 1991, 1999, 2002, 2006, and 2009). For the normally distributed mangroves data, Repeated Measures ANOVA with a Greenhouse-Geisser correction was employed to test the difference between the NDVI of area types over the seven years (alpha level p = 0.05) (Table 22.3). For non-normally distributed data for all land covers, savanna

Table 22.3 Normality of NDVI data was tested for (a) all land covers and then for the pixel clusters in each area type (b–d) in each year

Area type		Shapiro-Wilks		
		Statistic	*n*	*p*
(a) All land types				
1984	Dragon areas	.959	19	.559
	Forest parks	.932	11	.434
	National park	.915	27	.030
	Surrounding areas	.896	68	.000
1988	Dragon areas	.929	19	.169
	Forest parks	.937	11	.488
	National park	.800	27	.000
	Surrounding areas	.802	68	.000
1991	Dragon areas	.961	19	.584
	Forest parks	.828	11	.022
	National park	.945	27	.159
	Surrounding areas	.898	68	.000
1999	Dragon areas	.959	19	.551
	Forest parks	.870	11	.078
	National park	.764	27	.000
	Surrounding areas	.751	68	.000
2002	Dragon areas	.835	19	.004
	Forest parks	.826	11	.021
	National park	.941	27	.128
	Surrounding areas	.781	68	.000
2006	Dragon areas	.950	19	.394
	Forest parks	.869	11	.076
	National park	.819	27	.000
	Surrounding areas	.813	68	.000

Table 22.3 Cont.

Area type		Shapiro-Wilks		
		Statistic	n	p
2009	Dragon areas	.826	19	.003
	Forest parks	.912	11	.260
	National park	.846	27	.001
	Surrounding areas	.847	68	.000
(b) Mangroves				
1984	Dragon area	.991	3	.814
	National park	.925	8	.469
	Surrounding areas	.942	16	.379
1988	Dragon area	.994	3	.851
	National park	.882	8	.197
	Surrounding areas	.938	16	.321
1991	Dragon area	.933	3	.499
	National park	.947	8	.682
	Surrounding areas	.958	16	.631
1999	Dragon area	.797	3	.108
	National park	.854	8	.106
	Surrounding areas	.944	16	.401
2002	Dragon area	.991	3	.815
	National park	.941	8	.619
	Surrounding areas	.905	16	.095
2006	Dragon area	.974	3	.693
	National park	.907	8	.336
	Surrounding areas	.967	16	.786
2009	Dragon area	.922	3	.459
	National park	.941	8	.621
	Surrounding areas	.960	16	.656
(c) Savanna woodlands/woodland transition				
1984	Dragon areas	.956	9	.755
	Forest parks	.954	3	.586
	National park	.976	12	.960
	Surrounding areas	.944	20	.286
1988	Dragon areas	.906	9	.292
	Forest parks	.816	3	.152
	National park	.959	12	.768
	Surrounding areas	.975	20	.848
1991	Dragon areas	.916	9	.364
	Forest parks	1.000	3	.997
	National park	.956	12	.728
	Surrounding areas	.874	20	.014*
1999	Dragon areas	.967	9	.872
	Forest parks	.901	3	.390
	National park	.928	12	.358
	Surrounding areas	.891	20	.028*

(continued)

Table 22.3 Cont.

Area type		Shapiro-Wilks		
		Statistic	n	p
2002	Dragon areas	.899	9	.246
	Forest parks	.776	3	.059
	National park	.894	12	.134
	Surrounding areas	.958	20	.512
2006	Dragon areas	.930	9	.485
	Forest parks	.953	3	.584
	National park	.921	12	.296
	Surrounding areas	.836	20	.003★
2009	Dragon areas	.931	9	.489
	Forest parks	.981	3	.733
	National park	.927	12	.347
	Surrounding areas	.890	20	.027★
(d) Riparian and fringing savanna woodlands/woodlands				
1984	Dragon areas	.924	7	.502
	Forest parks	.880	8	.189
	National park	.886	7	.254
	Surrounding areas	.987	32	.963
1988	Dragon areas	.963	7	.847
	Forest parks	.893	8	.251
	National park	.875	7	.206
	Surrounding areas	.965	32	.384
1991	Dragon areas	.932	7	.567
	Forest parks	.809	8	.036★
	National park	.858	7	.145
	Surrounding areas	.981	32	.823
1999	Dragon areas	.958	7	.800
	Forest parks	.909	8	.348
	National park	.954	7	.770
	Surrounding areas	.963	32	.333
2002	Dragon areas	.605	7	.000★
	Forest parks	.877	8	.175
	National park	.894	7	.299
	Surrounding areas	.983	32	.880
2006	Dragon areas	.929	7	.539
	Forest parks	.866	8	.139
	National park	.937	7	.609
	Surrounding areas	.982	32	.848
2009	Dragon areas	.931	7	.558
	Forest parks	.932	8	.532
	National park	.918	7	.456
	Surrounding areas	.982	32	.851

Notes: ★
Normality was assessed via Shapiro–Wilks tests in SPSS v.19 (alpha level p = 0.05) (a) All land covers (mangroves, savanna woodlands/woodland transition, and Riparian and fringing savannah woodlands/woodlands).

woodlands/woodland transition and riparian and fringing savanna woodlands/woodlands, the Kruskal-Wallis Independent Samples test compared NDVI of area types within each year (alpha level p = 0.05).

Survey methods

Survey data were collected in six weeks of fieldwork from June 26th to August 7th, 2009. Fieldwork coincided with the Gambia's rainy season when village members living in the urban area return to the village to assist with rice farming, which improved the representativeness of the survey sample. The January 2009 Dumbuto census consisted of a list of households, and the gender and ages (0–30, 30+) of household members (Table 22.4). To select survey respondents, the census data were recorded in a SPSS data sheet and a random sample of 20 percent of the 553 village members aged 6 and above was selected via the "Select Cases" function. The survey was administered to 89[2] respondents, representing 16 percent of the village members aged 6 and above (Table 22.5). The representativeness of the Dumbuto random sample of the village population was determined via a Chi-square test. The oral survey included demographic questions of education, religious practice, and residence outside the village, followed by questions about the *ninkananka* (Table 22.6). The oral survey was translated into the local language, Mandinka. The translator, Lahmin Njie, was instructed to deliver the same translation of questions to each respondent, and the researcher's proficiency in Mandinka ensured continuity of translation.

Key informants were recommended from villages in the study region and government departments, and included local villagers, a hunter, wildlife and forest park managers, field guides to dragon areas, and a snake expert. The 18 key informant interviews were semi-structured and

Table 22.4 Dumbuto village demographics

Factor	Frequency	Percentage	Cumulative percentage
Gender			
Male	251	45.4	45.4
Female	302	54.6	100
Total	553	100	
Age			
6–14	177	32	32
15–29	156	28.2	60.2
30–49	123	22.2	82.5
50–69	82	14.8	97.3
70+	15	2.7	100
total	553	100	
Education			
None	248	44.8	44.8
Some < grade 5	106	19.2	64
Finished grade 5	22	4	68
Grade 5 < x < grade 9	72	13	81
Finished grade 9	51	9.2	90.2
Grade 9 < x < grade 12	18	3.3	93.5
Finished grade 12	36	6.5	100
Total	553	100	

Table 22.5 Dumbuto random sample demographics

Factor	Frequency	Percentage	Cumulative Percentage
Gender			
Male	41	47.7	47.7
Female	45	52.3	100.0
Total	86	100.0	
Age			
6–14	28	32.6	32.6
15–29	23	26.7	59.3
30–49	20	23.3	82.6
50–69	13	15.1	97.7
70+	2	2.3	100.0
total	86	100.0	
Education			
None	38	44.2	44.2
Some < grade 5	19	22.1	66.3
Finished grade 5	3	3.5	69.8
Grade 5 < x < grade 9	11	12.8	82.6
Finished grade 9	5	5.8	88.4
Grade 9 < x < grade 12	3	3.5	91.9
Finished grade 12	7	8.1	100.0
Total	86	100.0	

Table 22.6 Dumbuto survey questions and responses

Question	Frequency	Percentage
Where is the farthest away from Dumbuto that you have lived?		
Another village	35	39.3
I haven't lived outside	32	38.2
Urban area	14	21.3
Another West African country	1	1.1
Where is the farthest away from Dumbuto that people from your compound live?		
Europe or United States	41	51.7
Urban area	31	34.8
I don't have relatives outside	7	7.9
Another West African country	3	3.4
Another village	2	2.2
Do people from your compound live in Kombo (the urban area of the capital)?		
Yes	80	89.9
No	9	10.1
What Islamic studies have you completed?		
Village study	61	74.2
None	17	19.1
Advanced study	5	6.7
How many times a day do you pray?		
5	85	95.5
4	3	3.4
2	1	1.1

Table 22.6 Cont.

Question	Frequency	Percentage
How often do you go to the mosque?		
More than once a day	35	39.3
Daily	16	18.0
Fridays	16	18.0
Never	15	16.9
Multiple times a week	5	5.6
Special occasions	2	2.2
How often do you go into the bush?		
Daily	71	79.8
Never	7	7.9
Weekly	5	5.6
Rarely	4	4.5
Monthly	2	2.2
For what purpose do you go to the bush?		
Farming	69	77.5
Don't go to the bush	8	9.0
Herding cows	5	5.6
Firewood for household use	3	3.4
Plants/fruit	2	2.2
Park ranger	1	1.1
Takes lunch to farmers	1	1.1
Are there (bad) places in the bush you will not go?		
Yes, there are places I avoid	75	84.3
No, I will go everywhere	7	70
I don't go to the bush	8	9.0
I don't know the bad places	1	1.1
Why won't you go there?		
I'm afraid	54	60.7
I don't go to the bush	7	7.9
I don't avoid places	6	6.7
Elders/people told me not to go there	6	6.7
I don't have time	4	4.5
It is far	3	3.4
I stay to places I know	3	3.4
Dangerous animals/snakes	3	3.3
Dragons	2	2.2
I don't know	1	1.1
Have you heard of the *ninkananka*?		
Yes	84	94.4
No	5	5.6
Is the *ninkananka* real?		
Yes	72	80.9
I don't know	9	10.1
I haven't heard of the *ninkananka*	5	5.6
No	3	3.4

(*continued*)

Table 22.6 Cont.

Question	Frequency	Percentage
Do others believe it is real?		
Everyone believes it is real	34	38.2
Most people believe it is real	31	34.8
Some people believe it is real	11	12.4
I haven't heard of the *ninkananka*	5	5.6
Almost everyone believes it is real	3	3.4
No one believes it is real	2	2.2
I don't know	2	2.2
Many people believe it is real	1	1.1
How have you learned of it?		
Old men	46	51.7
Parents	17	19.1
Elders	12	13.5
I haven't heard of the *ninkananka*	5	5.6
In the bush	2	2.2
Book	2	2.2
Teacher	1	1.1
Local incident	1	1.1
Boys	1	1.1
How old were you when you first learned about it?		
7–14	51	57.3
0–6	19	21.3
15–29	9	10.1
I haven't heard of the *ninkananka*	5	5.6
30–49	4	4.5
50+	1	1.1
Do you know where the *ninkananka* lives?		
Yes	47	52.8
No	37	41.6
I haven't heard of the *ninkananka*	5	5.6
Have you been to that area?		
No	61	68.5
Yes	23	25.8
I haven't heard of the *ninkananka*	5	5.6
Do other people go to that area?		
No	33	37.1
Yes	23	25.8
Few people go	15	16.9
I haven't heard of the *ninkananka*	5	5.6
Most people go	4	4.5
People might enter accidentally	3	3.4
Some people go	2	2.2
Have you seen the *ninkananka*?		
No	82	92.1
I haven't heard of the *ninkananka*	5	5.6
Yes	2	2.2

Table 22.6 Cont.

Question	Frequency	Percentage
Do you know someone who has seen it?		
No	55	61.8
Yes	29	32.6
I haven't heard of the *ninkananka*	5	5.6
What happens if you see the *ninkananka*?		
You die	46	51.7
Sick or die	11	12.4
Some are lucky, but some die	9	10.1
If you are scared, you become sick and/or die	5	5.6
I haven't heard of the *ninkananka*	5	5.6
Nothing	3	3.4
I don't know	2	2.2
Chases you and you run away	2	2.2
If you wait to tell of it, then you will survive	2	2.2
You die if you are over 50 or if *ninkananka* sees you	1	1.1
You are scared, lucky if you see white *ninkananka*	1	1.1
You get sick	1	1.1
Have you heard stories about the *ninkananka*?		
Yes	59	66.3
No	25	28.1
I haven't heard of the *ninkananka*	5	5.6
Do you know what it looks like?		
Yes	64	71.9
No	20	22.4
I haven't heard of the *ninkananka*	5	5.6
Like a snake	49	55.1
Big	11	12.4
Like a lamp or a mirror	7	7.9
It has different colours	6	6.7
Like a python	4	4.5
Like a goat or sheep	4	4.5
It can change shape/animal	4	4.5
Like a crocodile	3	3.4
Starts small then gets big	3	3.4
It has big eyes	2	2.2
Like an insect	2	2.2
It has scales	2	2.2
It is green	2	2.2
It is brown	2	2.2
If you see a white one, it is lucky	2	2.2
It breathes fire from the mouth	1	1.1
It makes a noise like clanging metal	1	1.1
It has a crested head/horn	1	1.1
It has legs	1	1.1
Like a bird	1	1.1

Table 22.7 Correlation of demographic factors with belief in the *ninkananka*

Demographic factor	Fisher's exact value	p-value
Gender	4.46	0.20
Age	7.42	0.55
Living outside village	1.61	0.77
Relatives outside village	2.02	0.58
Prayer frequency	5.09	1.00
Mosque visit frequency	22.29	0.19
Purpose for bush visits	19.37	0.60
Education	16.61	0.01★
Islamic Studies	17.65	0.03★
Frequency bush visits	21.30	0.03★

Note: ★Significance level < 0.05.

described local encounters with the *ninkananka*, the impacts of the belief on forest management, and place-specific *ninkananka* information.

Limitations

The remote sensing methodology of the chapter presumes forest cover is an effective indicator of biodiversity conservation. However, the presence of forest patches alone does not indicate species are present. Redford (1991) described "empty forests" as seemingly "intact" forests with large trees defaunated of large mammals. Future work could measure the biodiversity of dragon forests. A remote approach was chosen for this study to track forest cover over a 25-year period, and out of respect for local beliefs. Indeed, despite stopping a respectful distance from the dragon areas to map their near-boundaries, our gun-toting hunter guides ran away on numerous occasions after hearing sounds or seeing an animal near a termite mound. Having completed his assistance of our study, one guide remarked he would never again venture so close to the dragon areas, not for any amount of money. Despite consistency in translator and translation, administering the survey via translation limits the researcher's understanding of the nuances of local cosmologies and beliefs. Belief in the *ninkananka* is an excellent example of this as it is such a secretive belief that people are hesitant to discuss it openly.

Results

Knowledge of the *ninkananka* and their whereabouts is gained on a need-to-know basis by community members; for example, a mother will point out a dragon area to her daughter on her daughter's first trip to the forest to collect firewood. Dragon areas are located within a variety of different management zones (including protected areas, forest parks, and surrounding areas) and are not acknowledged by the state or local institutions. Local people are extremely reticent to talk about the *ninkananka* for fear of negative supernatural repercussions (primarily, death) and regularly expressed concern for our safety in the field. When we initially asked community members about the mysterious forest patches, we received vague and non-committal responses; it was only after living in the rural community for six months and people learned we had biked through these areas that they expressed concern and warned us about the *ninkananka*.

The Dumbuto random sample was determined to be representative of the village population by gender (village: 45 percent male, 55 percent female; sample: 48 percent male, 52 percent female), age (village: 32 percent 6–14, 28 percent 15–29, 22 percent 30–49, 18 percent 50+; sample: 33 percent 6–14, 27 percent 15–29, 23 percent 30–49, 17 percent 50+), and education (village: 45 percent none, 23 percent some primary, 22 percent some junior secondary, 10 percent some senior secondary; random sample: 44 percent none, 26 percent some primary, 19 percent some junior secondary, 11 percent some senior secondary) (Chi-square test; $\chi^2 = 0.585$, df = 1, $p = 0.444$; $\chi^2 = 0.553$, df = 3, $p = 0.907$; $\chi^2 = 1.878$, df = 3, $p = 0.598$, respectively).

Descriptions of the *ninkananka* were highly variable among our key informants and respondents. Seventy-two percent of our respondents said they knew what it looks like and said it was: like a snake (55 percent) or python (5 percent), big (12 percent), like a lamp or mirror (8 percent), different colors (7 percent), able to change shape/animal (5 percent), like a goat or sheep (5 percent), crocodile (3.4 percent), able to change from small to big (3.4 percent), big-eyed (2.2 percent), like an insect (2.2 percent), scaled (2 percent), green (2 percent), brown (2 percent), white if you were lucky (2 percent), fire-breathing (1 percent), making a noise like clanging metal (1 percent), having a crested head/horn (1.1 percent), having legs (1.1), like a bird (1.1 percent). Informants told us that *ninkananka* often lived in hollowed trees or termite mounds, or in holes in the ground. Some of the dragon areas had such a prominent feature where the *ninkananka* was believed to reside.

When we asked a representative sample of 89 respondents in the village of Dumbuto,[3] "Is the *ninkananka* real?," 81 percent responded yes, 10 percent I don't know and 3 percent no; 6 percent had not heard of the *ninkananka*. No significant difference in belief is attributable to the demographic factors of gender, age, having lived outside the village, having relatives outside the village, frequency of prayer, frequency of mosque visits, or purpose of visits to the bush (Fisher's exact test, p > 0.05, all factors) (Table 22.7). However, belief in the *ninkananka* was negatively correlated with education (p = 0.01) (Table 22.8) and Islamic Studies[4] (p = 0.03) (Table 22.9)

Table 22.8 Belief in the *ninkananka* correlated with education level of respondents

Education		Is the ninkananka real?				
		No	Yes	I don't know	I haven't heard of ninkananka	Total
None	count	0	37	2	1	40
	% within education	0	92.5	5	2.5	100
	% within belief	0	51.39	22.22	20	44.94
Some primary	count	0	19	3	2	24
	% within education	0	79.17	12.5	8.33	100
	% within belief	0	26.39	33.33	40	26.97
Some junior secondary	count	1	11	1	2	15
	% within education	6.67	73.33	6.67	13.33	100
	% within belief	33.33	15.28	11.11	40	16.85
Some senior secondary	count	2	5	3	0	10
	% within education	20	50	30	0	100
	% within belief	66.67	6.94	33.33	0	11.24
Total	count	3	72	9	5	89
	% within education	3.37	80.9	10.11	5.62	100
	% within belief	100	100	100	100	100

Table 22.9 Islamic Studies correlated with belief in the *ninkananka*

Islamic Studies		Is the ninkananka real?				
		No	Yes	I don't know	I haven't heard of ninkananka	Total
None	count	0	17	0	0	17
	% within Islamic Studies	0	100	0	0	100
	% within belief	0	23.6	0	0	19.1
Village study	count	1	51	9	5	66
	% within Islamic Studies	1.5	77.3	13.6	7.6	100
	% within belief	33.3	70.8	100	100	74.2
Advanced study	count	2	4	0	0	6
	% within Islamic Studies	33.3	66.7	0	0	100
	% within belief	66.7	5.6	0	0	6.7
total	count	3	72	9	5	89
	% within education	3.4	80.9	10.1	5.6	100
	% within belief	100	100	100	100	100

Table 22.10 Frequency of trips to the bush correlated with belief in the *ninkananka*

Frequency of trips to the bush		Is the ninkananka real?				
		No	Yes	I don't know	I haven't heard of ninkananka	Total
Never	count	0	6	0	1	7
	% within bush frequency	0	85.7	0	14.3	100
	% within belief	0	8.3	0	20	7.9
Rarely	count	0	2	2	0	4
	% within bush frequency	0	50.0	50.0	0	100
	% within belief	0	2.8	22.2	0	4.5
Monthly	count	0	0	1	1	2
	% within bush frequency	0	0	50	50	100
	% within belief	0	0	11.1	20	2.2
Weekly	count	0	4	0	1	5
	% within bush frequency	0	80	0	20	100
	% within belief	33.33	15.28	11.11	40	16.85
Daily	count	3	60	6	2	71
	% within bush frequency	4.2	84.5	8.5	2.8	100
	% within belief	100	83.3	66.7	40	79.8
Total	count	3	72	9	5	89
	% within bush frequency	3.4	80.9	10.1	5.6	100
	% within belief	100	100	100	100	100

and positively correlated with frequency of trips to the bush (p = 0.03) (Table 22.10). When we asked, "Are there (bad) places in the bush you will not go?," 84 percent responded yes, there are places they avoid. To compensate for self-reporting error, we also asked the question "Do other people believe the *ninkananka* is real?" and 90 percent of respondents answered that other people believe the *ninkananka* is real, of which 73.4 percent responded that everyone, nearly everyone, or most people believe it is real. Key informants we interviewed who had lived in the capital city for many years and worked for international organizations expressed belief in the *ninkananka*.

The NDVI of dragon areas (0.35±0.01, 0.50±0.01, 0.34±0.01, 0.49±0.01, 0.64±0.01, 0.49±0.01, 0.52±0.01) was higher than the forest parks (0.31±0.01, 0.42±0.02, 0.35±0.01, 0.43±0.02, 0.64±0.01, 0.41±0.01, 0.51±0.01), national park (0.32±0.01, 0.43±0.01, 0.30±0.01, 0.45±0.01, 0.56±0.02, 0.42±0.01, 0.46±0.01), and surrounding areas (0.29±0.01, 0.38±0.01, 0.30±0.01, 0.37±0.02, 0.50±0.02, 0.36±0.01, 0.40±0.02) across most years, except in 1991 (when forest parks were 0.01 greener) and in 2002 (when forest parks were 0.003 greener) (Figure 22.3) (Table 22.11).

The NDVI of the area types was significantly different in all years (p = 0.002, 0.000, 0.008, 0.000, 0.000, 0.000, 0.000) (Table 22.12).

When we break the data down by land cover types, we find that in the Gambian mangroves, dragon areas exhibited the highest NDVI of the area types across all years (0.30±0.01, 0.55±0.01, 0.31±0.02, 0.46±0.02, 0.53±0.01, 0.49±0.02, 0.49±0.03), followed by the national park (0.27±0.02, 0.41±0.04, 0.27±0.02, 0.38±0.03, 0.45±0.03, 0.37±0.03, 0.38±0.03) and surrounding areas (0.19±0.02, 0.23±0.04, 0.18±0.02, 0.17±0.04, 0.22±0.05, 0.22±0.03, 0.22±0.04) (Figure 22.4).

Mean NDVI differed statistically significantly between area types (F(3.746, 22.257) = 5.607, p = 0.001). Post hoc tests using the Bonferroni correction revealed that mean NDVI differed statistically significantly between the national park and surrounding areas (p = 0.011) and between dragon areas and surrounding areas (p = 0.007) (Table 22.10).

In savanna woodlands/woodland transition, dragon areas exhibited the highest NDVI across all years except for 1991 (Figure 22.5).

The NDVI of the area types was significantly different in 1984 (p = 0.003), 1991 (p = 0.003), 1999 (p = 0.049), and 2002 (p = 0.039).

In riparian and fringing savanna woodlands/woodlands, the NDVI of dragon areas was highest in all years except for 1991 (Figure 22.6).

The NDVI of the area types was significantly different in all years except for 1991 (p = 0.006, 0.001, 0.000, 0.000, 0.000, 0.000) (Table 22.13).

Discussion

Like formally protected areas, dragon areas are significantly "greener" than unprotected surrounding areas. While this may indicate selection bias, that is, dragon areas may be identified by local people as areas comprised of different or denser vegetation than surrounding areas, the longitudinal nature of the study demonstrates that these areas have been effectively maintained over time. Importantly, dragon areas conserve forest in a landscape where formal protection is limited: only 2 percent of the Guinean forest-savanna mosaic ecoregion is protected and 6.7 percent of West Sudanian savanna is protected (World Wildlife Fund, 2007, 2008). These forest patches, like most sacred natural sites and other "informally" protected areas, do not appear in conservation maps, plans or networks. However, in the land use matrix beyond protected areas, sacred natural sites and other "informally" protected areas offer an endogenous

Table 22.11 N, mean and standard error of the mean NDVI for area types by land cover

		n	1984 Mean	s.e.m.	1988 Mean	s.e.m.	1991 Mean	s.e.m.	1999 Mean	s.e.m.	2002 Mean	s.e.m.	2006 Mean	s.e.m.	2009 Mean	s.e.m.
Mangroves																
Savanna woodland/ woodland transition	Dragon areas	2	0.30	0.01	0.55	0.01	0.31	0.02	0.46	0.02	0.53	0.01	0.49	0.02	0.49	0.03
	National park	8	0.27	0.02	0.41	0.04	0.27	0.02	0.38	0.03	0.45	0.03	0.37	0.03	0.38	0.03
	Surrounding areas	16	0.19	0.02	0.23	0.04	0.18	0.02	0.17	0.04	0.22	0.05	0.22	0.03	0.22	0.04
	Total	27														
Riparian and fringing savanna woodlands/ woodlands	Dragon areas	9	0.35	0.02	0.48	0.01	0.35	0.01	0.49	0.02	0.65	0.01	0.47	0.02	0.52	0.01
	Forest parks	3	0.31	0.01	0.44	0.02	0.37	0.01	0.48	0.01	0.59	0.01	0.43	0.01	0.49	0.01
	National park	12	0.34	0.01	0.43	0.01	0.31	0.01	0.48	0.01	0.61	0.02	0.41	0.01	0.48	0.02
	Surrounding areas	20	0.30	0.01	0.44	0.01	0.36	0.01	0.44	0.01	0.61	0.02	0.41	0.01	0.48	0.02
	Total	44														
	Dragon areas	7	0.38	0.01	0.50	0.01	0.34	0.01	0.52	0.01	0.67	0.02	0.51	0.01	0.54	0.00
	Forest parks	8	0.31	0.01	0.41	0.01	0.35	0.02	0.41	0.03	0.64	0.01	0.40	0.01	0.51	0.01
	National park	7	0.35	0.01	0.46	0.01	0.33	0.01	0.48	0.01	0.62	0.02	0.46	0.01	0.49	0.01
	Surrounding areas	22	0.33	0.01	0.42	0.01	0.32	0.01	0.43	0.01	0.57	0.01	0.40	0.01	0.44	0.01
	Total	44														

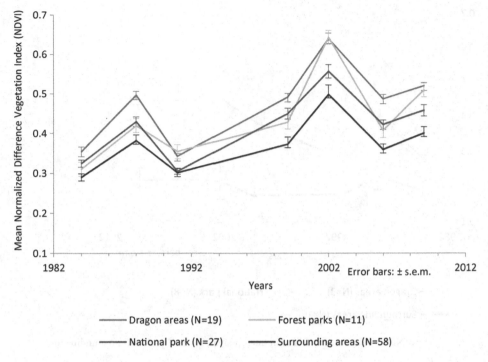

Figure 22.3 Mean NDVI of dragon areas, forest parks, national park and surrounding areas in all land covers (mangroves, savanna woodlands/woodland transition, riparian and fringing savanna woodlands/woodlands)

Table 22.12 Repeated measures ANOVA (1984, 1988, 1991, 1999, 2002, 2006, 2009) for NDVI of mangroves: pairwise comparisons of area types

(A) Area type	(B) Area type	Mean difference (A-B)	Std. Error	p^a	95% Confidence interval for difference^a	
					Lower bound	Upper bound
Dragon areas	National park	.086	.077	.825	-.112	.284
	Surrounding areas	.245*	.072	.007*	.060	.429
National park	Dragon areas	-.086	.077	.825	-.284	.112
	Surrounding areas	.159*	.049	.011*	.032	.286

Notes: Based on estimated marginal means.
(Dragon areas n = 3, National park n = 8, Surrounding areas n = 16, Total n = 27), alpha value p = 0.05* Pairwise comparisons.
a. Adjustment for multiple comparisons: Bonferroni

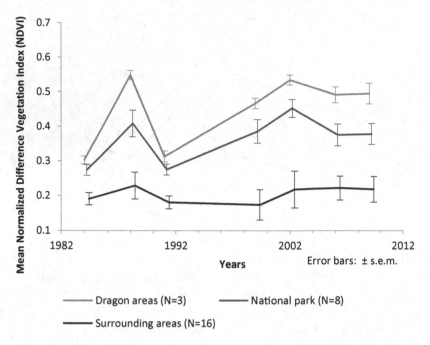

Figure 22.4   Mean NDVI of mangroves found in dragon areas, national park and surrounding areas

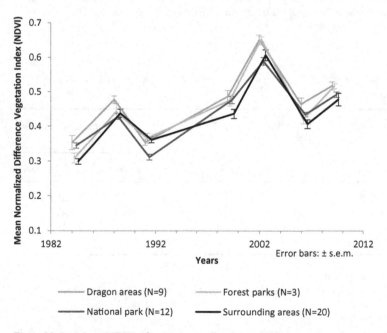

Figure 22.5   Mean NDVI of savanna woodlands/woodland transition found in dragon areas, forest parks, national park and surrounding areas

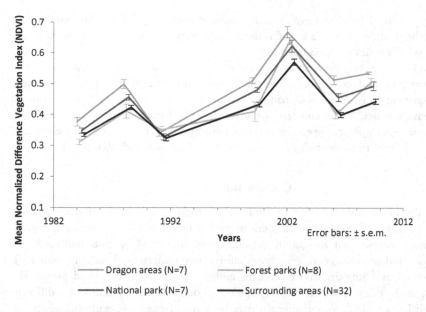

Figure 22.6 Mean NDVI of riparian and fringing savanna woodlands/woodlands found in dragon areas, forest parks, national park and surrounding areas

Table 22.13 Independent samples Kruskal–Wallis test for NDVI of area types in land covers (alpha level p = 0.05*, asymptotic significances displayed)

Year	All land covers (p)	Savanna woodlands/ woodland transition (p)	Riparian and fringing savanna woodlands/ woodlands (p)
1984	.002★	.003★	.006★
1988	.000★	0.055	.001★
1991	.008★	.003★	0.001
1999	.000★	.049★	.000★
2002	.000★	.039★	.000★
2006	.000★	0.195	.000★
2009	.000★	0.231	.000★

Notes: All land covers: Dragon areas n = 19, Forest parks n = 11, National park n = 27, Surrounding areas n = 58, Total n = 115; Savanna woodlands/woodland transition: Dragon areas n = 9, Forest parks n = 3, National park n = 12, Surrounding areas n = 20, Total n = 44; Riparian and fringing savanna woodlands/woodlands: Dragon areas n = 7, Forest parks n = 8, National park n = 7, Surrounding areas n = 22, Total n = 44).

approach to conservation widely acknowledged by local people, such as the respondents in our study (Verschuuren et al., 2010). While we found that 85.7 percent of people expressed belief in the *ninkananka*,[5] skeptical respondents had achieved higher levels of education or Islamic studies. Their skepticism does not necessarily affect local patterns of resource conservation as they may respect the majority view. Cultural psychologists Atran and Norenzayan (2004) attribute the resilience of supernatural beliefs to evolutionary forces affecting psychology. The mnemonic of a belief, i.e. the memorability of a belief, is the most powerful cognitive factor affecting the "cultural success" of a belief in a population.

Conservation planners may be reticent to acknowledge sites conserved by local beliefs that can change over time and specifically those that employ fear instead of support. However, adaptive management can address the changing nature of ecosystems as well as institutions and local beliefs (Scoones, 1999), and fear is not relegated to local belief systems but rather underscores some contemporary approaches in conservation management, for example, the use of armed poaching units to prevent illegal resource use (McClanahan and Wall, 2016).

Conclusion

Emphasis on the importance of current land tenure regimes and resilient management institutions may result in conservationists overlooking the presence and persistence of sacred groves, like dragon forests, that are quietly yet effectively conserved by custodians, even in the face of rapid land use change. Conservation efforts have underscored the importance of community buy-in and support, and custodians endogenously conserve sacred groves via local cosmologies. In cases of secretive, rarely-spoken-about beliefs, or cosmologies difficult to comprehend by outsiders, conservationists may be wary of engaging with custodians of dragon forests or similar sacred groves. However, this study demonstrates that in a state-owned common property regime, widespread belief in local cosmologies can conserve forest patches over time, as consistently as formally protected areas. This study also showed that local cosmologies can be syncretic with major world religions; in the Muslim village, 85.7 percent of respondents and the imam expressed belief in the dragon. Interviews with urban residents who had spent years away from the village but also expressed belief in the dragon, indicate that a local cosmology may be retained after separation from the location where it was formed.

In the case of sacred groves, the aim of conservationists' engagement with custodians is not to promote local beliefs beneficial to their mutual aims, but rather to recognize the conservation benefits that the sacred groves provide and to support the autonomy of custodians in management. For example, if sacred groves face threats from outsiders, such as the leasing of state land to resource extraction companies, conservationists can align with custodians to support their gaining land tenure. "Evil forests" and "sacred groves" are prevalent across Africa (Sheridan and Nyamweru, 2008), and similar "informally" protected sacred forests are found around the world. For example, there are nearly 35,000 Ethiopian church forests (Bongers et al., 2006), 80,000 Japanese Shinto shrine forests (Jinja Honcho, 2011), and 100,000–150,000 Indian sacred groves (Malhotra et al., 2001). It is estimated that the global land area informally conserved by custodians equals that of formally protected areas (ICCA Consortium, 2013). Our remote sensing analysis comparing these sites to formally protected areas in the same landscape over time indicates that the persistence of sacred groves is of interest and relevance for conservation planning and management beyond protected areas.

Notes

1 Fluctuations in NDVI between years were directionally similar in all land cover types.
2 There was a recording error for the demographic data for 3 of the 89 respondents.

3 Updated census (2009) reports total Dumbuto population as 697; our random sample of 89 is 13.5 percent of total population and is representative by gender, age and education.
4 Islamic studies are supplemental to classroom education, not an alternative form.
5 Calculated as a percentage of respondents who answered yes to "Is the *ninkananka* real?" (n = 72) out of the number of respondents who knew what the *ninkananka* is (n = 84).

References

Atran, S. and Norenzayan, A. 2004. Religion's evolutionary landscape: Counterintuition, commitment, compassion, communion. *Behavioral and Brain Sciences* 27, 713–730.

Babigumira, R., Angelsen, A., Buis, M., Bauch, S., Sunderland, T. and Wunder, S. 2014. Forest clearing in rural livelihoods: Household-level global-comparative evidence. *World Development* 64, S67–S79.

BBC News. 2006. Hunt for Gambia's mythical dragon. Available at: http://news.bbc.co.uk/1/hi/world/africa/5180404.stm (accessed 3 January 2012).

Bojang, L. 2001. Forestry Outlook Study for Africa (FOSA): the Gambi. In FAO (Ed.), *Forestry Sector Outlook Studies*. Rome: FAO.

Bongers, F., Wassie, A., Sterck, F. J., Bekele, T. and Teketay, D. 2006. Ecological restoration and church forests in northern Ethiopia. *Journal of the Drylands* 1, 35–44.

Cardinale, B. E. A. 2012. Biodiversity loss and its impact on humanity. *Nature* 489, 59–67.

Chavez Jr, P. S. 1988. An improved dark-object subtraction technique for atmospheric scattering correction of multispectral data. *Remote Sensing of Environment* 24, 459–479.

Defries, R. S., Rudel, T., Uriarte, M. and Hansen, M. 2010. Deforestation driven by urban population growth and agricultural trade in the twenty-first century. *Nature Geoscience* Advance Online Publication, 1–4.

Ellis, E. C. and Ramankutty, N. 2008. Putting people in the map: Anthropogenic biomes of the world. *Frontiers in Ecology and the Environment* 6, 439–447.

FAO Statistics Division 2015. CountrySTAT Gambia. Rome, Italy: FAO.

ICCA Consortium 2013. Indigenous peoples' and community conserved territories and areas (ICCAs). Available at: www.iccaconsortium.org/ (accessed May 22, 2014).

Jarlan, L., Tourre, Y. M., Mougin, E., Philippon, N. and Mazzega, P. 2005. Dominant patterns of AVHRR NDVI interannual variability over the Sahel and linkages with key climate signals (1982–2003). *Geophysical Research Letters* 32.

Jinja Honcho. 2011. Jinja Honcho: Association of Shinto Shrines. Available at: www.jinjahoncho.or.jp/en/ (accessed 3 January 2012).

Lambin, E. F. and Ehrlich, D. 1997. Land-cover changes in Sub-Saharan Africa (1982–1991): Application of a change index based on remotely sensed surface temperature and vegetation indices at a continental scale. *Remote Sensing of Environment* 61, 181–200.

Malhotra, K. C., Chatterjee, S., Gokhale, Y. and Srivastava, S. 2001. *Cultural and ecological dimensions of sacred groves in India*. New Delhi: Indian National Science Academy and Indira Gandhi Rashtriya Manav Sangrahalaya.

Mcclanahan, B. and Wall, T. 2016. 'Do some anti-poaching, kill some bad guys, and do some good': Manhunting, accumulation, and pacification in African conservation. In G. R. Potter, A. Nurse and M. Hall (Eds.), *The Geography of Environmental Crime*. Basingstoke: Palgrave.

Redford, K. 1991. The ecologically noble savage. *Cultural Survival Quarterly* 15, 46–48.

Scoones, I. 1999. New ecology and the social sciences: What prospects for a fruitful engagement? *Annual Review of Anthropology*, 28, 479–507.

Sheridan, M. and Nyamweru, C. (Eds.) 2008. *African Sacred Groves: Ecological Dynamics and Social Change*. Oxford: Oxford University Press.

Tyldum, G. 1982. GIS data. Available at: http://resourcepage.gambia.dk/gis_data.htm

UNEP-WCMC 2012. World database on protected areas. Available at: www.wdpa.org/ (accessed August 19, 2014).

United States Geological Survey 2012. EarthExplorer. Available at: http://earthexplorer.usgs.gov/.

Verschuuren, B., Mcneely, J., Wild, R. and Oviedo, G. 2010. *Sacred Natural Sites: Conserving Nature and Culture*. London: Earthscan.

Vogelmann, J. E., Helder, D., Morfitt, R., Choate, M. J., Merchant, J. W. and Bulley, H. 2001. Effects of Landsat 5 Thematic Mapper and Landsat 7 Enhanced Thematic Mapper Plus radiometric and geometric calibrations and corrections on landscape characterization. *Remote Sensing of Environment* 78, 55–70.

World Wildlife Fund 2007. West Sudanian savanna. In *Encyclopedia of the Earth*. Available at: www.encyclopedia.com/.../world-wildlife-fund

World Wildlife Fund 2008. Guinean forest-savanna mosaic. In *Encyclopedia of the Earth*. Available at: www.encyclopedia.com/.../world-wildlife-fund

23

FIRE, NATIVE ECOLOGICAL KNOWLEDGE, AND THE ENDURING ANTHROPOGENIC LANDSCAPES OF YOSEMITE VALLEY

Douglas Deur and Rochelle Bloom

While recognized globally as an iconic natural landscape and the centerpiece of a pioneering national park, Yosemite Valley is in truth a place cultivated by generations of Native caretakers. Many Native peoples' homelands converge in Yosemite Valley—including Southern and Central Miwok, Northern and Owens Valley Paiute, Chukchansi Yokuts, and Western Mono tribal communities, linked to both the western Sierra Nevada slope and to the deserts to the east. This amalgam of tribal communities together actively shaped this landscape in myriad ways. Prior to Euro-American resettlement, seasonal, patterned burning transformed the plant communities of the entire valley, sustaining open grassy meadows and California black oak (*Quercus kelloggii*) groves in locations that would otherwise have been overgrown in subalpine conifer forests. This burning, in addition to other types of active management—selective harvesting, pruning, replanting, and many other techniques—produced signature anthropogenic plant communities that have long sustained the food, material, medicinal, and spiritual needs of Yosemite Valley's Native residents. In spite of a century and a half of active displacement, Native peoples and aspects of their management practices endure, as do clear traces of human stewardship of the lands and plant communities of Yosemite Valley. This chapter summarizes these interconnected historical, cultural, and biological developments with special attention to the specific mechanisms used by Native American managers of the landscape, and the implications for both tribes and park managers today.

In this context, Yosemite Valley is a place with rich and enduring traditions of Indigenous Ecological Knowledge, manifesting in specific management practices that, in turn, leave discernible imprints upon the natural landscape. Moreover, as one of the world's first national parks, and a destination of enduring global significance, Yosemite presents us with a uniquely rich corpus documenting this knowledge and its manifestations within Native American resource practices and within the environments and humanized landscapes of Yosemite Valley.

DOI: 10.4324/9781315270845-26

Background and methodology

Our experience at Yosemite Valley demonstrates how IEK studies can be a highly productive historical and archival research endeavor, alongside ethnographic research with Indigenous Knowledge bearers. This is in part because documentation of Yosemite abounds within a nearly unbroken chronological succession of ethnographic, historical, agency, and travelers' accounts spanning the period from the mid-nineteenth century to the present. Today, the need to understand this knowledge and its implications has become relatively urgent. After a century and a half of fire suppression and other imposed changes, National Park Service staff confront rapid changes in Yosemite Valley vegetation, the decline of archetypal plant species and communities such as California black oak groves, the loss of iconic views to the encroachment of dense conifer stands, and a growing risk of catastrophic fire. These managers recognize that they must engage Indigenous Knowledge systems, and sometimes reintroduce Native management practices that extend from this knowledge—either by mimicking these traditions with modern analogues, or inviting Native harvesters back into the landscape to reestablish time-honored relationships between people and plants. With a growing sense of urgency, park managers recognize that Indigenous Ecological Knowledge is not a mere historical curiosity, but a living tradition that might yet provide compass points for future land and resource management for our common good.

In an attempt to understand the imprint of Indigenous Ecological Knowledge upon the Yosemite Valley landscape, we undertook certain research steps. Over a 15-year period, Deur carried out intermittent ethnographic interviewing and field visits with tribal members relating to traditional land and resource use (Deur 2007), following upon earlier ethnographic studies by researchers such as Bibby (1994) and Anderson (1988). Concurrently and in collaboration, drawing from the rich corpus of historical and ethnographic material relating to Yosemite Valley, Bloom oversaw the development of the Yosemite Ethnographic Database, which consists of roughly 12,800 individual entries describing traditional Native American use of Yosemite National Park lands and resources—the vast majority being within Yosemite Valley. Database entries incorporate data from a wide range of sources, including the Yosemite National Park Archives, the Yosemite Research Library, several university and museum collections, the California Digital Newspaper Collection, and numerous digital sources. National Council for Preservation Education (NCPE) interns working under the supervision of Bloom and Deur compiled and entered the assembled data from over 575 sources. This includes data derived from historical reports, early historical accounts written by visitors to Yosemite, ethnographies, ethno-ecological studies, Yosemite Nature Notes, oral histories, park notes from contemporary tribal events, archival materials, historical and contemporary newspaper articles, and more. These were supplemented with other materials, such as notes from tribal consultation meetings with park-associated tribes, and meetings with these tribes specifically on matters of traditional plant community management.

The resulting database provides a resource of rare detail and comprehensiveness, containing the vast majority of the written corpus regarding Native American communities' relationships to the Yosemite landscape across time. The database has become a tool for park managers seeking to understand the traditional uses of particular lands and resources within Yosemite, when used alongside their regular communications with contemporary park-associated tribal members. And this database has become a powerful tool for park-associated tribes attempting to document and understand their own practices in this unique montane environment. As a research tool, the database has allowed the authors to identify nearly all reported traditional land management practices employed by tribes both historically and today, through repeated and mutually reinforcing references in tribal oral accounts, written historical accounts, and other documentation. The picture provided by these combined accounts coheres with the findings of allied disciplines relating to Yosemite vegetation history and Yosemite Valley archaeology (Gassaway 2005, 2009; Gibbens and Heady 1964; Heady and Zinke 1978), while also illuminating working hypotheses for future vegetation and archeological studies. In the pages that follow, we summarize the specific forms of traditional resource management identified through this analysis.

Methods of traditional management in Yosemite Valley

Historically, and to a much lesser extent today, the Native American inhabitants of Yosemite Valley have employed a variety of techniques that materially enhance the availability of culturally preferred plant communities. While similar to other practices documented in California and the Great Basin, these techniques were uniquely adapted to fit the opportunities and constraints of the subalpine zone. Here, we identify specific techniques that appear consistently in the oral traditions and written historical accounts of the valley. These methods included anthropogenic burning, pruning and coppicing, clearing underbrush beneath trees, hand eradication ("weeding") of certain competing species, selective harvesting, smoking, "knocking" of dead wood from the tree, and other practices associated with both mundane activities and the spiritual beliefs of tribal communities traditionally associated with Yosemite.

Burning

Native peoples traditionally employed anthropogenic burning to clear Yosemite Valley of underbrush and maintain an open landscape (Commissioners to Manage the Yosemite Valley and the Mariposa Big Tree Grove 1891–92: 6–7; Ernst 1949:40). The fire regime in this region historically involved frequent, low-intensity surface fires set in a rotation to create a dynamic mosaic of plant communities in different stages of succession. Burns varied in size from localized spot fires, impacting only specific patches of plants, to clearing burns covering much larger areas (Gassaway 2009: 10; Taylor 2006: 9). Burning typically occurred in the fall after a few rains, with fires set at one- to several-year intervals (Anderson and Moratto 1996: 197–198).

Regular burning in the valley was guided by systems of Indigenous Ecological Knowledge pertaining to the timing of fires, as well as their location, required weather conditions, requisite scales, ignition points, and myriad other considerations. In turn, burning—guided by this understanding of fire, situated in place and accumulated knowledge across generations—created a variety of outcomes, including the production of culturally preferred plant communities that fostered Native sustenance, and other needs of Native communities. Burning promoted the dominance of certain plant species such as the black oak that were particularly suited to low-intensity surface fires, clearing competing vegetation and facilitating the germination and succession of new trees (Kuhn and Johnson 2008: 4; Reynolds 1959). Burning also stimulated the growth of several other culturally preferred plant species, caused the release of nutrients from accumulated biomass, and triggered the germination of fire- and heat-stimulated seeds (Anderson and Rosenthal 2015: 21; Reynolds 1959: 159–160; Stewart 2002: 294). Furthermore, Elders note that burning encouraged certain shrub species to regrow straighter stems following a fire, which produced high quality materials for producing baskets, spears, harpoons, canes, and a variety of other items (Anderson and Rosenthal 2015: 20, 24; Wickstrom 1987: 9). Vegetation in certain stages of regrowth after a fire were also ideal habitats for animals—producing forage and cover for the likes of mule deer, black-tailed jackrabbits, dusky-footed woodrats, California quail, greater roadrunners, and turkey vultures—all used for some combination of meat, furs, and feathers by Native peoples (Anderson 1988: 141; Anderson and Rosenthal 2015: 21, 24).

Burning created additional habitat-scale benefits as well. By producing clearings and creating a park-like open space in the meadows of Yosemite Valley, burning made acorn collection easier, eliminated hiding places for enemies, and reduced conifer encroachment on oaks, which thrive in open spaces (Grom and Camp 1959; Kuhn and Johnson 2008: 4; Martin 1996). Additionally, regular and systematic burning had the effect of reducing destructive wildfires by eliminating fuels such as underbrush, dead leaves, pine needles, and other debris (Bigelow 1904: 8; Kuhn and Johnson 2008: 4; Stoy 1890: 26) and assisted in controlling pests and pathogens that plagued some culturally preferred plant and animal species (Anderson and Rosenthal 2015: 22; Kuhn and Johnson 2008: 4). Burning may have limited the spread of root fungus, in part by keeping trees scattered and isolating the disease (Champion 1986; West 1986:2). Brush-burning also may have increased the availability of surface water and increased the flow of springs by reducing water uptake from less culturally significant shrub and tree species, while reducing moisture stress for those culturally preferred species that remained after repeat burning (Anderson and Rosenthal 2015: 22; Wickstrom 1987: 8).

Smoking

Exposure of plants to smoke, a byproduct of anthropogenic burning, provided ancillary benefits to culturally important plants, and represented an important and sometimes independent traditional management technique. Traditionally significant plants like chia, red mains, bluehead gilia, caterpillar Phacelia, and coyote tobacco (*Nicotiana attenuata* S. Watson) depend on exposure to smoke or charred wood for germination (Anderson and Rosenthal 2015: 15). A reported additional benefit of smoke is the decrease of insects that live and feed on food plants like sourberry, and ostensibly many others (ibid.: 22). Native consultants also share that smoke benefits the health of black oaks, contributing to increased acorn production (Anderson 1993: 189; Goode 2014: 6). Though references to this practice in Yosemite are few, Elders suggest that smoking may have been among many benefits of burning for the Yosemite Valley flora.

Pruning

Native peoples have traditionally pruned the ends of oak branches, allowing them to gather acorns early in the season before they were ripe enough to fall (Clark 1894: 15; NPS 2016b: 4). Particularly with oaks, pruning removes diseased limbs from trees and makes branches less likely to be broken by heavy winter snows or to combust during fires to the detriment of living trees (Clark 1894: 15; NPS 2014: 2). Pruning the tips of oak branches also encourages the growth of new branches and stimulates production of acorns (Anderson 2005: 139; Long et al. 2016: 63). Similarly, traditional harvesters have pruned other species, such as elderberry and manzanita, cutting off the flower or fruit-bearing portion of the branches to facilitate the growth of multiple fruiting stems (Anderson 1988: 132). Some sources suggest that pruning the tips of plants sustains the health, growth, and productivity of the plants generally, beyond the production of fruit- and flower-bearing stems (Anderson 2005: 139, 173; Stevens 1998: 31–32). Harvesters also prune various other plants, such as redbud, sourberry, and willow close to the ground or stem base, in order to produce the long, straight shoots ideal for use in basketry (Anderson 1988: 133; 2005: 319; Bates 1997; McCarthy n.d.: 26, 40).

Clearing underbrush

Individuals, families, and communities have traditionally raked leaf litter and cleared the ground around certain culturally preferred plant communities—especially black oak groves (Anderson 1988: 215; Bibby 1994: 57). This made access to culturally preferred plant species easier, while also serving to remove the buildup of underbrush and reduce potential competition and crowding by less prioritized species (Anderson 1988: 164–165). This practice also eliminated ladder fuels, reducing the odds of catastrophic fires that might harm culturally significant plant communities, especially black oak groves. Some attempt to remove sticks, litter, and underbrush has continued into modern times, being undertaken clandestinely or otherwise by Native harvesters, especially in black oak groves and other places where community harvests and investment are most concentrated.

Hand eradication

Native harvesters have often weeded competing plants from patches of culturally preferred species, or have engaged in the hand-removal of young trees in combination with anthropogenic burning to control brush and tree growth. Galen Clark (1894: 14–15) writes of Native harvesters he observed in the nineteenth century: "When the fires did not thoroughly burn over the moist meadows, all the young willows and cottonwoods were pulled by hand." Native peoples also weeded to clear the ground and improve gathering potentials (Bibby 1994: 57). Similar to the effects of clearing underbrush, weeding reduced the potential for competition among different plants and the crowding out of preferred plants (Anderson 1988: 164–165). Additionally, clearing underbrush and maintaining open meadows eliminated ladder fuels and reduced the likelihood of destructive crown fires, promoting light surface fires that only burned undergrowth while leaving oak groves and other preferred habitats relatively untouched (Martin 1996). This tradition of weeding competing species such as conifer trees has persisted, sometimes with NPS involvement and sometimes not, to protect culturally significant plant communities such as black oak groves.

Selective harvesting

Tribal members have often practiced selective harvesting, leaving parts of various tubers, bulbs, corms, rhizomes, and mushrooms within the soil when they harvested, so that remaining plants could regenerate (Anderson 1993: 93; 2005: 299; Anderson and Lake 2013: 58). Additionally, harvesters gather only the largest mushrooms, leaving smaller mushrooms to ensure the health and dispersal of culturally preferred mushroom species (Anderson 2005: 299; Anderson and Lake 2013: 58). Elders suggest that these practices are ancient, long predating Euro-American contact and calibrated by their long-term experiences as a community with plant resource abundance and scarcity. These selective harvesting traditions ensure that culturally significant plant communities will not be overexploited. Some suggest that selective harvesting may also preserve the genetic diversity of the species in various ways, such as by moderating the effects of human harvesters on the reproductive success of plant populations with distinctive and desirable genetic traits (Anderson and Lake 2013: 58).

Soil aeration

Harvesters loosened the earth in meadows as they dug edible bulbs with digging sticks, which also functioned to aerate the soil, reduce soil compaction, encourage bioactivity, and allow water to percolate deeper into drier soils (Anderson 2005: 173; Reynolds 1959: 189–190; Stevens 1998: 31–32). Together, these effects facilitate the development of larger and more numerous edible bulbs, the straighter growth of underground stems used for basketry, and the like. Systematic studies demonstrate that these traditional methods stimulate the growth of bulbs—working humus into the soil column, accelerating biotic action, improving aeration and decomposer access, and through other mechanisms (Anderson 2005: 173; Ortiz 1991: 5).

Knocking

"Knocking" is a management technique that occurs in conjunction with traditional gathering. Elders describe it as "a kind of massage for the tree, to give it energy and continue the relationship between the harvester and the tree" (Goode 2014: 6). Gatherers use knocking sticks to assist with removal of acorns from higher branches that cannot be reached from the ground, which encourages acorns to fall without damaging branches or bark (Anderson 1993: 174–175, 208; 2005: 141; Long et al. 2016: 44; NPS 2016c: 3). While knocking off acorns, or knocking trees generally, gatherers also successfully remove dead and diseased wood from the trees; in turn, this reduces the likelihood of disease or catastrophic fire that might affect the living tree, while also stimulating new growth (Martin 1996). Only healthier branches, with live and healthy cambium beneath the bark, remain after a thorough knocking. As one Native consultant explained, "Knocking wakes up the tree" (NPS 2014: 2).

Sowing

While detailed documentation is unavailable for Yosemite Valley, documentation regarding tribes associated with Yosemite shows that they also actively sowed certain plant propagules, most notably tobacco, by scattering seeds and scratching the ground with a stick in well-watered and burnt-over ground. People especially cultivated certain well-watered northern slopes or the inside of a burnt, rotten log to reduce moisture stress (Anderson 1988: 135; Barrett and Gifford 1933: 194). Accounts suggest that, through such practices, the tobacco grown in plots managed with these methods produced larger leaves and better flavor than untended or casually tended plots. Selective gathering and sowing enhanced these desired characteristics and provided valley residents with a regular supply of tobacco for use or trade (Anderson 1988: 160; Barrett and Gifford 1933:1 94).

Other methods and considerations

Tribal members today consistently stress the holistic nature of landscape management, and the ways in which management for one resource is inseparable from the management of whole ecosystems on the valley floor. As a consultant explained, "Native people did not manage one thing, they managed the whole landscape" (NPS 2016d: 3). Another consultant stressed the need to "look at the big picture. We gather the acorns, we hunt the deer. It's all connected" (NPS 2016b: 3). The various management techniques—gathering, hunting, and other ways of interacting with landscape, all contribute to ensuring the health of plants, animals, people, and the land (NPS 2014: 1; 2016d: 2).

Elders also report that, traditionally, the spiritual relationship and sense of interconnectedness between tribal members and culturally significant plant communities are also an essential part of sustaining the wellbeing of plants (NPS 2014: 1). Beyond mechanical management methods, harvesters sustain the relationship between plants and people through singing, dancing, and talking to plants and entire plant communities (NPS 2016b: 2). Traditional harvesters have understood that "giving thanks to the oak tree and other plants is essential for the continued health of the plant, and to the people who utilize the resource" (NFR 2019: 11). Even today, tribes hold acorn festivals to show thanks and ritually intervene to ensure the abundance of acorns. This is shaped, in part, by certain beliefs regarding the first acorn crop—for example, beliefs about how burning will affect fertility and about the importance of sharing some part of the harvest with animals (Deur 2007: 58; Moore 1985: 20; NPS 2014: 1). Each of these practices has certain material manifestations, such as leaving behind acorns for squirrels to cache and effectively replant, that unambiguously enhance plant output. Moreover, tribal members assert that showing such respects enhances plant output in ways that, as yet, defy the comprehension and methods of Western science.

The displacement of Native peoples and Native landscapes

The advent of Euro-American settlement in Yosemite Valley in the 1850s and 1860s greatly affected Native lifeways in Yosemite Valley. The removal of Native people from the valley and the suppression of fire and other traditional practices undermined both the integrity of the human communities and the biotic communities with which they were interdependent. Euro-Americans first entered Yosemite Valley in 1851 as part of the Mariposa Battalion, a militia unit formed to fight the Yosemite and Chowchilla Indians in the Mariposa War—one of many anti-Indian campaigns undertaken in the first years following the advent of the California Gold Rush and American occupation of California. The next several years were marked by interethnic hostilities, as an expanding population of Euro-American settlers entered the valley and lands adjacent. The astonishing scenery of the valley gained almost instant national attention. Tourism and settlement in Yosemite Valley began almost immediately after the first entry of the Mariposa Battalion. James Mason Hutchings had begun leading tourist parties into Yosemite Valley by 1855, while permanent Euro-American settlement of the valley began in 1859 (Greene 1987). The changes introduced by these developments quickly and dramatically impacted Native habitation patterns and traditional practices.

By 1864, a mere 13 years after the Mariposa Battalion first glimpsed the valley, U.S. Congress passed the Yosemite Park Act. This Act formally transferred Yosemite Valley and the Mariposa Grove of giant sequoias (*Sequoiadendron giganteum*)—including some of the largest trees on Earth—to the State of California under the management of a state board of commissioners. In 1890, the U.S. Congress designated the area surrounding Yosemite Valley and Mariposa Grove as a national park, initially administered by the U.S. Army. In 1906, a congressional act formally returned the valley and grove to the management of the federal government (Mancillas 2000: 5–7). Predating the development of the United States' National Park Service by a full ten years, this was a foundational park and served as a precedent for other national parks around the world.

These developments compelled many Native people to relocate among neighboring tribes. The NPS gradually consolidated those who remained into increasingly circumscribed areas with tightening limitations on who the agency permitted to remain (Mancillas 2000: 3). By the first decade of the twentieth century, Native families in Yosemite Valley were restricted to a single village site. When 1930s' Park Service regulations demanded reduced numbers of Indians in the park, the agency determined that some possessed a "moral right" to remain, and then removed the rest. The agency destroyed the one remaining village to allow for the construction of a park medical clinic, and relocated remaining residents to the "New Village," near the site of the historic village of *Wahogah* (Bibby 1994: 112; Greene 1987: 9). In 1953, the park adopted the Yosemite Indian Village Housing Policy, which called for the gradual eviction of all remaining Indians from the park (Mancillas 2000: 5–7). The NPS officially terminated the final Indian village in 1969, razing cabins in turn, as their residents lost their seasonal or full-time jobs in the park. Those who remained employed by the park were allowed temporary housing in other areas until they retired and were evicted—a process that continued through the 1990s. Even today, in communities surrounding the park, many tribal members recall residing in Yosemite Valley for decades of their younger lives (Bibby 1994: 113; George 2017; Spence 1999; Turek and Keller 1998).

Impacts on traditional management, cultural practices, and IEK

Alongside the physical displacement of Native peoples, traditional resource management was also limited or prevented—undermining myriad cultural, dietary, social, and economic practices of tribes while also negatively impacting the health and fecundity of the species under their care (Deur and James 2020). The traditional management practices described in this chapter entered a sharp decline immediately upon the entry of non-Native settlers and park managers into the valley. Under the period of state control (1864–1890), authorities sought to suppress fires and prohibited traditional burning; enforcement, however, was often disorganized and intermittent. Beginning in 1890, when Yosemite became a national park, federal troops served to aggressively suppress fires and police Native land managers under a host of new regulations (Rothman 2005: 16–19; Taylor 2006: 2). The last officially recorded instances of burning conducted by Yosemite's Native inhabitants occurred in the late nineteenth century (Ernst 1943: 59; YNP 2010: III-109). Meanwhile, official park reports documented the encroachment of brush and conifer trees, along with the loss of key vistas, as early as the 1870s as an outcome of emerging park restrictions on Native burning (Briggs 1882: 10–11; Gibbens and Heady 1964: 11). The Report of the Commissioners to Manage the Yosemite Valley and the Mariposa Big Tree Grove (1891–92: 6–7) detailed the changes to the valley following the suppression of anthropogenic burning and other traditional management practices:

> [T]he valley originally was a forest park, dotted with open meadows. Its Indian owners kept the floor clear of underbrush. It is known that besides the careful use of fire for this purpose they annually pulled up unnecessary shrubs and trees as soon as they sprouted. This protected the large trees from destruction by fire and left a free view of the walls, waterfalls, and beauties of the valley. Letting nature have her way in choking every vista with underbrush has obscured many of the finest views, has hastened the destruction of many fine old trees, especially the oaks, which, when crowded and starved by younger growth, yield to parasites and decay, and has increased the risk from fire.

In subsequent decades, writers and researchers continued to record the deteriorating landscape conditions characterized by diminishing meadows, conifer encroachment, deteriorating quantity and quality of native plant species, and obscured vistas (Ernst 1949; Gibbens and Heady 1964; Storer and Usinger 1963: 35). As Native harvesters sought to respond to these conditions, indications suggest that burning continued intermittently and on a small scale, beyond the gaze of park managers, in the less-developed areas on the edges of the valley into the early twentieth century (Gassaway 2005: 117; Deur 2007).

The redevelopment of the valley to accommodate tourist development only compounded these effects. In 1879, Galen Clark, Guardian for the State of California Yosemite Grant, blasted the El Capitan moraine in an attempt to reduce flooding, in order to facilitate visitor access to Yosemite Valley and enable agricultural use of the valley floor (Clark 1904: 396; Deur 2007: 52; Gibbens and Heady 1964: 15–17; Huber and Snyder 2007: 107–109). Historically, the moraine and the rock obstruction below Mirror Lake on Tenaya Creek served as rock barriers and contributed to the wetland hydrology of the valley floor (Deur 2007: 51–52; Matthes 1930). Researchers argue that the blasting of the moraine resulted in lowering the water table and, as a result, decreased seasonal flooding within the low riparian meadows (Huber and Snyder 2007: 107–109; Milestone 1978). Moreover, through much of the early twentieth century, park staff made other hydrological "improvements"—dredging and revetment of the Merced River, dredging sand from Mirror Lake, filling and developing wetlands, ditch construction in meadows to route pedestrian tourists, the removal of beaver dams and logjams, and other forms of flood control (Deur 2007: 52–53; Gibbens and Heady 1964: 5; Milestone 1978).

Together with fire suppression, these effects accelerated conifer encroachment into anthropogenic habitats of enduring concern to Native harvesters (Deur 2007). These changes diminished the size of riparian wetlands and seasonally-flooded meadows in Yosemite Valley and reduced the availability of culturally preferred species found disproportionately in these habitats (ibid.: 51–52). The lowered water table is cited as one of the primary explanations for the death of oak trees and the decrease in the number of acorns, as well as the encroachment of non-native invasive species into these increasingly dry meadow environments (NPS 2016c; Yosemite National Park 2011: 45). Riverine downcutting and bank hardening have also dramatically decreased the scale of alluvial sand deposits where the best fern and sedge roots for basketry grow (Deur 2007: 30, 53–54).

Several other practices associated with non-Native settlement, development, and tourism also contributed to the erosion of Yosemite Valley vegetation and soil. As early as 1851, with the entry of the Mariposa Battalion, livestock grazing began in Yosemite Valley. It steadily increased in the ensuing decades alongside the rise in settlers and, in time, park and concessionaires' livestock herds in support of tourist operations. Though grazing gradually declined as automobile use increased, and though it was discontinued in 1933, the practice had long-term impacts on the soil and vegetation of the valley. Livestock cropped, pulled, and trampled finer grasses; coarser ones remained and flourished. Compaction of the soil and subsoil within meadows obstructed the percolation of water, undermining the effects of traditional soil aeration by Native harvesters, causing soils to dry out faster (Gibbens and Heady 1964: 15). These changes created an environment favorable to the establishment of introduced species, as well as young conifers and woody shrubs in former meadows and riparian margins, while making it more difficult for certain culturally preferred native species to thrive (ibid.: 25).

Activities of park visitors have also contributed to the deterioration of certain plant species and the decline in traditional plant use in Yosemite Valley (Deur 2007: 56). The number of visitors to the park has increased exponentially over the decades. In 1906, the year that the federal government assumed management of Yosemite Valley, park managers recorded only 5,414 visitors. Visitation has increased almost inexorably ever since. By the 1990s, annual visitor numbers had already reached more than 4,000,000; the total exceeded 5,000,000 for the first time in 2016 (NPS 2019). Visitor foot traffic through meadows and oak groves has created damaging social trails, while picnicking, camping, and other activities have harmed native vegetation while also displacing remaining Native harvesters from time-honored gathering sites (Deur 2007: 53, 56; Gibbens and Heady 1964: 5, 15; Rothman 2005: 17). Visitors reportedly take acorns as souvenirs, which, over time, not only reduces available acorns for Native harvesters, but also reduces seedling recruitment in black oak groves (NPS 2016b). Similar small-scale picking or gathering of other plant materials by visitors also incrementally affects vegetation and reduces gathering opportunities for tribal members in heavily visited portions of the park. Moreover, tribal members suggest that the rising number of tourists and increasing park traffic have resulted in worsening pollution that adversely affects the health of gatherers, plants, and those who consume plants from the valley (NPS 2016a; 2016b; 2016c; Vasquez 2019: 33). Consultants describe concerns that the construction of the park infrastructure—such as road development, sewer lines, lawn maintenance, and possible soil contamination resulting from past park development—has also adversely impacted plant species (Anderson 2005: 321; NPS 2016d). Herbicides that the park sprays on invasive plant species, some harvesters suggest, also put gatherers at risk of health issues (Pfeiffer and Ortiz 2007).

Tribal members assert that the large number of visitors to Yosemite Valley results in crowds in and around traditional gathering areas. This undermines gathering and plant management activities that tribes traditionally carry out privately (Deur 2007: 15). Tribal consultants note that while some visitors are supportive of traditional gathering, they frequently ask questions, make comments, take photographs, or report activities to park employees when they are perceived to violate park rules (ibid.: 56). Visitors, in effect, continue to provide unwelcome surveillance and policing of Native activities, even when NPS policy has become more open to traditional management and harvests. Tribal members also commonly report feeling "like a tourist attraction" while trying to carry out culturally significant harvests in the valley. This is a challenge not only to the mechanics of plant gathering and management, but also undermines the ceremonial aspects of plant gathering, which are integral to the practice. In response, most tribal members who gather plants in Yosemite Valley make efforts to gather in locations not frequented by visitors, or at times when they are not present. As a result, the timing and locations of gathering have changed, becoming more dynamic and diffuse (Deur 2007; NPS 2016b, 2016c).

Native Elders generally agree that the effects on Yosemite Valley's anthropogenic landscapes have been dire. Black oak recruitment has diminished dramatically; undergrowth and competing species have quickly overtaken the groves; and riparian meadows, once characterized by a high degree of biodiversity and a mosaic of different plants, are increasingly supplanted by relatively uniform stands of conifers. In 2009, echoing the official findings of the park commissioners some 118 years prior, Yosemite National Park staff inventoried 181 scenic vistas and found that vegetation encroachment resulting from park policies had "completely obscured about one-third of the vistas, and partially obscured over half the vistas" (YNP 2010: vii). Consequently, the forcible elimination of Native harvesting and management from Yosemite Valley has had an adverse impact on more than Native peoples; the elimination of longstanding Native knowledge and practice from this landscape has served to erode the very same natural landscapes that park founders sought to protect through the creation of one of the world's most prominent national parks.

Yosemite's anthropogenic landscapes today

Today, certain keystone plant communities persist in diminished form on the landscape, such as in black oak groves, despite a century and a half of dramatic change. Such anthropogenic landscapes remain a potent locus of cultural meaning to modern tribes, and a focal point of ongoing National Park Service efforts to sustain these unique landscapes—managed with a growing recognition not only of their biological value, but also of their role as keystone cultural landscapes shaped by countless generations of Native knowledge-holders. The park currently conducts consultation and partners with seven tribes and tribal organizations with ancestral ties to Yosemite who continue to maintain connections to park lands and resources: the American Indian Council of Mariposa County (Southern Sierra Miwuk Nation); the North

Fork Mono Rancheria; the Tuolumne Band of Me-Wuk Indians; the Picayune Rancheria of Chukchansi Indians; the Mono Lake Kutzadika'a Tribe; the Bridgeport Paiute Indian Colony; and the Bishop Paiute Tribe. Together, these tribes and organizations continue to recount the Indigenous Ecological Knowledge of their ancestors and its implications for the past, present, and future management of Yosemite Valley landscapes. In recent consultation meetings and shared planning efforts, these descendants of Yosemite's original caretakers stress the importance of reintroducing the traditional management activities integral to caring for their ancestral land and maintaining the integrity of its natural and cultural landscapes (BIC 2019: 94–95; BPT 2019: 202; NFR 2019: 7; SSM-AICMC 2019).

The NPS has increasingly sought to incorporate Indigenous Ecological Knowledge and specific management practices (or their analogues) into modern land and resource management planning for Yosemite Valley. In recent years, the park has invited participation by tribal members in conducting prescribed burns, reintroducing regular fires to the valley floor (Kinoshita 2008). Within park natural resource planning, there has been growing and explicit acknowledgment of the role of Yosemite's traditional Native residents in the shaping of Yosemite Valley's landscape, vistas, habitats, and species composition. As such, management plans such as the Merced River Plan (YNP 2014) and the Yosemite Valley Scenic Vista Management Plan (YNP 2010) specifically acknowledge the contributions of Indigenous Ecological Knowledge to the landscape.

In this context, the authors have directed a multi-year effort to develop background documentation and monitoring protocols, addressing both cultural and culturally salient biological criteria in support of Merced River planning. As mandated by NPS policy and decisions in U.S. courts, Yosemite National Park must now assess California black oak as "ethnographic resources" that require protection and enhancement by virtue of their cultural significance. Together, we have developed protocols that incorporate Indigenous Ecological Knowledge and management strategies, using metrics to assess the integrity of these resources based on both biological and cultural criteria. As the NPS implements these protocols, Yosemite's anthropogenic plant communities will be increasingly monitored and brought into line with criteria reflecting the knowledge of many generations of Native land managers (Deur and Bloom 2018a; 2018b; 2018c).

In sum, Native peoples associated with Yosemite Valley have long practiced a range of traditional management techniques, contributing to a natural environment in which culturally preferred species thrived. Various methods such as burning, pruning and coppicing, clearing underbrush, weeding, selective harvesting, smoking, and "knocking" sustained traditional subsistence practices and fulfilled the caretaking responsibilities mandated by the Creator within Native oral tradition. Tribal members today explain that they "have been commissioned from the first time of our existence to be stewards of the land, knowing that this is how we will survive" (NFR 2019: 13). The techniques associated with this calling promoted the dominance of preferred species for use as food, medicines, and materials, improved the quantity and quality of species produced, produced open vistas, and decreased the threat of pests, pathogens, and conflagration. Indigenous Ecological Knowledge, and the management practices implied by this knowledge, created the park-like conditions of the valley as first encountered by Euro-Americans. This was no mere "wilderness" encountered by the likes of the Mariposa Battalion, but a fully humanized landscape—cultivated and under constant care —manifesting the long-held knowledge and the deepest cultural values of resident peoples, clearly inscribed upon the landscape.

The displacement of Native peoples has dramatically and adversely impacted both Native communities and the landscape of the valley and plant communities with which they are connected. In recent years, however, the National Park Service and the public at large have come to accept that the concept of a Yosemite "wilderness" was largely an invention of the Western mind. Conservation, as conceived by this classic model, was integral to the theme of Native displacement, as has been true in so many places around the world (Dowie 2009). To continue to embrace this fiction would be to court disaster in both cultural and natural domains.

Accordingly, the NPS has increasingly recognized the need to reintroduce traditional management values and methods to this contested landscape. Yosemite has begun incorporating the traditional practices of its first inhabitants into its management of the landscape, consulting with traditionally associated tribes, and welcoming their participation in unprecedented ways. In spite of this progress, the continuing decline in quantity and quality of keystone plant habitats and species indicates that more work lies ahead. Multidisciplinary research, cross-cultural dialogue, and management actions will be required if the anthropogenic landscapes of Yosemite Valley, and the people who long sustained them, will continue to thrive into the foreseeable future. The often painful lessons of the Yosemite experience provide a cautionary tale for other parks and protected areas that seek to remove or materially affect the activities of Native peoples within their boundaries in pursuit of conservation goals—especially when those Native peoples have been integral to the very natural order these parklands are meant to protect. So too, recent responses by the National Park Service, such as ambitious programs for tribal research and engagement, may be important touchstones for other parks and protected areas nationwide. Much as Yosemite National Park was pioneering as an early experiment in national park creation, so its modern engagement with tribes—carried out with varying degrees of success, with unusually high levels of national and international scrutiny—might also illuminate tentative new paths forward in the protection of the world's most spectacular and sensitive natural areas.

References

Anderson, M. K. 1988. *Southern Sierra Miwok Plant Resource Use and Management of the Yosemite Region: A Study of the Biological and Cultural Bases for Plant Gathering, Field Horticulture, and Anthropogenic Impacts on Sierra Vegetation.* Unpublished Master's thesis, Department of Wildland Resource Science, University of California, Berkeley.

Anderson, M. K. 1993. Indian fire-based management in the sequoia-mixed conifer forests of the Central and Southern Sierra Nevada. Report submitted to Yosemite Research Center, Yosemite National Park.

Anderson, M. K. 2005. *Tending the Wild: Native American Knowledge and the Management of California's Natural Resources.* Berkeley, CA: University of California Press.

Anderson, M. K. and F. K. Lake. 2013. California Indian ethnomycology and associated forest management. *Journal of Ethnobiology* 33(1): 33–85.

Anderson, M. K. and M. J. Moratto. 1996. Native American land-use practices and ecological impacts. In *Sierra Nevada Ecosystem Project: Final Report to Congress, Assessments and Scientific Basis for Management Options*, Vol. II. Davis:, CA: University of California, Centers for Water and Wildland Resources, pp. 187–206.

Anderson, M. K. and J. Rosenthal. 2015. An ethnobiological approach to reconstructing indigenous fire regimes in the foothill chaparral of the Western Sierra Nevada. *Journal of Ethnobiology* 35(1): 4–36.

Barrett, S. A. and E. W. Gifford. 1933. Miwok material culture: Indian life of the Yosemite region. *Bulletin of Milwaukee Public Museum* 2(4): 117–376.

Bates, C. 1997. Memo on traditional gathering areas in El Portal, Jan. 12. On file, Anthropology Office, RMS, Yosemite National Park, El Portal, CA.

Bibby, B. 1994. An ethnographic evaluation of Yosemite Valley: The Native American cultural landscape. Manuscript on file, Department of Anthropology, Yosemite National Park.

BIC (Bridgeport Indian Colony). 2019. Bridgeport Yosemite Paiutes: Who we are. In *Voices of the People: The Traditionally Associated Tribes of Yosemite National Park*. El Portal, CA: National Park Service, Yosemite National Park, pp. 177–213.

Bigelow, J., Jr., Maj. 1904. Report of the Acting Superintendent of Yosemite National Park in California to Secretary of the Interior. In *Annual Reports of the Department of the Interior for the Fiscal Year Ended*

June 30, 1904. (Box 979.447Y-11 ["Reports of Superintendents"], No Folder).Yosemite National Park Research Library.Washington, DC: Government Printing Office.

BPT (Bishop Paiute Tribe). 2019. Manahu. In *Voices of the People: The Traditionally Associated Tribes of Yosemite National Park*. El Portal, CA: National Park Service,Yosemite National Park, pp. 187–206.

Briggs, M. C. 1882. Report of the Commissioners to Manage the Yosemite Valley and the Mariposa Big Tree Grove. In *Biennial Report of the Commissioners to Manage the Yosemite Valley and the Mariposa Big Tree Grove, so extended as to include All Transactions of the Commission from April 19, 1880, to December 18, 1882*. Sacramento: State Office, J.D.Young, Supt. State Printing.

Champion, D. 1986. Disease threatens to kill all of Yosemite's evergreens. *San Francisco Chronicle*, 18 October.

Clark, G. 1894. Letter to the Hon. Board of Commissioners of the Yosemite Valley and Mariposa Big Trees Grove. Folder 68, Box 2, Craig Bates Collection. El Portal, CA: Yosemite Archives.

Clark, G. 1904. *Indians of the Yosemite Valley and Vicinity: Their History, Customs, and Traditions*.Yosemite Valley, California: G. Clark.

Commissioners to Manage the Yosemite Valley and the Mariposa Big Tree Grove. 1891–1892. Biennial Report of the Commissioners to Manage the Yosemite Valley and the Mariposa Big Tree Grove for the Years 1891–92. Reports of the Commissioners to Manage the Yosemite Valley and the Mariposa Big Tree Grove for the Years, Biennial Reports 1889-1904-1888. State Office, Sacramento. On File, Resources Management and Science Library, El Portal, CA.

Deur, D. 2007. *Yosemite National Park Traditional Use Study: Plant Use in Yosemite Valley and El Portal*. Pacific West Social Science Series No. 2007-001. Seattle, WA: USDI National Park Service.

Deur, D. and R. Bloom. 2018a. *A Review and Assessment of Ethnographic ORV Standards, Indicators, and Monitoring Protocols: Yosemite Valley, Yosemite National Park*. El Portal, CA, and Seattle, WA: USDI National Park Service,Yosemite National Park and Pacific Northwest CESU.

Deur, D. and R. Bloom. 2018b. *Yosemite Valley Traditional Use Plant Inventory and Threat Assessment: Yosemite Valley, Yosemite National Park*. El Portal, CA and Seattle, WA: USDI NPS, Yosemite National Park and Pacific Northwest CESU.

Deur, D. and R. Bloom. 2018c. *Black Oak Monitoring Protocols for the Merced Wild and Scenic River Ethnographic ORV*. El Portal, CA, and Seattle, WA: USDI NPS,Yosemite National Park and Pacific Northwest CESU.

Deur, D. and J. E. James, Jr. 2020. Cultivating the imagined wilderness: Contested Native American plant-gathering traditions in America's national parks. In N. J.Turner (Ed.), *Plants, People, and Places: The Roles of Ethnobotany and Ethnoecology in Indigenous Peoples' Land Rights in Canada and Beyond*. Montreal: McGill-Queens University Press.

Dowie, M. 2009. *Conservation Refugees: The Hundred-Year Conflict between Global Conservation and Native Peoples*. Cambridge, MA: MIT Press.

Ernst, E. 1943. Preliminary Report on the Study of the Meadows of Yosemite Valley. Unpublished Report. File No. 880-01.Yosemite National Park, CA. United States Department of the Interior, National Park Service.

Ernst, E. 1949.Vanishing meadows in Yosemite Valley. *Yosemite Nature Notes* 28(5): 34–40.

Gassaway, L. 2005. *Hujpu-St: Spatial and Temporal Patterns of Anthropogenic Fire in Yosemite Valley*. MA thesis, Anthropology/Archaeology, San Francisco State University, San Francisco, California.

Gassaway, L. 2009. Native American fire patterns in Yosemite Valley: Archaeology, dendrochronology, subsistence, and culture change in the Sierra Nevada. *SCA Proceedings* 22.

George, C. 2017. Decades after it was destroyed,Yosemite's last Native American village is returning. Fresno Bee, 15 December.

Gibbens, R. F. and H. F. Heady. 1964. The influence of modern man on the vegetation of Yosemite Valley. California Agricultural Experiment Station, Manual 36. University of California, Division of Agricultural Sciences.

Goode, R. 2014. Cultural burn.Tribal Chair North Fork Mono. Paper Presented to the Dinkey Collaborative, Dinkey Creek, California. On File,Yosemite National Park, RMS, Department of Anthropology.

Greene, L. W. 1987. *Yosemite: The Park and its Resources: A History of the Discovery, Management, and Physical Development of Yosemite National Park, California*. Denver, CO: National Park Service. Historic Resource Study series. Printed by Government Printing Office, Denver, 3 vols.

Grom, B. and C. Camp. 1959. Interview: Jack Leidig. Columbia College Oral History Series. Recorded 17 May 1959. Available at: http://apps.gocolumbia.edu/oralhistory/listen?96&t=yosemite&i=All&a=All&p=0,Yosemite National Park Archives, El Portal, CA. (accessed January 3, 2017).

Heady, H. F. and P. J. Zinke. 1978.Vegetational changes in Yosemite Valley. National Park Service Occasional Paper, no. 5. Washington, DC: Department of the Interior.

Huber, N. K. and J. Snyder. 2007. A history of the El Capitan moraine. In Yosemite Association, *Geological Ramblings in Yosemite*. California: Yosemite National Park, pp. 103–110.

Kinoshita, J. 2008. A legacy of fire: Looking back to look forward in Yosemite Valley. *Yosemite*, 70(2): 7–9.

Kuhn, B. and B. Johnson. 2008. Status and trends of black oaks (*Quercus kelloggii*) Populations and recruitment in Yosemite Valley (a.k.a. preserving Yosemite's oaks). Final report prepared for the Yosemite Fund. National Park Service, Yosemite National Park, Yosemite, CA. Available at: www.nps.gov/yose/learn/nature/upload/Status-Trends-Black-Oaks-2008.pdf, (accessed December 14, 2016).

Long, J. W., M. K. Anderson, L. Quinn-Davidson, R. W. Goode, F. K. Lake, and C. N. Skinner. 2016. Restoring California black oak ecosystems to promote tribal values and wildlife. General Technical Report. U.S. Department of Agriculture, Forest Service, Pacific Southwest Research Station, Albany, CA.

Mancillas, S. 2000. When the rocks came down. A history of the Indians of Yosemite Valley, 1916–1953. Unpublished Master of Arts thesis, Department of History, Sacramento State University, Sacramento.

Martin, G. 1996. Keepers of the oaks. *Discover Magazine* August. Craig Bates Collection (Box 2, Folder 68). El Portal, CA: Yosemite Archives.

Matthes, F. E. 1930. Geologic history of the Yosemite Valley. U. S. Geological Survey Professional Paper #160. USDI Geological Survey. Washington, DC: U.S. Government Printing Office..

McCarthy, H. n.d. Field guide to plants important to the Central Sierra Me-Wuk Indians with traditional uses, edited by T. Norton.

Milestone, J. F. 1978. The influence of modern man on the stream system of Yosemite Valley. Unpublished MA thesis. Department of Geography, San Francisco State University.

Moore, E. 1985. The eternal acorn: An ethnoarchaeological perspective. Unpublished Anthropology Paper for University Course (no university identified, possibly UCLA). On file, Anthropology Office, RMS. El Portal, CA: Yosemite Archives.

NFR (North Fork Rancheria of Mono Indians). 2019. Nikwa Nim (We are the people). In *Voices of the People: The Traditionally Associated Tribes of Yosemite National Park*. National Park Service, Yosemite National Park, pp. 1–19.

NPS (National Park Service). 2014. Notes taken at Yosemite Forum: Kat Anderson Talk, December 9, 2014. Yosemite National Park, California. On file in the Anthropology Office, RMS, Yosemite National Park.

NPS (National Park Service). 2016a. Meeting notes from Fourteenth Annual All-Tribes Meeting. Draft 1, June 30, 2016, Yosemite East Auditorium, Yosemite National Park, California. On file in the Anthropology Office, RMS, Yosemite National Park.

NPS (National Park Service). 2016b. Meeting notes from tribal meeting and site visit to discuss Black Oak Research Project with Dr. Douglas Deur. Draft 2, October 18, 2016 Yosemite East Auditorium, Yosemite National Park, California. On file in the Anthropology Office, RMS, Yosemite National Park.

NPS (National Park Service). 2016c. Meeting notes from Tribal Meeting to discuss Black Oak Research Project with Dr. Douglas Deur. Draft 2. October 21, 2016, American Indian Council of Mariposa County Headquarters, Mariposa, California. On file in the Anthropology Office, RMS, Yosemite National Park.

NPS (National Park Service). 2016d. Meeting notes from Tribal Meeting to discuss Black Oak Ethnographic Research Project with Dr. Douglas Deur. Draft 3. March 17, 2016, Yosemite Resources Management Science Building, El Portal, California. On file in the Anthropology Office, RMS, Yosemite National Park.

NPS (National Park Service). 2019. Yosemite NP (YOSE) Reports, Annual Park Recreation Visitation (1906-Last Calendar Year), National Park Service Visitor Use Statistics. Available at: https://irma.nps.gov/Stats/SSRSReports/Park%20Specific%20Reports/Annual%20Park%20Recreation%20Visitation%20(1904%20-%20Last%20Calendar%20Year)?Park=YOSE (accessed 25 October 2019).

Ortiz, B. R. 1991. *It Will Live Forever: Traditional Yosemite Indian Acorn Preparation*. Berkeley, CA: Heyday Books.

Pfeiffer, J. M. and E. H. Ortiz. 2007. Invasive plants impact California native plants used in traditional basketry. *Fremontia* 35(1): 7–13.

Reynolds, R. D. 1959. Effect of natural fires and aboriginal burning upon the forests of the Central Sierra Nevada. Master's thesis, Department of Geography, University of California, Berkeley.

Rothman, H. K. 2005. *A Test of Adversity and Strength: Wildland Fire in the National Park System*. Washington, DC: National Park Service, U.S. Department of the Interior.

Spence, M. D. 1999. *Dispossessing the Wilderness: Indian Removal and the Making of National Parks*. Oxford: Oxford University Press.

SSM-AICMC (Southern Sierra Miwuk Nation). 2019. Reflections on the past, visions for the future. In *Voices of the People: The Traditionally Associated Tribes of Yosemite National Park*. El Portal, CA: National Park Service, Yosemite National Park, pp. 21–51.

Stevens, M. 1998. The ethnobotany and distribution of white root (*Carex barbarae*). Folder 145, Box 3, Craig Bates Collection. El Portal, CA: Yosemite Archives.

Stewart, O. C. 2002. *Forgotten Fires: Native Americans and the Transient Wilderness*. Norman, OK: University of Oklahoma Press.

Storer, T. I. and R. L. Usinger. 1963. *Sierra Nevada Natural History: An Illustrated Handbook*. Berkeley, CA: The Regents of the University of California Press.

Stoy, W. H. 1890. Letter to the Honorable Secretary of the Interior. Biennial Report of the Commissioners to Manage the Yosemite Valley and the Mariposa Big Tree Grove for the Years 1889–90, 25–27. In Reports of the Commissioners to Manage the Yosemite Valley and the Mariposa Big Tree Grove for the Years, Biennial Reports 1889-1904-1888. Sacramento, CA: State Office.

Taylor, A. H. 2006. Fire History of Yosemite Valley. Final report for the Yosemite Fund through cooperative agreement 1443CA309701200 between the National Park Service and the Pennsylvania State University. On file. El Portal, CA: Yosemite RMS Library.

TBMI (Tuolumne Band of Me-Wuk Indians). 2019. Background and historic overview. In *Voices of the People: The Traditionally Associated Tribes of Yosemite National Park*. El Portal, CA: National Park Service, Yosemite National Park, pp. 147–185.

Turek, M. F. and R. H. Keller. 1998. *American Indians and National Parks*. Tucson, AZ: University of Arizona Press.

Vasquez, I. A. 2019. Evaluation of restoration techniques and management practices of Tule pertaining to eco-cultural Use. MS thesis, Humboldt State University.

West, L. 1986. The demise of Yosemite Valley's evergreens. *Yosemite*, 48(4): 2.

Wickstrom, C. K. R. 1987. Issues Concerning Native American use of fire: A literature review. Yosemite Research Center, Yosemite National Park, National Park Service, U.S. Department of the Interior.

YNP (Yosemite National Park). 2010. Scenic Vista Management Plan for Yosemite National Park. Environmental Assessment, July 2010. Yosemite National Park, National Park Service, U.S. Department of the Interior. Available at: www.nps.gov/yose/learn/management/upload/SVMP_YOSE_EA.pdf, (accessed 29 October 2019).

YNP (Yosemite National Park). 2011. 2010 Assessment of Meadows in the Merced River Corridor, Yosemite National Park. Resources Management and Science, Yosemite National Park, National Park Service, Department of the Interior.

YNP (Yosemite National Park). 2014. Merced Wild and Scenic River Final Comprehensive Management Plan and Environmental Impact Statement. National Park Service, U.S. Department of the Interior.

PART IV

Governance and equity

24

WHO BENEFITS?

Indigenous Environmental Knowledge (IEK) in multilateral biodiversity agreements

Wendy Jackson and Phil Lyver

Introduction

Indigenous Peoples and local communities (IPLCs)[1] globally play an important role in the protection and conservation of biodiversity. It is widely recognized that Indigenous Environmental Knowledge (IEK) systems reflect a deep knowledge of local biodiversity, and provide valuable insights and potential responses to the current global biodiversity crisis (Gómez-Baggethun, Corbera, and Reyes-Garcia 2013; IPBES 2013; Brondizio and Le Tourneau 2016). IPLCs frequently inhabit sparsely populated areas that are often hotspots of biodiversity and therefore possess IEK critical to understanding complex ecological processes and changes, e.g., relationships between wildlife populations and climate change (Mistry and Berardi 2016). Moreover, IEK may complement what is known through modern science, contribute to knowledge gaps, and interact with modern science to co-produce new knowledge (Bohensky and Maru 2011; Tengö et al. 2017).

It is estimated that IPLCs hold as much as two-thirds of the world's land area under customary community-based tenure systems, although governments recognize the rights of IPLCs only to a small proportion of these lands (Alden Wily 2011; Rights and Resources Initiative 2015). As a result, land degradation and intensification, forced evictions from traditional lands, poaching and over-utilization, urbanization of populations, and the effects of climate change continue to erode biological and cultural diversity (Maffi and Woodley 2012; Mowforth 2014; Tauli-Corpuz 2016). The loss of IEK associated with many of these processes, and declining connections to the natural environment, pose a significant issue for humanity. Loss of knowledge weakens the capacity of societies to respond to current and future environmental crises. Therefore, supporting the connection between IPLCs and their environments is as important as their engagement and contributions to confront ecological challenges.

Coordinated efforts from governments, non-governmental organizations, and multilateral instruments to halt or reverse biodiversity loss are vital. However, the impact of conservation policies and action on IPLCs needs to be evaluated ahead of implementation. Conservation policies and statements of intent can limit IPLCs from using land and resources according to their traditions, thereby decoupling or unduly harming the integral relationships IPLCs have with the environment (Lyver and Tylianakis 2017). As such, the development and implementation

DOI: 10.4324/9781315270845-28

of conservation authorities should be thoroughly examined with regard to their potential to divorce IPLCs from their lands and environments. If well designed, conservation policies and approaches may be able to engage IPLCs in problem identification and solutions, weave IEK into decision-making, and extend the spectrum of response options to critical social-environmental challenges.

The goals of this review are to survey one conservation authority, multilateral instruments, for indications of their capacity to interact with, and support IPLCs. We will also present case studies that illustrate the diverse manifestations of IEK within multilateral instruments. We examine the extent to which multilateral instruments have the approaches and procedures in place to effectively engage IEK, and whether or not IPLCs gain benefit from engaging with multilateral instruments. For the purposes of this review, we use Berkes' (2012) definition of Traditional Ecological Knowledge to define IEK as: "[the] dynamic body bodies of social-ecological knowledge, practice and belief, evolving by adaptive processes, grounded in territory, intergenerational and cultural transmission, about the relationship of living beings (including humans) with one another and with their environment". Such knowledge is characterized by diversity, local governance, contrasting world-views, historical-cultural contexts, knowledge protection and transmission mechanisms, and multi-scalar dimensions (Thaman et al. 2013).

Multilateral instrument processes for engaging IEK holders

To identify how IEK is integrated and engaged in multilateral processes, six biodiversity-related instruments were examined:

1. Convention on Wetlands of International Importance, especially as Waterfowl Habitat (Ramsar)
2. Convention Concerning the Protection of the World Cultural and Natural Heritage (WHC)
3. Convention on International Trade in Endangered Species of Wild Fauna and Flora (CITES)
4. Convention on the Conservation of Migratory Species of Wild Animals (CMS)
5. Convention on Biological Diversity (CBD)
6. Convention to Combat Desertification (CCD).

The primary objectives of the instruments are summarized in Table 24.1.

These instruments were chosen for two reasons: (1) all are recognized as the primary global agreements that constitute international law and policy on biodiversity;[2] and (2) examining numerous instruments creates a wider spectrum of analysis, providing more comparative examples of successes and challenges in engaging IEK and its holders.

An overview of the multiple pathways for engaging IEK in multilateral instruments is provided in Table 24.2. The mechanisms are listed along with an indication of their capacity to engage IEK and its holders.

1. *Little or no capacity* – no mechanism, or mechanism does not specify IEK or its holders.
2. *Potential capacity* – mechanisms exist, but provision for engagement is a peripheral aspect (e.g., brief mention of IPLCs in an otherwise comprehensive resolution).
3. *Evident capacity* – mechanisms exist, and engagement is mandated through direct and specific provisions.

Table 24.1 Primary objectives of six biodiversity-related multilateral instruments reviewed in this study

Instrument	Primary objective(s)
Convention on Wetlands of International Importance (Ramsar)	Conservation and wise use of wetlands and their resources
World Heritage Convention (WHC)	Prevent the loss, through deterioration or disappearance, of cultural and natural heritage of outstanding universal value
Convention on International Trade in Endangered Species (CITES)	Monitor and regulate trade in endangered species to ensure their long-term survival in the wild
Convention on the Conservation of Migratory Species (CMS)	Promote the conservation and effective management of migratory species of wild animals
Convention on Biological Diversity (CBD)	The conservation of biological diversity, the sustainable use of its components and the fair and equitable sharing of the benefits arising out of the utilization of genetic resources
Convention to Combat Desertification (CCD)	Combat desertification and mitigate the effects of drought in countries experiencing serious drought and/or desertification … with a view to contributing to the achievement of sustainable development in affected areas

Table 24.2 Multilateral instruments: overall view of capacity to engage IEK and its holders

Instruments	Ramsar	WHC	CITES	CMS	CBD	CCD
Entry into force	1975	1975	1975	1983	1992	1996
Number of parties (Feb. 2018)	169	193	183	126	196	197
Convention text	X	X	X	X *	+	+
Resolutions/decisions	+	X	+	+	++	+
Guidelines	+	+	+	+	++	X
Working groups	++	X	X	X	++	+
Strategic plans/frameworks	+	+	X	+	++	+
Evidence of meeting/workshop participation	+	+	+	+	++	+
Other mechanisms (e.g., national committees, multi-disciplinary panels, national implementation, etc.)	+	X	X	X	+	+

X = little or no capacity; + = potential capacity; ++ = evident capacity
* CMS itself does not contain provisions, although some "daughter" agreements do.

Table 24.2 also provides additional information about the instruments (entry into force and number of signatories).

Basic convention texts are one of the most fundamental methods for multilateral instruments to recognize a commitment to IPLCs. Convention text is legally binding for contracting parties, and therefore represents an obligation to establish approaches and mechanisms for engaging IEK and its holders. Of the six instruments examined, only two have such references to traditional knowledge, Indigenous Knowledge, or IPLCs: CBD (Articles 8, 10, 17, 18); and CCD (Articles 16, 17, 18).

All instruments have developed pathways for engaging IEK via resolutions, decisions, or guidelines. These mechanisms tend to be legally non-binding and hortatory in nature. Nonetheless, they can be significant in establishing global norms of implementation, as well as global best practice. Examples of these mechanisms include: Ramsar Convention's resolution VII.8 (*Guidelines for Establishing And Strengthening Local Communities' and Indigenous People's Participation in the Management of Wetlands*; Ramsar Secretariat 1999a); and CMS's *Guidelines on the Integration of Migratory Species into National Biodiversity Strategies and Action Plans (NBSAPs)*; CMS 2011) All but one agreement included reference to IEK (and its holders) in core documents such as strategic plans, e.g., the Strategic Plan for Biodiversity 2011–2020 (CBD 2010a; United Nations General Assembly 2011; see target 18); and the CMS Strategic Plan 2015–2023 (CMS 2014; see target 14); and the Ramsar Strategic Plan for 2016–2024 (Ramsar Secretariat 2015; see target 10). Half of the instruments have had working groups, task forces, or similar bodies tasked with addressing IEK or its holders in a direct manner. Similarly, IEK holders are able to participate in some meetings, workshops, and other events organized by the agreements. However, the rules of procedure governing the nature and extent of participation vary, depending on the instrument. IEK holders may be included on government delegations for official meetings, but it is difficult to ascertain the nature and extent of participation because IEK holders are generally not identified as such in meeting participant lists. Indeed, IEK and its holders may be provided the most opportunities to engage in multilateral instruments as they are implemented at the national level (see Tengö et al. 2017, for comments on "scale-crossing" actors).

IEK and multilateral instruments: a range of experiences

Given the variation among multilateral instruments in terms of their capacity to engage with IEK and its holders, it is no surprise that the experience of the holders varies from positive to negative. We view this diversity through three inter-related 'lenses': (1) opportunities for participation in multilateral instruments; (2) effective use of IEK in the multilateral instruments; and (3) benefits for IPLCs from multilateral instruments.

Opportunities for participation of IPLCs in multilateral instruments

Across the instruments, there was substantial variation in opportunities for IEK holders to participate. At the global scale, the CBD has the strongest mechanism, with a dedicated working group (known as the Working Group on Implementation of Article 8(j)). Specific activity related to Article 8(j) started in 1998; with meetings held regularly since then. IEK holders are active participants in the Working Group, whose decisions cover a wide range of issues relevant to biodiversity and IEK (called indigenous and local knowledge in the CBD system). Specific aspects of IEK are picked up in Article 8(j) workstreams, such as those on customary sustainable use and repatriation of indigenous and traditional knowledge. The Ramsar Convention has a Culture Working Group that can provide for IEK, but it is not a dedicated mechanism like the Article 8(j) Working Group. Participation of IEK holders in this Ramsar process is neither clear nor evident.

An attempt to create a similar body in the World Heritage system was unsuccessful. In late 2000, IPLCs from Australia, Canada, and New Zealand submitted a proposal to the World Heritage Committee to establish a World Heritage Indigenous Peoples Council of Experts (WHIPCOE) to: (1) serve as a network; (2) allow Indigenous Peoples' voices to be heard in protecting and promoting the world's natural and cultural heritage; (2) bring complementary Indigenous competencies and expertise; and (4) support best practice management and, upon request, make recommendations for improvements (*Report of the 25th Session of the WHC*; UNESCO 2001). The WHIPCOE proposal was discussed at the 2001 World Heritage Committee meeting, and was not approved. Reasons given for rejecting the proposal include: (1) perceptions that establishment of such a committee would challenge the sovereignty of States Parties; (2) unclear definitions of 'indigenous'; and (3) the ability of Indigenous Peoples to engage via other means, such as national delegations (Meskell 2013). The long-term impact of this failed initiative is not clear. The issue has not disappeared, and IPLC representatives active in the WHC regime have continuously advocated for a meaningful role for IPLCs (IWGIA 2012; Meskell 2013).

Where mechanisms do not exist at the global level, opportunities may exist at the domestic level. For example, management of World Heritage sites is a national or local-level activity, which may involve co-management by IPLCs (World Heritage Centre 2012). The Tongariro National Park in New Zealand was the first property to be inscribed on the World Heritage list under revised criteria describing cultural landscapes (World Heritage Convention 2015): "The mountains at the heart of the park have cultural and religious significance for the Māori people and symbolize the spiritual links between this community and its environment" (UNESCO 2018). With the cultural values of the property specified in the designation – and therefore a critical element of New Zealand's international obligations – any park or site management plan must ensure that management policies do not adversely impact on these values. In the case of Tongariro National Park, IEK is purported to be an integral part of the role and functions of management, including for "decision-making processes for use of cultural materials, the reintroduction of previously-present bird species, the consideration of concessions which may impact on cultural values or the development of further park guidelines or strategies" (Department of Conservation 2006: 41). IEK engagement in park administration has not been explicit, despite its role in risk management from volcanic activity in the park (see Gabrielsen et al. 2017). IEK was identified in recent Treaty settlement documents between the government and one of the local iwi linked to the Tongariro National Park (see Government of NZ 2018). The Deed of Settlement, which explicitly describes IEK linking one local iwi with physical, cultural, and spiritual elements of the land, should provide a solid basis for future IEK and IPLC engagement in management of the World Heritage site.

Effective use of IEK in multilateral instruments

Instruments with weak pathways for IEK do not necessarily prevent use of this knowledge in decision-making processes. CITES meetings provide a good example of the use and application of IEK in global negotiations. In 2013, the Inuit Tapiriit Kanatami attended CITES CoP16 to lobby against a US/Russia proposal to strengthen regulation of trade in polar bear (*Ursus maritimus*) specimens. IEK presented by the Inuit Tapiriit Kanatami suggested the conservation status of polar bears did not merit additional regulation, and moreover, the proposed protection would have adverse impacts on local livelihoods (Inuit Tapiriit Kanatami 2013). The Inuit Tapiriit Kanatami was active in the lead-up to and at the meeting, providing information to delegations and the public, and engaging with media. The proposal to strengthen regulation of trade in polar bear went to a vote, and did not receive enough support to pass. Ascribing rejection of the proposal to recognition of Inuit IEK would be speculative: there was alignment between the Inuit Tapiriit Kanatami and some countries who asserted the species did not meet the biological criteria for stronger regulations. However, other countries claimed that climate change is the primary threat to polar bears (cf. international trade), which may have been a bigger influence in countries voting against the proposal. Nonetheless, IEK was highly visible in the proposal and the ensuing debate (IISD-RS, 8 March 2013).

Effective use of IEK in the Ramsar system may be possible through their culture and wetlands work. In 2005, Ramsar Parties adopted resolution IX.21, recognizing wetlands as places where local communities and Indigenous Peoples have developed strong cultural connections and sustainable use practices (Ramsar Secretariat 2005). This resolution established the Ramsar Culture Working Group, which has evolved into the Ramsar Culture Network. In June 2019, Terms of Reference for the Network specify that one of the purposes of the Network is to encourage "normative alignment with international principles regarding cultural diversity, human rights and the rights of indigenous peoples and local communities as expressed in international law" (see Annex 3 of Ramsar Secretariat, 2019). A bio-cultural diversity thematic group has been established that specifies Indigenous Knowledge as it relates to wetlands, though it is too new to identify any successes in terms of use of IEK.[3] The Ramsar Secretariat has also published this resource "Learning from Experience: How indigenous peoples and local communities contribute to wetland conservation in Asia and Oceania".[4]

Benefits for IPLCs from engagement in multilateral instruments

IPLCs experience the *impacts* of decisions emerging from these instruments: however, a critical question is whether or not IEK holders derive any *benefit* from engagement with these instruments, or if they want to participate so as to derive benefit.

For example, there is only scant evidence (see Heinämäki et al. 2015) that guidelines produced by the CBD and relevant to IEK and its holders have had uptake; the three sets of guidelines are: *Akwé:Kon Voluntary Guidelines for the Conduct of Cultural, Environmental and Social Impact Assessments* (CBD 2004); the *Mo'otz Kuxtal Voluntary Guidelines on the Prior and Informed Consent of Indigenous Peoples and Local Communities for Accessing Their Knowledge, Innovations and Practices and for Fair and Equitable Sharing of Benefits* (CBD 2016); and the *Tkarihwaié:ri Code of Ethical Conduct to Ensure Respect for the Cultural and Intellectual Heritage of Indigenous and Local Communities* (CBD 2010b).

It may be that national- or local-level engagement provides a discernible positive impact for IEK holders. For example, the Upper Navua Conservation Area in Fiji – a Ramsar site – is located on land owned by traditional families. An eco-tourism company has leased the land, and the company expressly includes the families ("indigenous peoples") in their mission and operations (Rivers Fiji 2010; Bricker and Kerstetter 2016). Information on the Ramsar Secretariat website about the Upper Navua Conservation Area (UNCA) indicates that, "Villagers' centuries-old traditional knowledge of the river and its systems is seen as the foundation for the long-term preservation and sustainable use of the river and near-river resources" (Ramsar Secretariat 2014). Benefits for Indigenous local Fijians linked to the establishment of "white-water tourism" have been purported to include the facilitation of increased protection of the forest ecosystem from illegal logging, increased local employment, integration of traditional practices into community projects, philanthropic efforts, and community initiatives (e.g., environmental education programmes; Bricker and Kerstetter 2016).

Conversely, the East Rennell site in the Solomon Islands provides an example of where IEK holders have been visibly and vocally dissatisfied with World Heritage processes. This site was inscribed in 1998, and there was considerable interest by the international community in having the site listed because it was to be the first based on natural criteria under customary ownership and management (Smith 2011). It was observed that the East Rennell traditional communities had been supportive of World Heritage listing at the time of inscription, as they had expected small-scale eco-tourism activity to provide livelihood benefits (ibid.). However, such benefits did not emerge – in part because of civil unrest in the Solomon Islands between 1998 and 2005 (ibid.). It was suggested that a more likely cause was poor implementation of the WHC (ibid.: 596) by the Solomon Islands government, as well as by the provincial government. The issues were then exacerbated by a lack of accurate understanding of the WHC and UNESCO, and what World Heritage listing means for sites and communities. Challenges with the site continue, although there are efforts underway – which include the customary owners – to address them.[5]

There is an increasing awareness of the effects that the application of decisions from multilateral instruments has had on IPLC livelihoods (e.g., GS/OAS and Secretariat of CITES 2015a). While standards, certifications, and measures to address and mitigate potential effects have been considered, and in some instances applied (GS/OAS and Secretariat of CITES 2015b), the role and direct support for the protection and application IEK within these measures remains uncertain. The second of three pillars for CITES's standards for the sustainable use of species focuses on "[the] resource's social sustainability, wherein it is expected that trade generates benefits for poor rural communities, thereby improving their quality of life and strengthening their livelihoods while simultaneously respecting their cultural norms and traditional uses and practices". Assumptions are made that principles guiding the application of decisions by multilateral instruments will protect the human-environment relationships that contribute to maintaining the integrity of IEK. For example, changes in the behaviour of former American crocodile (*Crocodylus acutus*) hunters in response to CITES criteria for conservation have led to the recovery of regional populations in Colombia (Ulloa-Delgado and Sierra-Diaz 2012). Multilateral instrument interventions have also resulted in IPLCs contributing their IEK as part of alternate use opportunities such as ecotourism operations (ibid.; Bricker and Kerstetter 2016). The effectiveness of these programmes in augmenting IEK, however, is often not clear.

Discussion

This survey of multilateral instruments has led the authors to two conclusions. The first is there are very few examples of visible or credible engagement of IEK and its holders in the selected instruments. Despite a collective 25+ years working across these and other instruments (at multiple levels and in various capacities), we were unable to identify a sufficient sample size of useful and relevant case studies illustrating engagement of IEK and its holders. Second, where examples do exist, the experience of IEK holders is not necessarily positive.

In our view, one of the primary barriers of IEK engagement is the fundamental structure of the multilateral system. With these instruments designed to operate in a *state*-based system, the pathways of engagement are not straightforward for *non-state* actors (see, for example, Corell and Betsill 2008). For IPLCs (and others), the instruments can be impenetrable (Witter et al. 2015) or minimally productive (Marion Suiseeya 2014). Mauro and Hardison (2000) have described the challenge of IPLC engagement in multilateral processes, noting that historically IPLCs have sought to engage as both "sovereign" and "nations". However, the state-based (and state-run) system has not been equipped to engage with IPLCs on this basis, and the response has been to recognize IPLCs through discrete instruments peripherally related to global biodiversity policy.[6]

A related barrier may be the scale and applicability of commitments. Melick et al. (2012) have examined the CBD Aichi Biodiversity Targets and their applicability in Papua New Guinea. The authors point specifically to Aichi target 11 on protected areas.[7] It was noted that the focus of the target is "perceived as an alien concept with little relevance for many cultures" (ibid.: 344). Similarly, even where a target or process is deemed as relevant, implementation and benefits from implementation may not be understood in the same way. This was evident in the case of the East Rennell World Heritage site, arguably a case where instruments may be "mysterious and alienating" (Hill et al. 2011: 572), causing "deep frustration" (*Indigenous Peoples and the Ramsar Convention*; Ramsar Secretariat 1999b), in a system characterized by overall complexity of obligations (Chasek 2010: 9).

Another barrier that has been identified in other contexts is the "persistence of epistemological differences" (Mistry and Berardi 2016: 1275) between science and IEK. Although Intergovernmental Science-Policy Platform on Biodiversity and Ecosystem Services (IPBES) is not covered in this chapter, an analysis of its *modus operandi* has found that the way knowledge, meaning, and concepts are communicated (language, nomenclature, taxonomies, etc.) cannot be easily translated (Thaman et al. 2013). Furthermore, the authors of the report point to social structures that underpin IEK, where individuals may hold certain "domains" of knowledge, and access to that knowledge is regulated by particular customary protocols and rules (ibid.). These protocols and rules are not likely to be understood in a state-based system, which may have very different protocols for information access. Moreover, critiques suggest if IEK is removed from within its cultural context, it becomes something else, losing its relational identity, power, and effectiveness (Agrawal 1995; Nadasdy 1999).

This observation is linked to a related challenge associated with engagement of IEK and its holders: i.e. an underlying assumption (of the state-based system, and of Western science) that engagement is a one-way relationship based on the utility of IEK for the wider system. Reimerson observed:

> The inclusion of indigenous subjects seems to be based mainly on the possible contributions they can make to the objectives of the convention and the work of the state parties to the convention, not on their possible rights as peoples to self-determination and collective rights to land, water, and natural resources, or as possible parties in their own right.
>
> *(2013: 1004)*

While IEK has had clear utility in some situations (e.g., polar bear conservation measures; management of the Upper Navua Ramsar site), these cases exist in a longer historical context – often specified by IEK holders. An examination of written submissions by IPLCs to the CBD on matters related to Article 8(j) revealed that technical information was often supplemented by strategic content that made the connection between the CBD and wider processes intended to address issues such as Indigenous rights. Indeed, engagement between multilateral instruments and IEK holders without the critical context of historical injustices hinders the efficacy of that relationship.

This raises questions about the overall context of engaging IEK and its holders. In our view, it may be that the goal of multilateral instruments should not be to integrate IEK *per se*. As noted earlier and observed by Nadasdy (1999), integration may not be effective or even possible given the asymmetric power issues between holders and the state. We emphasize a specific goal for multilateral instruments should be to take a systems approach and implement policies and approaches that first secure and support the relationship of IPLCs with their traditional lands and biodiversity. This works towards ensuring benefits for IPLCs, and provides a more equitable space for knowledge exchange. As instruments evolve, there will be opportunities to create more sophisticated approaches and collaborative agendas for engaging IEK and its holders (e.g., IPBES's use of Mother Earth terminology, see Díaz et al. 2015). This evolution can be expedited through focused initiatives by the instruments to facilitate meaningful engagement by IEK holders, which in turn may provide more opportunities for "co-production of knowledge" (Berkes 2009, 151) or "multiple evidence-based approaches" for connecting diverse knowledge systems (Tengö et al. 2014). Thoughtfully designed multilateral instruments may provide mechanisms that support IPLCs to protect their lifestyles and apply their knowledge to the biodiversity crisis and accompanying socio-economic changes, albeit within their own cultural contexts.

Notes

1 This terminology is generally used by the multilateral biodiversity agreements following decision XII/12 taken at the 12th Conference of the Parties to the Convention on Biological Diversity. See Section F of www.cbd.int/doc/decisions/cop-12/cop-12-dec-12-en.pdf.
2 Other international processes exist (such as those under the International Union for Conservation of Nature, IUCN), but they have a stronger role in creating norms as opposed to establishing formal commitments.
3 See /www.ramsar.org/activities/bio-cultural-diversity.
4 See www.ramsar.org/fr/learning-from-experience-how-indigenous-peoples-and-local-communities-contribute-to-wetland.
5 See http://whc.unesco.org/en/news/1719, "Ministries and customary owners discuss future East Rennell World Heritage site".
6 For example: International Labour Organization Conventions 107 and 169 on Indigenous and Tribal Populations, adopted in 1957 and 1989, respectively; Declaration and International Convention on the Elimination of Any Form of Racial Discrimination, adopted in 1963; the UN Permanent Forum on Indigenous Issues, established in 2000; and the UN Declaration on the Rights of Indigenous Peoples, adopted in 2007.
7 See www.cbd.int/sp/targets/rationale/target-11/.

References

Agrawal, A. 1995. Dismantling the divide between indigenous and scientific knowledge. *Development and Change,* 26: 413–439.
Alden Wily, L. 2011. *The Tragedy of Public Lands: The Fate of the Commons Under Global Commercial Pressure.* Rome: International Land Coalition.
Berkes, F. 2009. Indigenous ways of knowing and the study of environmental change. *Journal of the Royal Society of New Zealand* 29(4): 151–156. http://dx.doi.org/10.1080/03014220909510568
Berkes, F. 2012. *Sacred Ecology,* 3rd edn. New York: Routledge.
Bohensky, E. L. and Y. Maru. 2011. Indigenous knowledge, science, and resilience: What have we learned from a decade of international literature on "integration"? *Ecology and Society* 16(4): 6. http://dx.doi.org/10.5751/ES-04342-160406
Bricker, K. S. and D. L. Kerstetter. 2016. Ecotourism and environmental management: A case study of a partnership for conservation. In P. Modica and M. Uysal (eds), *Sustainable Island Tourism: Competitiveness and Quality of Life.* Wallingford, UK: CABI, pp. 19–34.
Brondizio, E. and F-M. Le Tourneau. 2016. Environmental governance for all. *Science* 352: 1272–1273. DOI:10.1126/science.aaf5122.
Chasek, P. S. 2010. *Confronting Environmental Treaty Implementation Challenges in the Pacific Islands.* Pacific Islands Policy 6. Honolulu, HI: East West Center.
CBD (Convention on Biological Diversity). 2004. Akwé:Kon Voluntary Guidelines for the Conduct of Cultural, Environmental and Social Impact Assessment regarding Developments Proposed to Take Place on, or which are Likely to Impact on, Sacred Sites and on Lands and Waters Traditionally Occupied or Used by Indigenous and Local Communities. Montreal, Canada: Secretariat of the Convention on Biological Diversity.

CBD (Convention on Biological Diversity). 2010a. Strategic Plan for Biodiversity 2011–2020 and the Aichi Targets. Montreal, Canada: Secretariat of the Convention on Biological Diversity.

CBD (Convention on Biological Diversity). 2010b. Tkarihwaié:ri Code of Ethical Conduct to Ensure Respect for the Cultural and Intellectual Heritage of Indigenous and Local Communities. Montreal, Canada: Secretariat of the Convention on Biological Diversity.

CBD (Convention on Biological Diversity). 2016. Mo'otz Kuxtal Voluntary Guidelines for the Development of Mechanisms, Legislation or Other Appropriate Initiatives to Ensure the "Prior and Informed Consent", "Free, Prior and Informed Consent" or "Approval and Involvement", Depending on National Circumstances, of Indigenous Peoples and Local Communities for Accessing Their Knowledge, Innovations and Practices, for Fair and Equitable Sharing of Benefits Arising from the Use of Their Knowledge, Innovations and Practices Relevant for the Conservation and Sustainable Use of Biological Diversity, and for Reporting and Preventing Unlawful Appropriation of Traditional Knowledge. Montreal, Canada, Secretariat of the Convention on Biological Diversity. Decision XIII/18. Adopted by the 13th meeting of the Conference of the Parties.

CMS 2011. Guidelines on the Integration of Migratory Species into National Biodiversity Strategies and Action Plans (NBSAPs) and Other Outcomes from CBD COP10. UNEP/CMS/Resolution 10.18. Adopted by the Conference of the Parties at its Tenth Meeting, held in Bergen, Norway, 20–25 November 2011.

Corell, E. and M. Betsill. 2008. Analytical framework: Assessing the influence of NGO diplomats. In M. Betsill and E. Corell (eds), *NGO Diplomacy: The Influence of Nongovernmental Organizations in International Environmental Negotiations*. Cambridge, MA: MIT Press, pp. 19–42.

Department of Conservation 2006. *Tongariro National Park Management Plan: 2006–2016*. Turangi, New Zealand: Department of Conservation.

Díaz, Sandra et al. 2015. The IPBES Conceptual Framework: Connecting nature and people. *Current Opinion in Environmental Sustainability* 14: 1–16.

Gabrielsen, H., J. Procter, H. Rainforth, T. Black, G. Harmsworth, and N. Pardo. 2017. Reflections from an indigenous community on volcanic event management, communications and resilience. In *Advances in Volcanology*. Heidelberg: Springer.

General Secretariat of the Organization of American States / Secretariat of the Convention on International Trade in Endangered Species of Wild Fauna and Flora. 2015a. *Handbook on CITES and Livelihoods. Part 1: How to rapidly assess the effects of the application of CITES decisions on livelihoods in poor rural communities*. Washington: GS/OAS.

General Secretariat of the Organization of American States / Secretariat of the Convention on International Trade in Endangered Species of Wild Fauna and Flora. 2015b. *Handbook on CITES and Livelihoods. Part II: Addressing and mitigating the effects of the application of CITES decisions on livelihoods in poor rural communities*. Washington: GS/OAS.

Gómez-Baggethun, E., E. Corbera, and V. Reyes-Garcia. 2013. Traditional ecological knowledge and global environmental change: Research findings and policy implications. *Ecology and Society* 18(4): 72. http://dx.doi.org/10.5751/ES-06288-180472.

Government of NZ 2018. Ngāti Rangi and the Trustees of Te Tōtarahoe o Paerangi and the Crown: Rukutia Te Mana Deed of Settlement of Historical Claims. Available at: www.govt.nz/dmsdocument/7037.pdf (accessed 10 March 2018).

GS/OAS and Secretariat of CITES 2015. *Handbook on CITES and Livelihoods*. Washington, DC: General Secretariat of the Organization of American States/Secretariat of the Convention on International Trade in Endangered Species of Wild Fauna and Flora.

Heinämäki, L., T. M. Herrmann, and A. Neumann. 2015. Protection of the Culturally and spiritually important landscapes of Arctic Indigenous Peoples under the Convention on Biological Diversity and first experiences from the application of the Akwe:Kon Guidelines in Finland. In *The Yearbook of Polar Law VI*. Leiden: Brill, pp. 189–225.

Hill, R., L. C. Cullen-Unsworth, L. D. Talbot, and S. McIntyre-Tamwoy. 2011. Empowering Indigenous peoples' biocultural diversity through World Heritage cultural landscapes: A case study from the Australian humid tropical forests. *International Journal of Heritage Studies* 17(6): 571. http://dx.doi.org/10.1080/13527258.2011.618252.

IPBES 2013. Report of the international expert workshop on the contribution of indigenous and local knowledge systems to the Platform. Document IPBES/2/INF/1. Bonn, Germany: Intergovernmental Science-Policy Platform on Biodiversity and Ecosystem Services.

International Institute for Sustainable Development Reporting Services (IISD-RS) 2013. *CITES COP16 Highlights*, 21(78). Available at: http://enb.iisd.org/vol21/enb2178e.html. (accessed 18 March 2018).

IWGIA (International Work Group for Indigenous Affairs) 2012. International Expert Workshop on the World Heritage Convention and Indigenous Peoples. 21 September. Copenhagen, Denmark.

Inuit Tapiriit Kanatami. 2013. Polar bears, harvesting, and Inuit. See www.youtube.com/watch?v=oshrlHBrVRc (accessed 1 April 2017).

Lyver, P. O'B. and J. M. Tylianakis. 2017. Indigenous peoples: Conservation paradox. *Science* 357(6347): 142–143.

Maffi, L. and E. Woodley. 2012. *Biocultural Diversity Conservation: A Global Sourcebook*. London: Earthscan.

Marion Suiseeya, K. R. 2014. Negotiating the Nagoya Protocol: Indigenous demands for justice. *Global Environmental Politics* 14(3): 102–124.

Mauro, F. and P. D. Hardison. 2000. Traditional knowledge of indigenous and local communities: International debate and policy initiatives. *Ecological Applications* 10(5): 1263–1269.

Melick, D. R., J. P. Kinch, and H. Govan. 2012. How global biodiversity targets risk becoming counterproductive: The case of Papua New Guinea. *Conservation and Society* 10(4): 344–353.

Meskell, L. 2013. UNESCO and the fate of the World Heritage Indigenous Peoples Council of Experts (WHIPCOE). *International Journal of Cultural Property* 20: 155–174.

Mistry, J. and A. Berardi. 2016. Bridging indigenous and scientific knowledge. *Science* 352(6291): 1274–1275.

Mowforth, M. 2014. *The Violence of Development. Resource Depletion, Environmental Crises and Human Rights Abuses in Central America*. London: Pluto Press.

Nadasdy, P. 1999. The politics of TEK: Power and the "integration" of knowledge. *Arctic Anthropology* 36: 1–18.

Ramsar Secretariat 1999a. Guidelines for establishing and strengthening local communities' and indigenous peoples' participation in the management of wetlands. Adopted by the 7th Meeting of the Conference of the Contracting Parties to the Convention on Wetlands, held in San José, Costa Rica, 10–18 May 1999.

Ramsar Secretariat 1999b. Indigenous Peoples and the Ramsar Convention. Statement on Proposals about the Resolution Project No. 8, submitted at the Regional Workshop on Biodiversity Conservation and Traditional Knowledge in Relationship to the Implementing of Article 8(j) of the Biological Diversity Convention [sic]. See www.iisd.ca/ramsar/cop7/chiapas-ramsar-e.html.

Ramsar Secretariat 2005. Taking into account the cultural values of wetlands. Adopted by the 9th Meeting of the Conference of the Parties to the Ramsar Convention on Wetlands, held in Kampala, Uganda, 8–15 November 2005.

Ramsar Secretariat 2008. *Culture and Wetlands: A Ramsar Guidance Document*. Gland, Switzerland: Ramsar Secretariat.

Ramsar Secretariat 2014. *Upper Navua Conservation Area*. See www.ramsar.org/upper-navua-conservation-area (accessed 1 April 2017).

Ramsar Secretariat 2015. *Fourth Ramsar Strategic Plan 2016–2024*. Gland, Switzerland: Ramsar Secretariat.

Ramsar Secretariat 2019. Report of the Chair of the Scientific and Technical Review Panel, including draft work plan for 2019-2021. Gland, Switzerland: Ramsar Secretariat.

Reimerson, E. 2013. Between nature and culture: Exploring space for indigenous agency in the Convention on Biological Diversity. *Environmental Politics* 22(6): 992–1009.

Rights and Resources Initiative 2015. *Who owns the world's land? A global baseline of formally recognized indigenous and community land rights*. Washington, DC: RRI.

Rivers Fiji 2010. About us. See www.riversfiji.com/about-us. (accessed 1 April 2017).

Smith, A. 2011. East Rennell World Heritage Site: Misunderstandings, inconsistencies and opportunities in the implementation of the World Heritage Convention in the Pacific Islands. *International Journal of Heritage Studies* 17(6): 592–607. http://dx.doi.org/10.1080/13527258.2011.618253.

Tauli-Corpuz, V. 2016. Rights of Indigenous Peoples: Report of the Special Rapporteur of the Human Rights Council on the rights of indigenous peoples. United Nations General Assembly, A/71/229. New York: UN.

Tengö, M., E. S. Brondizio, T. Elmqvist, P. Malmer, and M. Spierenburg. 2014. Connecting diverse knowledge systems for enhanced ecosystem governance: the multiple evidence base approach. *Ambio* 43: 579–591.

Tengö, M., R. Hill, P. Malmer, C. M. Raymond, M. Spierenburg, et al. 2017. Weaving knowledge systems in IPBES, CBD and beyond: Lessons learned for sustainability. *Current Opinion in Environmental Sustainability* 26–27: 17–25. http://dx.doi.org/10.1016/j.cosust.2016.12.005.

Thaman, R., P .O'B. Lyver, R. Mpande, E. Perez, J. Cariño, and K. Takeuchi (eds) 2013. *The contribution of indigenous and local knowledge systems to IPBES: Building synergies with science.* IPBES Expert Meeting Report, Paris, France: UNESCO.

Ulloa-Delgado, G. and Sierra-Díaz, C. 2012. Conservation project for *Crocodylus acutus* of the Cispatá Bay with the participation of local communities in the municipality of San Antero-Department of Córdoba, Colombian Caribbean. Córdoba, Colombia: Regional Autonomous Corporation of Valleys of the Sinú and San Jorge CVS.

UNESCO 2001. Report of the 25th session of the WHC. Document CONF 208 XV.1–5. Paris, France: UNESCO.

UNESCO 2018. World Heritage List – Tongariro National Park. Case study: Tongariro. Available at: http://whc.unesco.org/en/activities/613/ (accessed 3 March 2018).

United Nations General Assembly 2011. Convention on Biological Diversity. Decision adopted by the United Nations General Assembly on 11 March 2011. UN document A/RES/65/161. New York: UN.

Witter, R., K. R. M. Suiseeya, R. L. Gruby, S. Hitchner, E. M. Maclin, et al. 2015. Moments of influence in global environmental governance. *Environmental Politics* 24(6): 894–912. DOI:10.1080/09644016.2015.1060036.

World Heritage Centre 2012. Managing Natural World Heritage. Paris, France: UNESCO.

World Heritage Convention 2015. Tongariro National Park. Available at: http://whc.unesco.org/en/list/421 (accessed 1 April 2017).

25

THE USE AND MISUSE OF IEK IN CONSERVATION IN VIETNAM

Pamela McElwee

Introduction

Environmental conservation projects have increasingly turned to Indigenous Environmental Knowledge (IEK) to inform practice, and IEK is often presented as a win-win solution to preserving nature and recognizing Indigenous Peoples' rights and contributions. However, less attention has been paid to failures to include IEK in conservation, including cases where highly selective readings of IEK have been used to justify conservation interventions that may actually harm local peoples by restricting their access to traditionally-used landscapes. This has been in the case in several conservation projects in Vietnam, where local Indigenous communities and their related knowledge practices have been acknowledged only partially, and in only highly selective ways, and the communities have experienced exclusion from protected areas as a result.

In Vietnam, the concepts of Indigenous Knowledge (IK) (*tri thức bản địa*), local knowledge (LK) (*tri thức địa phương*), traditional knowledge (TK) (*kiến thức cổ truyền*) and traditional environmental knowledge (TEK) (*kiến thức truyền thống sinh thái*) have been increasingly acknowledged, particularly given that over 13 percent of the country's population are identified as "ethnic minorities."[1] However, little work has assessed the contributions of these communities and their IEK to practical concerns of environmental management in Vietnam, aside from a few studies of ethnobotany (Hoàng Văn Sâm et al.. 2008; Whitney et al. 2016). This lack of attention to IEK is particularly notable within Western-funded conservation projects in biodiversity-rich areas of Vietnam. Projects have selectively deployed the idea of IEK in an essentialized form, if they have paid attention to it at all, such as focusing primarily on taxonomic naming practices or natural resources management found in customary law, with little attempt to understand the epistemological grounding from which IEK developed. Further, political restrictions on ethnic minority organizing and mobilization, along with government policies against and media denigration of many cultural practices, have already contributed to a weakening of IEK (McElwee 2004). The end result has been unsuccessful approaches to conservation that have confirmed exclusionary practices to keep communities out of resources protected with "fences and fines".

In this chapter I analyze several conservation initiatives in the central area of Vietnam, one of the most biologically and ethnically diverse areas of the country. The data come from fieldwork

DOI: 10.4324/9781315270845-29

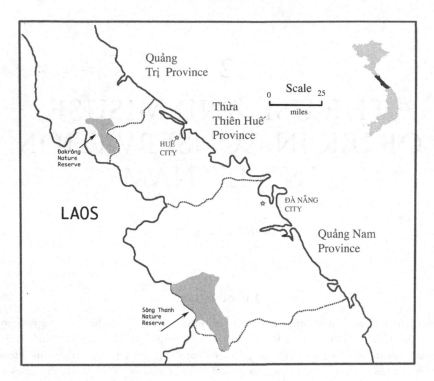

Figure 25.1 Location of Dakrông and Sông Thanh Nature Reserves

conducted in a 2004–2005 postdoctoral research project on environmental change in the Annamite highlands (primarily in Quảng Trị, Thừa Thiên Huế, and Quảng Nam Provinces) looking at the environmental practices of villages who shared similar forested environments. The methods for the research included focus group discussions, household surveys, key informant interviews, and participatory landscape walks in two nature reserves, the Sông Thanh Nature Reserve in western Quảng Nam and the Dakrông Nature Reserve in western Quảng Trị (see Figure 25.1). In both nature reserves, all forested areas falling within boundaries were no longer legally or freely accessible to local populations, despite being some of their ancestral lands, and increasing ranger activity was common at that time to begin to enforce these exclusionary rights.

Conservation projects in Vietnam

The Central Annamites of Vietnam have been a site of significant conservation interest in recent years, largely the result of the discovery in the 1990s of several new mammals previously undescribed by science, such as the saola (*Pseudoryx nghetinhensis*), the giant muntjac (*Megamuntiacus vuquangensis*), the Truong Son muntjac (*Muntiacus truongsonensis*), the Annamite Striped Rabbit (*Nesolagus timminsi*), and rediscovery of the Heude's warty pig (*Sus bucculentus*) (Sterling and Hurley 2005). Scientists involved in the species discoveries have posited that Vietnam's mountainous center was a "biotically unique region where primitive taxa, long extinct elsewhere, have been able to survive into the late 20th century" (Groves et al. 1997).

As a result of the attention to these endemic species discoveries and concerns that Vietnam's wildlife populations needed conservation protection from threats such as hunting and agricultural expansion, the World Wildlife Fund (WWF) named the Annamites a "Global 200 Ecoregion," a designation establishing it as one of the top 238 conservation sites in the world and a priority for expanded funding and conservation projects (Balzer et al. 2001). This process, however, was largely driven by expert scientists, rather than any local people, beginning the process of excluding IEK. The Global 200 project involved "over 1000 biogeographers, taxonomists, conservation biologists, and ecologists from around the world" (Olson et al. 2001, pp. 933–934), and in the case of the Indochinese region, local and Indigenous People were specifically not included. Attendees at a conference in Phnom Penh in 2000, where ecoregional maps were produced to determine where conservation projects were to be placed in the Annamites, included more than 80 people, the large majority of whom were Westerners and scientists. Only 13 Vietnamese, all of whom were either university or government biologists or working for a conservation NGO, attended (Baltzer et al. 2001).

More than $200 million in aid to conservation of wild animals and protected areas was pledged to Vietnam by foreign donors in the decade after these discoveries (1995–2005). With this money, the Ministry of Agriculture and Rural Development's Forest Protection Division (*Kiểm Lâm*) established several new protected areas in the Annamites in particular. Many of these areas were formerly owned and managed by provincial State Forest Enterprises, para-statal logging companies common throughout the socialist era in Vietnam (McElwee 2016). Once they were turned into nature reserves, there was supposed to be no more extractive activities of any kind, including no logging, no agriculture, no hunting, and no forest product collecting by local communities (McElwee 2002). In conjunction with the government demarcation of protected areas, foreign NGOs and donors used their financial influence to help set up management practices at these sites, including providing salaries for rangers, GIS mapping, and some alternative development activities. The most active organizations in the Annamites were the NGOs World Wildlife Fund (WWF) and Birdlife International, with financing for this work from the Global Environment Facility (GEF), USAID, US Fish and Wildlife Service, and the MacArthur Foundation. Several of these initiatives in the Annamites included:

- a WWF plan called the "Central Annamites Initiative" to establish ecoregional planning in the Annamites, and to "create a partnership of a broad range of stakeholders – from local communities to international organizations – working together to secure biodiversity conservation and sustainable development" (Nguyễn Lâm Thành 2003; Tordoff et al. 2003; Villemain et al. 2003). The project funded the development of a *Trường Sơn Conservation Action Plan*, a plan adopted by the Vietnamese government to coordinate conservation in the Annamites in which a series of new parks, some joined with protected areas on the other side of the border of Laos and Cambodia, would be located. The work also included threat assessments for two parks, the "paper park" Dakrông and a proposed Sông Thanh Nature Reserve (Lê Nho Năm et al. 2003), eventually leading to new management plans and funding to operationalize the reserves (Lê Trọng Trảiand Richardson 1999; Lê Nho Năm et al. 2004).
- a WWF project called "Green Corridor: conserving global conservation targets in a productive landscape," with funding from GEF to protect and maintain forests of Thừa Thiên Huế Province through conservation zonation plans and "participatory conservation agreements" (Averyanov et al. 2006).
- a WWF project, "Management of Strategic Areas for Integrated Conservation" (MOSIAC), to "protect the forests and rivers of global conservation significance" in

Nam Province, including developing the capacity of local communities and authorities to plan for, manage, and monitor natural resources to benefit conservation and
sustainable development (Hardcastle 2002; Dudley 2006).

• a Birdlife International project, "Conservation of Important Bird Areas in
  Indochina: Strengthening Site Support Groups to Conserve Critical Biodiversity,"
  with funding from the MacArthur Foundation, to improve local monitoring and
  involvement in protected areas, including Dakrông Nature Reserve in the central
  Annamites (Birdlife International in Indochina 2006).

Despite the different scales of the conservation projects and the organizations implementing
them, all these projects were aimed at increasing the enforcement of boundaries of nature
reserves, primarily through reducing the dependency of local communities on lands now
in parks and encouraging communities to respect and contribute to enforcement efforts.
The fundamental approach was to get local people to participate and "buy into" the idea
that they could no longer use protected areas for production or hunting. In order to achieve
this goal, the activities promoted included "awareness raising" of biodiversity conservation
through environmental education, posters, community meetings, and other formats, along
with some models for household economic development to substitute for income lost to the
park demarcation.

Despite this plethora of conservation projects, there was at the time a surprising lack of
attention to the social conditions under which these conservation projects began to operate, and
very little information about the communities affected. Throughout the time in which these
reserves were being set up, there was little focus on what human populations were using these
forest resources for and how they might be affected by conservation efforts. In addition to being
biologically diverse, much of the central Annamites has traditionally been home to several ethnic
minorities, including the Bru-Vân Kiều, Pa Cô (also known as Pa Hy), Tà Ôi, and Kà Tu. In the
gazetting of the new nature reserves, many of these communities were severely restricted in
terms of lands they could use for swidden cultivation for agriculture, as well experiencing loss of
access to the collection and use of forest products like timber, fuelwood, palms and rattans, and
wild animals.

The Central Annamites Initiative did commission several reports on the socioeconomics of
the area in the early 2000s; however, the consultants spent little time in local communities and
relied heavily on the collection of provincial and district data on economics and production,
rather than undertaking research on the livelihoods of the (mostly) ethnic minorities of the
conservation areas (Baker et al. 2000; Nguyễn Lâm Thành 2003; Villemain et al. 2003). It was
very common to see pejorative statements and generalizations in these reports when discussing
local populations; for example, one report noted:

> Income generation activities are varied, depending on the ethnic group and on a
> specific area. But most activities are simple, decentralized, self-serving and without
> bearing the hallmark of true goods production. Goods exchange is conducted using
> the barter system, and the equivalent price system is governed by definite unwritten
> law. The people tend to spend more than they earn, particularly on festivals and on
> worship. All of this contributes to starvation each year.
>
> *(Nguyễn Lâm Thành 2003: 36)*

Further, even when local knowledge was given attention in the early conservation planning
documents, the reports prepared gave the impression that while minority groups might have

had important IEK and local rules on resources use in the past, years of cooperatives and state decision-making, including policies to abolish "superstitious" practices, had rendered most minority traditions either obsolete or ineffectual and thus of little use to conservation practice (Huỳnh Thu Ba et al. 2003: 31).

Cherry picking IK/TEK for conservation

I turn now to examine how one particular conservation project approached IEK in its work around the newly established Sông Thanh nature reserve as a key site for biodiversity conservation. The project prepared a "Conservation Needs Assessment" to assess the role of the Indigenous People in the area, who were primarily from the Kà Tu ethnic group. The project noted that there was a need to focus on tasks such as establishing a "community outreach and conservation education team" at the park; to establish "Village Protection Teams" to get local communities, not just rangers, to enforce the park's borders; and to increase land allocation to communities in buffer zones, given that most valuable forests were now included in the park (Lê Nho Năm et al. 2003; Dudley 2006).

One approach to community outreach and protection of borders centered on getting villages to agree to what the project called "community '*hương ước*' agreements." The term *hương ước* comes from Vietnamese and means a village or community compact or set of customary laws. Supporting Vietnamese villages to readopt or rewrite traditional *hương ước* became a new central state policy in the 1990s and early 2000s to deal with increasing local social unrest and unhappiness with government corruption (Woodside 1999). Such *hương ước* were often written to include a mix of "party directives, state laws and local customs" (Vasavakul 1999). This interest in *hương ước* was extended to ethnic minority communities, with the assumption that similar modes of customary law must also exist, although there was little analysis of "how decisions are made and how a group's political structures are integrated with or in competition with Kinh-communist forms of political and legal organization and political culture" (Lempert 2001). The conservation projects in the central Annamites made the same assumption – that customary law must exist, and it could be codified – without assessing if Kà Tu customary law might differ from the form and content of a *hương ước* that derived from lowland Kinh (ethnic Vietnamese) cultures and communities.

To collect the data necessary to formalize these "community '*hương ước*' agreements", teams of park rangers were sent out to villages around the Sông Thanh Nature Reserve to collect basic information on "some main traditions" of each village. Responses on the forms were shared with me by rangers, and included answers like "at weddings, we have to have gongs, beef, and pigs" and "we have buffalo sacrifices at rice harvest time." (One sentence was often the extent of the information collected, as the survey form had a very small box in which to record the "main traditions.") The ranger teams then asked village heads about "any traditional laws still existing in the village." Again, responses were very basic, and included such statements as: "we have traditional laws on unwed mothers, where the village will investigate those who did wrong and require them to wed the girl." Other questions asked who the local "customary law" applied to, how it was applied, and how it was passed on and preserved.

The questions then moved to the practicalities of the customary laws as applied to forest protection, a key goal for the conservation project. The teams asked if there were "areas in the village or in the forest that from the past until now were protected and not encroached upon." Most Kà Tu respondents to the survey reported no to this question. In later discussions with Kà Tu villages, I found that communities did indeed have several areas of forests that they actively conserved. The reason they replied no in the survey was because the wording of the question implied the forests that were protected could not be used in any way, and the forests that the Kà Tu protected were used for burials, for harvest of certain ritual products, and for other social uses. But these did not fall within the purview of the survey done by the project, whose questions aimed at trying to find equivalents to "nature reserves" in Kà Tu traditions.

A final question on the survey asked, "if the traditional laws will continue to exist in the future," and at this point the disjuncture between the idealized systems of IEK that the project wanted to catalogue, preserve, and use, and the reality of the Vietnamese state's activities toward minorities in recent years was clear. One Kà Tu headman responded to the question: "They still will exist, but we've had to get rid of the traditional system of marriages, reduce the buffalo sacrifices, and only have gong ceremonies with pig and beef when we have the ability to do so." Even in the short response of the headman, it was plain to read the ironic fact that such customary practices that the project was asking about had been discouraged by the state in recent years. Getting rid of the objectionable parts of Indigenous Knowledge, including social systems of cross-cousin and arranged marriages as well as cere-monies involving sacrifices and belief in spirits, through national laws on marriage and the discouragement of certain ritual practices have been widespread in Vietnam since the socialist era first began (McElwee 2004).

The conservation project not only asked about traditions and customs that could be used for conservation, but also attempted to judge the scientific understanding of communities and their beliefs about forest management that might be compatible with the project's goals. The final section of the local survey done by rangers in Sông Thanh included a series of questions that were supposed to be asked of the village headmen and one or two others in the village to gauge the extent of local IEK and support for conservation activities. Ten statements were given, to which respondents were to agree or disagree:

1. Reducing the area of forest will reduce the diversity of species of animals living there.
2. Living next to the forest brings people many benefits.
3. The laws on protecting forests are equally applied to everyone.
4. Bats and birds help the forest recover after it is cut down.
5. If everyone understood about the problems deforestation causes then they wouldn't cut the forest.
6. There are no longer any tigers near the forest because they went to live someplace else.
7. Reducing the area of forest will reduce the number and kinds of animals that live there.
8. I understand the forest protection law and it has meaning for me and my family.
9. If I owned an area of forest I would cut it and use it for some other purpose.
10. We need to transfer our forests into a protected area.

Such statements tapped into a very limited perception of IEK. Only knowledge that might fit into Western science's view of what was important in conservation was asked about (i.e. whether or not communities recognized the importance of bat/bird migration for the dis-persal of seeds), or else agreement was sought between local and project views of conservation (i.e. following forest laws and making forests into protected areas). Yet there were numerous examples of IEK among the Kà Tu that were not tapped into in any of these quick surveys, because they did not fit into the predetermined boxes that the organizations had decided would be useful for conservation, which were mainly *customary laws* (narrowly defined as *hương ước*) and *locally protected forest areas*.

However, some of the most important aspects of IEK in Kà Tu communities fell outside of these neat boxes, and instead continued to involve beliefs and rituals regarding the spirit world, some of which may have practical utilitarian benefits for conservation and some which are less clearly beneficial. However, because beliefs in spirits seems less tangible and directly conservation-oriented than "customary law," attention to religious and spiritual practices has been virtually non-existent in conservation-oriented IEK studies and surveys in Vietnam. I note below one major area in which understanding of the spirit worlds that coexist with and inform the environments in which Kà Tu communities live are necessary to truly understand the realm of IEK: beliefs about spirits relating to animals and hunting practices.

Assessing IEK about animals and hunting in the Annamites

Hunting and forest product collecting have traditionally played an important role in the lives of all the Indigenous peoples in the Annamites. Among the Kà Tu, for example, hunting has long played a very important role not only in supplying meat for protein, but also in providing a way to express prowess and masculinity (Hickey 1993: 133). Hunting trophies collected by men would be displayed prominently in the communal house of the village (known as a *guơl*): as US researchers noted in the 1960s

> It is thought that the presence of buffalo skulls promotes the fertility of the land and prosperity of the village. According to Kà Tu belief, animals have a soul; thus, the buffalo's soul stays near the communal house where its skull is hung. In addition to buffalo skulls, the skulls and tails of wild animals are hung in the communal house.
>
> *(Schrock 1966: 363)*

Hunting trophies and visual markers for culturally important animals (buffalos, deer, and snakes chief among them) continue to decorate the outside and inside of modern *guơl* as well, many of which were rebuilt in the 2000s after a relaxation of state prohibitions of ethnic minority community spaces and longhouses (McElwee 2004). Figure 25.2 shows a reconstructed *guơl* rebuilt with state permission in 2002.

Souls of animals would reside in the communal house if their skulls were hung there (see Figure 25.3), and the spirits of ancestors who had died good deaths (that is, natural ones in old age) were also considered to be resident within the communal house, watching "over their descendants, protecting them from danger in the forest by warning them when evil spirits are nearby" (Schrock 1966: 364). Kà Tu were particularly afraid of spirits of people who had died a "bad" death, such as being devoured by a wild animal like a tiger, and it was necessary to ask good spirits to help guard against these bad spirits, who might bring misfortune or calamity. Such warning signs from good spirits might include

> peacock eggs in a path; a large tree uprooted across a trail; and the call, from the left side of the path, of a bird nesting in reeds. The evil spirits also have visible forms such as a tiger, a cobra hissing in the afternoon, and the flood waters causing a person to drown.
>
> *(ibid.: 364)*

Kà Tu customs traditionally placed great emphasis on the avoidance of these bad spirits, leading to a pantheon of omens and taboos, many of which continue to be followed today. Hornbills flying in certain directions, cobras crossing one's path, and encountering certain types of trees and plants as one goes into the forest are all taken as signs from deceased ancestors watching over the living. These omens led to specific animal taboos which the community would never hunt,

Figure 25.2 Kà Tu communal house (*gươl*), Tabhing commune, 2005

Figure 25.3 Animal skulls hanging inside a Kà Tu communal house, 2005

Figure 25.4   Painting of a saola in a Kà Tu communal house, Tabhing commune, 2005

including hornbills and many types of snakes. Taboo plants were still recognized among many Kà Tu communities as well, including staghorn ferns, which may not be disturbed. Understanding these links between good and bad spirits, their earthly manifestations, and how hunting or collection of plants would be an invitation for the bad spirits to descend is an example of the type of IEK that could have been collected for conservation purposes, but was not.

Rather, the project encouraged as part of their "conservation awareness raising" activities that Kà Tu communities should paint pictures of wild animals in their communal house, instead of hanging hunting trophies (Figures 25.4 and 25.5). In an ironic twist, the Kà Tu in Tabhing commune were given photos of endangered animals, like the saola pictured, to paint in their communal house rather than hang hunting trophies like boar skulls. However, in reality, the saola had at that time not been seen in south Nam Dong district by Kà Tu or by biologists, and this area was presumably outside its natural range. Nonetheless, the Kà Tu agreed to paint a picture of an animal *which they had never encountered* in the deeply symbolic spiritual space of their communal house, all the while suffering from restrictions on collecting important trophies of locally common and non-endangered animals like deer or wild pig.

Understanding the spirit world and the complicated series of omens and taboos is necessary to understanding why groups like the Kà Tu have community rituals that often involve the sacrifice of domesticated, rather than wild, animals. Nancy Costello, who has worked with Kà Tu in Vietnam and in Laos, has written:

> The whole fabric of Kà Tu society is intermeshed with the environment. The Kà Tu must live in harmony with the world around them, which includes other people, animals, birds, trees, stones, water, traditions and the many spirits. When this harmony

Figure 25.5 Picture of a hunting party painted inside a *guơl*, Tabhing commune, 2005

is disrupted, though the breaking of taboos and traditions, which displeases the many spirits, the correct relationship must be restored through ritual and sacrifice.

(*Costello 2003*)

When omens were not followed, or bad luck befell a village, domesticated animals such as buffalos were often sacrificed, along with common hunted animals, such as wild pigs. Such sacrifices were also made at times of plenty, such as after a good harvest, to thank the spirits for their blessings.

Yet conservation organizations who conducted research on hunting in the Annamites and its conservation consequences paid little attention to putting hunting in a larger cultural context; indeed, it was often biologists who conducted hunting surveys (Lê Trọng Trải et al. 2003), and they were often not aware of cultural factors in hunting. In fact, social practices of both hunting and of domesticated animal sacrifices were actively discouraged by the conservation organizations, who were opposed to hunting of nearly all kinds, even of common non-endangered animals like wild pigs that often destroy agricultural crops, if not kept in check. A total hunting ban within the communities surrounding Sông Thanh was being applied across the board when I visited in 2005, particularly in response to what the ranger stations reported to be breaches of regulations on wild animal protection. Some 23 people had been caught in violation of the hunting law, although statistics kept by the management board did not distinguish between local violators and those who came from afar to poach, and 309 kg of wild animals were confiscated in the previous year as part of the new interdiction efforts, which indicated to rangers that local communities were hunting in violation of park laws.

Yet according to a brief survey of 30 Kà Tu households done by students I was supervising in a research project, 70 percent of Kà Tu surveyed did not hunt or trap wild animals at all, and of the 30 percent who did hunt, none of the households hunted for the market, but for subsistence and mostly for crop protection. The main animals hunted were wild boar, deer, monkey, porcupine, civet, rat, and coucals, which were all predators of the crops in their swidden fields, and none of which were endangered or threatened species. Yet these practices of protecting fields were assumed to have as much impact on biodiversity as outsider poachers who came in and targeted highly endangered wild animals like tigers, elephants, or primates, as the across-the-board hunting restrictions made no distinction between the types of animals hunted and who collected them.

Further, the impact of hunting restrictions at the Sông Thanh was combined with laws forbidding the raising of buffalo by villages living along the park boundary, the concern being that buffalo were being left to graze freely in the park, which damaged park flora. This meant that Kà Tu had been unable to perform any sacrificial ceremonies for the past few years and had to substitute pigs or even chickens instead, something many Kà Tu regretted and which elders said was leading to a "loss of culture" among their youth. The changes in practices can be contrasted to Kà Tu on the other side of the Lao border; there, Yves Goudineau notes there has been a "revival of extravagant sacrifices of buffalos" thanks to increasing prosperity and lenient laws in Laos (Goudineau 2003). Such ritual revival is not, however, occurring on the Vietnam side in the research sites visited, largely as a result of conservation restrictions and declining prosperity as agricultural fields have been restricted to areas outside of protected forests.

Conclusion

As noted above, for some conservation organizations in Vietnam, the main role that Indigenous People can play as knowledge producers is to provide insight into customs that might mirror current conservation practice, such as areas of forest that might have been traditionally protected from use. These communities are rarely seen as having any other forms of IEK that might be useful for conservation, and certainly ideas about beliefs, rituals, and spirits have not been used in conservation practice. Yet it is clear that much of the IEK held by Kà Tu communities in the forms of omens or taboos might indeed indicate a particular attention to species numbers and populations and could be used as the basis for monitoring and for conservation (Colding and Folke 2001). After all, omens are so important that it would not be easy for someone to ignore a sighting of a cobra or to forget the number of times hornbills had flown over the village in recent years. Such knowledge could easily be tapped into in order to monitor animal populations in conservation assessments. Nearly every community with which I have worked in the Annamites could provide a relative sense of population numbers of common and uncommon fauna, as well as relative increases or decreases in these populations; similar knowledge has been documented in other areas of Vietnam (Cano and Tellaria 2013). However, these local communities have not been included in systematic monitoring for protected areas management. In no case of reviewing management plans for parks in the Annamites have I ever come across park managers who have turned to local communities, rather than foreign or university-trained conservation biologists, for information on biodiversity data to inform management plans.

In the rest of the world, IEK studies have gone even further and have linked taxonomic classifications and ethnobotany with deeper epistemological questions, such as how knowledge about ecological relationships or long-term ecological processes are generated and how they reflect local world-views (Mokuku and Mokuku 2004). Such a focus has included more attention to the idea of IEK as embodied skills and practices, as well as a focus on modes of transmission (Pearce et al. 2011; Thornton and Scheer 2012). These approaches have been practically non-existent in Vietnam, as have studies of the ways in which engaging religion, ritual, and belief could lead to better co-management of conservation zones.

This lack of understanding of or interest in IEK as a systematic world-view has had practical consequences for conservation activities in Vietnam. First, conservation projects have taken a piece-meal approach to their use of IEK, looking only for bits of IEK that they believed served their conservation purposes, with the result that very little local IEK actually ended up being used effectively in the process of creating and managing protected areas. Second, these projects predominantly relied on outsiders' views of which local community practices were acceptable IEK, and which ones were not, leading to superficial adoption of "community *hương ước* agreements" that were more reflective of Vietnamese culture than Kà Tu practices. Further, conservation projects often did not acknowledge the systematic discouragement and elimination of customary practices that Indigenous communities were subjected to by the state, and controlling such practices, like buffalo sacrifice, was even supported, which has contributed to an overall weakening of IEK in the region. Consequently, as a result of their lack of deeper engagement with IEK, conservation projects in Vietnam have seen Indigenous communities primarily as obstacles to their work, and exclusionary practices like hunting bans and resettlement out of protected areas have reinforced this tension (McElwee 2002; Boissiere et al. 2009). The lack of recognition of the role of local communities in conservation and the lack of voice these local communities have had in managing protected areas through locally applicable IEK is in contrast to other countries that are increasingly realizing people-centered conservation, often with much better and more effective results (Berkes 2007). It remains to be seen if these lessons will be learned in Vietnam, leading to more attention to and incorporation of IEK and Indigenous Peoples in the management and protection of biodiversity.

Note

1 The term Indigenous Peoples is considered controversial and not used within Vietnam.

References

Leonid V. Averyanov, Phan Kế Lộc, Nguyễn Tiến Vinh, Trần Minh Đức, Ngô Trí Dũng, Dưng Văn Thành, Lê Thaí Hưng, Nguyễn Tiến Hiệp, Phạm Văn Thế, Anna Averyanova and Jacinta Regalado.(2006). An assessment of the flora of the Green Corridor Forest Landscape, Thua Thien Hue Province, Vietnam. WWF Green Corridor Project Report #1, Hue.

Baker, J., McKenney, B., and Hurd, J., (2000). *Initial assessment of social and economic factors affecting biodiversity conservation efforts: Ecoregion-based conservation in the Lower Mekong sub-region.* Hanoi, Vietnam.

Baltzer, M., Dao, N. T., and Shore, R. (2001). *Towards a Vision for Biodiversity Conservation in the Forests of the Lower Mekong Ecoregion Complex.* Hanoi, Vietnam: WWF Indochina.

Berkes, F. (2007). Community-based conservation in a globalized world. *Proceedings of the National Academy of Sciences of the U.S.A.* 104(39), 15188.

Birdlife International in Indochina (2006). Conservation of Important Bird Areas in Indochina: Strengthening Site Support Groups to Conserve Critical Biodiversity. Final Review. Unpublished Report to the MacArthur Foundation. BirdLife International, Hanoi, Vietnam.

Boissière, M., Sheil, D., Basuki, I., Wan, M., and Le, H. (2009). Can engaging local people's interests reduce forest degradation in Central Vietnam? *Biodiversity and Conservation* 18(10), 2743–2757.

Cano, L. S. and Telleria, J. L. 2013. Local ecological knowledge as a tool for assessing the status of threatened vertebrates: A case study in Vietnam. *Oryx* 47(2), 177–183.

Colding, J. and Folke, C. (2001). Social taboos: "Invisible" systems of local resource management and bio-logical conservation. *Ecological Applications* 11(2): 584–600.

Costello, N. A. (2003). Kà Tu society: A harmonious way of life. In Y. Goudineau, (Ed.), *Laos and Ethnic Minority Cultures: Promoting Heritage*. Paris: UNESCO Publishing, pp. 163–176.

Dudley, N. (2006). *Ecoregional Conservation in a Priority Landscape in Vietnam: An Assessment of the WWF MOSAIC Project in Quang Nam Province, Central Vietnam*. Hanoi: WWF.

Goudineau, Y. (Ed.) (2003). *Laos and Ethnic Minority Cultures: Promoting Heritage*. Paris: UNESCO Publishing.

Groves, C. P., Schaller, G. B., Amato, G., and Khounboline, K. (1997). Rediscovery of the wild pig *Sus bucculentus*. *Nature* 386(6623), 335–335.

Hardcastle, J. (2002). *Opportunities for indigenous community management of forest resources in the central Truong Son uplands, Quảng Nam*. Hanoi, Vietnam: WWF.

Hickey, G. (1993). *Shattered World: Adaptation and Survival among Vietnam's Highland Peoples during the Vietnam War*. Philadelphia, PA: University of Pennsylvania Press.

Hoàng Văn Sâm, Baas, P., and Keßler, P. J. A. (2008). Traditional medicinal plants in Ben En National Park, Vietnam. Blumea: *Biodiversity, Evolution and Biogeography of Plants* 53(3), 569–601.

Huỳnh Thu Ba, Lê Công Uẩn, Vương Duy Quang, Phạm Ngọc Mậu, Nguyễn Ngọc Lưng, Nguyễn Quốc Dũng (2003). *People, Land and Resources in the Central Trường Sơn Landscape*. Hanoi, Vietnam: WWF Central Trường Sơn Initiative.

Lempert, D. (2001). Ethnic communities and legal pluralism: The politics of legal argument in market-oriented Communist Vietnam. *Legal Studies Forum* 25, 539–566.

Lê Nho Năm, Trần Văn Thu, Lê Trường Thọ, Lê Văn Đức, Đình Văn Hồng, Ngô Đình Khôi, Phạm Đức Khong (2003). *Assessment of Sông Thanh Nature Reserve Management Effectiveness*. Hanoi, Vietnam: WWF and Quảng Nam Forest Department.

Lê Nho Năm, Trần Văn Thu, Lê Trường Thọ, Lê Văn Đức, Đình Văn Hồng, Ngô Đình Khôi, Phạm Đức Khong, Lê Công Bé, Trần Văn Hoàng, Ngô Hoàng Hải Sơn, Nguyễn Đình Tuấn, Nguyễn Văn Lên (2004). Sông Thanh Nature Reserve Management Plan 2005–2010. Sông Thanh Nature Reserve Management Board, Quảng Nam Forest Protection Department, World Wide Fund for Nature.

Lê Trọng Trải and William Richardson (1999). A feasibility study for the establishment of Phong Điền (Thừa Thiên Huế province) and Dak Rông (Quảng Trị province) Nature Reserves, Vietnam. Hanoi, Vietnam: BirdLife International Vietnam Programme.

Lê Trọng Trải, Đặng Thăng Long, Phan Thanh Hà, and Lê Ngọc Tuấn (2003). *Hunting and Collecting Practices in the Central Trường Sơn Landscape*. Hanoi, Vietnam: WWF Central Trường Sơn Initiative

McElwee, P. D. (2002). Lost worlds and local people: Protected areas development in Vietnam. In D. Chatty and M. Colchester (Eds.), *Conservation and Mobile Indigenous Peoples: Displacement, Forced Settlement, and Sustainable Development*. Oxford: Berghahn Books, pp. 312–329.

McElwee, P. D. (2004). Becoming socialist or becoming Kinh? Government policies for ethnic minorities in the Socialist Republic of Viet Nam. In C. Duncan (Ed.), *Civilizing the Margins: Southeast Asian Government Policies for the Development of Minorities*. Ithaca, NY: Cornell University Press, pp. 182–213.

McElwee, P.D. (2016). *Forests Are Gold: Trees, People and Environmental Rule in Vietnam*. Seattle, WA: University of Washington Press.

Mokuku, T. and Mokuku, C. (2004). The role of indigenous knowledge in biodiversity conservation in the Lesotho Highlands: Exploring indigenous epistemology. *Southern African Journal of Environmental Education* 21, 37–49.

Nguyễn Lâm Thành (2003). *Socio-economic issues in the Central Trường Sơn landscape. Central Trường Sơn Initiative Report No. 2*. Hanoi, Vietnam: WWF Indochina.

Olson, D. M., Dinerstein, E., Wikramanayake, E. D., Burgess, N. D., Powell, G.V. N., et al. (2001). Terrestrial ecoregions of the world: A new map of life on Earth. *BioScience* 51(11), 933–938.

Pearce, T., Wright, H., Notaina, R., Kudlak, A., Smit, B., et al. (2011). Transmission of environmental knowledge and land skills among Inuit men in Ulukhaktok, Northwest Territories, Canada. *Human Ecology* 39(3), 271–288.

Schrock, J. L. (Ed.) (1966). *Minority Groups in the Republic of Vietnam*. Washington, DC: Headquarters, Dept. of the Army.

Sterling, E. J. and Hurley, M. M. (2005). Conserving biodiversity in Vietnam: Applying biogeography to conservation research. *Proceedings of the California Academy of Sciences* 56(9), 98–118.

Thornton, T. F. and Scheer, A. M. (2012). Collaborative engagement of local and traditional knowledge and science in marine environments: A review. *Ecology and Society* 17(3).

Tordoff, Andrew, Timmins, Robert, Smith, Robert & Mai Kỳ Vinh. (2003) A biological assessment of the Trường Son landscape. Central Trường Son initiative report no.1, WWF Indochina, Hanoi, Vietnam.

Vasavakul, T. (1999). Rethinking the philosophy of central-local relations in post-central-planning Vietnam. In M. Turner (Ed.), *Central-Local Relations in Asia-Pacific: Convergence or Divergence*. London: Palgrave. pp. 167–195.

Villemain, Aylette, Trần Kim Long, Christ, Herbert, Bạch Tấn Sinh, Nguyễn Thanh Hải, and Đỗ Đức Thọ (2003). *An assessment of development initiatives in the Central Trường Sơn landscape. Central Trường Sơn Initiative Report No. 3*. Hanoi: WWF.

Whitney, C. W., Min, V. S., Giang, L. H., Van Can, V., Barber, K., and Lanh, T. T. (2016). Learning with elders: Human ecology and ethnobotany explorations in northern and central Vietnam. *Human Ecology* 75(1), 71–86.

Woodside, A. (1999). The struggle to rethink the Vietnamese state in the era of market economics. In T. Brook and H. Van Luong (Eds.), *Culture and Economy: The Shaping of Capitalism in Eastern Asia*. Ann Arbor, MI: University of Michigan Press, pp. 61–77

26

INCLUDING INDIGENOUS AND LOCAL KNOWLEDGE IN THE WORK OF THE INTERGOVERNMENTAL SCIENCE-POLICY PLATFORM ON BIODIVERSITY AND ECOSYSTEM SERVICES (IPBES) GLOBAL ASSESSMENT

Outcomes and lessons for the future

*Pamela McElwee, Hien T. Ngo, Álvaro Fernández-Llamazares,
Victoria Reyes-García, Zsolt Molnár, Maximilien Guèze,
Yildiz Aumeruddy-Thomas, Sandra Díaz and Eduardo S. Brondízio*

Introduction

This chapter examines how the Intergovernmental Science-Policy Platform on Biodiversity and Ecosystem Services (IPBES) has approached the inclusion of Indigenous Knowledge in its work. Other science assessments have used terms like Indigenous Ecological Knowledge (IEK), traditional knowledge (TK), Indigenous Knowledge (IK), or Traditional Ecological Knowledge (TEK) (Mauro and Hardison, 2000; Ford et al., 2016); however, IPBES uses the concept of Indigenous and Local Knowledge (ILK), defined as "knowledge and know-how accumulated across generations, which guide human societies in their innumerable interactions with their surrounding environment" (IPBES, 2014), and we use this terminology in this chapter. IPBES has strongly promoted the idea that Indigenous Peoples and Local Communities (IPLCs) should be key stakeholders in global assessments, as they are both holders of knowledge about the natural world, as well as impacted by decisions made to manage nature by other stakeholders, such as governments and the business sector. As stakeholders, they can both contribute with their knowledge and have the right to inform policies that might directly affect their livelihoods.

DOI: 10.4324/9781315270845-30

Attention to both IPLCs and ILK in the work of IPBES has been made possible through a series of deliberate steps and decisions to promote the recognition of ILK in knowledge production and decision-making by IPLCs regarding sustainability, and we evaluate the successes and challenges of this approach. In this chapter we discuss what IPBES is; how it has been mandated to address ILK and other concerns of IPLCs; what the processes of inclusion have been in the work of the Global Assessment (GA) in particular; how ILK has informed the GA outcomes; and what lessons can be learned from the IPBES approach for other intergovernmental and/or scientific assessments.

What is IPBES?

At the global scale, an inclusive, legitimate and effective science-policy interface to provide assessments and advice is critical for decision-makers (Görg et al., 2014). Formally established in 2012 as an independent intergovernmental body with (presently) 137 member states, the Intergovernmental Science-Policy Platform on Biodiversity and Ecosystem Services (IPBES) is an "IPCC-like mechanism for biodiversity" (Larigauderie and Mooney, 2010) providing decision-makers with policy-relevant information from scientific assessments and strengthening knowledge foundations for the maintenance of biodiversity and ecosystem services. IPBES assessments, which are to be conducted regularly, focus on the status, trends and future of biodiversity and ecosystems, and the contributions of nature to people's lives, including through both regional and global trend reports as well as specific problem areas, such as pollination and land degradation. IPBES's mandate and work programme deliverables (such as assessments) were developed from requests submitted by governments, Multilateral Environmental Agreements (MEAs), UN bodies, and a range of stakeholders/decision-makers, which were reviewed and prioritized for the first work programme starting in 2014 to 2018. In addition to the assessment function, IPBES has three other functions: (1) capacity-building; (2) policy support and implementation; and (3) knowledge generation and communication. These functions are interconnected through various IPBES deliverables and tools.

One of the earliest and most fundamental guiding elements of IPBES is its "Conceptual Framework", which was developed over the course of two years by many scientists and stakeholders in an iterative and consultative process and was approved by member states in 2013 (Díaz et al., 2015a; Díaz et al., 2015b) (Figure 26.1). The framework supports the analytical work of IPBES by recognizing that biodiversity and ecosystems' contributions to people underpin every aspect of human development and sustainability. While Figure 26.1 is a highly simplified schema depicting complex relationships between people and nature through major social and ecological components (boxes) and their interactions (arrows), it has been a useful heuristic framework not only for explaining, but also for operationalizing, the work of IPBES. The Conceptual Framework encompasses a diversity of scientific disciplines, stakeholders, and knowledge systems, fundamentally recognizing that knowledge is established and validated in many ways across multiple scales (Díaz et al., 2015a; Díaz et al., 2015b; Hulme et al., 2011). The evolution of the framework has introduced new concepts, such as Nature's Contributions to People (NCP) (Díaz et al., 2018), as an alternative way to understand some of the concepts associated with the term "ecosystem services", which some IPLCs and country members of IPBES objected to as a "manifestation of nature's commodification" (Borie and Hulme, 2015: 493). Embracing a concept like NCP that can reflect both ecosystem services understandings and a more intrinsic worldview in which nature provides gifts to those who are appreciative (as is associated with many IPLCs' beliefs and practices) allows for a framing that is diverse and reflects these multiple ontologies (Díaz et al., 2018). While some concepts associated with scientific approaches and some concepts from ILK are in different places in the framework, which might

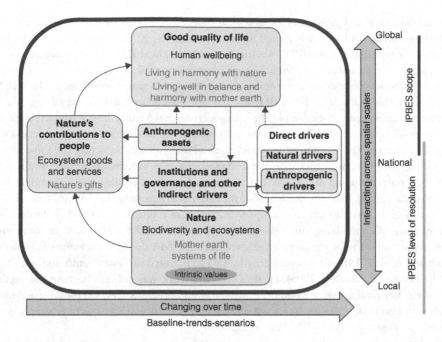

Figure 26.1 The IPBES Conceptual Framework
Source: Díaz et al. (2015a).

lead to continued divisions between science and ILK (Borie and Hulme, 2015), the framework was not meant to symbolize mutual exclusivity, but rather acknowledge different worldviews under common conceptual umbrellas.

Examples of and challenges in incorporating ILK into global science assessments

To date, many ILK issues have been under-represented in various intergovernmental processes and assessments (Ford et al., 2016). The inclusion or integration of ILK is challenging because often place-specific and oral-based forms of knowledge can be difficult to transpose into the written form more common in science (Berkes, 2008; Reid et al., 2006), and can lead to decontextualized information devoid of the cultural environment that establishes their meaning, and thus cannot represent the full "content or context of that knowledge" (Stevenson, 1996). Further, most conceptual frameworks and ontologies do not or cannot jointly recognize Western science approaches together with ILK understandings (Agrawal, 1995; Atran et al., 2001; Cruikshank, 2005; Huntington, 2000; Verran, 2001). Concepts in other knowledge systems include "living in harmony with nature" and "Mother Earth" and "nature's gifts", which are not easily reduced to transferable or generalizable content. The IPBES Conceptual Framework has made progress by attempting to recognize a diverse array of worldviews and knowledge systems into one framework which guides all IPBES assessments (Bohensky and Maru, 2011).

Reviewing the history of ILK in other global science assessments is challenging due to the inconsistent use of terms and their varying definitions (Usher, 2000). A notable development that may have been the impetus for the recognition of ILK in international processes and assessments was the establishment of the Convention on Biological Diversity (CBD) in 1993. Within the text of the Convention and its articles, namely, Article 8(j) and Decision IV/9,[1] the importance of connections between biological diversity and IPLC was stressed and highlighted. ILK and associated practices are considered important and relevant for achieving conservation goals and are treated with the same importance and relevance as any other form of knowledge in the implementation of the CBD.

Following this example in the CBD, ILK has since been included in several regional and international assessments. For example, the intergovernmental Arctic Climate Impact Assessment (ACIA), published in 2004, drew explicit attention to Arctic Indigenous Peoples and their knowledge systems. The assessment summary showed many visuals depicting the impact of climate change on Arctic Indigenous Peoples and used strategies of representation throughout the summary report (including in text, perspectives, quotes, and case studies, among others) (ACIA, 2004; Martello, 2008). In 2005, the Millennium Ecosystem Assessment (MA) was touted as one of "the first global assessments to attempt to incorporate multiple scales and multiple knowledge systems" (Gómez-Baggethun et al., 2013; Reid et al., 2006). However, in some views, the incorporation of multiple knowledge systems in the MA did not achieve its full potential (Sutherland et al., 2014), and was primarily limited to regional sub-assessments and not the full MA (Ericksen and Woodley, 2005). Further, the MA work was difficult because the methods to assess the key findings were based on a "standardized consensus view on biodiversity and on biodiversity knowledge" (Filer, 2009) and its conceptual framework was already deeply rooted primarily in Western science (Turnhout et al., 2016).

The Intergovernmental Panel on Climate Change (IPCC) has also recognized the value of including different knowledge systems into their assessments but has run into challenges. ILK often appears in grey literature or in non-written forms, which may downplay these contributions from the evidence base on which IPCC draws under their procedures (Raygorodetsky, 2011).[2] There has also been a lack of balance in ILK knowledge and focus among IPCC authors as only a small per cent had published any material on climate change and Indigenous Populations (Ford et al., 2012). Nonetheless, having authors with previous experience in ILK issues is not equivalent to having ILK incorporated into the assessment. To achieve that goal, procedures for inclusion should be explicit and emphasized (Turnhout et al., 2016). Similarly, although there is evidence that ILK has become more prominent over time in the IPCC Assessment Reports, coverage of ILK continues to be general in scope and limited in length, with its historical and contextual complexities being largely overlooked (Ford et al., 2016).

There are strong arguments in favor of incorporating multiple knowledge bases within a single assessment. The benefits of having recognition of ILK in global science processes include "a normative function that increases the legitimacy of the process; a substantive function that strengthens the knowledge base toward more appropriate solutions; and/or an instrumental function that supports more collaborative relationships and joint ownership of the knowledge produced" (Montana, 2017: 24). To some degree, IPBES continuing work in this area is aimed at showcasing the important roles ILK can play and highlighting how these benefits can be shared in other global assessments and in biodiversity governance and sustainability decision-making.

Operationalization of ILK in the IPBES work programme

IPBES innovative approach to the inclusion of ILK in assessments has been explicit from the beginning. Thus, the first work programme for IPBES agreed upon by member states included Deliverable 1(c) on "Procedures, approaches and participatory processes for working with indigenous and local knowledge systems". Even prior to the formal establishment of IPBES, there were several ad hoc and multi-stakeholder meetings and consultations to discuss the role of

ILK, among other issues. At the first ad hoc intergovernmental and multi-stakeholder meeting in 2008, a concept note called for improved dialogue between different knowledge systems regarding biodiversity and ecosystem services. At the third ad hoc intergovernmental and multi-stakeholder meeting in 2010, members adopted the Busan Outcome, which states that all the work of IPBES should recognize and respect the contribution of ILK to the conservation and sustainable use of biodiversity and ecosystems. In 2017, the member states of IPBES additionally approved an "Approach to recognizing and working with indigenous and local knowledge" (Annex II to decision IPBES-5/1), stating that the approach will be undertaken in line with the approved rules and procedures of IPBES and in accordance with internationally recognized rights of Indigenous Peoples and relevant commitments related to local communities.

As the first work programme has evolved over the past five years, IPBES has worked to incorporate ILK across all IPBES deliverables and functions. Several steps and procedures have helped consolidate the role ILK should play within IPBES. A 2013 international expert workshop on "The Contribution of Indigenous and Local Knowledge Systems to IPBES: Building Synergies with Science", held in Tokyo, aimed at examining the procedures and approaches for working with ILK within IPBES and the review and assessment of possible conceptual frameworks (Thaman et al., 2013). One recommendation was that an ILK Task Force on Indigenous and Local Knowledge Systems be set up within IPBES to "oversee the development of procedures and approaches for working with indigenous and local knowledge systems, including convening global dialogue workshops and developing case studies". The Task Force in the first work programme comprised 26 members from all regions, and currently includes 14 members, whose focus is to work on the implementation of objective 3(b) "Enhanced recognition of and work with indigenous and local knowledge systems" of the rolling work programme of IPBES up to 2030. Additionally, a Technical Support Unit (TSU) for ILK has been hosted at UNESCO in Paris since 2015, which helps to coordinate work and assist the Task Force with events and other activities.

ILK has been addressed specifically within the assessments conducted by IPBES through use of the Multiple Evidence Base (MEB) approach, which puts ILK on an equal footing to globally generated "science" (Hill et al. 2020). The MEB approach notes that there are "parallels where indigenous, local, and scientific knowledge systems are viewed to generate different manifestations of valid and useful knowledge" (Tengö et al., 2014). For example, there may be differences in spatial scale or temporal dimensions between different knowledge systems that can be combined in a process of triangulation and synergy (ibid.).

IPBES assessments address ILK on the basis of several key principles, as identified in Hill et. al (2020): respecting rights of IPLCs; supporting care and mutuality; strengthening IPLCs and their knowledge systems; and supporting knowledge exchange. The overall IPBES approach has four phases to recognize and work with ILK within assessments. Each of them addresses a specific set of conceptual, procedural and institutional challenges:

1. The first phase is the stage of developing the scoping document for any given assessment; this includes thinking through the broad policy-relevant questions that the assessment could answer involving issues of concern to IPLCs and which can be addressed through working with ILK.
2. The second phase refers to gathering a wide array of evidence and information, which can take on various forms outside of the formal peer-reviewed scientific literature.
3. The third phase consists of appropriately engaging IPLCs in the review of the various drafts of a specific assessment to make sure the assessment represents their views and knowledge systems.
4. Finally, the fourth phase involves knowledge sharing with appropriate IPLCs, so they can make use of the outcomes generated by assessments.

The rest of this chapter will detail some of the activities that the IPBES GA author team engaged in to recognize and work with ILK systems.

Processes for the inclusion of ILK in the Global Assessment

The Global Assessment (GA), which was designed to "assess the status and trends with regard to biodiversity and ecosystem services, the impact of biodiversity and ecosystem services on human well-being and the effectiveness of responses" (Annex I to decision IPBES-4/1), was accepted by member countries in May 2019 (Decision IPBES-7/1, Section II). The GA engaged different knowledge systems in a transparent, equitable and legitimate manner through a multi-pronged strategy to ensure that ILK was brought into the work on an equal footing as other knowledge systems. The process of operationalizing ILK within the GA included forming an ILK liaison group within the assessment; ensuring grey literatures and other forms of knowledge were assessed and included together with scientific literature; issuance of a global call for citations of ILK; and the holding of dialogues with ILK groups and other stakeholders (McElwee et al., 2020).

The ILK liaison group

The Global Assessment was prepared by 145 expert authors from 50 countries; most were natural scientists but, compared to previous assessments, the number of social scientists was considerable. To bring ILK into the assessment, special attempts were made during the selection of the coordinating and lead authors who wrote the report to include those with specializations in IPLC issues or working in areas that overlap with IPLC. In IPBES, three "knowledge-holder" categories were distinguished: (1) ILK-holders are members of IPLCs who possess and practise ILK; (2) ILK experts are scientists, NGO representatives, or practitioners who are members of IPLCs but at the same time have experience in science and/or policy (they may also be ILK-holders); and (3) experts on ILK who are not IPLC members but have experience in working with IPLCs and studying ILK. Among the GA authors were several ILK experts and around 20 experts on ILK.

These authors formed an "ILK liaison group" to help discussions on ILK and IPLC-related topics throughout the working meetings of the GA, and the group developed a guidance document to facilitate the consistent inclusion of ILK and topics highly relevant to IPLCs throughout the GA. Furthermore, cross-chapter story lines were developed for topics, including, for example, IPLCs and protected areas, how changes in NCP impact IPLCs, the impacts of global telecouplings on IPLCs livelihoods, and the contributions of IPLCs to urban systems. Another important role of the liaison group was to scale up experiences learnt from local case studies, so they might be used throughout the assessment.

ILK grey literatures and other forms of written texts

IPBES has made a concerted effort to consider forms of literature beyond peer-reviewed scientific literature. As such, authors were strongly encouraged to explore the importance of grey literature (e.g., technical reports, policy briefs, or case study compilations) as a vital complement to the peer-reviewed and scientific references. Although often not considered a scholarly form of publication, grey literature offers an opportunity to uncover important information on ILK and thus minimize bias in using comprehensive and balanced evidence (Thaman et al., 2013). For instance, reference documents compiled by Indigenous Peoples' Organizations can provide a more inclusive and representative picture of ILK in different parts of the world than the scientific literature published on IPLCs. Moreover, recent research suggests that over one third of new scientific conservation science documents are published in languages other than English, despite the assumption of English as the scientific "lingua franca" (Amano et al., 2016). These figures are arguably higher in the case of ILK literature, which is often unavailable in English, and published in local languages relevant to the IPLCs themselves. IPBES has placed a strong emphasis on overcoming language barriers through strategically creating multi-lingual author teams to avoid the problem of some ILK literature being potentially overlooked, as ignoring such non-English literature can contribute to biases in global understanding of ILK.

Solicited inputs through online calls

As part of the consultation strategy and to fuel the synthesis on ILK trends in the GA, an online call for contributions was launched in Spring 2017. The request sought to gather information on different types of inputs, including: (1) publications, data and knowledge (including scientific literature on ILK, reports, grey literature, or datasets); (2) IPLCs' networks and organizations; and (3) individual experts on ILK and ILK experts and holders. Submissions were called for within different broad and cross-cutting topics. The request was launched in English, Spanish, and French in an attempt to reach a wider audience. A publications database created from this call consists of nearly 1200 academic articles, reports, websites, and videos in 16 languages, some of which are Indigenous, and was made available in a searchable repository to the GA experts to facilitate inclusion in assessment chapters. The information received on organizations and experts was used to identify new aspects not yet covered in the assessment; it was also used to identify contributing authors (who wrote additional material on specific issues related to ILK for the different chapters).

Dialogues with ILK-holders and IPLCs

The GA launched eight Dialogue Workshops and consultations reaching over 250 people and allowing for constructive engagement with established networks of ILK-holders, following similar dialogues for other IPBES reports. IPBES has used both existing meetings where IPLC representatives are important interlocutors to hold side dialogues, as well as setting up task-designated dialogues to review ongoing assessments. ILK Dialogue Workshops were also held for the four regional assessments that were prepared in 2015–2018 and serve as a tangible resource for future assessments (Baptiste et al., 2017; Karki et al., 2017; Roué and Molnár, 2017; Roué et al., 2017). These side dialogues have included many activities, such as presenting the aims and goals of IPBES; soliciting additional ILK reports to be included in the assessment discussing and dialoguing with IPLC representatives for first-hand information on topics ranging from ILK to policies to support IPLCs; and requesting feedback on what aspects of IPBES reports might be particularly useful for IPLCs (see McElwee et al. 2020 for details). A total of eight ILK Dialogues were held for the GA, although funding to secure these meetings and ensure representation was a considerable challenge.

One such dialogue was held on the sidelines of a "Dialogue on Human Rights and Biodiversity Conservation" meeting, organized by the Forest Peoples' Programme, Swedbio at Stockholm Resilience Centre, and other local partners, held in Kenya in November 2017. The objective of the Dialogue was to identify and "suggest improvements to existing approaches, tools, and practices for ensuring that respect for human rights strengthens the ability to achieve conservation targets, and that securing conservation targets improves communities' ability to secure their human rights" (Malmer et al., 2018). Participants included representatives of several communities in Central and East Africa facing conservation conflicts with government authorities (including the Ogiek and Sengwer in Kenya, Batwa communities in Uganda and the Democratic Republic of Congo, and Maasai in Tanzania). It also included representatives of Hmong and Karen communities in Thailand and Aboriginal communities in Australia. Key issues discussed at the Kenya dialogue were:

- the interface and relation between scientific knowledge and ILK;
- how denigration of ILK contributes to the potential for displacement and conflict;
- opportunities and tensions surrounding the balance of human rights and biodiversity conservation;
- IPBES accountability and how the assessment work can be used by IPLCs to assist their efforts in recognition and rights.

Another dialogue was held at the tenth meeting of the Ad Hoc Open-ended Working Group on Article 8(j) and Related Provisions of the Convention on Biological Diversity in December 2017, in Montreal, Canada. This Dialogue was organized in close collaboration with the Secretariat of the Convention on Biological Diversity and the International Indigenous Forum on Biodiversity (IIFB). The main objective was to engage leading experts and representatives from IPLCs in the IPBES process by introducing IPBES and the GA and providing a space for IPLC representatives to discuss issues that challenge their communities and how these issues were or were not being addressed within the GA. Through this Dialogue, lead authors participating were able to improve drafts of the GA. The Dialogue participants were split into four different groups discussing, broadly, four different themes: (1) biodiversity changes and drivers; (2) IPLCs main concerns and priority areas; (3) potential policy pathways for IPLCs to conserve biodiversity; and (4) what options and opportunities identified could be important for IPLCs. Participants included 24 representatives from IPLCs from all global regions, 6 IPBES authors, 3 IPBES secretariat members, 3 CBD secretariat members, and 4 observers.

A dedicated ILK Dialogue Meeting in Helsinki, Finland, in June 2018, funded by the Government of Finland, was also held with Indigenous Peoples' representatives from across the Arctic to discuss their specific concerns surrounding issues addressed in the GA, including climate change impacts on IPLCs and the biodiversity they depend on. This meeting was organized in close collaboration with the Secretariat of the Arctic Council and the Indigenous Forum on Biodiversity (IIFB), as well as with the representative bodies of the six Indigenous Communities that hold a status of Permanent Participants at the Arctic Council. The Dialogue included a public seminar at the University of Helsinki with the aim of fostering critical, inter-disciplinary and evidence-based discussion on the importance of bridging diverse knowledge systems for Arctic sustainability.

Additional dialogues were held at the Annual Meeting of the UN Permanent Forum on Indigenous Issues in New York City in Spring 2017 and 2018, where the Permanent Forum formally endorsed the ILK strategy of IPBES; and side events at academic and practitioner conferences that focused on ILK and IPLCs, including at the Society for Ethnobiology in May 2017 in Montreal, the Community Conservation and Livelihoods conference in May 2018 in Halifax and the International Society of Ethnobiology meeting in Belém, Brazil in 2019.

How ILK has been incorporated into the Global Assessment

The IPBES GA writing team organization included lead authors working by chapters (six in total), each focused on a different theme laid out in an initial scoping document. Accordingly, ILK has been incorporated into the different chapters in different ways (IPBES, 2019). The first chapter defines the scope of the assessment and presents the IPBES Conceptual Framework. In addition to explaining other components of the assessment, Chapter 1 provides a global overview of who

IPLCs are, their populations and distribution, why they are important, and the lands and ecosystems that they manage which include global estimation of lands held and/or managed by IPLCs and their 'counter-mapping' efforts, as well as a summary table with dimensions around "Recognizing the Global Diversity of Indigenous Peoples and Local Communities." Thus, Chapter 1 sets the scene of the importance of IPLCs in relation to biodiversity and ecosystem services.

Chapter 2 is a large and complex chapter with a mandate to assess how direct and indirect drivers of biodiversity and ecosystem services cause change, what are the major trends in nature, how these changes in nature affect NCP, and finally how all these changes affect human well-being. ILK is brought into this chapter in diverse ways. Some sections prepared systematic reviews of peer-reviewed literature using search term-based reviews in Web of Science, Scopus, and Google Scholar. A limiting factor of these searches was the lack of proper key words to capture the complexity of the issues being analysed. For example, a section on the contributions of IPLCs to biodiversity, its management and protection used the terms *nature conservation, biodiversity, cultural landscape, biocultural diversity* and others to cover the ecological aspects of these contributions, but additional searches had to be run and a careful examination of the reference lists of the reviewed publications had to be made to find the most relevant local cases. These local cases often delivered important messages to the GA, but in some cases, the lack of knowledge up-scaling mechanisms hindered the assessment of local perceptions of status and trends of drivers, nature and NCP considerably. Grey literature (including films and other visuals) and discussions during the different Dialogue meetings proved highly relevant for understanding recent advances that had not yet reached the peer-reviewed literature. Key information conveyed in Chapter 2 includes data on amounts of biodiversity within landscapes managed by IPLCs; the trends in ecosystems whose biodiversity has co-evolved with and have been managed by IPLC; identification of the main drivers of change in IPLCs areas; and documentation of how these changes are influencing local livelihoods, NCP, and the ability of IPLCs to manage and conserve nature.

Chapter 3 assesses progress toward meeting major international objectives related to biodiversity and ecosystem services, and specifically the Aichi Biodiversity Targets (ABT) and the Sustainable Development Goals (SDGs). Key information conveyed by this chapter includes discussion of how IPLCs contribute to specific goals and targets; what factors help or hinder this contribution; and the type of recognition or benefits they receive from their contribution to these goals. The main strategy to address these issues was a peer-reviewed literature review. A group of 30 invited experts acting as "Contributing Authors" were asked to conduct a systematic literature review focusing on the role of IPLCs in each of the ABTs and SDGs. To ensure homogeneity, the chapter leadership team provided specific guidelines on how the literature review should be conducted and how to organize the information: Contributing Authors were asked to conduct a search in Web of Science with a first line capturing topics related to IPLC and ILK (e.g., *"indigenous communit*" OR "indigenous people$" OR "local communit*" OR aborigin* OR "traditional ecological knowledge" OR "TEK" OR "indigenous knowledge" OR "traditional management" OR "indigenous management" OR ILK*, and a second line with key words related to the ABT or SDG they were addressing. Then, the authors organized information from the literature review addressing the following topics: (1) what have been IPLCs contributions to achieve the ABT/SDG?; (2) how does progress (or lack of progress) in achieving the target affect IPLCs?; and (3) to what extent are IPLCs recognized, valued, and benefit from contributing to the target? Authors were asked to complement the text and the list of references with references and ideas from their own work, and to provide a case study to illustrate any of the three questions above. The text produced by the Contributing Authors (about 1,300 words each) was then sent to three or four experts on the topic for external review. Experts were encouraged to highlight any gaps in the topics or references covered, and the text was reviewed accordingly. Chapter 3 includes a condensed version of each text produced by this process, with the full text and the complete list of references appearing as an annex. One caveat of this type of analysis that, while comprehensive within existing literature, it does not capture well the views of the IPLCs themselves.

Chapters 4 and 5 deal with scenarios, and therefore had a major challenge in incorporating lessons from ILK in scenario development. The primarily use of ILK in these chapters has been to understand the potential impact of future pathways of change for IPLCs. Chapter 4, for example, notes that future development scenarios are likely to have significant impact on lands inhabited by IPLCs and their biodiversity. Key information in these chapters has included assessment of how social and environmental changes such as population growth or consumption impact IPLCs and challenges in reconciling the rights and needs of IPLCs with projected expansion in food, energy, water, and resource extraction to 2050.

Chapter 6 aims to identify options available to decision-makers in order to achieve the transformative change envisioned by the 2030 UN Agenda for Sustainable Development. The chapter specifically identifies policy options, including those that can be taken by IPLCs, and those options taken by others that might impact on IPLCs, in the context of sustainability decision-making. As part of these efforts, chapter experts conducted a comprehensive literature review addressing the effectiveness of several biodiversity-related policy instruments from the perspective of IPLCs, including protected areas, land tenure recognition, payments for ecosystem services, Reduced Emissions from Deforestation and Degradation (REDD), and sustainable wildlife management. The chapter explores the factors and conditions maximizing or undermining IPLCs contributions to biodiversity conservation, as well as how policy approaches might be reframed toward more inclusive environmental governance, including rights-based approaches, inclusive and adaptive governance, and reconfiguration of global economic processes. In particular, experts in this chapter focused on how to articulate knowledge co-production in sustainability decision-making and to promote recognition of ILK, thereby enhancing the legitimacy and effectiveness of environmental policies.

Conclusion: lessons learned for other science assessments

Even though IPBES is a relatively young global institution, its knowledge products and assessments have been favourably compared with other assessments like the IPCC and MA for its inclusivity and interdisciplinarity (Beck, 2014; Obermeister, 2017). Part of this engagement has included the attention to ILK and IPLCs outlined here. To achieve this inclusivity, it has clearly been fundamentally important that IPBES has recognized the usefulness and importance of ILK within its conceptual framework, a decision which has influenced all subsequent work products (Díaz-Reviriego et al., 2019).

Tengö et al. (2017) note that there are potentially five different steps through which bridging of knowledge systems can happen in assessments:

1. mobilization (the development of knowledge products through engagement);
2. translation (the adaptation of knowledge products across different systems and actors);
3. negotiation (interacting among knowledge systems to develop joint representations);
4. synthesis (shaping common knowledge toward a purpose);
5. application (using knowledge bases to make decisions).

Each of these approaches has been a part of the GA thus far, although the last step, application, remains particularly challenging, as there has not been adequate funding for planning post-GA follow-up activities although this aspect is anticipated to change within the next work programme of IPBES with more post-assessment follow-up activities.

In terms of mobilization, it has been a challenge to balance large-scale synthesis of ILK with the need for attention to contextualized knowledge, often represented through specific local case studies. The systematic literature reviews have been open to multiple sources of evidence beyond science, and the online calls for citations and case studies have been useful for opening up the process beyond initial experts' knowledge. Translation has also been a successful part of IPBES processes thus far, as the multiple dialogues have been designed to ensure exchanges between scientists and ILK-holders with translation processes moving both ways. However, by necessity, these dialogues have primarily involved ILK-holders and ILK experts already engaged in global fora (such as the CBD or academic conferences) and thus reaching ILK-holders still practising ILK and not already partnering with global institutions remains a challenge. Finding sufficient and sustained funding support is a critical limiting factor on ensuring IPLCs participation.

Negotiation can be seen in the development of the Conceptual Framework and the marriage of multiple ontologies within that schematic (Borie and Hulme, 2015). Part of negotiation also involves agreeing on the appropriate validation mechanisms between knowledge systems (Lofmarck and Lidskog, 2017), which remains an ongoing process for IPBES. Synthesis has been achieved in the GA and other assessments in that final work products reflect a common purpose to drive more careful management of biodiversity by multiple stakeholders. However, these syntheses have faced challenges as well: the diversity of ILK knowledge can become "flattened" and decontextualized through synthesis if not careful (ibid.) or be missing components when disassociated from institutions used to manage ILK on the ground (Tengö et al., 2017).

In the long term, it is hoped that the GA's emphasis on the inclusion of ILK and IPLCs will serve as a model for other assessments seeking to contribute to more respect and recognition of ILK, and for people as holders of this knowledge, as well as to more robust decision-making processes among a range of stakeholders.

Acknowledgments

We would like to thank all IPBES authors, particularly the ones participating in the ILK Liaison Group of the Global Assessment. We would also like to thank the other co-chair of the GA, Josef Settele. The TSU and the Task Force on ILK, along with the IPBES Secretariat, have also been essential in guiding this work. Thanks are also due to all the IPLCs who have participated in the work of IPBES, including key stakeholders like the members of the International Indigenous Forum on Biodiversity and the UN Permanent Forum on Indigenous Issues. The funders and supporters of the IPBES Dialogues also deserve our thanks, including the Finnish Ministries of the Environment and Foreign Affairs, Swedbio at Stockholm Resilience Centre, International Indigenous Forum on Biodiversity, Natural Justice, IUCN, University of Helsinki, Helmholtz Centre for Environmental Research and the Forest People's Programme.

Notes

1 Article 8(j) states the need to do the following:

> Subject to its national legislation, respect, preserve and maintain knowledge, innovations and practices of indigenous and local communities embodying traditional lifestyles relevant for the conservation and sustainable use of biological diversity and promote their wider application with the approval and involvement of the holders of such knowledge, innovations and practices and encourage the equitable sharing of the benefits arising from the utilization of such knowledge, innovations and practices.

2 The General Guidance Note on the Use of Literature in IPCC Reports and Appendix A to the Principles Governing IPCC Work states that "Contributions should be supported as far as possible with references from the peer reviewed and internationally available literature, and with copies of any unpublished material cited."

References

ACIA 2004. *Impacts of a Warming Arctic: Arctic Climate Impact Assessment.* Cambridge: Cambridge University Press.

Agrawal, A. 1995. Dismantling the divide between indigenous and scientific knowledge. *Development and Change* 26: 413–439.

Amano, T., González-Varo, J. P. and Sutherland, W. J. 2016. Languages are still a major barrier to global science. *PLoS Biology* 14(12): e2000933.

Atran, S., Medin, D., Lynch, E. et al. 2001. Folkbiology doesn't come from folkpsychology: Evidence from Yukatek Maya in cross-cultural perspective. *Journal of Cognition and Culture* 1: 3–42.

Baptiste, B., Pacheco, D., Carneiro da Cunha, M., and Díaz, S. (eds.) 2017. *Knowing Our Lands and Resources: Indigenous and Local Knowledge of Biodiversity and Ecosystem Services in the Americas.* Knowledges of Nature 11. UNESCO: Paris.

Beck, S., Borie, M., Chilvers, J. et al. 2014. Toward a reflexive turn in the governance of global environmental expertise: The cases of the IPCC and the IPBES. *GAIA* 23(2): 80–87.

Berkes, F. 2008. *Sacred Ecology.* New York: Routledge.

Bohensky, E. L. and Maru, Y. 2011. Indigenous knowledge, science, and resilience: What have we learned from a decade of international literature on "integration"? *Ecology and Society* 16(4): 6.

Borie, M. and Hulme, M. 2015. Framing global biodiversity: IPBES between Mother Earth and ecosystem services. *Environmental Science & Policy* 54: 487–496.

Cruikshank, J. 2005. *Do Glaciers Listen? Local Knowledge, Colonial Encounters, and Social Transformation.* Victoria, BC: University of British Columbia Press.

Díaz, S., Demissew, S., Carabias, J. et al. 2015a. The IPBES Conceptual Framework: Connecting nature and people. *Current Opinion in Environmental Sustainability* 14: 1–16.

Díaz, S., Demissew S., Joly, C., Lonsdale, W. M., and Larigauderie, A. 2015b. A Rosetta Stone for nature's benefits to people. *PLoS Biology* 13(1): e1002040.

Díaz, S., Pascual, U., et al. 2018. Assessing nature's contributions to people. Recognizing culture, and diverse sources of knowledge, can improve assessments. *Science*, 359(6373): 270–272. DOI:10.1126/science.aap8826.

Díaz-Reviriego, I., Turnhout, E. & Beck, S. (2019) Participation and inclusiveness in the Intergovernmental Science–Policy Platform on Biodiversity and Ecosystem Services. Nat Sustain 2, 457–464.

Ericksen, P. and Woodley, E. 2005. Using multiple knowledge systems: benefits and challenges. In Millennium Ecosystem Assessment (ed.), *Ecosystems and Human Well-being: Multiscale Assessments*, vol. 4. Washington, DC: Island Press.

Filer, C. 2009. A bridge too far: The knowledge problem in the Millennium Assessment. In J. Carrier and P. West (eds), *Virtualism, Governance and Practice: Vision and Execution in Environmental Conservation.* Oxford: Berghahn Books, pp. 85–111.

Ford, J. D., Cameron, L., Rubis, J., Maillet, M., Nakashima, D. et al. 2016. Including indigenous knowledge and experience in IPCC assessment reports. *Nature Climatic Change* 6(4): 349–353.

Ford, J. D., Vanderbilt, W., and Berrang-Ford, L. 2012. Authorship in IPCC AR5 and its implications for content: Climate change and indigenous populations in WGII. *Climatic Change* 113(2): 201–213.

Gómez-Baggethun, E., Corbera, E., and Reyes-García, V. 2013. Traditional ecological knowledge and global environmental change: Research findings and policy implications. *Ecology and Society* 18(4): 72. http://doi.org/10.5751/ES-06288-180472.

Görg, C., Spangenberg, J. H., Tekken, V., Burkhard, B., Thanh Truong, D., et al. 2014. Engaging local knowledge in biodiversity research: Experiences from large inter- and transdisciplinary projects. *Interdisciplinary Science Reviews* 39(4): 323–341. http://doi.org/10.1179/0308018814Z.00000000095.

Hill, R., Adem, C., Alangui, W., Molnár, Z., Aumeeruddy-Thomas, Y.,...& Xue, D. 2020. Working with Indigenous, local and scientific knowledge in assessments of nature and nature's linkages with people. *Current Opinion in Environmental Sustainability* 43: 8-20.

Hulme, M., Mahony, M., Beck, S., Görg, C., Hansjürgens, B., et al. 2011. Science-policy interface: Beyond assessments. *Science* 333(6043): 697–698.

Huntington, H. P. 2000. Using traditional ecological knowledge in science: Methods and applications. *Ecological Applications* 10: 1270–1274.

IPBES 2014. Initial elements for an approach towards principles and procedures for working with indigenous and local knowledge systems proposed for use by the Intergovernmental Science Policy Platform on Biodiversity and Ecosystem Services. Bonn: IPBES/2/INF/1.Add.1.

IPBES 2019, Global assessment report of the Intergovernmental Science-Policy Platform on Biodiversity and Ecosystem Services, Brondízio, E. S., Díaz, S., Settele, J., Ngo, H. T. (eds). IPBES secretariat, Bonn, Germany.

Karki, M., Hill, R., Xue, D., Alangui, W., Ichikawa, K. et al. (eds) 2017. *Knowing Our Lands and Resources: Indigenous and Local Knowledge and Practices Related to Biodiversity and Ecosystem Services in Asia.* Paris: UNESCO/IPBES.

Larigauderie, A. and Mooney, H. 2010. The Intergovernmental Science-Policy Platform On Biodiversity And Ecosystem Services: Moving a step closer to an IPCC-like mechanism for biodiversity. *Current Opinion in Environmental Sustainability* 2(1–2): 9–14.

Lofmarck, Erik and Lidskog, Rolf. (2017) Bumping against the boundary: IPBES and the knowledge divide. *Environmental Science & Policy* 69, 22–28.

Malmer, P. et al. 2018. *Global Dialogue on Human Rights and Biodiversity Conservation – Eldoret Kenya.* Stockholm: Swedbio & Stockholm Resilience Center.

Martello, M. 2008. Arctic indigenous peoples as representations and representatives of climate change. *Social Studies of Science* 38(3): 351–376.

Mauro, F. and Hardison, P. D. 2000. Traditional knowledge of indigenous and local communities: International debate and policy initiatives. *Ecological Applications* 10: 1263–1269.

McElwee, P.D., Fernández-Llamazares, A., Aumeeruddy-Thomas, Y., Babai, D., Bates, P., Galvin, K., Guèze, M., Liu, J., Molnar, Z., Ngo, H., Reyes-García, V., Roy Chowdhury, R., Samakov, A., Shrestha, U.B., Díaz, S. and Brondizio, E. "Integrating indigenous and local knowledge (ILK) into large-scale ecological assessments: The experience of the IPBES global assessment." *Journal of Applied Ecology* 57: 1666–1676.

Montana, J. 2017. Accommodating consensus and diversity in environmental knowledge production: Achieving closure through typologies in IPBES. *Environmental Science & Policy* 68: 20–27.

Obermeister, N. 2017. From dichotomy to duality: Addressing interdisciplinary epistemological barriers to inclusive knowledge governance in global environmental assessments. *Environmental Science & Policy* 68: 80–86.

Raygorodetsky, G. 2011. Why traditional knowledge holds the key to climate change. Tokyo: United Nations University. Available at :http://unu.edu/articles/global-change-sustainable-development/why-traditional-knowledge-holds-the-key-to-climate-change.

Reid, W., Berkes, F., Wilbanks, T. J. and Capistrano D. 2006. Bridging scales and knowledge systems: Concepts and applications in ecosystem assessment. In Millennium Ecosystem Assessment (ed.), *Ecosystems and Human Well-being: Multiscale Assessments.* Washington, DC: Island Press.

Roué, M., Césard, N., Yao, Y. C. A. and Oteng-Yeboah, A. (eds) 2017. *Knowing Our Lands and Resources: Indigenous and Local Knowledge of Biodiversity and Ecosystem Services in Africa.* Knowledges of Nature 8. Paris: UNESCO.

Roué, M. and Molnár, Z. (eds) 2017. *Knowing Our Land and Resources: Indigenous and Local Knowledge of Biodiversity and Ecosystem Services in Europe & Central Asia.* Knowledges of Nature 9. Paris: UNESCO.

Stevenson, M. G. 1996. Indigenous knowledge in environmental assessment. *Arctic* 49(3): 278–291.

Sutherland, W. J., Gardner, T. A., Haider, L. J., and Dicks, L. V. 2014. How can local and traditional knowledge be effectively incorporated into international assessments? *Oryx* 48(1): 1–2.

Tengö, M., Brondizio, E. S., Elmqvist, T., Malmer, P. and Spierenburg, M. 2014. Connecting diverse knowledge systems for enhanced ecosystem governance: The Multiple Evidence Base approach. *Ambio: A Journal of the Human Environment* 43(5): 579–591. http://doi.org/10.1007/s13280-014-0501-3.

Tengö, M., Hill, R., Malmer, P. et al. 2017. Weaving knowledge systems in IPBES, CBD, and beyond: Lessons learned for sustainability. *Current Opinion in Environmental Sustainability* 26–27: 17–25.

Thaman, R., Lyver, P., Mpande, R., Perez, E. et al. (eds) 2013. *The contribution of Indigenous and Local Knowledge Systems to IPBES: Building synergies with science.* IPBES Expert Meeting Report. Paris: UNESCO.

Turnhout, E., Dewulf, A. and Hulme, M. 2016. What does policy-relevant global environmental knowledge do? The cases of climate and biodiversity. *Current Opinion in Environmental Sustainability* 18: 65–72.

Usher, P. J. 2000. Traditional ecological knowledge in environmental assessment and management. *Arctic* 53(2), 183–193.

Verran, H. 2001. *Science and an African Logic.* Chicago: University of Chicago Press.

27

INDIGENOUS KNOWLEDGE, KNOWLEDGE-HOLDERS AND MARINE ENVIRONMENTAL GOVERNANCE

Suzanne von der Porten, Yoshitaka Ota and Devi Mucina

Introduction

Scholars and practitioners interested in protecting and sustaining marine environments have demonstrated increasing interest in Indigenous Knowledge (IK) systems[1] since at least the late 1980s and early 1990s (World Commission on Environment and Development, 1987). This interest was in part driven by the broad, international recognition of the need for revised or new approaches to all resource management (Bryant and Wilson, 1998; Latulippe, 2015), and in particular within fisheries and marine sciences where the lack of ecosystem-based management was increasingly seen as a driver of the global fisheries crisis (Jentoft, 2000; Mulrennan, 2012). UN Conventions such as Rio+20 (United Nations General Assembly, 1992) supported an opportunity for a change in perspective in international political fora such that it recognized Indigenous Knowledge in resource management and sustainability (Mulrennan, 2012). Indigenous Knowledge systems, which have been retained and (re-)cultivated to various degrees over millennia in local marine environments, are now frequently discussed in fisheries and marine management literature. For example, in the 2000s, the trend grew with academic and policy writing on resource management focusing on or including IK (Breton-Honeyman et al., 2016). These developments were occurring during the same era within marine scholarship where ideas of fisheries management were, to some extent, being merged with Indigenous Ecological Knowledge (Aswani and Hamilton, 2004; Johannes et al., 2000; Thornton and Maciejewski Scheer, 2012). The trend has continued through to today where hundreds of scholarly articles discussing IK can be found within the single niche in marine management (Breton-Honeyman et al., 2016).

While the scholarly discussion of Indigenous Knowledge in marine environmental governance is now widespread from both natural and social science perspectives, the rationale given by authors for the inclusion or integration of Indigenous Knowledge into non-Indigenous marine management systems or science is highly variable. Not all of these scholars cite motivations for interest in IK for the sake of benefitting Indigenous Knowledge-holders, but rather, in many cases, for reasons that knowledge integration will benefit resource planning, science or other processes. For example, Teixeira et al. (2013: 241) cite cost-effectiveness and community buy-in as supporting reasons for including TEK:

DOI: 10.4324/9781315270845-31

Analyses of overlap and proximity showed that TEK is relatively cost-effective and accurate for largescale benthic surveys, especially as a starting point for planning oceanographic surveys. Moreover, including TEK in the planning stage of MPAs may increase communities' participation and understanding of the costs and benefits of the new access and fishing effort regulations.

Additional reasons for scholarly interest[2] in Indigenous Knowledge transcend simply cataloging knowledge (Horowitz, 2015) and include its integration into legislative processes (Gadamus et al., 2015), building non-Indigenous Knowledge and understanding about the environment (Heaslip, 2008; von der Porten, de Loë, et al., 2016), conserving marine species (Aswani and Hamilton, 2004; Breton-Honeyman et al., 2016; Laidler, 2006), providing information for targeted marine research(Teixeira et al., 2013), non-Indigenous managers being obligated to consider it (Nadasdy, 2005; Breton-Honeyman et al., 2016), improving adaptive fisheries management (Latulippe, 2015), and filling data gaps in scientific or geographic knowledge (Teixeira et al., 2013). The rationale for the "inclusion" of Indigenous Knowledge is often rooted in the belief by progressive resource managers (Weiss et al., 2013) and administrators that "research of the greatest value can be obtained through mixing existing disciplines" (Hviding, 2003: 50). Indeed, the tenuous line between the integration and co-optation of IK has been well documented in both marine governance settings (Fernandez-Gimenez et al., 2006) and non-marine environments (Nadasdy, 2005).

It is not uncommon for resource management scholars to discuss IK in an apolitical way, as just another source of data to be accumulated by science. This apolitical view is expedient because it leaves out the much more complicated, relevant and urgent issues about the Indigenous People who hold that knowledge, and the rights and title to the traditional homelands and oceans upon which that knowledge was/is practiced. These unresolved issues center around the inherent and legal rights of Indigenous Knowledge-holders and their lands and oceans. In this chapter, we make a case for why applications of IK to marine governance must be considered within the context of Indigenous realities and Indigenous-led priorities. First, we explain how the incorporation of IK into sustaining marine environments is de facto political and is thus only relevant to marine governance insofar as Indigenous Knowledge-holders and land-holders are legitimately empowered to implement their knowledge. Second, we discuss the merits of greater deference to Indigenous-led re-implementation of Indigenous Knowledge systems, many of which provide a mechanism for directing people on how to protect and interact with the environment. Finally, we argue that unresolved issues regarding the inherent and legal rights of Indigenous Knowledge-holders and their lands and oceans, cannot be disaggregated from discussions of the applications of Indigenous Knowledge.

The political nature of IK

In addition to the academic and practical interest in the integration of Indigenous Knowledge into marine governance by scholars and policy-makers, there is mounting international political momentum in each of the fields of oceans governance, Indigenous Knowledge, and the rights of Indigenous Peoples. A recent example of an international political commitment on oceans governance is the United Nations Sustainable Development Goal (SDG) 14, to "[c]onserve and sustainably use the oceans, seas and marine resources for sustainable development" (United Nations, 2017: 1). The Sustainable Development goals, adopted by more than 150 world leaders in 2015 (United Nations Development Programme, 2015), have the potential to considerably impact future national-level policy-making with regards to oceans governance (Singh et al.,

2016). While Indigenous People are not explicitly discussed within the targets and indicators for SDG 14, language such as indicator 14.B.1 relating to countries' recognition and protection of "access rights for small-scale fisheries"(United Nations, 2017: 1) does pertain to Indigenous coastal peoples. Small-scale fisheries have been further supported at an international level under the UN Food and Agriculture Organization's *International Guidelines on Securing Sustainable Small-Scale Fisheries* (Food and Agriculture Organization of the United Nations, 2017). Other discussions within the United Nations pertain directly to Indigenous Knowledge. Specifically, the Convention on Biological Diversity (CBD) Article 8(j) includes recognition by the international community of "the close and traditional dependence of many indigenous peoples and local communities on biological resources ... [and] ... the contribution that traditional knowledge can make to both the conservation and the sustainable use of biological diversity ..."(Convention on Biological Diversity, 2016: 2).

This international consideration of Indigenous Knowledge in the context of biological diversity is mirrored in the United Nations Declaration on the Rights of Indigenous Peoples (UNDRIP) (United Nations, 2008). The UNDRIP, adopted by the UN General Assembly in 2007, similarly recognizes that "respect for indigenous knowledge, cultures and traditional practices contributes to sustainable and equitable development and proper management of the environment" (ibid.: 2) and that "Indigenous peoples have the right to maintain, control, protect and develop their ... traditional knowledge ..." (ibid.: 11). In concert, these international fora, declarations and agreements, among others (Hanich and Ota, 2013), demonstrate the political impetus behind some of the international focus on Indigenous Knowledge.

Indigenous Knowledge is also "political" in another sense of the word, beyond the realm of international policy-making. Politics, sometimes defined as the "competition between competing interest groups or individuals for power and leadership" (Merriam-Webster 2017), is inherent to the contested and colonized lands upon which many Indigenous coastal people live. These same coastal lands and oceans have attracted development and resource extraction including tourism, shipping, pipelines, and industrial fisheries. In many instances, Indigenous Peoples, communities, and knowledge-holders have been dispossessed of some or all of their traditional homelands and oceans upon which this knowledge was gathered and has been/ is applied (Huseman and Short, 2012; Memon and Kirk, 2012; Hall, 2013; Youdelis, 2016; Abbott, 2016). In the many cases where these lands and oceans have not been repatriated to Indigenous Peoples, the integration of IK into academe, policy-making, science, and conservation is thus rendered less meaningful. The reason that the dispossession of land/ocean is integral to discussions of Indigenous Knowledge, is similar to the reason why dispossession is integral to reconciliation. In their discussion of reconciliation, Rigby 2001 states:

> There were two friends, Peter and John. One day Peter steals John's bicycle. Then, after a period of some months, he goes up to John with outstretched hand and says: 'Let's talk about reconciliation.'
>
> John says, 'No, let's talk about my bicycle.'
>
> 'Forget about the bicycle for now,' says Peter. 'Let's talk about reconciliation.'
>
> 'No,' says John. 'We cannot talk about reconciliation until you return my bicycle.'

Replacing *reconciliation* with *Indigenous Knowledge* here, in cases where land/oceans dispossession has been effected but not resolved, is problematic for reasons similar to reconciliation. The intention of requoting the bicycle metaphor is not to trivialize land dispossession in a way that makes it synonymous with bicycle theft. Rather, the metaphor demonstrates how discussions of integration of IK, like reconciliation, can neither be tidily separated from the legacies and politics of land/ocean dispossession nor from Indigenous Knowledge-holders themselves.

Given the outstanding jurisdictional questions about lands/oceans, as well as matters of tenure, competition, circumscription, degradation, the role of Indigenous-Knowledge-holders becomes central to how IK relates to marine environmental governance. Instead of focusing on how disembodied IK could potentially contribute to non-Indigenous processes of marine environmental governance, we turn our focus to Indigenous Knowledge-holders themselves. The importance of knowledge-holders is indeed discussed in marine governance scholarship. In this scholarship, key issues include: how to correctly attribute contributions and authorship of Indigenous Knowledge-holders (Breton-Honeyman et al., 2016), the direct involvement of knowledge-holders in decision-making (Fernandez-Gimenez et al., 2006), and the effect of Indigenous Knowledge-holders on policy-making (Gadamus et al., 2015). However, Indigenous inclusion in marine policy-making could go further to accommodate the reality that Indigenous Peoples (and knowledge-holders) are "increasingly asserting their rights to *primary* roles in policy-and decision-making" on their traditional lands and oceans (von der Porten, Lepofsky, et al., 2016: 68). Michi Saagiig Nishnaabeg scholar Leanne Betasamosake Simpson (2004: 379–380) articulates how environmental Indigenous Knowledge-holders view such policy-making in the context of contested colonial jurisdiction:

> Indigenous leaders and knowledge holders are invited by state governments to share their Traditional Knowledge in national and international forums such as the Convention on Biological Diversity, yet those same state governments continue to actively participate in the destruction of the land and knowledge while categorically ignoring Indigenous jurisdiction over those lands. The hypocrisy is not lost on Traditional Knowledge holders. This is colonialism in action.

This outstanding issue of land/ocean dispossession is inextricable from Indigenous Knowledge and the holders of that knowledge in the context of marine governance. Rather than pulling Indigenous Knowledge and knowledge-holders into existing and future processes of marine governance, we must first, as suggested by Simpson (2004), name and challenge the forces which threaten Indigenous Knowledge and knowledge-holders.

Relational knowledge

One of the major concerns of this Handbook, and of environmental governance more broadly, is how Indigenous (Environmental) Knowledge can generate practical responses to contemporary environmental problems, at both the local and the global scales. The particular environmental problems that pertain to ocean resources are well-known and include: habitat destruction, pollution, overharvesting of marine species, and climate change (Laidler, 2006; Norman, 2012; Gadamus et al., 2015). Indigenous Knowledge, held by peoples who have managed local marine environments for millennia, is a part of the cultural capital which has supported and does support the avoidance or mitigation of many these problems (Turner et al., 2000; Dove, 2006; Turner and Clifton, 2009; Green et al., 2010). Indigenous Knowledge systems thus reflect a system to interpret the surrounding environment with a lengthier track record than scientific knowledge. Consequently, the knowledge offers a diverse source to look for answers for socio-ecological approach within IK systems.

However, in the context of environmental marine governance, we contribute to this aspect of the debate on IK by highlighting the point that resource management and environmental governance are not really about managing resources nor governing the environment. By and large, environmental marine governance, such as policy-making, regulation, decision-making and management are created to modify the behavior of people, because it is we who harvest, consume, extract, sell, trade, depend on, and/or protect the threatened resources. It is thus notable that Indigenous Knowledge systems are relational and embedded in sociality and the managing of human protocols, solidarity, equity, and behavior in relation to each other and the environment. This perspective of building protocols for solidarity between people is particularly important when it comes to resources with contested rights and capacity to access, such as fish stocks. The depth of Indigenous Knowledge is important to consider in the context of new and emerging ocean spaces, such as the increasingly ice-free Arctic Ocean where Indigenous rights have yet to be legally protected. Although it has been argued that, in some contexts, knowledge cannot prevent people from allowing a resource to be effectively destroyed (Foale, 2006: 134), the widely-held understanding is that Indigenous environmental knowledge systems could be vital for informing the social aspects of governance in relation to these dynamic ocean resources (Heaslip, 2008).

Thus, rather than viewing IK as a component of something that can be used as one of two components in reformulated or hybridized knowledges (Fish et al., 2016) (e.g., bridging Indigenous Knowledge and marine ecology), greater deference could simply be given to Indigenous Knowledge-holders to lead the development and implementation of their knowledge systems on their traditional homelands and oceans. In the realm of developing a marine conservation program, Drew (2005: 1288) argues that Indigenous Knowledge "is not about a one-time extraction of information" but rather "its use presents the opportunity for a long-term collaboration and development of information". However, Indigenous Knowledge systems encompass information that has been and is being developed over millennia (Mucina, 2013), that provide direction for Indigenous People relating to the natural resources in those environments (Aswani and Hamilton, 2004; Berkes and Turner, 2006; Foale, 2006; Carter and Hill, 2007; Davidson-Hunt and Michael O'Flaherty, 2007). In coastal settings, those Indigenous Knowledge systems historically included protocols for how coastal people should relate to the environment through, for example, ceremony, codes of conduct, systems of chieftainships, how harvesting was done, and how much was harvested (Aswani and Hamilton, 2004; Laidler, 2006; Ota, 2006; Menzies and Butler, 2007; Turner et al., 2013). This direction for people contributed (and still contributes) to the ethic of sustainability and conservation embedded in many Indigenous ways of living (Menzies and Butler, 2007).

Contemporary examples of neo-traditional Indigenous Knowledge being developed in marine environments exist as well (Gerkey, 2016), and these too provide direction for people relating to environments. For example, in the development of the Tla-o-qui-aht's land use plan for the Haa'uukimun Tribal Park on the Pacific coast of Canada, the Tla-o-qui-aht Nation created a zone named qwa siin hap or "leave as it was", which is governed by strict conservation standards, restricting human activities there to things like research and education (Murray and King, 2012). According to Murray and King (ibid.: 390), the lands defined by this zone have sacred meaning to the Tla-o-qui-aht Nation, as they are the people's place of origin. Similarly, in the Pacific, Hawaiians have created modern adaptations of traditional principles related to marine resources that pertain specifically to self-restraint in resources use and caring for marine species (Poepoe et al., 2003). These examples illustrate a view of Indigenous Knowledge as a world-view that serves as a reference point for Indigenous Peoples to relate to the land and oceans upon which their language, identity, culture, and way of life were developed. This stands in contrast to the view of Indigenous Knowledge as data that can be quantified, documented, digitized, and/or transferred to non-Indigenous processes and decision-making.

We present these examples of Indigenous Knowledge systems to provide direction to people so that the academic conversations about Indigenous Knowledge may make a substantial shift from a focus on the knowledge itself to a more instrumental and pragmatic focus on the people – both Indigenous Knowledge-holders and the people to whom this knowledge provides direction. As Simpson (2004) suggests, the documentation or digitization of IK, though seemingly benign, in fact denigrates the nature of IK by separating it from the land, depersonalizing it, and stripping it of its dynamism and fluidity. Non-Indigenous marine practitioners, policy-makers, and planners are visitors to Indigenous coastal lands and oceans, not hosts. Non-Indigenous individuals proceeding with the question of how to manage resources, and, in some cases, dictating the terms of how and what knowledges are folded into processes and procedures of resource management (Nadasdy, 2005), are bypassing an inherent problem of more primary need: scholars and practitioners with an interest in coastal Indigenous Knowledge systems should begin to focus primarily on finding ways to advocate for the leadership of Indigenous Peoples and knowledge-holders to reinstate their Indigenous Knowledge systems in their traditional homelands and oceans.

Advocacy of this kind should support Indigenous-led resurgence (von der Porten, Lepofsky, et al., 2016), a concept put forth by Indigenous governance scholars (Borrows, 2002; Alfred and Corntassel, 2005; Coulthard, 2008; Simpson, 2011; Corntassel, 2012), and would thus have the potential to redirect efforts to document Indigenous Knowledge to efforts to protect Indigenous lands, oceans, and "the Indigenous processes for the transmission of Indigenous Knowledge to younger generations" (Simpson, 2004: 380). Unlike research which supports the idea that Indigenous Knowledge and wisdom be integrated into broader (non-Indigenous) management regimes (Heaslip, 2008), this shift in focus would instead support the decolonization of knowledge and Indigenous People (Hviding, 2003), and divert further colonization of Indigenous Knowledges. This shift would be of particular importance in coastal Indigenous settings where Indigenous Knowledge plays a central role in marine resource management and coastal social dynamics (Cisneros-Montemayor et al., 2016). In sum, the support by non-Indigenous practitioners and scholars of Indigenous Peoples to reinstate their lands and Indigenous Knowledge practices, and direct Indigenous and non-Indigenous Peoples in how to relate to and interact with the land, would mark a more decolonizing approach to the debate surrounding the role of IK in environmental marine governance. Further, this approach holds the potential to also maintain the autonomy of differing processes of knowledge and knowledge production (Weiss et al., 2013) and moves beyond decision-makers simply "including" Indigenous Knowledge-holders in environmental marine governance (Heaslip, 2008).

Going forward

Given the interest in IK for marine environmental governance – and given the above-discussed issues of coastal lands/oceans dispossession and the role of knowledge in managing people – we offer a concrete way forward on how future marine scholarship and international and local marine environmental governance can align with Indigenous leadership on knowledge use.

Marine scholars and policy-makers are often focused on the integration of Indigenous Knowledge for reasons such as the decolonization of decision-making processes (Hviding, 2003), the improvement of interdisciplinary marine environmental governance, or generating practical responses to contemporary marine environmental problems. Here we suggest that, for any one of these goals, greater attention be concentrated on listening to the historical and political reality of Indigenous Knowledge-holders (Hoyle, 2011). Indigenous Knowledge, according to a Kluane First Nation member, is "not really knowledge at all … [but].. more a way of life" (Nadasdy, 2003: 52). Viewing IK as a way of life, we argue that "documenting this knowledge" is in fact not the "first step", nor is the step of "translating indigenous knowledge and science into forms that are mutually intelligible, in ways that make it accessible to decision-makers"(Berkes et al., 2007: 159) necessarily a first priority for Indigenous Peoples. Though important, we argue that these efforts are secondary to repatriating lands/oceans to Indigenous Peoples and Knowledge-holders so that they can better restore this way of life, the vital source for Indigenous environmental knowledge, and lead the decision-making in their coastal homelands. Before IK is, if ever, folded into patriarchal-colonial structures, including any non-Indigenous coastal environmental management processes, Indigenous Peoples must first be afforded self-determination within their own laws, customs, and ways of each Indigenous nation in a way that is free of external subordination (Tully, 1995). Like other scholars debating Indigenous Knowledge from a critical perspective, that is, that IK is "embedded in uneven, colonial relations of power"(Latulippe, 2015, p.121), we argue for the importance of prioritzing Indigenous rights/power to implement their knowledge systems over other non-Indigenous efforts to apply or incorporate IK.

The focus on Indigenous-managed coastal lands and oceans that we suggest here has two potential benefits. The first is decolonization and justice for Indigenous Peoples. The case for the repatriation of traditional homelands for Indigenous Peoples around the world, coastal or otherwise, has been made (Borrows 1997; Corntassel, 2003; Laidler, 2006; Norman, 2012). This need is true in the realm of Indigenous Knowledge and coastal resources management which has become an important site of mobilization and resistance for Indigenous Knowledge-holders and leaders who have been and are advocating for Indigenous control over Indigenous territories and knowledge to promote decolonization and justice in the coexistence of Indigenous and non-Indigenous Peoples (Simpson, 2004). The restoration of lands and oceans to Indigenous People would be a step in the direction of decolonization and justice, and would support Indigenous resistance and mobilization. For Indigenous coastal peoples, for whom seafood makes up a much greater portion of subsistence, trade, and economic livelihoods than their non-Indigenous counterparts (Pauwelussen 2015; Cisneros-Montemayor and Ota, 2016), the repatriation of coastal lands and oceans is truly vital to their way of life. The second benefit of deferring to Indigenous Peoples to lead environmental marine decision-making is that Indigenous Knowledge-holders can implement their knowledge without having to translate it into another form, process, or structure. Ostensibly, this deference has potential to be the most practical and streamlined approach to "applying" Indigenous Knowledge to contemporary marine environmental problems: Indigenous Peoples need not translate their knowledge and non-Indigenous practitioners need not try to apply Indigenous Knowledge to non-Indigenous processes. Rather, future efforts and research can focus on returning Indigenous lands and oceans to Indigenous Peoples so that they can again lead the protection, and now restoration, of the habitat, species, and systems within them, and maintain and continue their practice of Indigenous Knowledge systems now and in the future. Finally, it is important to the marine context to consider the realities of climate change, species migration, and the interconnectedness of oceans. For this reason, it is worth considering how Indigenous nations will integrate their own knowledge systems with the knowledge systems of Indigenous and non-Indigenous Peoples with whom they are increasingly interconnected. It is our contention that where Indigenous Peoples are afforded the right to self-determination and the appropriate resources to make decisions about their coastal lands, each Indigenous nation will address this quandary in a way that suits the needs of their nation.

Notes

1 Indigenous Knowledge systems can even more accurately be described as *relational Indigenous Knowledge systems* particularly where traditionally, harvesting requires reciprocal actions and where intimate connections to environments are elemental to human survival. For the sake of brevity and the diversity of Indigenous circumstances, we use the terms *Indigenous Knowledge* and *Indigenous Knowledge systems* in this chapter.
2 The focus of literature review for this chapter was on scholarship pertaining to marine/oceans research that addressed the subject of Indigenous (Ecological/Traditional) Knowledge.

References

Abbott, J. 2016. The neo-colonization of Central America. *New Politics* XVI(1): 1–8.

Alfred, T. and Corntassel, J. 2005. Being indigenous: Resurgences against contemporary colonialism. *Government and Opposition* 40(4): 597–614.

Aswani, S. and Hamilton, R. J. 2004. Integrating indigenous ecological knowledge and customary sea tenure with marine and social science for conservation of bumphead parrotfish (*Bolbometopon muricatum*) in the Roviana Lagoon, Solomon Islands. *Environmental Conservation* 31(1): 69–83.

Berkes, F., Berkes, M. K. and Fast, H. 2007. Collaborative integrated management in Canada's North: The role of local and traditional knowledge and community-based monitoring. *Coastal Management* 35(1): 143–162.

Berkes, F. and Turner, N. J. 2006. Knowledge, learning and the evolution of conservation practice for social-ecological system resilience. *Human Ecology* 34(4): 479–494.

Borrows, J. 1997. Wampum at Niagara: The Royal Proclamation, Canadian legal history, and self-government. *Aboriginal and Treaty Rights in Canada: Essays on Law, Equality and Respect for Difference*, pp. 155–172. Available at: http://scholar.google.com/scholar?hl=en&btnG=Search&q=intitle:Wamp um+at+Niagara:+The+Royal+Proclamation,+Canadian+Legal+History,+and+Self-Government#0.

Borrows, J. 2002. *Recovering Canada: The Resurgence of Indigenous Law*. Toronto, Canada: University of Toronto Press.

Breton-Honeyman, K., Furgal, C. M., and Hammill, M. O. 2016. Systematic review and critique of the contributions of traditional ecological knowledge of beluga whales in the marine mammal literature. *Arctic* 69(1): 37–46.

Bryant, R. L. and Wilson, G. A. 1998. Rethinking environmental management. *Progress in Human Geography* 22(3): 321–343.

Carter, J. L. and Hill, G. J. E. 2007. Critiquing environmental management in indigenous Australia: Two case studies. *Area* 39: 43–54.

Cisneros-Montemayor, A. M. et al. 2016. A global estimate of seafood consumption by coastal indigenous peoples. *PLoS ONE* 11(12).

Cisneros-Montemayor, A. M. and Ota, Y. 2016. Indigenous marine fisheries: A global perspective. In D. Pauly (Ed.), *Global Atlas of Marine Fisheries: A Critical Appraisal of Catches and Ecosystem Impacts*. Washington, DC: Island Press.

Convention on Biological Diversity 2016. Article 8(j) and Related Prohibitions: Draft decision submitted by the Chair of Working Group II. Cancun, Mexico. Available at: www.cbd.int/doc/c/c823/28e7/2b8302f5a527ab8b81c6aa7d/cop-13-l-14-en.pdf.

Corntassel, J. 2003. Who is indigenous? "Peoplehood" and ethnonationalist approaches to rearticulating indigenous identity. *Nationalism and Ethnic Politics* 9(1): 75–100.

Corntassel, J. 2012. Re-envisioning resurgence: Indigenous pathways to decolonization and sustainable self-determination. *Decolonization: Indigeneity, Education & Society* 1(1): 86–101.

Andrew Rigby, Justice and Reconciliation: After the Violence (Boulder: Lynne Rienner Publishers 2001), p. 142.

Coulthard, G. 2008. Beyond recognition: Indigenous self-determination as prefigurative practice. In L. Simpson (Ed.), *Lighting the Eighth Fire: The Liberation, Resurgence, and Protection of Indigenous Nations*. Winnipeg, Canada: Arbeiter Ring Publishing, pp. 187–204.

Davidson-Hunt, I. J. and Michael O'Flaherty, R. 2007. Researchers, indigenous peoples, and place-based learning communities. *Society & Natural Resources*, 20(4): 291–305.

Dove, M. R. 2006. Indigenous people and environmental politics. *Annual Review of Anthropology* 35(1): 191–208.

Drew, J. A. 2005. Use of traditional ecological knowledge in marine conservation. *Conservation Biology* 19(4): 1286–1293.

Fernandez-Gimenez, M. E., Huntington, H. P. and Frost, K. J. 2006. Integration or co-optation? Traditional knowledge and science in the Alaska Beluga Whale Committee. *Environmental Conservation* 33(4): 306.

Fish, R. D., Church, A. and Winter, M. 2016. Conceptualising cultural ecosystem services: A novel framework for research and critical engagement. *Ecosystem Services* 21(B): 208–217.

Foale, S. 2006. The intersection of scientific and indigenous ecological knowledge in coastal Melanesia: Implications for contemporary marine resource management. *International Social Science Journal* 58(187): 129–137.

Food and Agriculture Organization of the United Nations 2017. International Guidelines on Securing Sustainable Small-Scale Fisheries [SSF Guidelines]. Voluntary Guidelines for Securing Sustainable Small-Scale Fisheries in the Context of Food Security and Poverty Eradication. Available at: www.fao. org/fishery/ssf/guidelines/en (accessed May 10, 2017).

Gadamus, L. et al. 2015. Building an indigenous evidence-base for tribally-led habitat conservation policies. *Marine Policy* 62: 116–124. http://dx.doi.org/10.1016/j.marpol.2015.09.008.

Gerkey, D. 2016. The emergence of institutions in a post-Soviet commons: Salmon fishing and reindeer herding in Kamchatka, Russia. *Human Organization* 75(4): 336.

Green, D., Billy, J. and Tapim, A. 2010. Indigenous Australians' knowledge of weather and climate. *Climatic Change* 100(2): 337.

Hall, R. 2013. Diamond mining in Canada's Northwest Territories: A colonial continuity. *Antipode* 45(2): 376–393.

Hanich, Q. and Ota, Y. 2013. Moving beyond rights-based management: A transparent approach to distributing the conservation burden and benefit in tuna fisheries. *International Journal of Marine and Coastal Law* 28(1): 135–170.

Heaslip, R. 2008. Monitoring salmon aquaculture waste: The contribution of First Nations' rights, knowledge, and practices in British Columbia, Canada. *Marine Policy* 32(6): 988–996.

Horowitz, L. S. 2015. Local environmental knowledge. In T. A. Perreault, G. Bridge, and J. McCarthy (Eds.), *The Routledge Handbook of Political Ecology*. New York: Routledge.

Hoyle, D. 2011. Cameroon: Listening to indigenous peoples. *Nature* 474(7349): 36.

Huseman, J. and Short, D. 2012. "A slow industrial genocide": Tar sands and the indigenous peoples of northern Alberta. *The International Journal of Human Rights* 16(1): 216–237.

Hviding, E. 2003. Opening up? Reclaiming a plurality of knowledges. *The Contemporary Pacific* 15(1): 43–73.

Jentoft, S. 2000. The community: A missing link of fisheries management. *Marine Policy* 24(1): 53–60.

Johannes, R. E., Freeman, M. and Hamilton, R. J. 2000. Ignore fishers' knowledge and miss the boat. *Fish and Fisheries* 1: 257–271.

Laidler, G. J. 2006. Inuit and scientific perspectives on the relationship between sea ice and climate change: The ideal complement? *Climatic Change* 78: 407–444. http://dx.doi.org/10.1016/j.marpol.2015.09.008.

Latulippe, N. 2015. Situating the work: A typology of traditional knowledge literature. *AlterNative* 11(2): 118–131.

Memon, P. A. and Kirk, N. 2012. Role of indigenous Māori people in collaborative water governance in Aotearoa/New Zealand. *Journal of Environmental Planning and Management* August: 37–41.

Menzies, C. R. and Butler, C. F. 2007. Returning to selective fishing through indigenous fisheries knowledge: The example of K'moda, Gitxaała territory. *American Indian Quarterly* 31(3): 441–464.

Merriam-Webster 2017. Politics. In Merriam-Webster Dictionary. Available at: www.merriam-webster. com/dictionary/politics.

Mucina, D. D. 2013. Ubuntu orality as a living philosophy. *The Journal of Pan African Studies* 6(4): 18–36.

Mulrennan, M. E. 2012. Indigenous Knowledge in marine and coastal policy and management. *Ocean Yearbook* 27: 89–119.

Murray, G. and King, L. 2012. First Nations values in protected area governance: Tla-o-qui-aht Tribal Parks and Pacific Rim National Park Reserve. *Human Ecology* 40: 385–395.

Nadasdy, P. 2003. *Hunters and Bureaucrats: Power, Knowledge, and Aboriginal-State Relations in the Southwest Yukon*. Vancouver, BC: UBC Press.

Nadasdy, P. 2005. The anti-politics of TEK: The institutionalization of co-management discourse and practice. *Anthropologica* 47(2): 215–232.

Norman, E. S. 2012. Cultural politics and transboundary resource governance in the Salish Sea. *Water Alternatives* 5(1): 138–160.

Ota, Y. 2006. Fluid bodies in the sea: An ethnography of underwater spear gun fishing in Palau, Micronesia. *Worldviews: Environment, Culture, Religion* 10(2): 205–220.

Pauwelussen, A. P. 2015. The moves of a Bajau middlewoman: understanding the disparity between trade networks and marine conservation. *Anthropological Forum* 25(4): 1–21. http://dx.doi.org/10.1080/00664677.2015.1054343.

Poepoe, K., Bartram, P., and Friedlander, A. 2003. The use of traditional Hawaiian knowledge in the contemporary management of marine resources. In *Putting Fishers' Knowledge to Work: Conference Proceedings.* University of British Columbia, pp. 328–339.

Simpson, L. 2004. Anticolonial strategies for the recovery and maintenance of Indigenous Knowledge. *American Indian Quarterly* 28(3/4): 373–384.

Simpson, L. 2011. *Dancing on Our Turtle's Back: Stories of Nishnaabeg Re-creation, Resurgence, and a New Emergence.* Winnipeg, Manitoba: Arbeiter Ring Publishing.

Singh, G. G. et al. 2016. How are national biodiversity strategies and action plans contributing to Sustainable Development Goals? Paper presented at United Nations Development Programme: Indigenous Peoples and Local Communities Day, COP 13, Convention on Biological Diversity; Rio Conventions Pavilion, December 8. Cancun, Mexico.

Teixeira, J. B. et al. 2013. Traditional Ecological Knowledge and the mapping of benthic marine habitats. *Journal of Environmental Management* 115: 241–250. http://dx.doi.org/10.1016/j.jenvman.2012.11.020.

Thornton, T. F. and Maciejewski Scheer, A. 2012. Collaborative engagement of local and traditional knowledge and science in marine environments: A review. *Ecology and Society* 17(3).

Tully, J. 1995. *Strange Multiplicity: Constitutionalism in an Age of Diversity.* Cambridge: Cambridge University Press.

Turner, N. J., Berkes, F. and Dick, J. 2013. Blundering intruders: Extraneous impacts on two indigenous food systems. *Human Ecology* 41(4): 563–574.

Turner, N. J. and Clifton, H. 2009. 'It's so different today': Climate change and indigenous lifeways in British Columbia. *Global Environmental Change* 19(2): 180–190.

Turner, N. J, Ignace, M. and Ignace, R. 2000. Traditional Ecological Knowledge and wisdom of Aboriginal Peoples in British Columbia. *Ecological Applications* 10(5): 1275–1287.

United Nations 2008. United Nations Declaration on the Rights of Indigenous Peoples United Nations Declaration on the Rights of Indigenous Peoples. (March). New York: UN.

United Nations 2017. Sustainable Development Goal 14. Available at: https://sustainabledevelopment.un.org/sdg14.

United Nations Development Programme 2015. World leaders adopt Sustainable Development Goals. Available at: www.undp.org/content/undp/en/home/presscenter/pressreleases/2015/09/24/undp-welcomes-adoption-of-sustainable-development-goals-by-world-leaders.html (accessed March 14, 2017).

United Nations General Assembly 1992. Report of the United Nations Conference on Environment and Development, Rio de Janeiro. New York: UN.

von der Porten, S., de Loë, R. C. and McGregor, D. 2016. Incorporating indigenous knowledge systems into collaborative governance for water: challenges and opportunities. *Journal of Canadian Studies* 50(1): 214–243.

von der Porten, S., Lepofsky, D., et al. 2016. Recommendations for marine herring policy change in Canada: Aligning with Indigenous legal and inherent rights. *Marine Policy* 74: 68–76. Available at: http://linkinghub.elsevier.com/retrieve/pii/S0308597X16301464.

Weiss, K., Hamann, M. and Marsh, H. 2013. Bridging knowledges: Understanding and applying indigenous and western scientific knowledge for marine wildlife management. *Society & Natural Resources* 26(3): 285–302.

World Commission on Environment and Development 1987. *Our Common Future.* Oxford: Oxford University Press.

Youdelis, M. 2016. "'They could take you out for coffee and call it consultation !'": The colonial antipolitics of Indigenous consultation in Jasper National Park. *Environment and Planning A* 48(7): 1374–1392.

28

INCORPORATING SOCIAL-ECOLOGICAL SYSTEMS INTO PROTECTED AREA NETWORKS

Territories and areas conserved by Indigenous Peoples and local communities (ICCAs) in Sabah, Malaysian Borneo

Ashley Massey Marks, Paul Porodong and Shonil A. Bhagwat

Introduction

Protected areas comprise over 12 per cent of the Earth's terrestrial surface (UNEP-WCMC, 2012) and the Aichi Targets of the Convention on Biological Diversity (CBD) aim to expand this coverage to 17 per cent by 2020 (Secretariat of the Convention on Biological Diversity, 2011). Increasing the coverage of protected area networks is challenging, however, as the 87 per cent of the terrestrial surface beyond protected areas is a mosaic of land uses (Chazdon et al., 2009). Chazdon et al. propose a research agenda in these landscapes, noting: "Designing successful conservation strategies requires an understanding of how and why local residents manage their landscapes and adapt to environmental changes" (ibid.: 147).

In the matrix of land uses beyond protected areas, community-based conservation (CBC) devolves natural resource management from national governments to local communities (Horwich and Lyon, 2007). "Community" is a simplistic term used to denote self-regulating groups of natural resource users; the value of this term has been debated (Agrawal and Gibson, 1999). A systematic review of 146 CBC studies across 40 countries presents four definitions of CBC success that are alternatives to standard area-based protection targets: (1) attitudinal (positive changes in views of conservation goals); (2) behavioural (decreased off-take); (3) ecological (improved outcomes for the habitat or species of interest); and (4) economic (variety of livelihood benefits) (Brooks et al., 2012) (Table 28.1).

The multi-dimensional description of successful community-based conservation counterbalances conservation's focus on total protected area as the primary metric of conservation success. There is an increasing recognition that the success of conservation efforts should be measured not only by the size of the area conserved, but also by the interaction between local

DOI: 10.4324/9781315270845-32

Table 28.1 Factors increasing the likelihood of successful community-based conservation: results of a systematic review of 146 studies across 40 countries

Type of success	Definition of success	Factors that increase likelihood of success
Attitudinal	Positive changes in views of conservation goals	Project creates or develops social capital
		Communities participate in project initiation, establishment and daily management
		Equity of benefits across community members
Behavioural	Decreased off-take	Project designs develop individual and institutional capacity
		Smaller communities
		Local engagement in cultural traditions and governance
Ecological	Improved outcomes for the habitat or species of interest	Local engagement in cultural traditions and governance
		Project designs develop individual and institutional capacity
		Communities participate in project initiation, establishment and daily management
Economic	Variety of livelihood benefits	Project designs develop individual and institutional capacity
		Communities with tenure rights

Source: Brooks et al. (2012).

communities and the environment. Brooks et al.'s systematic review indicates that if community members perceive an equity of benefits, for example then their perceptions of conservation goals may be improved. Smaller communities have been found to be more likely to decrease off-take, a metric of successful conservation behaviour. Custodians who are active in governing the area and practising their cultural traditions lead to ecological success for the target habitat or species of interest. Finally, land tenure can provide livelihood benefits, such as agricultural security, leading to economic benefits. Porodong et al. (2011: 337) note: "Success of conservation efforts should not only be measured by the size of area we manage to conserve but also by how far we are willing to recognize the significance of the interaction between local communities and the environment". Pungetti et al. (2012: 2) concur, arguing that conservation efforts should "only expect to achieve optimum results when undertaking an approach that involves and fully integrates cultural perspectives in nature conservation".

In 2008, the World Commission on Protected Areas amended the governance types for protected areas to include Indigenous and community conserved areas and territories (Dudley, 2009). Territories and areas conserved by Indigenous Peoples and local communities (ICCAs) are "natural and/or modified ecosystems containing significant biodiversity values, ecological services and cultural values, voluntarily conserved by Indigenous peoples and local communities, both sedentary and mobile, through customary laws or other effective means" (IUCN, 2003). The prevalence and extent of ICCAs are unknown, however, the ICCA Consortium formed during the 2008 World Conservation Congress estimates their extent to equal that of protected areas (nearly 13 per cent of the Earth's terrestrial surface) (ICCA Consortium, 2013).

A wide variety of sites qualify as ICCAs with their own distinctive local contexts, however, the ICCA Consortium website notes three defining characteristics:

1) A strong relationship exists between indigenous or local communities (sedentary or mobile) and their physical environment (such as a given ecosystem, habitat, resource or species) as a result of cultural, social, economic and other reasons.
2) The concerned indigenous peoples or local community plays a key role in making decisions about the management of the ecosystem, area or species. The community possesses (in law or in practice) the power to make and enforce key management decisions regarding the territory and resources.
3) The voluntary management decisions and efforts of the concerned community lead to, or at least are well in the process of leading to, the conservation of biodiversity, habitats, species, ecological functions and associated cultural values, regardless of the original management objectives as perceived by the community.

(ibid.)

The ICCA Consortium website's first defining characteristic of ICCAs describes social-ecological systems as: "A strong relationship exists between indigenous or local communities (sedentary or mobile) and their physical environment (such as a given ecosystem, habitat, resource or species) as a result of cultural, social, economic and other reasons" (ibid.). Berkes and Turner (2006) attribute the development of conservation to two different conceptual mechanisms: the ecological understanding model and the depletion crisis model. The ecological understanding model roots conservation in the slow accumulation of knowledge describing the human-environment relationship, including observations of changes in ecosystems and natural cycles, lessons from the past and other environments, observations of predator-prey interactions and migration cycles, monitoring resource use and ecosystem engineering by humans (Turner and Berkes, 2006). The depletion crisis model posits that conservation behaviour is developed following a realization of resource scarcity. In this model, conservation is often the result of a resource crisis caused by human action, such as overharvesting (Berkes and Turner, 2006).

Gunderson and Holling's (2001) adaptive cycle illustrates transformations of social-ecological systems over time, from resource exploitation (r) to conservation (k) (Figure 28.1).

The shift from conservation (K) to release (Ω) is linked by a single long arrow and indicates a rapidly changing situation (e.g. disturbance via "small and fast processes"). Once the effects of

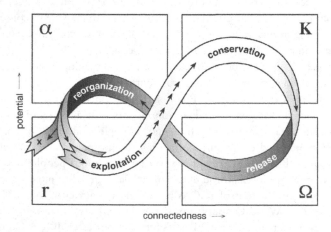

Figure 28.1 Transformations in social-ecological systems described by the adaptive cycle

Source: Gunderson and Holling (2001) *Panarchy*, edited by Lance H. Gunderson and C.S. Holling: Figure 2-1 (page 34). Copyright © 2002 Island Press. Reproduced by permission of Island Press, Washington, DC.

a disturbance are observed in the system in the release phase (Ω), there is potential for reorganization of the system (α) towards a new trajectory of exploitation (r) to conservation (K), or towards an alternative state (x). Berkes and Turner (2006) similarly describe this transformation with their ecological depletion model, where conservation occurs as a reaction to the observation of a sudden depletion of a resource.

Conversely, Berkes and Turner's (2006) ecological understanding model mirrors the multiple small arrows in the adaptive cycle between exploitation (r) and conservation (K), which represent a slowly changing situation (e.g. accumulation of ecological memory and social memory via "large and slow" processes). Ecological memory is described by Bengtsson et al. (2003: 389):

> For ecosystems to reorganize after large-scale natural and human-induced disturbances, spatial resilience in the form of ecological memory is a prerequisite. The ecological memory is composed of the species, interactions and structures that make ecosystem reorganization possible, and its components may be found within disturbed patches as well in the surrounding landscape.

McIntosh (2000: 142) notes that social memory

> links the symbolic reservoir, that is, the authoritative set of symbols, metaphors, and myths that derive from core values (deep-time and unarticulated) fundamental to a society's perception of the meaning of space and of its own place in nature and social action.

and can be "likened to a calculus that allows society to reinvent responses to the environment in the face of environmental change". Social-ecological memory is one form of adaptive governance of ICCAs, with the "remember" function linking alternative states of the adaptive cycle (Figure 28.2).

Here we ask, how can social-ecological processes beyond protected areas inform conservation policy and practice on a landscape scale?

The study area

The third largest island in the world (743,000 sq. km), Borneo is one of the most biodiverse places on Earth. Over 700 tree species have been recorded in a 10-hectare plot – the number of tree species found in Canada and the United States combined (WWF Germany, 2005). Malaysian Borneo is comprised of the two states of Sarawak and Sabah, with the small country of Brunei between them in the northern part of the island and Indonesian Borneo to the south. Sabah covers 73,631 sq. km, less than 10 per cent of the island of Borneo (Reynolds et al., 2011). A variety of protected areas including forest reserves, virgin jungle reserves, wildlife reserves and state parks, cover over 13,500 km², or over 18 per cent of Sabah, and Sabah plans to expand the protected network to cover 30 per cent of the land mass over the next ten years (Sario, 2013) (Figure 28.3).

However, Borneo is undergoing widespread conversion of forests to oil palm plantations; oil palm expansion accounted for 86 per cent of deforestation in Malaysian Borneo from 1995–2000 (WWF Germany, 2005) and the state of Sabah had near-complete clearance by 2010 of forest beyond Permanent Forest Reserves and State Parks (Reynolds et al., 2011).

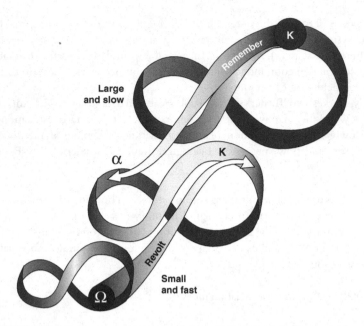

Figure 28.2 Social–ecological memory accumulated through "large and slow" processes and innovation developed through "small and fast" processes can shift an adaptive cycle to an alternative state ("remember" or "revolt")

Source: Gunderson and Holling (2001) *Panarchy*, edited by Lance H. Gunderson and C.S. Holling: Figure 3-10 (page 75). Copyright © 2002 Island Press. Reproduced by permission of Island Press, Washington, DC..

The population of Sabah doubled from 1.34 million in 1987 to 3.12 million in 2010, with a population density around 40 people per sq. km (ibid.). Seventy per cent of Sabah's population live in rural areas and conserve forests in a myriad of locally developed, culturally specific ways (Tongkul, 2002). Sabah has high rates of cultural diversity – Indigenous People comprise 60 per cent of Sabah's population, with over 30 ethnic groups speaking 50 languages and 80 dialects (Tongkul, 2002). Sabah is exploring the potential integration of territories and areas conserved by Indigenous Peoples and local communities (ICCAs) with its fragmented protected area network, as demonstrated by the Sabah ICCA Review (Majid Cooke and Vaz, 2011; Vaz and Agama, 2013). In this study, we present survey results on natural resource use, regulations and community-based conservation of respondents from 24 villages across Sabah and employ three case studies to contextualize these findings. Finally, we present a framework for conservation engagement using social-ecological systems theory (Folke et al., 2005) to make recommendations for the successful integration of ICCAs with protected area networks.

Methods

Fieldwork was conducted in Sabah, Malaysian Borneo, from October 2010 to April 2011, following a 6-week pilot trip in Sabah and Sarawak from June to August 2010. The research employed key informant interviews, oral and written questionnaires, and participant observation on-site and at conservation planning and capacity building workshops. Access was facilitated by researchers from Universiti Malaysia Sabah and the non-governmental organizations Global Diversity Foundation (GDF) and Partners of Community Organisations (PACOS).

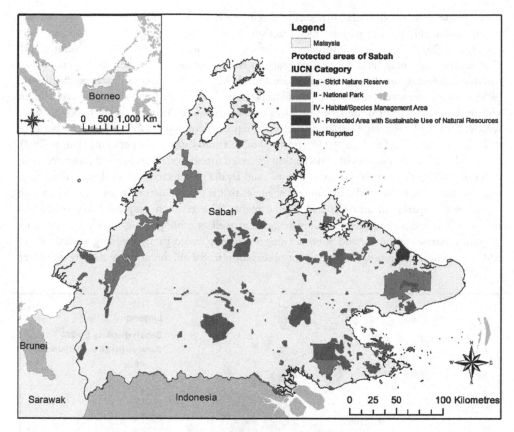

Figure 28.3   Protected areas in Sabah, Malaysia by IUCN category

Source: Global Administrative Areas (2012), UNEP-WCMC (2012).

PACOS serves as an umbrella organization for community organizations in 116 villages in 23 of the 25 Sabah districts and holds quarterly capacity-building workshops (i.e. Joint Technical Sessions) for community representatives. In addition to participant observation over eight months, we interviewed and surveyed participants at two PACOS workshops and an additional training (Joint Technical Session, 18–19 November 2010; paralegal training for interior communities, 23–25 November 2010; and Joint Technical Session, 16–17 February 2011). Employing a Malay translator, we piloted an oral questionnaire with two respondents at the first PACOS workshop and an improved version with another five respondents at the second PACOS workshop. We further improved the structure and clarity of the questionnaire, decreased its length, and reformatted it as a written questionnaire in Malay with the assistance of our translator (SI 5-1 and 5-2).

We distributed 49 written questionnaires on the first day of the third workshop and 38 questionnaires were completed and returned to us on the second day. Some representatives did not return to the workshop for the second day due to other obligations, accounting for some of the incomplete surveys. Once the surveys were handed in, we distributed one-page brief follow-up surveys for respondents who had indicated their village had a current or former village forest reserve, spirit forest or *tagal/bombon* resource management system, respectively. *Tagal* (or *bombon*) is a customary natural resource management system featuring spatial or temporal regulations on resource use in Sabah. In Kadazan, *tagal*, translated into English as "prohibition", is one way that people sustainably use natural resources, practising *gompi-guno*, meaning "protect and use" (Vaz and Agama, 2013). In Kadazandusun, *bombon* means "the controlled river" (Vun and Yung, 2010). Due to the extremely high cultural diversity in Sabah, we use "spirit forests" to encompass forests where ancestors or other "spirits" are found. We coded responses in statistical software SPSS v. 21 and removed duplicate respondents randomly. Duplicate respondents occurred when villages had sent two or more representatives to the workshop. After duplicate responses were removed, we had completed questionnaires from representatives of 24 villages, or 20.7 per cent of PACOS villages in Sabah (n = 116), from 10 of 25 rural districts (40 per cent) (Figure 28.4).

We identified cases studies of conservation of sacred forests and rivers beyond protected areas in Sabah: village forest reserves, "spirit forests", and *tagal*. The selected case studies each featured high profile interactions with government agencies that regulated natural resource use in Sabah (i.e. Forest Department and Fisheries Department). The research employed key informant interviews, scientific and grey literature reviews, as well as ethnographic tools including participant observation on-site and at conservation planning and capacity building workshops. In addition to over six months of participant observation in Sabah, the primary data for the three

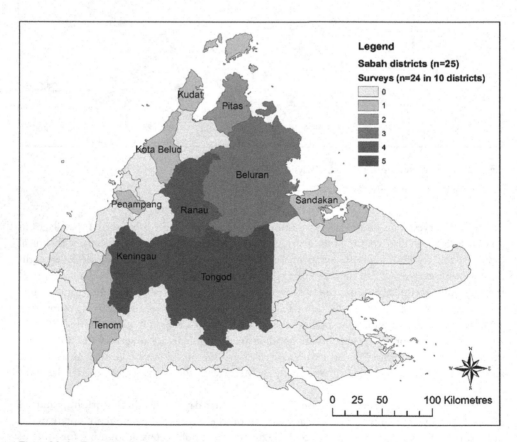

Figure 28.4 Districts of Sabah (n = 25)

Note: Survey respondents represented 24 of 116 PACOS villages (20.7 per cent) in 10 of 25 districts (40 per cent).

Source: Global Administrative Areas (2012).

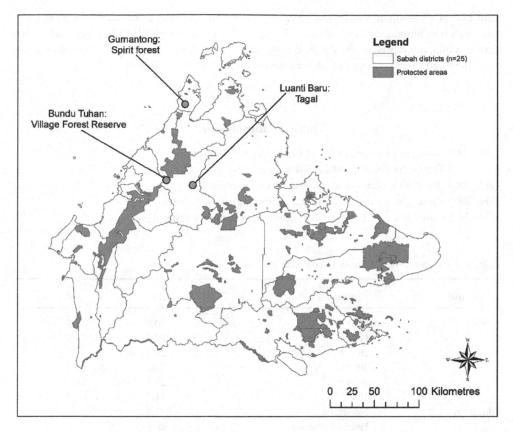

Figure 28.5 Case studies of the conservation of sacred forests and rivers in Sabah: Bundu Tuhan village forest reserve, Gumantong spirit forest and Luanti Baru *tagal*

Source: Global Administrative Areas (2012), UNEP-WCMC (2012).

case studies on village forest reserves (Bundu Tuhan), "spirit forests" (Gumantong) and *tagal* (Luanti Baru) came from site visits and key informant interviews on 2–4 December 2010, 6 March 2011 and 23 January 2011, respectively (Figure 28.5).

The primary data for the case study of Gumantong comes from an interview with Porodong Mogilin, Native Chief Representative of Matunggong Native Court in Bavanggazo, Kudat, and meetings of community leaders from the 13 villages surrounding Gumantong. Dr. Paul Porodong facilitated access and served as a Rungus translator in Kudat. A well-respected native of Bavanggazo, one of the villages surrounding Gumantong, his close ties with the village and interviewees enabled frank discussion and minimized reporting error due to access and control issues and power inequalities between researcher and interviewees.

Limitations

The reporting of natural resource management strategies by PACOS village representatives and the three case studies may not form a representative sample of villages in Sabah. In terms of survey responses, the familiarity of respondents with the forms of natural resource management in their home areas may vary. Thus, it is important to note that questionnaire results may suffer from bias of omission and additional research in the home villages across Sabah may uncover other forms of biocultural conservation. The context of the quarterly PACOS meetings helped limit this bias; the representatives' regular representation of their villages at these meetings may have improved their awareness of conservation activities in their village.

Results

Questionnaire results

Table 28.2 shows the characteristics of the survey respondents.
Table 28.3 shows the characteristics of the survey respondents' villages.
Table 28.4 shows the natural resource use survey responses.
Table 28.5 shows the natural resource regulations survey responses.
Table 28.6 shows the community-based conservation survey responses.

Table 28.2 Characteristics of survey respondents (n = 24)

Characteristic	Response	(%)	N
Gender	Male	70.8	17
	Female	29.2	7
Age	20–29	29.2	7
	30–39	29.2	7
	40–49	29.2	7
	50+	11.5	3
Education	No school	42.1	1
	Finished primary	20.8	5
	Finished secondary	66.7	16
	Finished form 6[1]	8.3	2
Ethnicity	Dusun putih, Dusun sokid, Kimaragang, Kedayan, Lundayeh, Makieng, Mangkaak, Murut, Nunuk Ragang, Tindal	4.2 each	1 each
	Dusun, Liwan, Lobu, Rumanau, Rungus, Sungai, Tombonuo	8.3 each	2 each
Religion	Christian	95.8	23
	Anglican	25	6
	Protestant Church in Sabah	8.3	2
	Roman Catholic	29.2	7
	Seventh Day Adventist	4.2	1
	Borneo Evangelical Church	25	6
	Gerakan Perkabaran Injil	4.2	1
	Muslim	4.2	1
Highest community role	Chairman of board	45.8	11
	Religious leader	25	6
	Village member	29.2	7

Note:
1 Form 6 is a 1.5-year pre-university course undertaken upon completion of secondary school.

Table 28.3 Characteristics of survey respondents' villages (n = 24)

Characteristic	Response	(%)	N
Access to village	Paved road	12.5	3
	Dirt road	87.5	21
Schools in village	Primary school	79.2	19
	Secondary school	8.3	2
Population	<250	12.5	3
	250–499	41.7	10
	500–749	33.3	8
	750–999	12.5	3
Ethnic groups	Bajau	4.2	1
	Bisaya	4.2	1
	Bugis	4.2	1
	Bundu	4.2	1
	Dusun	8.3	2
	Dusun Putih	4.2	1
	Dusun Sokid	4.2	1
	Kedayan	4.2	1
	Kimaragang	4.2	1
	Liwan	12.5	3
	Lobu	8.3	2
	Lundayeh	4.2	1
	Makaing	4.2	1
	Mangkaak	4.2	1
	Murut	8.3	2
	Nunuk Ragang	4.2	1
	Rumanau	12.5	3
	Rungus	8.3	2
	Sinobu	4.2	1
	Sungai	12.5	3
	Tausug	4.2	1
	Tidong	4.2	1
	Tindal	4.2	1
	Tombonuo	4.2	1
Religion	Christian	100	24
	Anglican	33.3	8
	Borneo Evangelical Church	25	6
	Gerakan Perkabaran Injil	4.2	1
	Protestant Church in Sabah	20.8	5
	Roman Catholic	20.8	5
	Seventh Day Adventist	8.3	2
	Muslim	25	6
	Animist	4.2	1

Table 28.4 Natural resource use survey responses (n = 24)

Question	Response	(%)	N
Why gather natural resources	For own use	100	24
	To sell at local markets	37.5	9
	To sell to middlemen	4.2	1
Where natural resources gathered	Forests	95.8	23
	Rivers	100	24
	Caves	37.5	9
	Sea	8.3	2
	Backyard farms	8.3	2
What natural resources gathered in forests	Wood for building	95.8	23
	Firewood	95.8	23
	Medicinal plants	91.7	22
	Fruits, vegetables and roots	87.5	21
	Hunting animals	91.7	22
What natural resources gathered in rivers	Fish	100	24
	Snails	70.8	17
	Shrimp	8.3	2
	Tortoises	4.2	1
	Eels	4.2	1
	Frogs	4.2	1
What natural resources gathered in caves	Bird nests	16.7	4
	Hunting animals	33.3	8
Land tenure of forests where natural resources are gathered	State land	12.5	3
	Forest Department land	41.7	10
	Native Reserve	29.2	7
	Native Title	20.8	5
	Native Customary Rights	83.3	20
	Pending land claim	8.3	2

Table 28.7 shows the village forest reserves survey responses.
Table 28.8 shows the "spirit forests" survey responses.
Table 28.9 shows the *tagal/bombon* survey responses.

Case studies

The three case studies of village forest reserves, "spirit forests" and *tagal* are located in the matrix beyond the protected areas in Sabah in three distinct local contexts. As expected, considering the high levels of diversity in Sabah's landscape (cultural, linguistic and biological), the case studies vary in a plethora of ways, *inter alia* ethnicity, religion, village size and population, ecology, and forms of natural resource use. However, in all three of the case studies, resources conserved via biocultural conservation have been targeted by government agencies that regulate natural resource use, leading to local engagement, and in some cases, conflict. Government engagement has raised the profile of these case studies as potential ICCAs. We consider their environmental histories and apply social-ecological systems theory to develop a framework and recommendations for the inclusion of sacred forests and rivers in protected area networks as ICCAs.

Table 28.5 Natural resource regulations survey responses (n = 24)

Characteristic	Response	(%)	N
Spatial or temporal rules for natural resource management	Yes	41.7	10
	No	45.8	11
	Unknown	4.2	1
Who makes spatial and temporal regulations for natural resource use?	Ketua Kampung (village head)	33.3	8
	JKKK (village committee)	20.8	5
	Religious leader	8.3	2
	Government	8.3	2
	Ancestors, inherited beliefs or customary law (*adat*)	8.3	2
Over the course of respondent's lifetime, regulations perceived to	Have stayed the same	37.5	9
	Have become stricter	8.3	2
	Have become less strict	16.7	4
Penalties exist for breaking regulations	Yes	45.8	11
Penalties enforced	Yes	45.8	11

Table 28.6 Community-based conservation survey responses (n = 24)

Conservation type	Time period	(%)	N
Village forest reserves	Current	66.7	16
	Past	70.8	17
"Spirit forests"[1]	Current	37.5	9
	Past	54.2	13
Tagal/bombon[2]	Current	33.3	8
	Past	16.7	4
Water catchment areas	Current	20.8	5
	Past	12.5	3
Village *adat* or regulations by JKKK	Current	8.3	2
None	Current	12.5	3
	Past	8.3	2

Notes: 1
2 *Tagal/bombon* is a resource management system that employs customary spatial and/or temporal regulations; it is a local form of taboo.

Village Forest Reserve case study: Bundu Tuhan, Ranau

In 1968, boundary markers were placed for the Bundu Tuhan Native Reserve (approximately 1,263 ha), officially declared by the Chief Minister of Sabah in 1983 (Doolittle, 2001). This land included and surrounded the village and did not overlap with Kinabalu Park. However, in 1984, the Forest Department gazetted Tenompok Forest Reserve, with boundaries overlapping the Bundu Tuhan Native Reserve (Majid Cooke and Vaz, 2011). The community of Bundu Tuhan has prevented the Forest Department from measuring or mapping the contested area. The gazettement of Bundu Tuhan's Native Reserve legally fended off appropriation of land by external forces and incorporated a greater area than could have been procured by Native

Table 28.7 Village forest reserves survey responses (n = 24)

Characteristic	Response	(%)	N
Size (acres)	20	4.2	1
	30	4.2	1
	200	8.3	2
	300	8.3	2
	500	4.2	1
	1,200	4.2	1
	2,000	4.2	1
	5,000	4.2	1
	6,000	4.2	1
	10,000	4.2	1
Land tenure	Forest Department land	16.7	4
	Native Reserves	20.8	5
	Native Title	12.5	3
	Native Customary Rights	50	12
	Water catchment	12.5	3
Ecosystem services	Gathered natural resources	8.3	2
	Water source	70.8	17
Establishment	Pre-British rule	41.7	10
	WWII–Independence	8.3	2
	Independence–1980	8.3	2
	1980–2000	8.3	2
	2000–present	4.2	1
Clearance	British rule–WWII	4.2	1
	WWII–Independence	4.2	1
	Independence–1980	4.2	1
	2000–present	12.5	3
Why cleared	Religious change	21	5
	Converted to agriculture or government projects	8.3	2

Title (individual land titles as opposed to communal). However, as the political economy of the region shifted, community members in Bundu Tuhan expressed frustration with land being locked up in the Native Reserve and unable to serve as credit or to be sold for profit (Doolittle, 2001). The Forest Department has presented a co-management proposal to the community, but community members express concern that the Forest Department might change the reserve allocation from class I to class II to allow logging (Majid Cooke and Vaz, 2011).

In 2005, community members observed that game populations were decreasing in their forest and a nearby community renowned for its vegetables, Kundasang, was suffering from soil dryness due to decreased river flow from the loss of its forest. Committees were formed to develop rules and policies for the Native Reserve using customary law (*adat*), including *Majlis Permuafakatan* (Community General Assembly), *Jawatankuasa Pemegang Amanah Hutan Simpan Bundu Tuhan* (Native Reserve Board of Trustees) and *Jawatankuasa Bombon* (conservation committees for forest and rivers) (Majid Cooke and Vaz, 2011) (Figure 28.6).

Bundu Tuhan has expressed interest in recognizing the ~60 per cent of the Native Reserve conserved as a *hutan simpan* (protected forest) as a community-managed ICCA (Majid Cooke and Vaz, 2011).

Table 28.8 "Spirit forests" survey responses (n = 24)

Characteristic	Response	(%)	N
Size (acres)	3	4.2	1
	10	4.2	1
	15	4.2	1
	500	4.2	1
	1,000	4.2	1
	2,000	4.2	1
	10,000	8.3	2
Land tenure	Forest Department land	8.3	2
	Native Reserves	8.3	2
	Native Title	12.5	3
	Native Customary Rights	50	12
	Agricultural land, hunting areas and water catchment	4.2	1
Ecosystem services	Gathered natural resources	8.3	2
	Water source	45.8	11
Establishment	Pre-British rule	29.2	7
	British rule–WWII	4.2	1
	WWII–Independence	4.2	1
	Independence–1980	8.3	2
	1980–2000	4.2	1
State of belief	No one believes	20.8	5
	A few believe	16.7	4
	Most believe	12.5	3
Clearance	British rule–WWII	4.2	1
	2000–present	12.5	3
Why cleared	Religious change	21	5
	Lost land to Forest Department	4.2	1

Table 28.9 *Tagal/bombon* survey responses (n = 24)

Characteristic	Response	(%)	N
Reason	Fish harvest	29.2	7
	Tourism	8.3	2
	Protect spawning grounds of fish	4.2	1
Fish harvest frequency	Twice a year	4.2	1
	Annually	12.5	3
	Every 2 years	8.3	2
	Every 3 years	4.2	1
Establishment[1]	1980–2000	4.2	1
	2000–present	29.2	7

Note:

1 None of the respondents said that the *tagal/bombon* had ended.

Figure 28.6 Bundu Tuhan committee to develop rules and policies for their Native Reserve using customary law, December 2010

Source: Majid Cooke and Vaz (2011).

Spirit Forest case study: Gumantong, Kudat

In Kudat, northern Malaysian Borneo, the Rungus people conserve *puru*, patches of forest approximately 1 hectare in size and inhabited by *rogon* ("spirits"). Five villages surround the highest hill and large *puru*, Gumantong (Figure 28.7).

"Dancing animals" are believed to reside on the mountaintop. These *kopizo* ("omens") are said to goad the interloper into laughing, and then dying upon his arrival at home (Porodong, 2010; Mogilin, 2011) (Figure 28.8).

This spiritual edict conserved the mountaintop in the first half of the twentieth century. However, the effectiveness of this form of resource regulation was threatened by outsiders, specifically Iban members of a British survey team. The Iban are an Indigenous ethnic group from the neighboring Malaysian state of Sarawak. While surveying the mountaintop, the Iban hunted animals, and the local Rungus believed they had hunted the "dancing animals" out of Gumantong's *puru*. Following their conversion to Christianity, the local Rungus people began to clear their *puru* for cropland. As a consequence of clearing these forest patches, the water table dropped to the point where the villages became reliant on the government for their water supply.

Although Kudat was formerly a mosaic of mature and fallow secondary rainforest, today the landscape of Kudat is primarily a monoculture of *Acacia mangium* due to its widespread planting in the 1980s by the Sabah Forestry Development Authority (SAFODA) for pulp production (Turnbull, Midgley and Cossalter, 1998, cited in Porodong, 2010: 24–25). The spread of Acacia

Figure 28.7 The mountaintop of Gumantong, Kudat, northern Sabah

was enabled by the pervasive use of fire in swidden agriculture, as fire catalyzes the germination of buried *Acacia mangium* seeds. *Acacia mangium* has also been shown to out-compete native species such as *Melastoma* (Osunkoya, Farah and Rafhiah, 2005, cited in Porodong, 2010: 24–25). At the end of the century, the Forest Department arrived to clear Gumantong's mountaintop of native vegetation to plant the fast-growing exotic, *Acacia mangium* (Kothari, 2006). When protesting against the proposed clearing of Gumantong for the planting of *Acacia mangium*, the communities surrounding Gumantong expressed concern that the exotic species would dry up their water source. By that point, the community had observed the connection between the *puru* and water access, and they vehemently protested against the Forest Department's planned clearing of the mountaintop. Their action attracted the attention of the United Nations Development Programme (UNDP) on Climate Change, and the Rungus villages formed a partnership. In 2007, the Forest Department gazetted a 590-hectare area including Gumantong as a Forest Reserve Class 1 (Watershed) without informing the village chiefs or native court chiefs representing the 13 communities and 3,000 villagers. Again the villages protested against the action, registering a complaint with the Chief Minister of Sabah and proposing Gumantong be recognized as a Native Forest Reserve, a form of territory or area conserved by Indigenous Peoples and local communities (ICCA) that would provide land tenure (Sabah Publishing House, 2011).

Tagal case study: Luanti Baru

With modernization and urbanization impacting rural village communities in Sabah in the 1960s, harvesting practices shifted to include the use of explosives in fishing and the widespread logging of forests silted and polluted rivers. In 1997, a village in Penampang district successfully

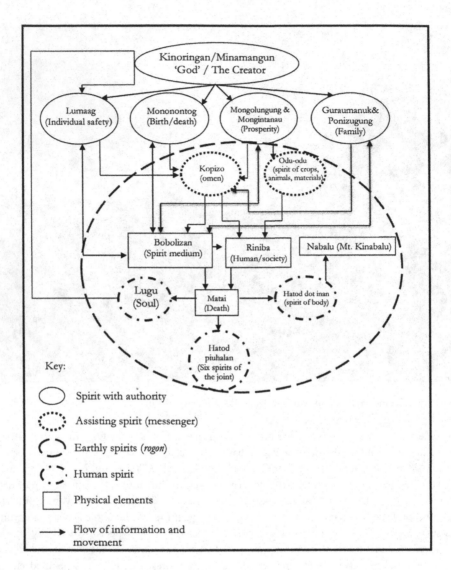

Figure 28.8 The Rungus Spirit World

Source: Reproduced from Porodong (2010).

reinstated their customary *tagal* system to reintroduce spatial and temporal fishing regulations (Vaz and Agama, 2013). As news of the improved condition of Penampang's Babagon River spread, so did renewed interest in the revitalization of *tagal* systems in communities across Sabah.

In 2002, the Sabah Fisheries Department ran a "Love your river" campaign. The village of Luanti Baru's river was full of litter, people washed their cars in it and it had become polluted and silted from logging upstream (Er et al., 2012). Community member Jeffrin Majangki remembered the many fish in the river of his childhood, and began work with his community to restore the river's fish population. The village approached the Fisheries Department for support to clean up the river; all rivers in Sabah were technically under governmental control. Resulting from   that collaboration, village, district and state-level committees were formed.

The zoning system adopted was a traffic light scheme, with red zones being no-take, yellow zones sometimes deemed fishable by the oversight committee, and green zones always fishable. Once large numbers of fish returned to the area, a tourism project was developed for fish massage, which was so successful that the community restricted fishing in all zones (Er et al., 2012) (Figure 28.9).

Tourists visit the river and wade in, submerging their feet. Fish food is sprinkled on the water to attract fish and the fish come to eat the fish food, but also to take nibbles of dead skin on the feet. This "natural spa" is especially popular with domestic tourists enjoying a weekend trip outside of the city. Luanti Baru received 50,000 tourists in 2013 alone (*Daily Express*, 2014). Tourism funds are managed by a committee and fund community projects, such as children's education. One challenge of *tagal* is that improved management practices in one section of the river do not necessarily buffer effects upstream. Even in the highly successful fish tourism project, logging upstream creates massive silt deposits that can stymie tourism until the river clears. The positive side of this challenge is that communities are incentivized to come together to advocate for more river-friendly practices upstream, which can clean up entire rivers instead of one zone.

The Sabah Fisheries Department conducted research on the effectiveness of *tagal*, and in 2003, *tagal* was codified in Sabah law in the Sabah Inland Fisheries and Aquaculture Enactment. Some villages with *tagal* projects had not formerly practised *tagal*; its meaning has expanded to encompass a range of community-based fisheries projects, regardless of the links to customary law. Several villages reportedly made annual tourism revenues between RM30,000 and RM50,000 (*Borneo Today*, 2016). *Tagal* systems have been (re)introduced by the Fisheries Department in 554 tagal centres across 223 rivers in Sabah to date and delete "with plans to add another 600 zones by 2016". Replace (Daily Express, 2014) citation here with (Daily Express, 2019). Add to Bibliography: Daily Express, 2019. Proper centre for tagal operators opens in Ranau. 25 October. *Tagal* has even been promoted in the neighbouring Malaysian state of Sarawak, where different ethnic groups with different customary practices live (Vun and Yung, 2010; *Borneo Today*, 2016).

Discussion

Fragmented protected area networks seek to increase connectivity and incorporate the conservation practices of local communities in the human-modified landscapes comprising 87 per cent of the Earth's terrestrial surface (Chazdon et al., 2009). Countries are identifying territory or area conserved by Indigenous Peoples and local communities (ICCA) to augment their protected area networks and ICCAs have been recognized as a form of IUCN protected area governance (Dudley, 2009). We surveyed natural resource use, regulations and community-based conservation in bioculturally diverse Sabah, Malaysian Borneo, and identified factors impacting the successful integration of ICCAs into protected area networks.

Our survey indicates that rural villages in Sabah employ *inter alia* village forest reserves (66.7 per cent), "spirit forests" (37.5 per cent), *tagal/bombon* (33.3 per cent), and water catchment systems (20.8 per cent) to conserve their natural resources. Conservation is practised to conserve ecosystem services including water sources (70.8 per cent) and gathering natural resources (8.3 per cent), which we also observed in the case studies (e.g. Gumantong's water catchment area and Luanti Baru's fish massage tourism), for aesthetic/moral reasons (e.g. to protect heritage sites in ancestral lands) and for spiritual reasons as in "spirit forests" (37.5 per cent), (e.g. Gumantong's "dancing animals" or *rogon* ("spirits") in *puru*). Our research indicates that the conservation of sacred forests and rivers is not static, but rather can be changed by events such

Figure 28.9 *Tagal*, customary spatial and temporal fishing regulations, (a) are painted on a river rock as a reminder to villagers in Buayan; (b) Luanti Baru's *tagal* is hailed as one of the most successful examples of revitalization of *tagal* in Sabah; (c) harvesting of fish has been abandoned to support a fish massage tourism project, which received 50,000 tourists in 2013 alone

Sources: (a) Photo: Nasiri Sabiah; (c) *Borneo Today* (2016).

as religious conversion (the reason cited for village forest or spirit forest clearance by 21 per cent of respondents, e.g. Gumantong's clearance of *puru* upon conversion to Christianity), modernization (e.g. Luanti Baru's introduction of plastics and car washing polluting the river) and the influence of outsiders on natural resource use (e.g. Gumantong's Iban hunting of "dancing animals", Bundu Tuhan's observance of Kundasang's soil aridity following forest clearance, and the adoption of *tagal* by communities who had not traditionally practised it). Communities can respond to these shifts by practising adaptive management of the social-ecological system, such as (re)protecting the Gumantong hilltop by collaborating with the UNDP climate change programme.

In the Gumantong case study, the influence of outsiders, i.e. changing hunting practices and religious conversion, disturbed the social-ecological system. The clearance of *puru* decreased the level of the water table, shifting the social-ecological system from conservation (K) to release (Ω). The input of a government water supply system was the first stage of reorganization (α) and then a "revolt" occurred as communities banded together to protect their largest *puru* in order to conserve their watershed (K) in an alternative state. Inherent in this adaptive process was the recognition of the link between forest clearance and the drop in the water table, which enabled the "small and fast" shift towards conservation of Gumantong (Figure 28.10).

Adaptive governance enabled the means of conservation (the *how* and *why*) to shift to an alternative adaptive cycle, from "spirit forests" ("dancing animals" and *rogon*) to water catchments providing ecosystem services (water conservation). This is evident in our survey, where we found the decreased use of village forest reserves (70.8 per cent in the past to 66.7 per cent in the present) and "spirit forests" (54.2 per cent in the past to 37.5 per cent in the present) over time, but also an increase in the use of water catchment areas (12.5 per cent in the past to 20.8 per cent in the present) and *tagal/bombon* (16.7 per cent in the past to 33.3 per cent in the present).

Religious conversion does not in itself result in the clearance of sacred forests, as occurred with the conversion to Christianity and the clearance of *puru*. While none of our survey respondents self-identified as Animist and only one respondent noted Animism was practised in his or her village, 37.5 per cent of respondents said they had "spirit forests", 12.5 per cent said most people believe in the "spirit forests", 16.7 per cent said a few people believe in them and 8.3 per cent said their ancestors, inherited beliefs or customary law (*adat*) set regulations for natural resource use. In other parts of Southeast Asia, conservation of sacred forests has been maintained in the face of religious conversion, with the reasons underpinning conservation shifting to aesthetic/moral reasons (Keyes, 1996). In Indonesia, Protestant Christianity incorporated many of the structures of local *adat* governance systems (Cooley, 1967). Although we do not observe similar syncretism of area-based conservation measures in Sabah, religious leaders were cited as rule-makers regarding natural resource use in 7.7 per cent of our survey villages and faith leaders convened community conservation meetings in Lokos, a village we observed over the course of our research. Collaboration with religious leaders who interpret environmental stewardship as part of their religious teachings has been successful in many contexts (Palmer and Finlay, 2003; Hitzhusen and Tucker, 2013) and can retain the functionality of the social-ecological system despite the disturbance of religious conversion. The motivations underpinning conservation can be reframed to accommodate the new belief system, for example protecting forest patches as a moral duty as opposed to the veneration of ancestral "spirits". The process of religious conversion occurred at the village level in Sabah, and the degree of acceptance of animist beliefs and integration with the conversion religion varied by missionary, even within the same branch of religion. Therefore, the clearance of the *puru* by the Rungus cannot be attributed solely to their religious conversion, but may be due to the interplay of multiple simultaneous shifts, including the method of religious conversion, modernization, and increased market access.

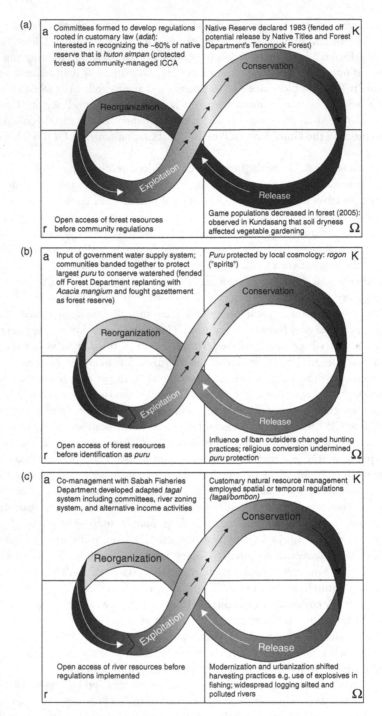

Figure 28.10 The case studies of (a) a village forest reserve; (b) "spirit forest"; and (c) *tagal/bombon* in Sabah pass through the phases of the adaptive cycle

Source: Modified from *Panarchy* edited by Lance H. Gunderson and C.S. Holling: Figure 3-10 (page 75). Copyright © 2002 Island Press. Reproduced by permission of Island Press, Washington, DC.

As communities reorganized their social-ecological systems to address polluted river zones, traditional forms of natural resource management (*tagal/bombon*) were revitalized and adapted to the current system. An observed shift in the ecological system to an undesirable state can spur conservation actions ("revolt" developed via "small and fast" processes) or formerly successful interventions can be revitalized and adapted (the "remember" function developed via "large and slow" processes).

In addition, it is important to note that social-ecological systems (here we focus on ICCAs) do not occur in a vacuum, but rather observation of other communities' experiences (i.e. other adaptive cycles) can spur conservation action before an undesirable state, such as a depletion crisis, is reached. In our Sabah case studies, this is observed in Bundu Tuhan's observation of Kundasang and in the rollout of *tagal* to river zones across Sabah (e.g. only 16.7 per cent of our survey noted *tagal/bombon* systems were in place in the past, but 33.3 per cent of respondents had current *tagal/bombon* systems). Participatory capacity-building workshops bringing together community representatives with a range of experiences, such as those we observed with PACOS and the Global Diversity Foundation, can facilitate social-ecological knowledge exchanges that foster "communal memory", that is, cumulative "social memory" at the community level.

The stage of the adaptive cycle for identifying ICCAs to integrate into the protected area network is conservation (K). Formal recognition, such as co-management with a governmental agency or the determination of an IUCN protected area category and inclusion in the World Database on Protected Areas, would ideally contribute to the process indicated by the multiple small arrows linking exploitation (r) to conservation (K). As such, formal recognition would be one of many social processes accumulated over time to conserve the resource base. However, it must be noted that formally recognizing an area as an ICCA can also act as a disturbance to the system, shifting it from conservation (K) to release (Ω) (Dove et al., 2011). Negative social impacts of imposing conservation on local communities have included the restriction of land use and the loss of management rights (West and Brockington, 2006). Even in cases where land use and management rights are unchanged, formally describing an area as conserved may alter local perceptions of rights and ownership (Pathak, 2006). The process of recognizing and supporting ICCAs must acknowledge that communities may have had negative experiences or hold preconceptions of formal conservation and must include safeguards to ensure the autonomy of local custodians (Kothari, 2006).

The United Nations Declaration of Principles on the Rights of Indigenous Peoples (UNDRIP) says:

> Indigenous Nations and Peoples are entitled to the permanent enjoyment of their aboriginal ancestral historical territories. This includes air space, surface and subsurface rights, inland and coastal waters, sea ice, renewable and non-renewable resources, and the economies based on these resources.
>
> *(United Nations, 2007)*

If ICCAs include the shift in land tenure from the government to Indigenous Peoples, UNDRIP principles state that the custodians have decision-making power, including the rights to forest clearance, mineral extraction, etc. Shifting decision-making to the community level for conservation therefore does not ensure that conservation will always be their choice, but, rather, that they will be the ones making that choice. The concern that custodians will choose economic development through mineral extraction over conservation has stymied the full transfer of decision-making power to communities in favour of co-management or participatory government-led decision-making processes. The large majority of the world's biodiversity

Figure 28.11 Commercial oil palm plantation as seen from the A4 roadway in Sabah. Forest clearance on the steep hillside has led to erosion and siltation of the water below

is found in Indigenous Peoples' areas and territories, a fact that is oft cited to demonstrate Indigenous Peoples' successful environmental management to date (Stevens, 2014). It is also important to note that protected areas are not immune to similar threats such as mineral extraction. Protected Area Downgrading, Downsizing, and Degazettement (PADDD) is widespread; a study of protected areas in Africa, Asia, Latin America and the Caribbean from 1900–2010 found that natural resource extraction (oil and gas, forestry, mining, industrial agriculture, industrialization and infrastructure) accounted for 37.5 per cent of the 282 PADDD events (Mascia et al., 2014). Protected area networks are also at risk during periods of national instability. Following the 2009 coup d'état in Madagascar, illegal rosewood harvesting in Masoala and Marojejy National Parks spiked, with $130,000,000 of illegal exports occurring from January to April alone (Schuurman and Lowry, 2009).

Some communities in Sabah have expressed interest in the ICCA concept as a means to circumvent their inability to acquire tenure of farmland with slopes greater than 25 degrees. Sabah's Forest Enactment 1968 (Sabah En. 268) forbids the clearance of vegetation on slopes greater than 25 degrees, barring community land claims for agricultural use of their steep ancestral lands. However, Sabahans perceive this environmental regulation to be loosely enforced when it comes to commercial interests (Figure 28.11).

Enforcement relies on 13 officers for over 300 projects and fines (maximum RM20,000) are significantly less than the short-term economic benefits of non-compliance (Wyn, 2013). This is an important distinction: land tenure may be a prerequisite for successful community-based conservation (Brooks et al., 2012) but is not in itself sufficient. ICCAs that exhibit the attributes of successful community-based conservation espoused by Brooks et al. may be more resilient to disturbances. Of particular note to protected area network planners are the factors that increase the likelihood of ecological success (that is, improved outcomes for the habitat or species of interest): (1) project designs develop individual and institutional capacity; (2) smaller communities; and (3) local engagement in cultural traditions and governance (Chazdon et al., 2009; Brooks et al., 2012).

Local governance comprises the second defining characteristic of ICCAs:

> The concerned indigenous peoples or local community plays a key role in making decisions about the management of the ecosystem, area or species. The community possesses (in law or in practice) the power to make and enforce key management decisions regarding the territory and resources.
>
> *(ICCA Consortium, 2013)*

The survey showed that regulations of natural resource use were set by the Ketua Kampung (village head) (33.3 per cent), the JKKK (village committee) (20.8 per cent), religious leaders (8.3 per cent), ancestors, inherited beliefs or customary law (*adat*) (8.3 per cent) and by the government (8.3 per cent). Our research found that *how* and *why* a community conserves its resources changes over time; 8.3 per cent of survey respondents perceived natural resource use regulations to have become stricter and 16.7 per cent perceived them as becoming less strict over the course of their lifetimes. Just as protected area management has shifted towards adaptive management of ecological systems, protected areas that include people should also include adaptive governance of social systems (Scoones, 1999; Dove et al., 2011). ICCAs are inclusive of systems where conservation has developed over time through an ecological understanding model. The ICCA Consortium describes such systems, noting:

> The voluntary management decisions and efforts of the concerned community lead to, or at least are well in the process of leading to, the conservation of biodiversity, habitats, species, ecological functions and associated cultural values, regardless of the original management objectives as perceived by the community.
>
> *(2013)*

Communities may engage with the ICCA recognition process to gain land tenure, and ICCAs developed through the ecological understanding model may require additional facilitation and support. Successfully integrating ICCAs into the protected area network a few at a time could provide landscape-scale rewards if the "proof of concept" engages neighbouring communities, as we found in the case studies in Luanti Baru (*tagal*) and Bundu Tuhan (village forest reserve).

Conclusion

As alternative existing forms of conservation in the landscape, territories and areas conserved by Indigenous Peoples and local communities (ICCAs) are of interest as conservation expands into the mosaic of land uses beyond protected areas. ICCAs take a myriad of forms, from "spirit forests" to village forest reserves, church forests and seasonal or temporal river resource regulations. The breadth of local conservation approaches and their dynamic nature over time can be challenging for conservation planning. However, sacred forests and rivers can make valuable contributions in a new paradigm of community conservation – one that acknowledges conservation rooted in ecological understanding, supports conservation in the face of social and ecological changes, and recognizes the autonomy of local custodians in the process. Just as the adaptive management of protected areas formerly viewed as static ecological systems has become best practice, so will adaptive governance of sacred forests and rivers by local communities beyond protected areas.

References

Agrawal, A. and Gibson, C. C. 1999. Enchantment and disenchantment: The role of community in natural resource conservation. *World Development* 27, 629–649.

Bengtsson, J., Angelstam, P., Elmqvist, T., Emanuelsson, U., Folke, C. et al. 2003. Reserves, resilience and dynamic landscapes. *Ambio: A Journal of the Human Environment* 32, 389–396.

Berkes, F. and Turner, N. J. 2006. Knowledge, learning and the evolution of conservation practice for social-ecological system resilience. *Human Ecology* 34, 479–494.

Borneo Today 2016. Sabah Fisheries Department willing to develop tagal system in Sarawak. Available at: www.borneotoday.net/sabah-fisheries-department-willing-to-develop-tagal-system-in-sarawak/ (accessed 11 November 2016).

Brooks, J. S., Waylen, K. A. and Borgerhoff Mulder, M. 2012. How national context, project design, and local community characteristics influence success in community-based conservation projectsv *Proceedings of the National Academy of Sciences of the U.S.A.* 109, 21265–21270.

Chazdon, R. L., Harvey, C. A., Komar, O., Griffith, D. M., Ferguson, B. G., et al. 2009. Beyond reserves: A research agenda for conserving biodiversity in human-modified tropical landscapes. *Biotropica* 41, 142–153.

Cooley, F. L. 1967. Allang: A village on Ambon Island. In T. Koentjaraningra (ed.), *Villages in Indonesia*. Ithaca, NY: Equinox Publishing.

Daily Express 2011. Matunggong villagers unhappy over the gazetting of 590ha. 15 November.

Daily Express, 2019. Proper centre for tagal operators opens in Ranau. 25 October.

Daily Express 2014. Tagal system for 600 more river zones in Sabah. 19 January.

Doolittle, A. A. 2001. From village land to "native reserve": Changes in property rights in Sabah, Malaysia, 1950–1996. *Human Ecology* 29, 69–98.

Dove, M. R., P. E. Sajise, and A. A. Doolittle (eds) 2011. *Beyond the Sacred Forest: Complicating Conservation in Southeast Asia*. Durham, NC: Duke University Press.

Dudley, N. 2009. The links between protected areas, faiths, and sacred natural site. *Conservation Biology* 23, 568–577.

Er, A. C., Selvadurai, S., Lyndon, N., Chong, S. T., Adam, J. H. et al. 2012. The evolvement of tagal on eco-tourism and environmental conservation: A case study in Kampong Luanti Baru, Sabah. *Advances in Natural and Applied Sciences* 6, 61–64.

Folke, C., Hahn, T., Olsson, P. and Norberg, J. 2005. Adaptive governance of social-ecological systems. *Annual Review of Environment and Resources* 30, 441–473.

Global Administrative Areas 2012. GADM database of Global Administrative Areas, version 2.0. Available at: www.gadm.org

Gunderson, L. H. and Holling, C. S. 2001. *Panarchy: Understanding Transformations in Human and Natural Systems*. Washington, DC: Island Press.

Hitzhusen, G. E. and Tucker, M. E. 2013. The potential of religion for Earth stewardship. *Frontiers in Ecology and the Environment* 11, 368–376.

Horwich, R. H. and Lyon, J. 2007. Community conservation: practitioners' answer to critics. *Oryx* 41, 376–385.

ICCA Consortium. 2013. Indigenous peoples' and community conserved territories and areas (ICCAs) . Available at: www.iccaconsortium.org/ (accessed 22 May 2014).

IUCN 2003. Vth World Parks Congress Recommendations. Durban, South Africa: IUCN.

Keyes, C. F. 1996. Being Protestant Christians in Southeast Asian worlds. *Journal of Southeast Asian Studies* 27, 280–292.

Kothari, A. 2006. Community conserved areas: towards ecological and livelihood security. *Parks* 16, 3–13.

Majid Cooke, F. and Vaz, J. 2011. *The Sabah ICCA Review: A Review of Indigenous Peoples and Community Conserved Areas in Sabah*. Kota Kinabalu: Global Diversity Foundation.

Mascia, M. B., Pailler, S., Krithivasan, R., Roshchanka, V., Burns, D. et al. 2014. Protected area downgrading, downsizing, and degazettement (PADDD) in Africa, Asia, and Latin America and the Caribbean, 1900–2010. *Biological Conservation* 169, 355–361.

Massey, A., Bhagwat, S. A. and Porodong, P. 2011. Beware the animals that dance: Conservation as an unintended outcome of cultural practices. *Society, Biology and Human Affairs* 76, 1–10.

Mcintosh, R. J. 2000. Social memory in Mande. In R. J. Mcintosh, *The Way the Wind Blows: Climate, History, and Human Action*. New York: Columbia University Press.

Mogilin, P. 2011. The Puru of Gumantong Bukit, Interviewed by Ashley Massey. 5 March 2011, Bavanggazo, Kudat.

Palmer, M. and Finlay, V. 2003. *Faith in conservation: New approaches to religions and the environment*. Washington, DC: World Bank.

Pathak, N. 2006. Lessons learnt in the establishment and management of protected areas in South Asia. Available at: http://cmsdata.iucn.org/downloads/cca_npathak.pdf (accessed 31 December 2011).

Porodong, P. 2010. Rungus death rituals and customary laws. In: H. J. M. Saidatul Nornis (ed.) *Death Rituals and Customary Laws in Sabah*. Kota Kinabalu, Malaysia: Universiti Malaysia Sabah.

Porodong, P., Lunkapis, G. J. and Sarbi, F. 2011. A brief note on conservation, swidden agriculture and indigenous community living in periphery of Imbak Canyon Conservation Area. Sabah: Geology, Biodiversity and Socio-economic Environment Department.

Pungetti, G. and Oviedo, G. 2012. *Sacred Species and Sites: Advances in Biocultural Conservation*. Cambridge: Cambridge University Press.

Reynolds, G., Payne, J., Sinun, W., Mosigil, G. and Walsh, R. P. D. 2011. Changes in forest land use and management in Sabah, Malaysian Borneo, 1990–2010, with a focus on the Danum Valley region. *Philosophical Transactions of the Royal Society B: Biological Sciences* 366, 3168–3176.

Sario, R. 2013. Increase in Sabah land conservation areas. *The Star Online*, 18 October

Schuurman, D. and Lowry, P. P., II. 2009. The Madagascar rosewood massacre. *Madagascar Conservation & Development* 4, 98–102.

Scoones, I. 1999. New ecology and the social sciences: What prospects for a fruitful engagement? *Annual Review of Anthropology* 28, 479–507.

Secretariat of the Convention on Biological Diversity 2011. Aichi Biodiversity Targets. Available at: www.cbd.int/sp/targets/default.shtml (accessed 13 August 2013).

Stevens, S. 2014. *Indigenous Peoples, National Parks, and Protected Areas: A New Paradigm Linking Conservation, Culture, and Rights*. Tucson, AZ: University of Arizona Press.

Tongkul, F. 2002. *Traditional Systems of Indigenous Peoples of Sabah, Malaysia: Wisdom Accumulated through Generations*. Penampang: PACOS Trust.

Turner, N.J. and Berkes, F. 2006. Coming to understanding: Developing conservation through incremental learning in the Pacific Northwest. *Human Ecology* 34, 495–513.

UNEP-WCMC 2012. World Database on Protected Areas. Available at: www.wdpa.org/ (accessed 10 August 2014].

United Nations 2007. Declaration on the Rights of Indigenous Peoples. Available at: www.un.org/esa/socdev/unpfii/documents/DRIPS_en.pdf (accessed 13 November 2016).

Vaz, J. and Agama, A. L. 2013. Seeking synergy between community and state-based governance for biodiversity conservation: The role of Indigenous and Community-Conserved Areas in Sabah, Malaysian Borneo. *Asia Pacific Viewpoint* 54, 141–157.

Vun, J. L. Y. and Yung, D. C. C. 2010. River preservation: Bombon. Paper presented at Borneo Research Council Conference. Borneo Research Council, Curtin University of Malaysia, Sarawak.

West, P. and Brockington, D. 2006. An anthropological perspective on some unexpected consequences of protected areas. *Conservation Biology* 20, 609–616.

WWF GERMANY 2005. *Borneo: Treasure island at risk*. Frankfurt: WWF.

Wyn, K. T. 2013. *Malaysia: Illegalities in Forest Clearance for Large-scale Commercial Plantations*. Washington, DC: Forest Trends.

INDEX

Printed in the United States
by Baker & Taylor Publisher Services

Printed in the United States
by Baker & Taylor Publisher Services